DAT

INTELLIGENCE AND LEARNING

NATO CONFERENCE SERIES

I   Ecology
II   Systems Science
III   Human Factors
IV   Marine Sciences
V   Air—Sea Interactions
VI   Materials Science

III HUMAN FACTORS

INTELLIGENCE AND LEARNING

Edited by
Morton P. Friedman
University of California, Los Angeles
Los Angeles, California

J. P. Das
University of Alberta
Alberta, Canada

and
Neil O'Connor
MRC Developmental Psychology Unit
London, England

Published in cooperation with NATO Scientific Affairs Division by

PLENUM PRESS · NEW YORK AND LONDON

Library of Congress Cataloging in Publication Data

Nato Conference on Intelligence and Learning, York, Eng., 1979.
 Intelligence and learning.

 (NATO conference series: III, Human factors; v. 14)
 "Proceedings of a NATO Conference on Intelligence and Learning, held July 16-20,
1979, in York, England."
 Includes index.
 1. Intellect—Congresses. 2. Intelligence levels—Congresses. 3. Learning Psychology
of—Congresses. I. Friedman, Morton P. II. Das, Jagannath Prasad. III. O'Connor, Neil.
IV. Title. V. Series.
 BF431.N377 1979   153   80-28692
 ISBN 0-306-40643-8

Proceedings of a NATO Conference on Intelligence and Learning,
held July 16—20, 1979, in York, England

© 1981 Plenum Press, New York
A Division of Plenum Publishing Corporation
233 Spring Street, New York, N.Y. 10013

Printed in the United States of America

PREFACE

This volume contains the Proceedings of an International Conference on Intelligence and Learning held at York University, England, on July 16-20, 1979.

The conference was made possible with the support and assistance of the following agencies: NATO Scientific Division, specifically the Human Factors panel, was the major sponsor of the conference. Special thanks are due to Dr. B. A. Bayraktar, who helped organize the conference. Special appreciation is also expressed for the support of the University of York where the conference was held, the University of Alberta, the University of California, Los Angeles, the Medical Research Council, especially its Developmental Psychology Research Unit in London, and the British Council. The conference was jointly directed by J. P. Das and N. O'Connor. The directors appreciate the assistance in administrative matters of Patricia Chobater and Emma Collins of the University of Alberta.

The Editors of the Proceedings acknowledge and appreciate the following individuals who assisted in the production of the volume at the University of California, Los Angeles: Francine Gray, Janet Koblen and Richard Russell. Special thanks go to Keith Felton, who prepared the final manuscript, and Carol Saro, who assisted the editors and prepared the indexes.

Morton P. Friedman

J. P. Das

Neil O'Connor

CONTENTS

Section 10.   CROSS-CULTURAL APPROACHES

Section 11.   INDIVIDUAL DIFFERENCES AND COGNITION

Section 12.   MENTAL RETARDATION AND LEARING DISABILITIES

INTRODUCTION AND OVERVIEW

Morton P. Friedman   J. P. Das   Neil O'Connor
University of California University of Alberta MRC Developmental
Los Angeles,   Alberta,   Psychology Unit
California U.S.A.   Canada   London, England

The traditional approach to intelligence has been a psychometric one which has emphasized the study of abilities. Recently, alternative conceptions of the nature of intelligence have been proposed: the developmental and structural models of Piaget and others, biological theories and information processing models. An international conference on intelligence and learning was organized to critically review these changes in the field. It brought together some of the leading researchers and promising young workers who represent contemporary approaches to intellectual behavior. This book is a result of that conference. We think it will provide a sample of research and thinking relating intelligence to major psychological processes. An added feature of the book is the discussion of the implications of recent research in intelligence for fields such as reading, cross-cultural psychology and cognitive psychopathology.

The organization of the book follows roughly the organization of the conference. Section 1 contains the conference keynote lecture by W. K. Estes and several special papers on theory and application. Sections 2, 3, and 4 are mainly concerned with the theoretical nature of intelligence. Piagetian approaches are considered in Sections 5 and 6. Sections 7, 8, and 9 deal with cognitive approaches, and also contain some applications to reading. Cross-cultural approaches are covered in Section 10. Sections 11, 12, and 13 consider individual differences and pathologies of intelligence. Sections 14 and 15 deal with information processing approaches to intelligence.

INTELLIGENCE AND LEARNING

W. K. Estes

Rockefeller University

New York, New York, U.S.A.

Given the title of this volume, some of the questions one should expect to be at issue are surely: What has been, what is, and what should be the relationship between learning and intelligence? Are the referents of the two terms identical? Are they, rather, related like two sides of a coin? Or do they perhaps refer to levels of intellect or intellectual function?

As a first step toward clarifying our ideas, it may be useful to partition the problem. Thus I propose to examine these questions with reference to several different relationships: First, interactions between the fields of study or research traditions bearing on intelligence and on learning, second, the correlation between measures of intelligence and learning, and third, conceptual relationships between intelligence and learning that should be significant in theories of either or both.

The Research Traditions of Intelligence and Learning Theory

In Figure 1 I have provided some materials for a synoptic look at the development of research and theory in these fields in longitudinal section. The time line along the bottom is intended to cover nearly a century, running from the mid-1880s to the present time. The names, most of which will be highly familiar, have been inserted at points roughly corresponding to notable developments in research or theory associated with the individuals. In the band representing intelligence, it will be apparent that the upper strand has to do with measurement and the lower strand with considerations of structure and the search for factors or components.

The fact that the study of intelligence has been quite sharply compartmentalized from the study of learning over most of the history of these disciplines is perhaps attributable to three factors--the

3

Intelligence

Tests   Binet   Terman   Wechsler

Factors &   Spearman   Thorndike   Thurstone   Guilford   Sternberg
Structure   Carroll

Learning

Assoc.   Ebbinghaus   Thorndike   Robinson   McGeoch   Underwood
                                                                       Hunt
                                                                       Simon & Feigenbaum

Learning   Pavlov   Tolman   Skinner   Estes   Rescorla
& Beh. Th.   Guthrie Hull   & Wagner

Interp. ment. def.   Sidman   Ellis
                                                                       Zeaman & House

Learning-to-learn   Harlow

Perceptual learn.   Lashley   Hebb   Gibson

Cognitive Psy.

Framework   James   Bartlett   Piaget   Broadbent   Neisser

Info. Proc.   Miller   Atkinson   Anderson
                                                                & Shiffrin & Bower

Org. of Memory   Tulving   Kintsch
                                                                          Collins & Quillian

Cog. Perform.   Sternberg   Posner
                                                                            Chase & Clark

|1880   1905   1930   1955   1980

Figure 1. A sampling of names of investigators associated with major
developments in the study of intelligence, learning theory,
and cognitive psychology, chronology running from left to
right as indicated by the time line at the bottom. A representative published work of each investigator is included in
the Reference list.

almost complete reliance of investigators of intelligence on correlational, those of learning on experimental methods, the uneven theoretical development of the two fields, and the need for a conceptual
bridge between them.

The predominantly correlational approach to intelligence over
many decades seems a natural consequence of the fact that in the
early period intelligence was almost without question taken to be a
trait, with the task of research being to find ways of measuring
this characteristic of the individual rather than to analyze intellectual performance.   Nonetheless, the concept of intelligence might
not have evolved in such uniform isolation from the methods and
accumulating results of research on learning had it not been for
the exceedingly primitive state of learning theory in the early 1900s.

Some years later Thorndike (1926), who was personally responsible for much of the development of learning theory during the first
quarter of this century, made a Herculean effort to bring intelligence
and learning within a single theoretical framework, with the basis for
both intelligence and learning ability being localized in an ensemble
of actual or potential connections in the cortex.   This effort was
rather more influential on research and practice than its scientific
merits warranted in my estimation (Estes, 1974).   Perhaps one of
Thorndike's most important contributions was to make it clear that
meaningful theoretical rapprochement between intelligence and
learning would have to wait on further development of both fields.

A body of systematic doctrine that might be termed learning
theory only began to take form about the middle of the period
covered by Figure 1.   And even then, there was no place for a
concept of intelligence in the psychology of human learning of the
association-functional tradition, represented in the first row under
learning theory, nor in the conditioning and reinforcement theories
associated with Pavlov (1927), Tolman (1932), Hull (1943), Skinner
(1938), and their intellectual descendants.   Trait-oriented concepts
were not at home in these theories, and the theories were for
several decades too closely tied to problems of detailed prediction
of behavior of laboratory subjects to provide much contribution
toward the understanding of human intellectual functioning.   The
one exception perhaps was Harlow's (1949) concept of learning set,
which quickly outgrew its early ties with discrimination learning in
monkeys and generated what has proved to be an important body of
research on learning-to-learn, with special reference to the mentally
retarded (see for example, Estes, 1970).

Over the time period we are considering, a slowly accelerating
but ultimately significant shift in the focus of research on intelligence from sheer measurement of ability to problems of dealing constructively with the mentally retarded set the stage for some important spinoffs of the behavioral learning theories, beginning in

the 1960s with the work of Sidman and Stoddard (1966) and other followers of Skinner on the shaping of behavior of the mentally retarded by reinforcement procedures and the work of Zeaman and House (1963) and Ellis (1963, 1970) on the application of concepts of Hull's learning theory to the interpretation of aspects of mental deficiency.

Over the same period during which the behavioral learning theories evolved and ultimately began to find application to problems of mental retardation, another current of thought in learning theory that was less dominated by behaviorism and operationism and more hospitable to the interweaving of concepts of learning and perception steadily gained influence (Hebb, 1949; Lashley, 1942). However, a gap remained between the main lines of research on learning and intelligence that began to be filled out in the 1960s with the emergence of a cognitive psychology broad enough in outlook and methods to encompass or interact with contemporary learning theories on the one hand and contemporary approaches to the measurement and interpretation of intelligence on the other.

To be sure cognitive psychology was not new in the 1960s; in fact its general philosophy and some of its enduring central concepts had been laid down by William James before 1900. However, methods for incisive experimental attacks on aspects of cognition other than learning were slow to develop; it is hard to identify notable theoretical contributions for several decades following William James (1890), although there was a steady accumulation of results on specific subtopics, well reviewed by Woodworth (1938). Contemporaneously Piaget's approach appeared and grew in influence, and, though foreign in outlook to experimental psychology, helped set the stage for the almost explosive developments in the 1960s when converging intellectual inputs from Piagetian theories of cognitive development, the computer revolution, and the rise of psycholinguistics gave rise to cognitive psychology as we now know it (Estes, 1978).

Although the Conference represented in this volume was entitled "Intelligence and Learning," it seems to me that it no longer makes sense to discuss interrelationships of intelligence and learning without consideration of the third member of the triumvirate, cognitive psychology. To be sure the three research traditions and the concepts associated with them overlap in various aspects, but nonetheless they are relatively distinct facets of intellectual function and each needs full consideration. There is doubtless room for debate over definitions, but usage of the three principal terms in today's literature seems to me reasonably consistent. The study of learning and learning theory bear on the development of skills and the acquisition of knowledge, with primary concern for the course and conditions of acquisition. Intelligence has primarily to do with the measurement of intellectual abilities, conceptualiza-

tion of the way abilities are organized, and the identification of
the abilities implicated in various kinds of intellectual tasks. Cognitive psychology is concerned primarily with the products of learning, that is the way knowledge is organized and accessed in the
memory system, and with the mental operations by means of which
intellectual tasks are actually accomplished.

With these working definitions and our overall picture of the
combined field in mind, I should like now to turn to two more
specific problems, first the interrelationships between intelligence
and learning abilities and, second, the interactions of both kinds
of abilities with the structures and processes contributing to intellectual performance.

The Relationship Between Intelligence and Learning Ability

The long-standing and widely held supposition that the inter-
relationship of intelligence and learning ability must at the least
be very close doubtless has its origins in the fact that the first
major contribution to intelligence testing, the Binet-Simon scale,
was produced in response to the commissioning of those investigators to find a way of identifying children "unable to profit, in
an average measure, from the instruction given in ordinary schools"
(Binet and Simon, 1905, p. 9). The supposition might, further,
seem to be strongly fortified by the fact that the validity of intelligence scales has been most commonly defined in terms of school
progress or the ability to profit from school instruction. In the
minds of the originators of the Binet-Simon scale, however, the
picture of their creation was quite different. These investigators
were not simply early "human engineers" carrying out a practical
assignment, but major theoretical psychologists of their time, quite
capable of debating with William James (as witness numerous articles
by Binet and James in early issues of the Psychological Review).
Binet and Simon conceived their scale, not as a measure of a single
trait that might be termed intelligence, but rather as a classifier
of "diverse intelligences" (Binet and Simon, 1905, p. 40). They
proposed equating intelligence with judgment, considered memory
to be quite independent of judgment and tried to keep their scale
free of tests in which a child might succeed by "rote learning."

The theoretical ideas of these investigators did not become as
well known as the tangible product of their efforts however, and
when the scales were revised by Terman (1916) for what proved to
be extremely widespread use in American schools, the focus was
almost entirely on diagnosing a child's inability to profit from instruction or ability to accelerate in the schools. In the course of a later
revision (McNemar and Terman, 1942) the nature and interrelationships of the various subtests were examined in detail and it proved,
contrary to the intention of Binet and Simon, that the subtests that
would be regarded as measures of memory correlated as highly with

measures of mental age as the reliabilities would permit.   The authors
concluded that "any reasonable allowance for these effects [overlap,
correlated errors] will lead to the conclusion that "memory" as determined by the items of a "memory" nature in the New Revision is not
very different from the general intelligence being measured by the
scale as a whole" (McNemar and Terman, 1942, p. 150).   This close
identification of intelligence and learning ability was by no means
peculiar to McNemar and Terman.   In the 1940 Yearbook on intelligence, for example, Freeman expressed the view that "intelligence,
then, is the ability to learn new acts or to perform new acts that
are functionally useful" (NSSE Yearbook, 1940, p. 18).

A long history of attempts to accrue empirical evidence concerning relationships between learning abilities and other aspects
of intelligence have on the whole provided more support for the
original ideas of Binet and Simon than for the conclusions of their
successors.   These efforts began in the early 1900s with the correlational studies of relations between laboratory tasks, many of them
designed to test memory or learning, and measures or criteria of
intelligence.   A review of these by Spearman (1904) assessed the
results as uniformly negative, concluding with the rather acid
comment, "The most curious part of the general failure to find
any correspondence between the psychics of the Laboratory
and those of Life is that experimental psychologists on the
whole do not seem in any way disturbed by it."

Continuing efforts over many decades yielded only a little
more by way of positive relationships.   Substantial efforts by Woodrow (1940) and other studies reported by Munn (1954) yielded only
low and at most barely significant correlations between measures of
IQ and laboratory measures of learning.   By the 1960s the measures
of learning had perhaps gained something by way of reliability, ad-
mitting correlations with IQ in the .20's and .30's, and in the case
of paired-associate learning, a bit closer in content to such aspects
of intelligence as vocabulary acquisition, some correlations as high
as .45-.60.   A critical and analytic review by Zeaman and House
(1967) made the point that many of the low correlations may have
resulted from restricted ranges of IQs entering into the correlations.   With this methodological defect allowed for, they conclude
that there is at least a significant positive relationship, for subjects of equal mental age, between IQ and measures of verbal
learning.

The checkered history of attempts to characterize the relationship between learning ability and other aspects of intelligence is
typical of research efforts that proceed for long periods with little
theoretical direction.   Some reasons for the variability of the correlational results and their continuing refractoriness to coherent
interpretation may be found in a consideration of the interactions
of abilities with the processes bearing on intellectual functioning,
to which we now turn.

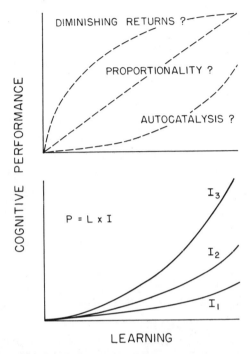

Figure 2. In the upper panel are shown several possible relationships between cognitive performance and learning and in the lower panel the hypothesized interaction between learning and intelligence (I).

Interactions of Processes and Abilities

First, let us ask how, in a general way, do intelligence and learning interact in the determination of intellectual performance. Some principal possibilities are sketched in the upper panel of Figure 2, with some measure of cognitive performance (in arbitrary units) on the vertical axis and some measurement of amount of learning on the horizontal axis. One possibility, illustrated by the upper curve, is a diminishing returns relationship. On this idea some learning would be essential to enable cognitive performance of any reasonable degree of efficiency, but beyond that the amount of learning would rapidly become less important, and other variables, presumably those subsumed under intelligence, would be the main determiners of individual differences in performance. A second possibility, indicated by the middle function, is proportionality, that is constant proportional contributions of the two factors at all levels. A third possibility I have termed autocatalysis, meaning a positively accelerated relationship in which increasing amounts of learning yield products of increasing value for the mediation of

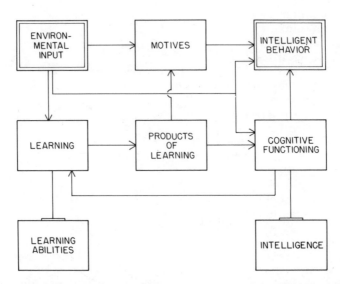

Figure 3. Schematic representation of interactions among the processes and abilities contributing to intelligent behavior.

cognitive performance. I think the question marks in the figure are highly appropriate, but my own reading of the literature, together with theoretical considerations that will be illustrated in the remainder of this paper, lead me to opt for the positively accelerated function as the best bet on the evidence we have.

Proceeding on this working hypothesis, I have sketched in the lower panel my surmise as to the way degree of intelligence, to the extent that this variable proves distinguishable from learning ability, would modify the contribution of learning to cognitive performance, the function being a multiplicative one. The specific form should not be taken seriously, of course, beyond the point of signifying that, in general, effort put into producing a given increment in learning should be expected to produce increasingly large increments in amount or quality of cognitive performance the greater the intelligence of the individual doing the learning. To set the stage for more fruitful and detailed discussion of these somewhat global concepts I will proceed to discuss Figure 3, which lays out a set of relationships among various aspects of learning, intelligence, and cognition that follow from the hypotheses suggested by a review of many years of research on both intelligence and learning.

The principal concepts I see entering into the global conception of intelligence will be seen to include intelligent behavior and the various kinds of internal and external determiners that enter into its prediction and modification. I will assume that all investigators conceive intelligence to be important, not just as an abstract property of an organism, but as a characterization of or determiner of behavior that aids the individual to adjust to his or her environment. Thus for a start I will take the class of dependent variables we are concerned with, intelligent behavior, to comprise, as Charlesworth (1976) puts it, adaptive behavior that is regulated by cognitive functions. By cognitive functions I refer to such activities as perceiving relationships, comparing and judging similarities and differences, coding information into progressively more abstract forms, classification and categorization, memory search and retrieval.

Having available in one's repertoire various cognitive rules and operations is necessary, but not sufficient for intelligent behavior, however; it is necessary for them to be activated in problem situations. Thus, although the fact has often slipped from attention, it has been recognized from the time of Binet, and perhaps first strongly emphasized by Lewin (1940), that motives must be considered on a par with the more intellectual determiners of intelligent behavior. The relevant motives must not be identified solely, or perhaps even most importantly, with simple biological drives and the like. Rather, they must be understood as organized components of the cognitive system, incorporating products of earlier learning and entering into cognitive function in ways that still demand elucidation (Bower, 1975).

Looking at the base of the structure shown in Figure 3 one may see that I am inclined to make some fairly definite assumptions about the role of abilities. I recognize that some people believe that all individual differences in intellectual behavior can be traced to differences in products of learning and thence to differences in opportunities to learn during individuals' earlier histories. There is certainly no harm in that viewpoint being pushed to the limit by investigators who wish to do so. However, it seems to me that all we know about individual differences in intellectual function and in learning points, rather, to the idea that both rates of learning and capabilities of employing the products of learning depend on abilities, that is characteristics of individuals, which, if not innate, are determined by events that occur early in developmental histories and that have not to date been successively identified.

As I have indicated in my extremely brief thumbnail review of research on relationships among various kinds of intellectual abilities, I think hardly any hypothesis one might hold at present could be firmly ruled out on the basis of solid evidence. Nonetheless, from my own subjective reading of the research results, together with more general theoretical considerations, I prefer to proceed on the working hypothesis of two relatively distinct clusters--one that might be termed learning abilities and the

other, which I have tagged intelligence for short, abilities that
pertain to the utilization of cognitive operations in problem
situations. The reason, in part, is not so much that anything
prevents us from classifying the two kinds together if we choose,
as that it seems more fruitful to distinguish them in theory and
leave it an empirical problem to determine their interrelationships.
It will be noted that in labelling the lower right-hand box I have
followed Binet and Simon rather than the consensus of most subsequent work, which has tended to equate the concept of intelligence with the conglomerate of all kinds of abilities that bear
on intellectual performance. Thus it might be better to think
of that box as being relabelled "information processing abilities."

The rather intricate pattern of interactions brought out by
the schema in Figure 3 has a number of implications with regard
to problems of measuring abilities. For one thing the schema
points up a fact that has been recognized by many thinkers in this
field, but still often fades from attention, namely that appraisals
of intelligence, or of either learning or information processing abilities taken separately, always involve indirect inference. The behavior we tap when we give tests or scales of intelligence falls
in the dependent variable box at the upper right of the diagram
and must <u>always</u> be assumed to depend on all of the other factors
portrayed. Thus to measure any one component it is necessary
either to hold all of the others constant, which may often be
impossible of realization, or to understand the interactions well
enough to partial out the effects of components other than the
one that is being measured.

With regard to the two main types of abilities, the problems
are somewhat asymmetric, with in general the information-processing
abilities being somewhat less difficult to appraise separately. One
reason is that intelligence tests tap performance during a short
interval of time within which the amount of learning that goes on may
be assumed negligible. The products of previous learning are
always important, but these may sometimes be handled by allowing
different amounts of previous time and training for different individuals in order to produce a common background of knowledge relevant
to the test. On the other hand, when one is attempting to test
learning ability, behaviour must necessarily be followed over a longer
period of time and one must contend with the important feedback
loop from cognitive functioning to learning, which means that cognitive functions that themselves depend on information processing
abilities influence the course of learning.

We noted above that from the time of Spearman it has been
well known that measures of learning abilities obtained from laboratory tasks typically exhibit low and variable correlations both with
measures of intelligence and with criteria of intelligent behavior,
and usually low intercorrelations among themselves. It will be ap-

parent from the theoretical schema, however, that one is not
justified in a logical inference from these observed results to the
conclusion that the abilities being measured are largely independent
of each other and of either intelligent behavior or school learning.
The network of interrelationships implies that each laboratory
test used in an attempt to get at some constituent of learning
ability calls on some pattern of cognitive operations to carry out
a given task and has its unique requirements with regard to
products of previous learning, both in kind and in degree, that
are prerequisite to the performance called for. Thus the low correlations commonly observed among laboratory tasks used to measure learning abilities may simply reflect variation in contexts
rather than independence of the abilities.

These considerations concerning context become particularly
important as hypothesized learning processes and the abilities they
depend on become incorporated into models for various types of
intellectual performance. To illustrate the point, consider current
models for task situations as diverse as paired-associate learning
(Crothers and Suppes, 1967), problem-solving (Gilmartin, Newell
and Simon, 1976), and comprehension during reading (Kintsch and
Vipond, 1978), in all of which short-term memory for verbal items
such as letters or digits is assumed to be an important constituent.
Now, I don't know that anyone has done so, but it seems a foregone
conclusion tht if anyone decides to correlate scores on digit span
tests with rate of paired associate learning, skill in problem solving,
or reading ability that depends on comprehension from text, the
correlations will prove to be near zero. From these hypothetical,
but I am sure obtainable, results, I would not want to assume that
the models were wrong, but rather that the rationale for such correlational studies is faulty. The correlations must be expected to
be low because the combination of factors with which the hypothesized memory capacity must interact in the test situation is quite
different from the combinations that must be operative in criterion
situations. Thus, to make progress toward determining to what
degree performance on any of these criterion tasks might be related
to greater or lesser short-term memory capacities, one must proceed
to develop ways of testing short-term memory for relevant material
in the context of the criterion task. This kind of measurement could
not be expected to be easy, for in each instance it will need to be
carried out within the framework of a model that represents the
important interactions between the ability in question and other factors
in the task situation. However, taking account of the current
progress toward functional models in a number of cognitive domains,
the goal may no longer be out of reach.

It may be noted that the problem of separating effects of
ability from effects of context are somewhat similar to those that
have been encountered, and to a considerable extent solved, in
signal detectability theory where the corresponding problem is

separating the effects of signal strength or discriminability from those of response bias. In the case of signal detectability a useful approach has been that of the choice model of Luce (1963). In that approach the stimuli presented in the detection situation, say S_1 and S_2 in a simple case of two alternatives and the correct responses to them, R_1 and R_2, can be taken to denote the rows and columns of a matrix, with the cells of the matrix indicating the strength of each response resulting from presentation of each stimulus.

	R_1	R_2
S_1	x	ηy
S_2	ηx	y

In Luce's model, parameter η denotes similarity or confusability between the two stimuli. Thus x is the strength of R_1 to stimulus and ηx the generalized strength of R_1 in the presence of S_2 resulting from the similarity of the two stimuli. For larger sets of stimuli the matrix takes the same form and the model provides a way for evaluating the similarity parameter from experimental data.

Turning to the problem, closer to our present interests, of dealing with the determiners of performance in simple learning tasks, we could portray relationships between tasks, say task A and task B in the simplest case, in a matrix analogous to that of the signal detection problem:

	Task A	Task B
Task A	u	ηv
Task B	ηu	v

Here the upper left and lower right cells can be taken to represent performance on Task A and Task B, respectively, following practice on the same task. Entries in the lower left and upper right would denote performance on either task following practice on the other. It would be assumed that performance depends on ability modified multiplicatively by a factor representing the degree of utilization of resources (relevant products of learning and cognitive operations) and that practice affects the utilization of resources but not the basic ability. Hence initial and transfer scores could perhaps be analyzed by methods somewhat akin to those of the choice model in order to permit evaluation of the parameter η, here denoting the similarity in context (that is the similarity or overlap in resources required) for the two tasks. And again for a larger number of tasks the matrix would take the same form, just as in the case of the signal detection problem. I do not wish to press the analogy too far, but the idea of bringing more of the methodology of experimental psychology to bear on the study of intelligence may be worth more serious exploration.

An important asymmetry between the measurement of learning ability and of information processing abilities has to do with the sheer speed of technical development. Whereas the latter has been the subject of a steady and cumulative research effort since the beginning of the century, with commensurate progress in solving the technical problems of measurement, either the same is not true with regard to learning ability or the literature has escaped me entirely. The idea of measuring learning abilities by simple laboratory tasks, or their equivalents embedded in intelligence scales, had its start in a period predating anything we would recognize as learning theory and in the context of an extremely simplistic and severely limited conception of learning as a rather homogeneous associative process. Within learning theory, that limited view has given way to broadened conceptions that take account of a major distinction between slow and fast learning, and, correspondingly, long and short-term retention of the products of learning. This distinction was not apparent at the time of Thorndike, nor even in the learning and reinforcement theories of the Tolman-Hull period.

To my knowledge it was Hebb (1949) who first brought together and organized the evidence for the prolonged and slow learning processes underlying, for example, the development of the ability to recognize sensory patterns and the growth of syntactical competence. This form of learning is to be distinguished from that studied in most laboratory tasks, which involves an almost instantaneous restructuring of products of earlier learning. A child, or even an animal, may very quickly learn a discrimination or concept requiring, say, the categorization of red triangles versus blue circles, but only if the test has been preceded by a long period of learning to discriminate colors and objects of differing forms. Very recently the work of a few investigators, for example LaBerge (1976), Shiffrin and Schneider (1977), has shown under well controlled laboratory conditions that even in adults long periods of slow learning may be required to change performance from relatively slow processing with heavy demands on attention to highly efficient processing that is relatively attention free.

Independently, but in close parallel, one finds that in much current research on memory a clear distinction is made between episodic memory, the memory (often short-term) for particular experiences or episodes and semantic memory, the accumulated products of learning with regard to language and verbal concepts (e.g., Tulving, 1968).

However these distinctions have yet to be effectuated in the measurement of learning abilities. So far as I know all of the tasks used to measure learning ability involve brief samples of activity that could not possibly begin to assess the rates at which slow learning occurs in different individuals, either the very prolonged learning that occurs outside the laboratory in relation to pattern perception and language or even the shorter term but still prolonged learning that is now effectively studied in some laboratory investigations (e.g., Crothers and Suppes, 1967; Friedman et al., 1964; Shiffrin and Schneider, 1977). Again, tests calculated to get at memory ability, except for the problem of context already mentioned, seem to do well enough at appraising abilities related to episodic memories over short time intervals and to assess the current state of important segments of semantic memory, but none so far as I know have yet been addressed to assessing the rate at which semantic memories are formed. Thus the possibility remains open that some of the complaints about lack of validity of extant laboratory tests of learning ability could be materially met by theoretically directed research.

Intelligence in Learning

Although the relation between learning and intelligence is usually conceived in terms of learning as one of the preconditions for intelligent behavior, the feedback loop whereby information processing abilities and cognitive operations influence the course of learning is beginning to be appreciated. A concrete example of this "backward" path of influence is given by some results shown in Figure 4 for an unpublished experiment carried out in my laboratory. College student subjects learned two successive lists of paired-associate items in a simulated vocabulary learning situation, the stimulus members of items being consonant-vowel-consonant trigrams and the response members ordinary English words. A novel feature was that at the point when a subject first recalled the response member of an item, he was asked to indicate how the correct answer had come to mind: (1) simply by rote, (2) by memory for the episode of the previous paired presentation, or (3) by utilization of a perceived relationship between the stimulus and the response word (either in sound or in visual pattern). Examples of type 2 would be remembering where in the list (following what other item) the item occurred or remembering the visual appearance of the printed response word (as by a visual image) on the previous trial. Examples of type 3 would be noticing that the stimulus member of an item is a syllable of the response member or that it rhymes with a synonym of the response member.

The data plotted in the main portion of Figure 4 shows the course of learning in terms of proportions of items that had been correctly recalled by the end of each trial on each of the two replications of the experiment. These curves exhibit the usual learning-to-learn effect from List 1 to List 2. More interesting

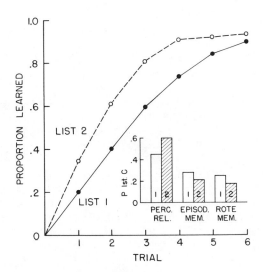

<u>Figure 4</u>. Results of a simulated vocabulary learning experiment
              described in the text.

is the inset, in which the heights of the bars show the proportions of instances in which the first correct recall of an item on
List 1 or List 2 fell into each of the three categories. We see
that the proportion of cases in which recall arose from a perceived
relationship increased appreciably from List 1 to List 2 whereas
the proportions of recalls falling in the other two categories
decreased. In another analysis, it was found that the probability
of a later failure after the first correct recall of an item was only
.02 if the first recall depended on a perceived relationship but was
.12 if the first recall involved episodic memory and .17 if the first
correct recall fell in the "rote memory" category.

     It seems clear that even as apparently simple a form of learning
as acquiring discrete verbal associations can occur in distinguishably
different ways, which can well be categorized as more or less intelligent and which implicate quite different cognitive processes. This
conclusion has been developed more fully, together with many relevant
empirical analyses by Greeno, James, Carlton, and Polson (1978).

     Once sensitized, I could see evidence of operation of the factor
of perception of relevant relationships in other learning situations
that I had habitually conceptualized solely in terms of stimulus-response
associations and I began to feel that this observation brings together
a number of otherwise relatively unrelated findings. One of these,
for example, has to do with the reinterpretation of Thorndike's
classical results on "belongingness" in terms of a conception of <u>open</u>

versus closed tasks (Nuttin and Greenwald, 1968). In this classification, a closed task is one in which there is no inherent reason why
reinforcement contingencies obtaining on a particular trial should
extend beyond it; in other words, remembering what happened on
one trial conveys no necessary information about what should happen
on others. In contrast, an open task is one in which there is reason
to expect carryover of contingencies from one trial or one task to
another. I found it congenial to reformulate this conception in terms
of an individual's perception of relationships among tasks, and in
doing so found that the idea extended quite fruitfully to situations
somewhat different from those Nuttin had dealt with (Estes, 1972).
Later the same idea was extended with some success by K. W. Estes
(1976) to the interpretation of individual differences in children's
discrimination learning. This line of thinking, though starting from
somewhat different origins, seems to mesh quite well with the emphasis of Zeaman and House (1963, 1978) on the role of attention in
discrimination learning, these approaches jointly supporting the
general idea that individual differences in speed of learning have
much to do with the employment of attentional and perceptual
processes.

Some Conclusions

This brief review of the interactions between research disciplines having to do with intelligence and learning suggests that there
is reason to hope for a more fruitful relationship in the future than
has characteristically obtained in the past. At the same time we
have perhaps pointed up more in the way of outstanding problems
than of substantial results. The problems fall into three natural
categories.

(1) The relation between intelligence and learning ability.
Many efforts addressed to this problem over some eight decades have
yielded few convincing results and a general impression of very low
correlation between learning abilities and measures of other aspects
of intelligence. However the correlations have always been based on
measures of learning taken from performance on brief laboratory
tasks, and it cannot be presumed that these are significantly related
to the slow and long-term forms of learning that occur outside of
the laboratory and yield the products of learning that are so important to intellectual functioning. I am afraid we have to conclude
that trying to answer questions concerning the relationship between
learning and other intellectual abilities is nearly as premature today
as it was in the early 1900s, owing to the lack of progress toward
the effective measurement of learning ability.

At the same time, it appears that research on the measurement
of the intellectual abilities generally associated with the term intelligence reached a point of diminishing returns a number of decades
ago; though there has been continuing refinement of technical

methods for test construction, progress has remained essentially asymptotic with regard to problems of predicting intellectual functioning outside of testing situations. An important reason suggested by the present analysis is continuing overdependence on the concept of context-free ability tests and consequent lack of analysis of the interactions and contexts. A glimmer of hope for the future is perhaps to be found in some current efforts to embed concepts of ability within information processing models. The exploitation of such augmented models in research may conceivably further both the measurement of abilities and the understanding of the ways in which abilities and the products of learning influence performance.

(2) The role of learning in intellectual performance. Although learning and memory were regarded as lower-order mental functions scarcely related to intelligence by the early developers of intelligence scales, there has been substantial progress over the years toward appreciating the role of learning in intellectual performance, notable advances being associated with the extended analyses of discrimination learning in the mentally deficient by Zeaman and House (1963, 1967) and Gagne's (1968) conceptualization of the dependence of intellectual performance on the cumulative products of learning. Once again, though, diminishing returns are apparent after a burst of activity, perhaps in this case because the development of learning theory itself has been in the doldrums during the recent period of enthusiasm for the newer specialties of cognitive psychology and information processing models. Healthy development of research in the broad field of intelligence may depend rather critically on the correction of this imbalance and the implementation of new developments of learning theory, taking more effective account of individual differences and the distinctions between fast and slow learning.

(3) The role of intelligence in learning. We have seen that throughout the history of research on learning and on intelligence the prevailing view of the interaction between these aspects of mental function has been one-sided, learning and the products of learning being conceived as prerequisites for intelligent performance in such activities as problem solving. However the last decade or so of intensive research in cognitive psychology and in information processing may have set the stage for a new wave of effort aimed toward redressing the balance. Theoretical analysis suggests that the role of information processing operations in learning may be just as important as the role of the products of learing in information processing. A new learning theory that takes account of this side of the interaction might prove to be quite different in form from the learning theories it was possible to conceive during the period in which the efforts of the great systematists from Thorndike to Hull evolved the concept of learning theory as we now understand it and might be more relevant to problems of intellectual function.

References

Anderson, J. R., and Bower, G. H. Human Associative Memory.
 Washington, D.C.: V. H. Winston, 1973.
Atkinson, R. C., and Shiffrin, R. M. Human memory: A proposed
 system and its control processes. In K. W. Spence and J. T.
 Spence (Eds.), The Psychology of Learning and Motivation.
 Vol. 2. New York: Academic Press, 1968. Pp. 89-195.
Bartlett, F. C. Remembering. Cambridge: Cambridge University
 Press, 1932.
Binet, A., and Simon, T. The Development of Intelligence In Chil-
 dren. Baltimore: Williams and Wilkins, 1916.
Bower, G. H. Cognitive psychology: An introduction. In W. K.
 Estes (Ed.), Handbook of Learning and Cognitive Processes.
 Vol. 1. Hillsdale, NJ: Erlbaum Associates, 1975. Pp. 25-80.
Broadbent, D. E. Perception and Communication. Oxford: Pergamon
 Press, 1958.
Carroll, J. B. Psychometric tests as cognitive tasks: A new
 "structure of intellect." In L. B. Resnick (Ed.), The Nature
 of Intelligence. Hillsdale, NJ: Erlbaum Associates, 1976.
 Pp. 27-56.
Charlesworth, W. R. Human intelligence as adaptation: An etho-
 logical approach. In L. B. Resnick (Ed.), The Nature of
 Intelligence. Hillsdale, NJ: Erlbaum Associates, 1976. Pp.
 147-168.
Chase, W. G., and Clark, H. H. Mental operations in the compar-
 ison of sentences and pictures. In L. Gregg (Ed.), Cognition
 in Language and Memory. New York: Wiley, 1972.
Collins, A. M., and Quillian, M. R. How to make a language user.
 In E. Tulving and W. Donaldson (Eds.), Organization of Memory.
 New York: Academic Press, 1972. Pp. 309-351.
Crothers, E., and Suppes, P. Experiments in Second-Language
 Learning. New York: Academic Press, 1967.
Ebbinghaus, H. Memory. 1885. Translated by H. A. Ruger and
 C. E. Bussdenius. New York: Teachers College, 1913.
Ellis, N. R. Handbook of Mental Deficiency. New York: McGraw-
 Hill, 1963.
Ellis, N. R. International Review of Research in Mental Retarda-
 tion. 1970, 4.
Estes, K. W. An information-processing analysis of reinforcement
 in children's discrimination learning. Child Development,
 1976, 47, 639-647.
Estes, W. K. New perspectives on some old issues in association
 theory. In N. J. Mackintosh and W. K. Honig (Eds.),
 Fundamental Issues in Associative Learning. Halifax, Nova
 Scotia: Dalhousie University Press, 1969. Pp. 162-189.
Estes, W. K. Learning Theory and Mental Development. New York:
 Academic Press, 1970.
Estes, W. K. Reinforcement in human behavior. American Scientist,
 1972, 60, 723-729.

Estes, W. K.   Learning theory and intelligence. American Psychologist, 1974, 29, 740-749.

Estes, W. K.   On the organization and core concepts of learning theory and cognitive psychology. In W. K. Estes (Ed.), Handbook of Learning and Cognitive Processes. Vol. 6. Hillsdale, NJ: Erlbaum Associates, 1978. Pp. 235-292.

Freeman, F. N.   The meaning of intelligence. In the National Society for the Study of Education's 39th Yearbook, Intelligence: Its Nature and Nurture. Bloomington, Ill.: Public School Publications, 1940.

Friedman, M. P., Burke, C. J., Cole, M., Keller, L., Millward, R. B., and Estes, W. K.   Two-choice behavior under extended training with shifting probabilities of reinforcement. In R. C. Atkinson (Ed.), Studies in Mathematical Psychology. Stanford: Stanford University Press, 1964.

Gagne, R. M.   Contributions of learning to human development. Psychological Review, 1968, 75, 177-191.

Gibson, E. J.   A systematic application of the concepts of generalization and differentiation to verbal learning. Psychological Review, 1940, 47, 196-229.

Gilmartin, K. J., Newell, A., and Simon, H. A.   A program modeling short-term memory under strategy control. In C. Cofer (Ed.), The Structure of Human Memory. San Francisco: W. H. Freeman and Co., 1976. Pp. 15-30.

Greeno, J. G., James, C. t., DaPolito, F., and Polson, P. G. Associative Learning: A Cognitive Analysis. Englewood Cliffs, NJ: Prentice-Hall, 1978.

Guilford, J. P.   The Nature Of Human Intelligence. New York: McGraw-Hill, 1967.

Guthrie, E. R.   The Psychology of Learning. New York: Harper, 1935.

Harlow, H. F.   The formation of learning sets. Psychological Review, 1949, 56, 51-65.

Hebb, D. O.   The Organization of Behavior: A Neuropsychological Theory. New York: Wiley, 1949.

Hull, C. L.   Principles of Behavior. New York: Appleton-Century-Crofts, 1943.

Hunt, E. B.   Concept Learning: An Information Processing Problem. New York: Wiley, 1962.

James, W.   Principles of Psychology (2 vols.). New York: Holt, 1890.

Kintsch, W.   The Representation of Meaning in Memory. New York: Wiley, 1974.

Kintsch, W., and Vipond, D.   Reading comprehension and readability in educational practice and psychological theory. In L.-G. Nilsson (Ed.), Perspectives on Memory Research. Hillsdale, NJ: Erlbaum Associates, 1978. Pp. 329-366.

LaBerge, D.   Perceptual learning and attention. In W. K. Estes (Ed.), Handbook of Learning and Cognitive Processes. Vol. 4. Hillsdale, NJ: Erlbaum Associates, 1976. Pp. 237-274.

Lashley, K. S. An examination of the continuity theory as applied
  to discriminative learning. Journal of General Psychology, 1942,
  26, 241-265.
Lewin, K. Intelligence and motivation. In the National Society for
  the Study of Education's 39th Yearbook, Intelligence: Its
  Nature and Nurture. Bloomington, Ill.: Public School Pub-
  lishing Co., 1940.
Luce, R. D. Detection and recognition. In R. D. Luce, R. R.
  Bush, and E. Galanter (Eds.), Handbook of Mathematical
  Psychology. Vol. 1. New York: Wiley, 1963. Pp. 103-190.
McGeoch, J. A. The Psychology of Human Learning. New York:
  Longmans and Green, 1942.
Miller, G. A. The magical number seven, plus or minus two: Some
  limits on our capacity for processing information. Psychological
  Review, 1956, 63, 81-97.
Munn, N. L. Learning in children. In L. Carmichael (Ed.), Manual
  of Child Pschology, 2nd Ed. New York: Wiley, 1954. Pp. 374-
  458.
Neisser, U. Cognitive Psychology. New York: Appleton-Century-
  Crofts, 1967.
Nuttin, J. R., and Greenwald, A. G. Reward and Punishment in
  Human Learning. New York: Academic Press, 1968.
Pavlov, I. Conditioned Reflexes. London: Oxford University
  Press, 1927.
Piaget, J. The Origins of Intelligence in Children. New York:
  International Universities Press, 1952.
Posner, M. I. Chronometric Explorations of Mind. Hillsdale, NJ:
  Erlbaum Associates, 1978.
Rescorla, R. A., and Wagner, A. R. A theory of Pavlovian
  conditioning: Variations in the effectiveness of reinforcement
  and non-reinforcement. In A. H. Black and W. R. Prokasy
  (Eds.), Classical Conditioning II. New York: Appleton-
  Century-Crofts, 1972.
Robinson, E. S. Association Theory To-day. New York: The
  Century Co., 1932.
Shiffrin, R. M., and Schneider, W. Controlled and automatic
  human information processing: II. Perceptual learning,
  automatic attending, and a general theory. Psychological
  Review, 1977, 84, 127-190.
Sidman, M., and Stoddard, L. T. Programming perception and
  leaning for retarded children. In N. R. Ellis (Ed.), Inter-
  national Review of Research in Mental Retardation, 1966, 2,
  141-208.
Simon, H. A., and Feigenbaum, E. A. An information-processing
  theory of some effects of similarity, familiarization, and meaningfulness in verbal learning. Journal of Verbal Learning and
  Verbal Behavior, 1964, 3, 385-396.
Skinner, B. F. The Behavior of Organisms: An Experimental
  Analysis. New York: Appleton-Century-Crofts, 1938.
Spearman, C. "General intelligence" objectively determined and

measured. American Journal of Psychology, 1904, 15, 201-293.

Sternberg, R. J. Intelligence, Information Processing, and Analogical Reasoning: The Componential Analysis of Human Abilities. Hillsdale, NJ: Erlbaum Associates, 1977.

Sternberg, S. High-speed scanning in human memory. Science, 1966, 153, 652-654.

Terman, L. M. The Measurement of Intelligence. Boston: Houghton Mifflin, 1916.

Thorndike, E. L. The Measurement of Intelligence. New York: Teachers College Press, 1926.

Thorndike, E. L. Human Learning. New York: Appleton-Century-Crofts, 1931.

Thurstone, L. L. The Nature of Intelligence. New York: Harcourt, Brace and World, 1924.

Tolman, E. C. Purposive Behavior in Animals and Men.. New York: Appleton-Century, 1932.

Tulving, E. Theoretical issues in free recall. In T. R. Dixon and D. L. Horton (Eds.), Verbal behavior and General Behavior Theory. Englewood Cliffs, NJ: Prentice-Hall, 1968. Pp. 2-36.

Underwood, B. J. Are we overloading memory? In A. W. Melton and E. Martin (Eds.), Coding Processes in Human Memory. Washington, D.C.: V. H. Winston, 1972. Pp. 1-23.

Wechsler, D. Cognitive, conative, and non-intellective intelligence. American Psychologist, 1950, 5, 78-83.

Woodrow, H. Interrelations of measures of learning. Journal of Psychology, 1940, 10, 49-73.

Woodworth, R. S. Experimental Psychology. New York: Holt, 1938.

Zeaman, D., and House, B. J. The role of attention in retardate discrimination learning. In N. R. Ellis (Ed.), Handbook of Mental Deficiency. New York: McGraw-Hill, 1963. Pp. 159-223.

Zeaman, D., and House, B. J. The relation of IQ and learning. In R. M. Gagne (Ed.), Learning and Individual Differences. Columbus, Ohio: Charles E. Merrill, 1967. Pp. 192-212.

Zeaman, D., and House, B. J. Interpretation of developmental trends in discriminative transfer effects. In A. D. Pick (Ed.), Minnesota Symposia in Child Psychology, 1974, 8, 144-186.

RECENT ISSUES IN THE DEVELOPMENTAL APPROACH

TO MENTAL RETARDATION

Edward Zigler and David Balla

Yale University

New Haven, Connecticut, U.S.A.

Abstract

A two-group approach to the range of intellect was explained to account for irregularities in the "normal" IQ curve. Organically retarded persons would be represented by one curve at the lowest end of the distribution. Familial retarded persons would be grouped with the rest of the population--their lower IQs considered a part of the normal variation dictated by the diversity of human genetic inheritance. The extreme environmental approach to mental retardation was summarized, as were the difference and general-developmental positions. Behavioral differences between mildly retarded and nonretarded persons of the same MA were explained in terms of environmentally-based motivational differences, including such factors as social deprivation, expectancy of success, optimal reinforcers, outerdirectedness, and institutionalization.

The field of mental retardation continues to be plagued by myths and fallacies. For example, look at the typical introductory textbook chapter on mental retardation. Here we inevitably find a graph of the normal curve for intelligence. There is also some arbitrary cutoff point, usually IQ 70, and it is implied that everybody below that point is retarded and everybody above it is not. Thus, we give students the impression that mental retardation is a homogeneous phenomenon for which we can expect to find some single underlying cause. But there are myriad known causes of retardation and many more as yet undiscovered, and the behavior of retarded persons is no more homogeneous than is that of any random group of individuals.

Actually the normal IQ curve which we hold in such high esteem has some basic problems. First of all, the distribution of IQ scores in every population that has been studied turns out not to be bell-shaped at all (e.g., Penrose, 1963). It deviates from symmetry in two ways, both of which are important to our thinking about mental retardation. For one, there are many more cases below IQ 50 than we would predict from our basic polygenic formulation. This bulge at the lower IQ levels has led several theorists (e.g., Penrose, 1963; Zigler, 1966) to assert that a major step in our understanding of retarded persons would be to adopt a two-group approach to mental retardation. Rather than viewing intelligence as a single curve representing a single population, we should try to envision two curves representing two populations. The curve at the lower end of the distribution would represent retarded persons with known anatomical or physiological defects. This organically retarded group has a mean IQ of approximately 35 and a range from 0 to about 70.

The IQ curve of the rest of the population is almost symmetrical and encompasses IQs from approximately 50 to 150. We have argued that this range probably reflects the genetic variation of our species. That is, people are destined to be different, and human traits will always have a distribution with some persons considerably above and some well below the mean. From an evolutionary point of view, such variation is in fact desirable. Where do organically retarded persons fit into this polygenic explanation of intelligence? They would appear to represent persons with a wide range of genetic potential whose intellectual expression was altered by some major and usually identifiable physiological problem.

The two-group approach to intelligence raises some serious issues concerning mildly retarded persons who have no evidence of organic involvement. They are sometimes called cultural-familial retarded, sometimes endogenous retarded, and, in official terminology, those suffering from retardation due to psychosocial disadvantage. This group comprises between 65 and 75 percent of all retarded individuals. We know enough about labeling theory and the phenomenon of stigmatization (see Mercer, 1973) to lead us to believe that there should be a better term to describe what seems to be the lower portion of the normal distribution of intelligence. Zigler (1977) previously suggested that no child with an IQ above 50 be labeled retarded, because the social services that follow cannot compensate for the harm done by being branded with the mental retardation label. The problem, of course, is that such an action would immediately reduce the number of retarded persons by a huge percentage. But at least then we would be talking about the two to three million individuals with IQs below 50--the most seriously afflicted--for whom we could expend the bulk of our professional efforts. This does not mean that the other group would be of no interest to us. It simply means that we would no longer refer to them as retarded. We need some term in the area of intelligence that is analogous to the term "short" when we speak of height.

Let us return for a moment to the second IQ curve mentioned which describes the intelligence of the majority of the population. Here again there is a notable deviation from symmetry. There seem to be too many cases in the 70 to 100 IQ range, with the excess shading into the mildly retarded levels. It is here that interactionists and environmentalists can take their stand. They would explain that every genotype is capable of producing a range of phenotypes depending on the individual's experiences. Although behavior geneticists do not agree on the reaction range of intelligence, let us assume high heritability and put it in the neighborhood of, say, 20 points. This means that there could be a 20-point difference in IQ between identical genotypes which experience the very best and the very worst environments. From this point of view, the excess of cases in the lower IQ range means that there are a great number of children in our society who experience very poor environments. These adverse conditions combined with a genetic predisposition have thus placed more individuals in the mildly retarded IQ ranges than is dictated by the nature of our population's gene pool.

We in the mental retardation area, as in psychology in general, have been in the throes of a more extreme environmentalism for over a decade. We have heard very knowledgable people assert that if we surround the child with the right experiences and/or arrange the reinforcement contingencies properly, we could do away with the problem of mild mental retardation altogether. This sort of belief in the infinite plasticity of the human organism has been widely popularized. Some time ago middle-class parents read in a magazine that they could raise their child's IQ by 20 points. About then a fine intervention program was heralded for no other reason than that it could increase IQ scores by a point a month. In fact, 13 years ago when the Head Start program was started, we acted as if we believed that six weeks of nursery school could produce dramatic cognitive changes and somehow immunize the child from the effects of all kinds of future adverse experiences. This sort of optimism is simply unwarranted. We know that it is extraordinarily difficult to change the life outcome of a child. Furthermore, there must be a limit to the reaction range of intelligence, and this limit cannot be altered by some relatively small intervention.

We do not mean to say that environment is unimportant, but the extreme environmental position troubles us for several reasons.

Consider the anxiety that it must create in parents. What do parents think when they learn that they missed the latest, supposedly IQ-enhancing activity, such as putting a mobile over their child's crib? Or what are they feeling when they put their children into nursery school because they see it as a first step in a brilliant career? If we find this sort of anxiety in the parents of nonretarded children, what kind of anxiety can we expect to haunt parents of retarded children?

There is still another danger in the extreme environmental position--that undue optimism will eventually breed undue pessimism. As a lesson from history the mental retardation field began with the mental orthopedics movement (a rather therapeutic-sounding term). Such great thinkers as Itard and Seguin developed a variety of interventions which they believed would enable retarded persons to become productive and independent members of society. The state institutions in the U.S. were started with such educational and therapeutic goals in mind. What happened was that these goals were not reached. This disappointment led to a widespread belief that absolutely nothing could be done for retarded individuals, and the history of mental retardation entered its darkest phase. Retarded persons were segregated into large state schools far removed from population centers so that they would not mingle with the rest of society. Mild mental retardation was seen as a primary social menace and blamed for most criminality, illegitimacy, and whatever ills might befall society. Sterilization laws were passed in the majority of states. While we certainly do not mean to imply that such a state of affairs will occur again, we do believe that if the claims of extreme environmentalists are not fulfilled, the hardwon gains to improve the quality of life for retarded persons may be in jeopardy.

There is another approach, antithetical to the extreme environmental position, which has enjoyed considerable popularity in the mental retardation area, especially among basic research workers. We have labelled a group of these theories defect or difference approaches (reviewed by Zigler, 1966, 1969). What these theories have in common is that they view cultural-familial retarded persons as inherently different from those who are not retarded. According to these theories, at every level of development, there must be some difference or defect in the retarded person's physiological or cognitive structure. These hypothesized differences are believed to produce differences in behavior, even when retarded and nonretarded individuals have the same mental age.

There are a variety of difference positions. An early one which has had considerable impact on the training and treatment of retarded persons was proposed by Lewin and Kounin. They took the common observation that retarded individuals often display perseverative and stereotyped behavior and developed a

theory of cognitive rigidity to explain the difference which characterizes individuals exhibiting mental retardation. Others have
asserted that retarded individuals do not effectively use verbal
means to guide their behavior; that is, they have a verbal
mediation deficit. Still other proposals as to the difference which
afflicts retarded persons include deficits in short-term memory,
attention, or information processing.

The various difference approaches thus typically deal with a
narrow segment of human functioning. As such, none could constitute a comprehensive theory capable of explaining the behavior
of retarded persons. An advantage of the difference positions,
though, is that they provide quite specific areas for intervention.
Indeed, when efforts have been made to remediate specific cognitive deficiencies, the results have been quite encouraging. For
example, after Butterfield, Wambold, and Belmont (1973) demonstrated that retarded individuals do not effectively use rehearsal
strategies in short-term memory problems, they went on to teach
them how to use these strategies. This instruction greatly enhanced short-term memory performance. It is somewhat ironic
that intervention efforts that have their theoretical origins in
difference or defect positions have resulted in findings that
cognitive structure is considerably more plastic than originally
implied.

In opposition to the difference approaches to the study of
mental retardation, we have long espoused a general developmental position (e.g., Balla and Zigler, in press; Zigler, 1969).
Stated most simply, this view holds that the behavior of familial
retarded persons is governed by the same principles which apply
to the behavior of nonretarded persons. The only difference
would be that retarded children have a slower rate of cognitive
development and attain a lower final limit. The emphasis here on
similarities is consistent with the view that the intelligence of
mildly retarded persons falls within the normal variation dictated
by our gene pool. Consequently, retarded and nonretarded
persons of equivalent MA would be expected to perform cognitive
tasks in much the same way. In a comprehensive review of
Piagetian research relevant to the developmental-difference controversy, Weisz and Zigler (1979) found strong support for the
cognitive-developmental position. The only exceptions were
findings from studies which included institutionalized and/or
organically retarded individuals.

It is interesting to note that, though poles apart, the extreme
environmental and difference approaches share a common feature--
they emphasize cognitive factors in behavior to the almost total
exclusion of other factors which are known to be most important.
Behavior is never an inexorable readout of cognitive processes
alone. Researchers in the area of mental retardation seem in

such awe of the cognitive deficit of retarded individuals that they
have ignored other factors which influence everyone's performance.

We have argued (e.g., Balla and Zigler, in press) that there
are three classes of determinants of behavior for everyone, be
they retarded or nonretarded. The first is formal cognition,
including those processes that people like Piaget, Bruner, Vigotsky,
and Werner have studied for many years. These cognitive processes include such factors as memory, reasoning, and abstractive
abilities. The second class of determinants involves achievements.
A person may have a perfectly intact cognitive system, but without
particular experiences that person will not readily be able to do
certain things. Of course we are referring here to the process-
content distinction that has been much discussed in psychology.
These achievement factors are almost totally determined by experience.

The third class of factors includes motivational determinants
of behavior. We said before that we did not mean to imply that
environment is unimportant to intellectual behavior. Perhaps the
best support we can give that statement is to hold up the 20-or-so
years of work that our group at Yale has done to determine how
environmentally caused motivational factors influence what a
person can or cannot do. We are convinced that such personality
features underlie many of the behaviors that mildly retarded
persons exhibit. Yet we have repeatedly found that certain
motivational variables which can hamper performance are not intrinsic to mental retardation. They appear also in nonretarded
individuals who have experienced the same deprivation and failure
that have riddled the lives of so many retarded persons. We thus
believe that specific motivational and emotional states are key
determinants of behavior, and that these states arise from certain
experiences which are common but not limited to the lives of
retarded individuals. We will briefly mention some of these personality influences.

One of the common background features of cultural-familial
retarded persons is that they come almost exclusively from the
lower socio-economic groups. While many parents from the lowest
SES are just as adequate as parents from any other SES level, it
is clear that many mildly retarded children experience extremely
adverse environments while growing up. This history of social
deprivation has been found to pervade many aspects of the child's
behavior. For example, it has been associated with decreased
behavior variability and increased verbal dependency (Balla,
Butterfield, and Zigler, 1974). Of special significance is the fact
that social deprivation leads to a heightened motivation to interact
with adults. This repeated finding (e.g., Balla et al., 1974;
Zigler and Balla, 1972) seems congruent with the common observation that retarded individuals actively seek attention and affection,

and this in turn seems related to the overdependency which they frequently exhibit. With a slight shift in terminology we might conclude that a general consequence of social deprivation is overdependency. We cannot place enough emphasis on the role of over-dependency in the behavior of retarded persons. We have come to believe that, given some minimal intellectual level, the shift from dependence to independence is the single most important factor that would enable retarded persons to become self-sustaining members of society.

In keeping with the general developmental progression from helplessness and dependency to autonomy and independence, we have found both retarded and intellectually-average children of higher MAs to be less motivated for social reinforcement than children of lower MAs (Zigler and Balla, 1972). However, at each MA level the retarded children were more responsive to social reinforcement than their nonretarded peers. The relation between social deprivation and this need for social reinforcement was strongest for the youngest retarded group. This suggests that the younger the child, the more his or her behavior depends on social interactions within the family. Perhaps as the child grows older and interacts with a broader spectrum of socializing agents, motivation for social reinforcement becomes less determined by the quality of family experiences. This view is certainly consistent with the fact that with increasing age, the child's personality is much more influenced by peers, teachers, and other nonfamily socializing agents.

We should mention that there is a controversy over whether social deprivation leads to an atypical desire for interaction with adults or to apathy and withdrawal. Indeed, the retarded person's reluctance and wariness to reciprocate with adults has often been commented upon. Although seemingly inconsistent, experimental work has suggested that social deprivation can lead to both positive and negative attitudes toward adults. We have found that retarded individuals with a history of severe social deprivation are more wary than less deprived individuals (e.g., Balla, Kossan, and Zigler, 1976), and that those institutionalized at an older age are more wary than those institutionalized when younger (Balla, McCarthy, and Zigler, 1971). Thus, excessive wariness is not an inexorable consequence of institutionalization, but it can become quite longstanding if the preinstitutional deprivation persists for some length of time.

Another common trait of retarded persons is their low expectancy of success and high expectancy of failure. These expectancies are believed to stem from the fact that retarded people frequently encounter tasks with which they are intellectually ill-equipped to deal. The extent of feelings of failure in retarded individuals has been well documented (Cromwell, 1963). A clear example comes from a series of studies by MacMillan and colleagues (e.g., MacMillan and Keogh, 1971). An experimenter prevented children from finishing several tasks and then asked why the tasks were not completed. The retarded children consistently blamed themselves, whereas the non-retarded children used a variety of excuses to place the responsibility on others rather than themselves.

Our studies of expectancy of success have often used a three-choice discrimination-learning task where two choices are never reinforced and one is rewarded only part of the time. Children who expect success learn the proper choice more slowly, because they are busy formulating strategies which will result in 100 percent success (e.g., Gruen and Zigler, 1968; Kier, Styfco, and Zigler, 1977). Retarded children and others who expect failure learn quickly because they are content with being right just some of the time. To determine if these findings might be explained by cognitive rigidity, we also employed intense success and failure preconditions (Ollendick, Balla, and Zigler, 1971). We found that failure resulted in a low expectancy of success, while positive experiences raised expectancy of success. The impact of this finding is highlighted by another report (Zeaman and House, 1963) that retarded persons who experienced a series of failures became unable to solve simple learning problems that they previously mastered easily. In a more life-like school situation, Gruen, Ottinger, and Ollendick (1974) found that retarded children from mainstreamed classrooms had lower expectancies of success than those from segregated special education classes--presumably because the mainstreamed children were exposed to a greater amount of failure.

Social learning experiences acquired fairly early in life also appear to influence a child's motivation for particular rewards. For example, familial retarded children seem less responsive to intangible reinforcement than are intellectually-average children. Work, in this area (reviewed by Havighurst, 1970) is of particular importance since intangible incentives are most frequently offered in real life. Studies have shown that retarded and lower SES children may perform better on a variety of tasks if their reward is something tangible. Children from middle SES homes generally do better with intangible rewards such as being told they are correct--and this has been found for Down syndrome children as well as for those of average IQ (Byck, 1968). However, we should note that studies of optimal reinforcement have not had clear-cut results. We have found middle SES children to be responsive to both intangible and tangible rewards, and interestingly, upper SES children to switch concepts more readily inter-tangible rather than intangible reinforcement condition (Zigler and Unell, 1962).

As in the case of particular reinforcers, the strength of the effectance motive may be different for retarded and nonretarded persons. Work in this area owes much to White's (1959) formulation that using one's cognitive resources to their fullest is intrinsically gratifying and thus motivating. The desire to be effective shows up in behavior such as curiosity, exploration, and a willingness to take up challenges and attempt problem-solving. We have found retarded children to be less motivated by a need to be effective than are nonretarded children (Harter and Zigler, 1974). But here again experience is important. This lack of effectance motivation was particularly pronounced for retarded persons living in institutions. Thus, although retarded children on the average may value being correct or effective less than middle SES children on the average, the crucial factor is not membership in a particular social class or intellectual level per se, but rather the particular social learning experiences.

Another behavioral trait we have found to be characteristic of retarded individuals is their outerdirectedness (see Balla et al., 1976). It has been observed that retarded children are very sensitive to cues provided by an adult and are highly imitative. Of course children at lower levels of cognitive development should be more outer-directed than those at higher levels. With relatively limited experience and cognitive resources, reliance on cues from others to guide behavior is in fact adaptive. However, either too little or too much imitation can be a negative psychological indicator. If the child never imitates an adult, it may be that he or she has come to mistrust adults and thus cannot profit from their guidance. Excessive imitation can indicate a distrust of one's own abilities. Some intermediate level of imitation is viewed as a positive developmental phenomenon, reflecting the child's healthy attachment to adults and responsivity to cues from adults which can be helpful in problem-solving.

In general, we have found that outerdirectedness decreases with higher mental age. This has been found for children of average IQ as well as for retarded children whether institutionalized or not (e.g., Balla, Styfco, and Zigler, 1971; Zigler and Yando, 1972). However, presumably because of their histories of failure, retarded children are more outerdirected than nonretarded children of the same MA. (Excessive outerdirectedness has also been found in non-retarded children following induced failure experiences.) It seems reasonable to expect that children who have an environment adjusted to their developmental level will be less imitative than children in an environment where they are confronted with their intellectual shortcomings and experience considerable failure. Indeed, we have found that noninstitutionalized retarded children rely more on external cues on certain tasks than do retarded children living in institutions (e.g., Achenbach and Zigler, 1968). The school setting of retarded children living at home can make a difference too. In one study (Lustman, Balla and Zigler, 1977) we discovered a small group of children who were active nonimitators, and they were all from a self-contained special education classroom where failure was apparently a nonexistent word.

No discussion of motivational factors in retarded persons would be complete without special mention of the effects of institutionalization. Many of the studies reported in the mental retardation literature have compared institutionalized retarded and noninstitutionalized nonretarded individuals. Thus, there is a recurring ambiguity in interpretation: Do these studies inform us about the effects of intellectual deficit, institutionalization, or some interaction of these factors?

There is little question that, at least before the advent of small community-based facilities, the prevalent position was that institutions had extremely negative and monolithic effects on development. There was certainly support for this view in both the psychological and sociological literature. In our eagerness to blame institutions for everything, we hardly noticed some findings that institutions can also have beneficial effects. There have been scattered reports in several studies of overall increases in IQ following institutionalization (e.g., Balla and Zigler, 1975; Clarke, Clarke, and Reiman, 1958). Increasing length of institutionalization has also been associated with greater behavior variability and autonomy in problem solving, and with decreased verbal dependency and imitation (Balla et al., 1974; Yandon and Zigler, 1971).

Some of the most revealing research on institutional effects has concerned their relation to social deprivation. Indeed, institutionalization has often been considered the epitomy of a life of deprivation. In one longitudinal study (Zigler and Williams, 1963) we found that after three years of institutional experience, residents became more responsive to social reinforcement. However, this increase was related to the extent of preinstitutional social deprivation. Institutionalization was less depriving for persons from very deprived backgrounds than for those from relatively good homes. In contrast to these findings, we found retarded residents to become less responsive to social reinforcement over the three years of another study (Zigler, Balla, and Butterfield, 1968). Persons from relatively good homes demonstrated a smaller decrease in this responsiveness than did persons from poorer homes. These inconsistent findings appeared to be due to differences in quality of the institutions studied. The institution in the first study was apparently depriving, while the one in the second study had practices which ameliorated the effects of preinstitutional deprivation.

Extreme deprivation, however, does not go away so readily. In a follow-up of individuals in the second study, we found the effects of preinstitutional deprivation were still in evidence after six years of institutional experience (Balla and Zigler, 1975). Organically retarded persons who came from homes characterized by marital discord, mental illness, and/or child abuse were more responsive to social reinforcement for all six years than those who had been less deprived. In another longitudinal study (Zigler, Butterfield, and Capobianco, 1970), we found discernible effects of severe preinstitutional deprivation even after ten intervening years of institutional experience. We cannot overemphasize the importance of these findings. It seems that social deprivation experiences become part of the personality structure of the individual and forever mediate his or her interactions with the environment.

Our work has taken us to so many institutions that we could not help noticing striking differences among them. Thus began cross-institutional studies. Butterfield and Zigler (1965) found that even a large central institution could provide a home-like atmosphere and be less depriving to residents than a facility with the locked-ward atmosphere which stereotypes institutions in our minds. We also examined how several institutional demographic variables might affect residents (Balla et al., 1974). These included such things as size, number of residents per living unit, cost per resident per day, employee turnover rate, and numbers of direct care and professional personnel per resident. Over the course of 2½ years in one study, we found that in all four institutions we investigated, the residents showed considerable psychological growth. Somewhat surprisingly, none of the objective characteristics of the institutions was found to be related to the residents' motivational traits. The one exception was size, in that residents of the largest institution were more responsive to social reinforcement. So we did another comparison of the behavior of persons residing either in large central institutions or in small regional centers (Balla et al., 1976). This time we found no differences. This was another surprise, since the average size of the largest institutions was over 1,600, while the regional centers averaged only 111. The number of aides per resident and the cost per day were twice as high in the regional centers, and the proportion of professional staff was almost six times as great. Simply increasing cost or staff or the fact of placement in a small regional center did not seem, in and of themselves, to ensure greater behavioral competency.

Of course the findings we emphasize here are just to point out that large institutions are not necessarily synonymous with the diminution of life. We all know that institutions can have serious detrimental effects on their residents. What we are saying is that their effects do not have to be bad, nor are they necessarily related to the aspects of institutions that we most often blame. We have come to believe that the question of the effects of institutionalization is a very complex one, and that it cannot be answered without first considering several factors. These include the characteristics of the retarded person such as age, gender, and diagnosis, the nature of the person's preinstitutional life experiences, and the nature of the institution both in demographic and social/psychological terms.

In conclusion, we assert that the total body of evidence concerning motivation and the retarded person is of considerable importance. We think that many of the reported differences between retarded and intellectually-average children of the same MA are a result of motivational and emotional differences that reflect variations in experiential histories. This is not to say that we believe the cause of cultural-familial mental retardation can be explained in terms of motivation. The cognitive functioning of retarded persons unquestionably has a profound effect on their behavior. The crucial questions are just how great is this influence and how does it differ across tasks with which retarded people are confronted? We would like to think that if we could change the motivational stance of many retarded persons, they would have a better chance to become self-sustaining members of society rather than be consigned to a life of dependency and neglect.

References

Achenbach, T., and Zigler, E. Cue-learning and problem-learning strategies in normal and retarded children. Child Development, 1968, 3, 827-848.

Balla, D., Butterfield, E. C., and Zigler, E. Effects of institutionalization on retarded children: A longitudinal cross-institutional investigation. American Journal of Mental Deficiency, 1974, 78, 530-549.

Balla, D., Kossan, N., and Zigler, E. Effects of preinstitutional history and institutionalization on the behavior of the retarded. Unpublished manuscript, Yale University, 1976.

Balla, D., McCarthy, E., and Zigler, E. Some correlates of negative reaction tendencies in institutionalized retarded children. Journal of Psychology, 1971, 79, 77-84.

Balla, D., Styfco, S. J., and Zigler, E. Use of the opposition concept and outerdirectedness in intellectually-average, familial retarded, and organically retarded children. American Journal of Mental Deficiency, 1971, 75, 663-680.

Balla, D., and Zigler, E. Preinstitutional social deprivation and responsiveness to social reinforcement in institutionalized retarded individuals: A six-year follow-up study. American Journal of Mental Deficiency, 1975, 80, 228-230.

Balla, D., and Zigler, E. Mental retardation. In M. Hersen, A. Kazdin, and A. Bellack, (Eds.), New perspectives in abnormal psychology. New York: Oxford University Press, in press.

Butterfield, E. C., Wambold, C., and Belmont, J. M. On the theory and practice of improving short-term memory. American Journal of Mental Deficiency, 1973, 77, 654-669.

Butterfield, E. C., and Zigler, E. The influence of differing institutional social climates on the effectiveness of social reinforcement in the mentally retarded. American Journal of Mental Deficiency, 1965, 70, 48-56.

Byck, M. Cognitive differences among diagnostic groups of retardates. American Journal of Mental Deficiency, 1968, 73, 97-101.

Clarke, A. D. B., Clarke, A. M., and Reiman, S. Cognitive and social changes in the feebleminded--three further studies. British Journal of Psychology, 1958, 49, 144-157.

Cromwell, R. L. A social learning approach to mental retardation. In N. R. Ellis (Ed.), Handbook of mental deficiency. New York: McGraw-Hill, 1963.

Gruen, G., Ottinger, D., and Ollendick, T. Probability learning in retarded children with differing histories of success and failure in school. American Journal of Mental Deficiency, 1974, 79, 417-423.

Gruen, G., and Zigler, E. Expectancy of success and the probability learning of middle-class, lower-class, and retarded children. Journal of Abnormal Psychology, 1968, 73, 343-352.

Harter, S., and Zigler, E. The assessment of effectance motivation in normal and retarded children. Developmental Psychology, 1974, 10, 169-180.

Havighurst, R. J. Minority subcultures and the law of effect. American Psychologist, 1970, 25, 313-322.

Kier, R. J., Styfco, S. J., and Zigler, E. Success expectancies and the probability learning of children of low and middle socio-economic status. Developmental Psychology, 1977, 13, 444-449.

Lustman, N. M., Balla, D., and Zigler, E. Imitation in institutionalized and noninstitutionalized retarded children and in children of average intellect. Unpublished manuscript, Yale University, 1977.

MacMillan, D. L., and Keogh, B. K. Normal and retarded children's expectancy for failure. Developmental Psychology, 1971, 4, 343-348.

Mercer, J. R. Labelling the mentally retarded. Berkeley: University of California Press, 1973.

Ollendick, T., Balla, D., and Zigler, E. Expectancy of success and the probability learning of retarded children. Journal of Abnormal Psychology, 1971, 77, 275-281.

Penrose, L. S. The biology of mental deficiency. London: Sidgwick and Jackson, 1963.

Weisz, J. R., and Zigler, E. Cognitive development in retarded and nonretarded persons: Piagetian tests of the similar sequence hypothesis. Psychological Bulletin, 1979, 86(4),

831-851.

White, R. W. Motivation reconsidered: The concept of competence. Psychological Review, 1959, 66, 297-333.

Yando, R., and Zigler, E. Outerdirectedness in the problem-solving of institutionalized and noninstitutionalized normal and retarded children. Developmental Psychology, 1971, 4, 277-288.

Zeaman, D., and House, B. J.   The role of attention in retardate discrimination learning. In N. R. Ellis (Ed.), Handbook of mental deficiency.   New York: McGraw-Hill, 1963.

Zigler, E.   Research on personality structure in the retardate. In N. R. Ellis (Ed.), International review of research in mental retardation (Vol. 1). New York: Academic Press, 1966.

Zigler, E. Developmental versus difference theories of mental retardation and the problem of motivation. American Journal of Mental Deficiency, 1969, 73, 536-556.

Zigler, E. Dealing with retardation. Science, 1977, 196, 1192-1194.

Zigler, E., and Balla, D. Developmental course of responsiveness to social reinforcement in normal children and institutionalized retarded children. Developmental Psychology, 1972, 6, 66-73.

Zigler, E., Balla, D., and Butterfield, E. C. A longitudinal investigation of the relationship between preinstitutional social deprivation and social motivation in institutionalized retardates. Journal of Personality and Social Psychology, 1968, 10, 437-445.

Zigler, E., Butterfield, E. C., and Capobianco, F. Institutionalization and the effectiveness of social reinforcement: A five- and eight-year follow-up study. Developmental Psychology, 1970, 3 255-263.

Zigler, E., and Unell, E. Concept-switching in normal and feeble-minded children as a function of reinforcement. American Journal of Mental Deficiency, 1962, 66, 651-657.

Zigler, E., and Williams, J. Institutionalization and the effectiveness of social reinforcement: A three-year follow-up study. Journal of Abnormal and Social Psychology, 1963, 66, 197-205.

Zigler, E., and Yando, R. Outerdirectedness and imitative behavior of institutionalized and noninstitutionalized younger and older children.   Child Development, 1972, 43, 413-425.

REACTION TIME AND INTELLIGENCE

Arthur R. Jensen

University of California

Berkeley, California, U.S.A.

Abstract

Measurements of various parameters derived from different reaction time (RT) paradigms are found to be correlated with psychometric measurements of general mental ability. Such RT-derived measurements, when combined in a multiple regression equation, predict some 50 percent or more of the variance in IQ or g. This relationship of IQ or g to RT parameters indicates that our standard IQ tests tap fundamental processes involved in individual differences in specific knowledge, acquired skills, or cultural background.

This article reviews the main currents in research on the relationship of reaction time (RT) to general intelligence and other psychometric mental abilities.

The first conclusion we can draw with confidence is that RT parameters in a variety of paradigms are significantly related to scores on standard tests of intelligence and other psychometric abilities. As I have noted elsewhere (Jensen, 1979), the study of RT as a measure of mental ability got off to a bad start in the early history of psychology, for a number of reasons, largely due to psychometric naivete and inadequate statistical methods. Modern investigators have been more successful in finding substantial and replicable relationships between RT and IQ.

Correlation coefficients between RT and IQ are not as impressive or as consistent as are mean differences in RT between different criterion groups selected on the basis of IQ or other psychometric indices of ability. Correlations between RT and IQ can be generally characterized as fairly low. But in the entire literature on RT and IQ there are virtually no correlations on the "wrong" side of zero. Most rs fall in the range from 0 to -.50, with a mode in the -.30's. A correlation of -.50 is about maximum. It is theoretically important to understand the causes of this apparent low correlation ceiling. But there is no doubt that the present evidence overwhelmingly rejects the null hypothesis. This is true of simple RT as well as choice RT (also termed discriminative or disjunctive RT). Both simple and choice RT are negatively correlated with IQ.

Mean differences in RT (or in various parameters of RT) between criterion groups selected for differences in ability as measured by psychometric tests or scholastic performance always give more clearly impressive evidence of a relationship between RT and general ability than the correlation coefficient. The mean RT difference between criterion groups is often of at least the same magnitude as the mean IQ difference between the groups, when the mean differences in RT and IQ are both expressed in standard deviation or σ units. We have found that borderline retarded young adults, with a mean IQ of about 70, differ from university students about 6σ on Raven's Matrices. These groups differ about $7\,\sigma$ (σ of the university students) in mean RT. University students compared with academically less highly selected students of the same age in a two-year vocational college differ about $1\,\sigma$ in scholastic aptitude scores; in mean RT they differ $1.2\,\sigma$ in terms of the vocational college σ and 1.9σ in terms of the universityσ.

From the standpoint of psychometrics, I think the most important conclusion from all the RT research is that it proves beyond reasonable doubt that our present standard tests of IQ measure, in part, some basic intrinsic aspect of mental ability and not merely individual differences in acquired specific knowledge, scholastic skills, and cultural background. The RT parameters derived from typical procedures cannot possibly measure knowledge, intellectual skills, or cultural background in any accepted meaning of these terms. Yet these RT parameters show significant correlations with scores on standard tests of mental ability and scholastic achievement and show considerable mean differences between criterion groups selected on such measures.

Three Basic RT Paradigms

There are three distinct and basic paradigms in RT research. Each paradigm measures different facets of information processing speed, and each has shown a relationshp to psychometric variables. I shall refer to these paradigms by the names of the three psychologists who initiated them.

The Hick paradigm measures the linear increase in RT to visual or auditory stimuli as a function of the amount of information (measured as bits=$\log_2$ of the number of stimulus alternatives) conveyed by the reaction stimulus, but involves no need to access either short-term or long-term memory (STM or LTM). The classical experiment contrasting simple and two-choice RT is the simplest example of the Hick paradigm, involving 0 and 1 bit of information, respectively.

The Sternberg (1966) paradigm presents the subject with a small set of digits (or letters) followed immediately by a single "probe" digit to which the subject responds "yes" or "no" as to whether the probe was or was not included in the set. The S's RT or decision time in pressing the "yes" or "no" key involves speed of scanning STM, and RT increases as a linear function of the number of items in the set, unlike the Hick phenomenon, in which RT increases as a linear function of the logarithm (to the base 2) of the number of stimulus alternatives.

The Posner (1969) paradigm contrasts discriminative ("same" versus "different") RTs to pairs of stimuli which are the same or different either physically or semantically. For example, the letters AA are physically the same, whereas Aa are physically different but semantically the same. When Ss are instructed to respond "same" or "different" to the physical stimulus, RTs are faster than when Ss must respond to the semantic meaning. The physical discrimination is essentially the same as classical discriminative RT, but RT in the semantic discrimination involves access to semantic codes in LTM, which takes considerably more time than physical discriminative RT. The difference between semantic and physical RT thus measures access time to highly overlearned semantic codes in LTM. Interestingly, Hunt (1976) and his co-workers have found that this measurement is especially related to verbal ability as measured by the Scholastic Aptitude Test (SAT-V) in university students.

Typical Findings

Posner Paradigm. Figure 1 shows the results of a study by Hunt (1976) using the Posner paradigm with groups of university students scoring high or low on the SAT-Verbal. AA represents the physical identity choice (same-different) RT task; Aa represents the semantic identity task. University students require on the average about 75 milliseconds more time to respond to Aa than to AA types, which is the time taken by semantic encoding of the stimulus. Two features of Figure 1 are particularly interesting in relation to findings from the Sternberg and Hick paradigms: (1) the high and low groups on SAT-V show a mean difference in RTs even on the physical, nonsemantic identity task, which is essentially just a form of classical two-choice discriminative RT;

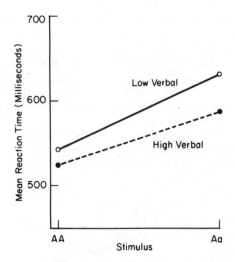

Figure 1. Time required to recognize physical or semantic identity
of letter pairs by university students who score in the
upper (high) or lower (low) quartile on the SAT-Verbal.
(After Hunt, 1976, Table 1, p. 244.)

and (2) the mean RTs are all greater than 500 milliseconds, which
is appreciably slower than the RTs of university students in the
Hick paradigm, even for RT to three bits (i.e., eight stimulus
alternatives) of information, which has a mean RT of 350 to 400
msec.   Because the times needed for physical discrimination
between extremely familiar stimuli and for accessing simple, highly
overlearned semantic codes in LTM are in excess of the RTs to
three bits of information in the Hick paradigm, it suggests that
performance in our Hick paradigm does not depend on discriminating anything as difficult as familiar letters or accessing anything
in LTM.   The average RT difference between A̲A̲ and A̲a̲ (i.e.,
semantic encoding time) of 75 msec for Hunt's university students
is exactly the same as the difference in RT between 0 and 3 bits
of information in our Hick paradigm with university students.

Sternberg Paradigm.   Figure 2 shows Sternberg STM-scan
RTs for groups of fifth and sixth grade children with moderate
and high IQs, from a study by McCauley et al. (1976).   The
intercepts and slopes of the moderate and high IQ groups both
differ significantly.   Stanford University students given a comparable Sternberg task (Chiang and Atkinson, 1976) show much lower
intercepts (about 400 msec) but show about the same slope (i.e.,
a scan rate of 42 msec per digit in target set) as the high IQ
children (with a scan rate of 40 msec per digit), whose IQs (with
a mean of 126) are probably close to the IQs of the Stanford
students.   The moderate IQ group has a significantly greater
slope (i.e., slower STM scanning rate) of 58 msec per digit. IQ

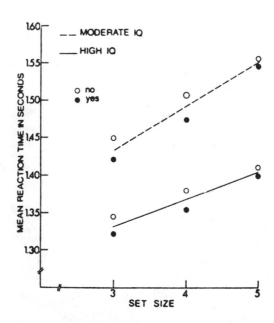

Figure 2. Mean RTs for correct "yes" and "no" (i.e., presence or
absence of probe digit in target set) for moderate IQ (95
or below, X=88) and high IQ (115 or above, X=126) fifth
and sixth grade children. The equations for the two lines
are: moderate IQ RT = 1265 + 58s, and high IQ RT = 1210
+ 40s, where RT is in milliseconds and s = number of
digits in the target set. (From McCauley et al., 1976.)

would appear to be more crucial than mental age for short-term
memory scan rate. This has interesting implications for scanning
and rehearsal of information in STM to consolidate it into LTM.
In terms of such a model, and in view of the observed differences in scan rates as a function of IQ, it should seem little wonder
that high IQ persons in general know more about nearly everything than persons with low IQs. Snow, Marshalek, and Lohman
(1976) were able to "predict" the intercepts and slopes of the
Sternberg memory scan paradigm for individual Stanford students
with multiple R's of .88 and .70, respectively, using scores on
several psychometric tests (in addition to sex). The intercept
and slope parameters of the Sternberg scan, on the other hand,
predicted each of four factor scores derived from a large battery
of psychometric tests with R's between .33 and .56. SAT-Verbal
and SAT-Quantitative scores were predicted with R's of .54 and
.21, respectively. Remember, we are dealing here with the quite
restricted range of ability in Stanford University students.

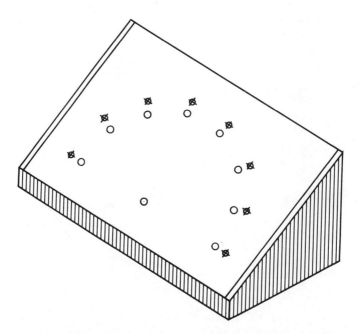

Figure 3. Subject's console of the reaction time-movement time apparatus. Push buttons indicated by circles, green jeweled
lights by circled crosses. The "home" button is in the
lower center.

     Hick Paradigm.   This is an elaboration of simple and choice
RT. Hick (1952) discovered that RT increases linearly as a function of log$_2$ of the number of choices or stimulus alternatives -- a
phenomenon now known as Hick's Law. I have been doing studies
of this paradigm, using an apparatus show in Figure 3. (It is
described in more detail by Jensen and Munro, 1979.) The S
places his index finger on the "home" button, a "beep" warning
signal is sounded for 1 second, and after a random interval of 1
to 4 seconds one of the green lights goes on. The S must turn
off the light as fast as possible by touching the button adjacent
to it. The time between the light's going on and removal of the
S's finger from the home button is the RT. The interval from
release of the home button to turning out the light is the movement time (MT). Templates can be placed over the console to
expose any number of light/button alternatives from 1 to 8. We
have most often used 1, 2, 4, and 8 alternatives, corresponding
to 0, 1, 2, and 3 bits of information. Following instructions and
several practice trials, Ss are usually given 15 trials on each

number of alternatives (60 trials in all) in a single session lasting about 20 minutes.

To insure that RT is in fact related to intelligence, I have sought correlations between RT parameters and IQ in criterion groups selected from every available level of the IQ distribution, ranging from the severely retarded (with IQs of 15 to 50), to the mildly retarded and borderline (IQs 50 to 80 or so), to average and bright school children and average young adults, and to university students with IQs above the 95th percentile of population norms. We have now tested nine such groups totalling about 800 persons. Without exception, groups differing in mean IQ also differ very significantly in the expected direction in a number of RT (and also MT) parameters. Also, within every group we have tested, the RT parameters are significantly correlated with IQ, with all correlations in the theoretically expected direction, mostly ranging between about .20 and .50. Many of these findings have been described elsewhere (Jensen, 1979; Jensen and Munro, 1979).

We describe an individual's RT performance in the Hick paradigm in terms of three parameters: the slope of the linear regression of RT on bits, the intercept of the regression line, and the intraindividual variability over trials, which is indexed by the root mean square of the variances among trials within bits. (We have also used the slope of the regression of the standard deviation among trials, as a function of bits.) Individual differences in all of the RT parameters are positively intercorrelated. Other investigators, too, have found a positive correlation between intercepts and slopes in the Sternberg paradigm (Dugas and Kellas, 1974; Snow et. al, 1976; Oswald, 1971). Moreover, all these parameters are negatively correlated with g. At first I expected that intercepts, which represent simple RT, and hence involve little or no information processing, would not be correlated with IQ. I was wrong; intercepts are negatively correlated with IQ, although within fairly homogeneous criterion groups the correlations are often too small to be significant and are almost invariably smaller than the correlations of slope and intraindividual variability with IQ. Figure 4 shows the intercepts and slopes of RT data from seven criterion groups. None of the regression lines except that of the severly retarded group shows a significant nonlinear trend.

Intraindividual Variability. Surprisingly little attention was ever given to intraindividual variability in RT in the older literature. Yet it is this aspect of individual differences in RT that seems to be the most profoundly related to intelligence level, as has been frequently noted by investigators of RT in the mentally retarded (Berkson and Baumeister, 1967; Baumeister and Kellas, 1968a, 1968b, 1968c; Liebert and Baumeister, 1973; Wade, Newell, and Wallace, 1978; Vernon, 1979). The negative correlation

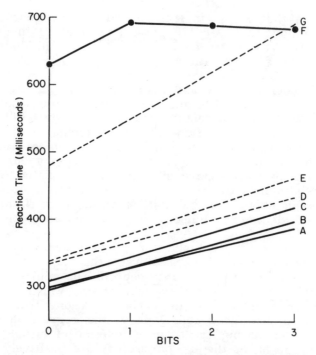

Figure 4. RT as a function of bits, illustrating Hick's law and dif-
        ferences in intercepts and slopes, for diverse groups vary-
        ing in age and intelligence: A - university students, B -
        ninth grade girls, C - 6th graders in a high SES-high IQ
        school, D and E - white and black, respectively, male vo-
        cational college freshmen with approximately equal scholas-
        tic aptitude scores, F - severely mentally retarded young
        adults (mean IQ 39), G - mildly retarded and borderline
        young adults (mean IQ 70). (From Jensen, 1979.)

between intraindividual variability in RT and IQ is found within
every level of intelligence, from the severely retarded to university students.

    I have looked more closely at this phenomenon in our data
by rank ordering each S's RTs from the shortest to the longest
in 15 trials. (The 15th rank is eliminated to get rid of possible
outliers.) Figure 5 shows the means of the ranked RTs of 46
mildly retarded (IQ 70) and 50 bright normal (IQ 120) young
adults each given 15 trials on simple (0 bit) RT. Note that even
on the fastest trial (rank 1) the retarded and normal Ss differ by
111 msec. In fact, the normal Ss' slowest RT (rank 14) is 32 msec

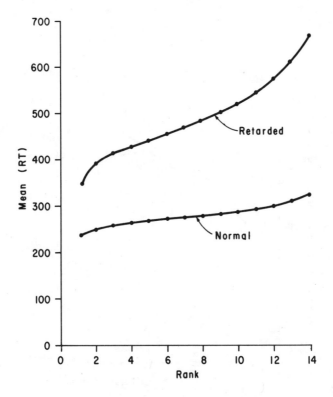

Figure 5. Mean simple RT plotted after ranking RTs on 15 trials
from the fastest to the slowest trial (omitting the 15th
rank) for retarded and normal S̲s.

shorter that the retardates' fastest RT. In case anyone might
think these are trivial differences, let us look at then in terms of
standard deviation or σ units, i.e. (normal RT minus retarded
RT)/σ, as shown for simple RT in Figure 6 for σ differences
based on both normal and retarded σ units. The fastest simple
RT of retardates and normals differs 1.2σ in terms of the retardates' σ units and 4.8 σ in terms of the normals' σ units.

The fact that even the fastest RTs of the retarded S̲s are
slower than the RTs of normals, even for simple RT, suggests
that the difference is at some very basic, one might almost say
neural, level and not at any very complex level of information
processing. Possibly even simpler responses might show reliable
speed differences related to general intelligence.

Combining RTs in the Hick, Sternberg, and Posner Paradigms.

If RT and the derived parameters in the three different

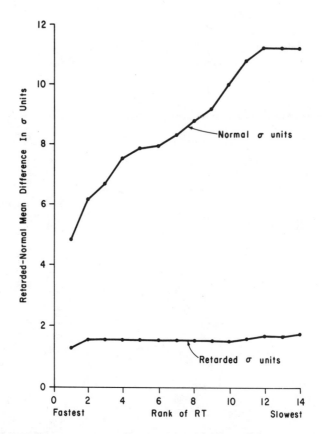

Figure 6. Difference in simple RT between retarded and normal Ss,
         expressed in both normal and retardate σ units, with
         RTs for 15 trials ranked from fastest to slowest.

paradigms reflect different processes, involving stimulus encoding,
scanning of STM, and retrieval of semantic codes in LTM, all of
which are probably involved in arriving at the correct answers to
the relatively complex items used in ordinary intelligence tests,
we should expect that an optimally weighted combination of RT
measurements derived from all three paradigms should show a
much more substantial correlation with mental test scores than
measurements derived from any one RT paradigm.   This is exaclty
what Keating and Bobbitt (1978) found.   Three RT-derived measures were obtained on each S: (1) choice RT minus simple RT
(Hick paradigm), (2) semantic minus physical same/difference RT
to letter pairs (Posner paradigm), and (3) slope of RT on set
size with sets of 1, 3, or 5 digits (Sternberg paradigm).   The
multiple R of these three measurements with Raven scores of 60
school children in grades 3, 7, and 11 was .59, .57, and .60, in

the three grades, respectively. I imagine that still higher correlations would be obtained if intraindividual variability were taken into account and if the correlations were corrected for attenuation using the between days test-retest stability coefficients. The average intercorrelation among the three paradigm measures was only .27, indicating that they are tapping different processes as well as sharing some variance in common.

The burning question is this: Will it be possible to discover a small number of such basic processes, measurable by means of RT, that will yield parameters which, in an optimally weighted combination, will "account for" practically all of the true g variance in psychometric tests of mental ability? Might not differentl y weighted combinations of a few process measurements based on RT also account for the variance in the so-called group factors involved in verbal, quantitative, and spatial abilities? This is what we must try to find out. Whatever the outcome may be, the effort will be amply rewarded by the gain in our theoretical understanding of the nature of mental abilities, to say nothing of the potential for practical applications should it turn out that most of the variance in complex mental abilities now measured by psychometric tests can be accounted for in terms of a number of RT parameters in a few fundamental paradigms.

References

Baumeister, A. A., and Kellas, G. Reaction time and mental retardation. In N. R. Ellis (Ed.), International Review of Research in Mental Retardation, Vol. 3. New York: Academic Press, 1968. (a)

Baumeister, A. A., and Kellas, G. Distribution of reaction times of retardates and normals. American Journal of Mental Deficiency, 1968, 72, 715-718. (b)

Baumeister, A. A., and Kellas, G. Intrasubject response variability in relation to intelligence. Journal of Abnormal Psychology, 1968, 73, 421-423. (c)

Berkson, G., and Baumeister, A. A. Reaction time variability of mental defectives and normals. American Journal of Mental Deficiency, 1967, 72, 262-266.

Chiang, A. and Atkinson, R. C. Individual differences and interrelationships among a select set of cognitive skills. Memory and Cognition, 1976, 4, 661-672.

Dugas, J., and Kellas, G. Encoding and retrieval processes in normal children and retarded adolescents. Journal of Experimental Child Psychology, 1974, 17, 177-185.

Hick, W. On the rate of gain of information. Quarterly Journal of Experimental Psychology, 1952, 4, 11-46.

Hunt, E. Varieties of Cognitive Power. In L. B. Resnick (Ed.) The Nature of Intelligence. Hillsdale, N.J.: Erlbaum, 1976. P. 237-259.

Jensen, A. R. g: Outmoded theory or unconquered frontier?
  Creative Science and Technology, 1979, 2, 16-29.
Jensen, A. R., and Munro, E. Reaction time, movement time, and
  intelligence. Intelligence, 1979, 3, 121-126.
Keating, D. P., and Bobbitt, B. Individual and developmental
  differences in cognitive processing components of mental
  ability. Child Development, 1978, 49, 155-169.
McCauley, C., and Dugas, J., and Kellas, G., and DeVellis, R. F.
  Effects of serial rehearsal training on memory search. Journal
  of Educational Psychology, 1976, 68, 474-481.
Oswald, W. D. Über Zusammenhange zwischen Informationgescwin-
  digkeit, Alter und Intelligenzstruktur beim Kartensortieren.
  Psychologische Rundschau, 1971, 22, 197-202.
Posner, M. I. Abstraction and the process of recognition. In
  G. H. Bower and J. T. Spence (Eds.) The Psychology of
  Learning and Motivation (Vol. 3). New York: Academic Press,
  1969, 43-100.
Snow, R. E., Marshalek, B., and Lohman, D. F. Correlation of
  selected cognitive abilities and cognitive processing parameters: An explanatory study. Technical Report No. 3., Aptitude Research Project, School of Education, Stanford University, December, 1976.
Sternberg, S. High speed scanning in human memory. Science,
  1966, 153, 652-654.
Vernon, P. A. Reaction time and intelligence in the mentally retarded. Unpublished paper, 1979.
Wade, M. G., Newell, K. M., and Wallace, S. A. Decision time
  and movement time as a function of response complexity in
  retarded persons. American Journal of Mental Deficiency,
  1978, 83, 135-144.

INTELLIGENCE AND LEARNING: SPECIFIC AND GENERAL HANDICAP

N. O'Connor and B. Hermelin

Medical Research Council

London, England

Abstract

It is difficult to ignore the value of normative psychometrics and the resultant concept of intelligence in the study of groups of low IQ. However, such an approach ignores the advances made through the study of cognitive processes in the subnormal. Such studies generate dynamic hypotheses which the psychometric approach does not, although the linear information flow assumptions characteristic of the latter are questionable on neuropathological grounds. In consequence we sought an alternative strategy.

The neuropsychological model is attractive but presents problems in the study of children because of the compensatory mechanisms common in a developing organism. We therefore chose our examples of "localised" injury from the "peripherally" handicapped, i.e. the congenitally blind and deaf. Such groups were compared with groups with central neuropathology such as the severely subnormal. Absence of a modality was found to lead to alternative strategies also occurred in the centrally handicapped. Comparisons are made and the reason for similarities and differences are discussed.

Introduction

The first combination of the two words "intelligence" and "learning" and the concepts they represent was made by Binet. It is problematical whether he used the words as virtual synonyms because intelligence had formerly meant knowledge, new information, or what is learned. So the two words might have been seen as

related in the sense of substantive and gerund. Certainly,
Binet's construction of his early tests included a strong element
of information. One part of his test was based on information
taught in schools and might well figure these days in any terminal
test of scholastic progress arranged by school years. It is this
aspect of Binet's original test which has received most attention
subsequently, his physiological tests such as, for example, tests
of two-point threshold having been allowed to disappear. These
latter Binet introduced because he had observed a connexion
between economic, physiological and cognitive deprivation. It was
this aspect of his work which appealed to Galton, Pearson and
Spearman because of their involvement, sometimes extreme, in the
Genetic Reform movement.

In consequence, Spearman developed his notion of a hierarchical structure of intelligence with its general and specific
components. By general, of course, we understand the positive
correlation of performance across many tasks and by specific we
understand particular tasks or abilities not especially correlated
with others. Spearman and Binet originally demonstrated a positive
correlation between different scholastic abilities and it is essentially this scholastic ability which has come to be thought of as
intelligence, although there is a sense in which the definition is
circular except in so far as it assumes the relative permanence of
the scholastic skill. However, without the positing of a physiological connexion, the concept of intelligence has proved both
stable and sterile. Stable in the sense that measurable development
like the development of height, gives rise to few dramatic surprises
once its relative course is determined and in the absence of
serious physiological insult; sterile in the sense that so far as
subnormality or early cognitive handicap is concerned, perhaps
because of the circularity of derivation mentioned above, intelligence levels would appear to explain both everything and nothing.
Commonly, as with Binet, low intelligence correlates with failure
to learn, largely because that is how it is defined. The specification of intelligence therefore has little explanatory value unless
it can be substantially defined in independent physiological or
other terms.

So far as subnormality is concerned, this has been quite
hard to accomplish for a number of reasons. Extensive damage to
the brain and the nervous system from birth or soon after has
the peculiar effect of retarding all aspect of learning and not
selectively damaging specific functions. Extensive damage has
this stunting effect to such a degree that exceptional selective
damage is rare and so called receptive childhood aphasia or dys-
phasia, a case in point, is assumed to result from specific bilateral injury in a generally undamaged nervous system. Such
specific cases, apart from occurring so very rarely that authenticated cases are found less frequently than 4 per 10,000 live

births, are almost unique examples of their kind. Even autistic children have intelligence levels about 30 points below average in some 66% of cases and those with above normal IQs or very high IQs are a group of very great rarity indeed. To the best of my knowledge, research has not yet identified 100 cases in Great Britain. Clearly, then, if specific deficits are hard to find in children, the techniques of neuropsychology, so effective with adults with developed nervous systems, are inappropriate with children.

It is perhaps not surprising therefore that few psychologists working in this area have successfully discovered a good methodology for the study of handicap. Disturbed by the unproductive character of intelligence test results they have in recent decades been attracted by models of information processing. These models which have been largely linear and successive in character have led them to hypthesize an explanation of learning or processing failure which no longer needed to depend on a failure of general ability--a failure which appeared to generate no hypotheses--but could be seen as a widespread failure, retardation or stunting of learning which could be accounted for in terms of a break in the learning chain. Such breaks for instance were envisaged by Zeaman and House (1963) as attentional, i.e. selective, as short term memory weakness by Ellis (1963) or later as a weakness of rehearsal (1970), as a secondary signalling system failure by Luria (1961) and as a cross-modal coding deficit by O'Connor and Hermelin (1963).

Most of these approaches tried to account for general learning failure in terms of a specific deficit and had the advantage of appearing to have a bearing on the learning process by appearing to explain it. In many ways therefore this approach was an advance on the measurement of intelligence as an explanatory paradigm. Unfortunately, it also has its weaknesses. These are chiefly that the model so useful in the neuropsychology of adults is inapplicable with children, especially severely handicapped children. The concept of a broken chain is inadequate as an explanation for overall learning failure, primarily because pathological and psychological findings indicate strongly that not just one link in the chain is damaged, but all links.

Another objection to the concept of a successive chain is that the chain is so interlinked both forward and backward, that the motion of a successive direction for boxes in a flow diagram, must be seen as a useful but misguided conception.

A caveat must be inserted at this point because it would be wrong to give the impression that measuring intelligence is a waste of time. Nor must one conclude that all those experimenters including ourselves, who attempted to explain learning deficit in terms of the breakdown in one part of a flow diagram were also squandering effort. Many of them have made very useful and intriguing contributions to our understanding of cognitive processes, and still do. Nor is it incorrect to compare mongols and non-mongols as so many have done.

It must also be noted that we are aware that not all subnormality is severe subnormality and that the models which we wish to discuss and illustrate apply only to some subnormals and not necessarily to those mildly subnormal children who may have no detectable damage to their central nervous systems.

However, although those who explained subnormality in terms of defective intelligence could be criticised for circularity, they might win points because they generally show a delayed development in all subjects, admittedly with variations, but not with very great variations of standard deviations.

The concept of islets of intelligence has not gained ground even among those working on autism. The positive correlation between scores on IQ subtests continues to be one of the most reliable findings of cognitive tests, just as a low mental age is reflected in all such subtests with subnormals.

Some further discussion of the experimental approach is necessary because of the type of argument which we have advanced ourseved at different times (O'Connor and Hermelin, 1963) namely that there is in fact an apparent sparing of some functions by the general pathology. For example, we have argued that although some have claimed long term memory deficits in the subnormal, we did not find them, although some input defects were noted. Does not this argue for differential handicap? The simple answer is yes. Differential handicap occurs, but within the limit of mental age level. The differences found are often of the order found among normals and called individual differences. Psychology must one day account for them and is far from doing so but we believe their level of operation need not lead us to modify our present approach.

Thus the experimental model clearly has both weaknesses and strengths. The linear model concept breaks down learning or information processing into connected and less arbitrarily determined components than does the intelligence model. At the same time, it tends to ignore the strengths of this model in so far as it (the latter) compares its components on a normative basis which the experimentalists have so far not systematically attempted.

The problem we have proposed therefore can be restated more succinctly. There are objections to the use of intelligence as an explanatory structure in relation to learning because it is to some extent circular and therefore sterile. But there are also problems in the experimental approach because it ignores what intelligence testing has taken into account, the comparative and normalised structure of population statistics. It also ignores the strangely general effect of early brain damage which retards all aspects of learning except in a limited number of children such as developmental aphasics, where other explanations of a neurological kind must be taken into account.

Clearly, therefore, there would be good reason to combine the strengths of both methods, but the appropriate paradigm has not yet occurred to anyone. We hope that now that we have attempted to state what we think is the problem someone will come up with the solution.

An Interim Approach

However, in the interim, our own thinking led us to pursue a neuropsychological approach which began as an attempt to compare the effects of specific injuries or lesions with the effect of more general disabilities. The foundation for exploring this possibility was the model of information processing and learning which we developed as an explanatory model to help visualize the information acquisition process some years ago in anticipation of the work on subnormal perceptual, mnemonic and encoding functions. Our report of this work was published in our monograph "Speech and Thought in Severe Subnormality." However, there have been many subsequent models most of which follow a simple consecutive pattern. The assumption of nearly all these models is linear processing but with varying feedback or feed-forward links. One inference from this set of conditions is that an ineffective box in any part of the line of functions will block acquisition in the subsequent boxes in whole or in part. However, we faced, as we said, the problem of testing a model which we trusted only in part because any part of the flow diagram could be defective and perhaps in the subnormal, all could be defective.

One solution was theoretically and practically quite simple. We wished to compare the effect of specific lesions with that of general lesions in the developing nervous system. Our chosen solution therefore was to select children on the one hand who were known to suffer from specific lesions and known not to suffer from general lesions, and compare their performance on certain tasks with children of known general deficit who appeared not to have the specific deficits characteristic of other groups. In other words, we compared blind and deaf children on the one hand with subnormal and low IQ autistic children on the other hand. From time to time also we introduced normal children of matched mental age. Most children were aged between 10 and 12 years of age.

We had other aims beside the aim of comparing the effect of specific (or peripheral) with general lesions in the developing nervous system. We wished also to compare the effect of specific and general lesions on the manipulation of spatial and temporal qualities at this stage of development. Some notion of the way we worked can be given by describing some experiments carried out in these two areas. At the same time where appropriate the relevance of these findings to the issue of specific and general lesions will be indicated.

Experiments with Spatial Organisation

To begin with experiments on space, it can be said that the purpose of the exposition essentially will be to show how the absence of a sensory input sometimes results in an alternative encoding procedure if the modality of input is primary or appropriate but sometimes does not if it is not. Another aim, incidental to this, will be to indicate which kinds of operations are specific to one modality and which not.

The first experiment is one which we carried out to pursue an interesting observation of Attneave and Benson (1969). They noted that an interchange of hand location apparently had no effect on position sense when finger ends on each hand had been successfully and randomly stimulated by touch both before and after hand reversal. All this took place in the presence of sight using adults subjects. What would happen in its permanent or temporary absence? We simplified the experiment to involve two fingers of each hand with the two hands on the table one in front of the other. In the learning phase of the study children were taught to respond with certain words whenever appropriate fingers were touched. After a criterion performance had been achieved, the hands were simply reversed. All groups who performed both phases of the task were either blind or blindfolded with the exception of two sighted groups. The results are of interest to us because they show how effectively the deprivation of sight robs even those with a lifetime of visual experience of the characteristic method of encoding noted even after reversal by those using sight as well as touch - as shown by Attneave and Benson (1969) as well as one of our own results.

The results illustrate two consequences of specific deficits in children and perhaps adults. They show that encoding processes in two separate modalities follow different rules even when concerned with one dimension, the dimension of spatial order in this case. They also illustrate the consequences of specific deprivation, namely that at least in this case, as sight would appear to be essential to a certain aspect of spatial ordering, deprivation of vision results in an alternative kind of coding.

TABLE 1
Frequency of Responses after Hand Reversal

Group	Finger Response	Location Response	Random Response
10 Seeing normal children	158	239	3
10 Seeing autistic children	124	251	25
10 Blindfold adults	398	0	2
10 Blindfold normal children	297	85	18
10 Blind children	276	116	8

The distinctive character of these two forms of encoding can also be found in another experiment. This experiment was concerned with shape or form. Two shapes fixed to a background were presented tactually to a subject who was asked to feel them blind and one after another. When he had felt them he was asked to decide whether they would form a square if pushed together. The decision was recorded. Some pairs of shapes would form a square if pushed together, some pairs would not and yet others needed to be mentally rotated as well as pushed together to produce a square. The subjects were blind or age-matched blindfold, or sighted. The task is illustrated in Figure 1. The results are presented in Table 2.

The rotation effect, which is notable in the case of sight, does not occur in the tactile modality. However, the most notable finding is the clear lack of difference between the blind and blindfold groups, i.e. transfer from visual experience does not occur as the total error scores reveal. The change to a "new" modality of presentation in the case of the blindfold obviously leads to a new encoding behaviour and the blind seem to have acquired little greater skill from a long experience of the tactile appreciation of form.

The inferences to be drawn from these two spatial experiments

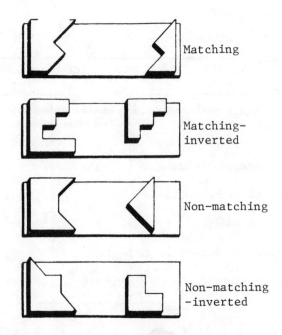

Matching

Matching-
inverted

Non-matching

Non-matching
-inverted

Figure 1. Forms for mental manipulation

TABLE 2
Shapes
Total Error Scores by Groups and Presentations

	Rotated	Unrotated	Totals
Blind	139	124	263
Blindfold	154	159	313
Sighted	38	15	53
Totals	331	298	629

might be firstly that spatial appreciation and manipulation inheres in the modality of vision and cannot be entirely recoded into an alternative modality. Notions of order and of shape despite any assumptions we might make concerning their interchangeability between modalities, apparently do not interchange easily. Under what conditions would a transfer of an appropriate ability in vision occur, if it could occur at all? This question which we presented to ourselves could not be answered completely rationally but some errors could be avoided. Order could hardly be transferred, nor shape, as we knew from the two previous studies. Rotated shapes were however no problem for touch, although we knew from Shephard and Metzler's (1971) work that they were a problem for vision. We decided to explore the allied question of mirror imagery where neither shape, nor order, nor in fact rotation was involved. Mirror images cannot be achieved by rotation, not can they be achieved strictly by superimposition. The most appropriate word to describe the form of spatial agreement which we hit on is the word symmetry or better still the geometric term congruence. Perhaps an even better term but a might literal is the German term "Klapp Symmetrie." We hit on the notion of congruent differentiation by considering the very organs which are specialized for touch, i.e. the hands. We also considered the many varied tests which Henry Head invented to test neurological normality. In a number of these tests the subject sits opposite the examiner and must imitate his gestures. One element which is subject to error is cross-lateral imitation.

A variety of such considerations led us to choose the following task. We decided to present a single plastic hand, either a left hand or a right hand, to the blind or blindfold subject to feel. His task was to say whether it was a left hand or a right hand.

The question was whether this strange task, like the two previous ones, would once more demonstrate lack of transfer. As, of course, we do not place much store by the visual discrimination of right and left hands, transfer might not be expected any more for this task than for the other two. We therefore presented to congenitally blind subjects and to blindfold controls, a single plastic hand in six separate orientations and they were required to judge by touch whether it was a right hand or a left.

The results were quite different from those of the other experiments and are given in Table 3. Error scores were significantly lower in the case of the blindfold than in the case of the blind. We assume that this was because the sighted were able to transfer the visual experience which they had acquired but of course this is an assumption.

TABLE 3
Hands
Total Error Scores by Groups and Orientations

	Up	Down	Right	Left	Towards E	Away From E	Totals
Blind	39	39	39	38	44	46	245
Blindfold	19	26	23	22	13	26	129
Sighted	10	15	16	7	9	22	79
Totals	68	80	78	67	66	94	453

The upshot of these kinds of experiments of which I have been able to describe only a few, is that in many situations involving key dimensions of spatial perception such as shape and order, coding into touch follows different rules from coding into vision and in these two cases it seems as if transfer from vision does not occur. Specific defects therefore are liable to lead to entirely different encoding methods to achieve the same intended aims.

Experiments in Temporal Ordering

It could be shown that a somewhat similar situation emerges in relation to the appreciation of time by specifically handicapped congenitally deaf children. The reason why time, i.e. duration and temporal order were chosen as the variable to be explored through studies with the congenitally deaf is because of the literature showing a strong association between auditory verbal input and the sense of time. Authors such as Hirsh, Bilger, and Deatherage (1956) and Savin (1967) have drawn attention to this phenomenon. Frankenhaeuser (1959) has also more systematically shown how auditorily filled time seems longer than unfilled time.

For this kind of reason we considered that encoding of duration and temporal succession was likely to be different or perhaps handicapped in congenitally deaf as compared with hearing children.

One of our first experiments devised to explore this area and to help us to work with deaf subjects was an experiment carried out with deaf, blind, normal and subnormal subjects in which all groups were taught to discriminate between two durations, of two seconds and six seconds respectively. The discrimination in which subjects were asked to appreciate two successive tactile stimuli and then judge whether they were the same or different was presented in the form of a rotary probe to the left hand. Subjects who learned the task to criterion were then asked to transfer the discrimination from touch to another appropriate modality, vision in the case of the deaf and hearing in the case of the blind. The control groups were allotted appropriately to blind and to deaf transfer conditions. Before transfer all subjects except the deaf were asked to verbalize the principle of solution of the learning task. All succeeded.

However, no transfer succeeded and nearly all subjects were unsuccessful in either the visual or the auditory discrimination of two stimuli of similar duration, taking as long to learn these differentiations as they had in the original tactile task. Once again, this time in temporal discrimination, the specificity of the task to modalities seemed to have been demonstrated.

Another experiment which we conducted at this time was concerned with the ordering of events in time as distinct from duration. Language concerning ordering is slow to develop but experiments on temporal ordering have generally shown that it is a distinctive skill independent of event recall. Conrad (1965) has shown this to be so. Our own study began with the question of how deaf children would store, memorize and recall digits. We presented three digits in the first instance to the subjects at the approximate limit of their digit span and in fact as three digits. These were in our first study presented both visually to deaf children and to controls and auditorily to blind children and controls. They were always presented in an order which was incongruent with a left to right order and the subject was asked to watch (or listen to) the numbers and when they had finished to say, or indicate, which was the middle one.

The results were very clear cut. All subjects whether deaf or normally hearing, given a visual presentation, chose the visually middle digit irrespective of presentation order and all subjects whether blind or normally sighted chose the successively middle digit when presented with incongruent auditory material. Once again in this instance the encoding processes seemed to be modality specific.

Another experiment with similar results is of interest. In this study a series of stimuli was presented twice. Sometimes the two series would be identical and sometimes different. The series could vary in length and could be visual or auditory. In addition, they could be of a Morse code type, for example two successive patterns such as long, short, long, long, short followed by the same or a different series, emitted from one source, or they could be demarcated by being emitted from two sources, e.g. right, left, right, right, left followed by the same or a different series.

Irrespective of the length of the series we can characterize the experiment as involving light and sound and one source or two sources of emission. Of course, deaf subjects and controls could see visual signals and blind and control subjects would hear auditory ones. Briefly, results clearly established that heard stimuli series were best judged from one source and visual series from two, as predicted.

Over 40 trials, deaf, blind, normal and subnormal children aged about 13 years, and of normal IQ except for the 15 year old subnormals (IQ 70) gave results showing that auditory stimuli led to more correct recognitions when the sequences were temporally structured (Morse type signals from one source) than when they were spatially structured, i.e., from two sources. The reverse was true of visual signals. Once, again, the specificity of modality encoding was demonstrated irrespective of level of intelligence in this case, or of type of handicap.

Another case in which this phenomenon was observed in relation to temporal encoding was in an experiment very similar to the "middle" experiment with three spatially incongruent digits presented visually. In this case subjects were not asked which was the middle one of a series of three, but were asked to wait until the presentation was finished and then to either recall the three digits or recognize them from three alternatives. In this experiment deaf children always recalled the left to right order of the visual presentation and the hearing always recalled the successive order. In this instance a specific deficit involved an alternative form of encoding. These examples will serve to illustrate our experimental method and an evaluation of this material can now be made.

An Evaluation of the Experiments

The first thing which should be done before drawing any general conclusions is to say something about the relationship between specific and general handicap as revealed from the results of the experiments presented. Our success in comparing specific with general deficits has been limited in part because experiments take time, especially with children and we have also faced a number of basic problems which required attention. Perhaps in summary one can say that our experiments show that specificity of encoding tends to be modality bound but in the case of speech its absence as an encoding medium can occur for several reasons, either specific or peripheral on the one hand or general or central on the other. In either case an alternative method of encoding may be elected by the subject. An example will make this clear. In the figure presented below, results of the recall of three digits visually presented are depicted. The figure suggests that whereas normal children generally opt for a temporal recall order and the deaf for a spatial recall order, other centrally handicapped subjects may, in this respect, resemble the deaf more than the hearing, even when their own hearing is intact. Although the mechanisms underlying the two similar encoding phenomena may be quite different, the alternative coding techniques, for example in the severely subnormal and the mentally handicapped autistic children would appear to be identical. In fact, subsequent experimentation has shown us that the use of language in thinking creates a sub-division in the subnormal between those using words in communication only and those able to use words as mental tools. The barrier between the two groups is indicated by a verbal IQ score around the 60 point level.

Other examples of the similarity in the encoding response among specifically and generally handicapped children can be seen in the failure of transfer in the durational judgment experiment, although in this experiment control groups are limited and the weakness may be of wider denotation, not necessarily applying only to the handicapped. Therefore our solution to the problem of how to study general cognitive handicap must be admitted to be only weakly established. Naturally, it seems to us to deserve further exploration but at this stage we can claim only limited success.

In some other respects, however, our findings seem to us to be of considerable interest, especially in the area of sensory specific encoding and consequent processing. To summarize our findings in a manner relevant to the problems raised in the introduction, it is best to summarize some of the conclusions which appear in a recent book where our experiments have been reported in more detail.

One inescapable conclusion from our experiments is that information is frequently processed in terms of the sensory modality of input providing this modality is appropriate. Secondly, processing frequently occurs in appropriate modalities and these tend to be visual in the case of spatial dimensions and auditory for temporal dimensions. Quite obviously the absence of these modalities deprives people of the capacity to manipulate material in the appropriate dimensions in the same way as those not handicapped and alternative encoding techniques are adopted.

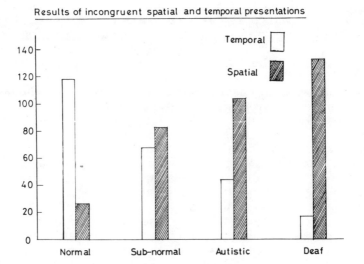

Figure 2. Results of digit recall by groups with different handicap

However, our evidence also suggests that at least in relation to the severe language incapacity associated with subnormality, concepts of temporal order may be differently handled, or processed in an alternative fashion closely resembling the methods common in the deaf. It can also be said that when deprived of sight, even normal children resort to a non-Euclidian and therefore developmentally earlier stage of spatial conceptualization, as in the four finger hand reversal experiment.

There is thus some evidence that sensory deprivation can have consequences in lowering, in a developmental sense, the level of processing of a given sensory input. There is also evidence, again limited, that general handicap especially involving verbal or "abstract" encoding can sometimes have similar effects. We think therefore that we can claim to have shown some ways in which specific and general deficits can resemble each other, without committing the solecism of assuming specific neurological or flow diagram deficits and without neglecting the general nature of cognitive deficits as shown by tests of intelligence. Naturally, as before, we want to emphasize the tentative character of this statement but we feel that the method is free from some of the obvious weaknesses associated with the more naive experimental model while leading to some interesting if unpopular conclusions.

References

Attneave, F. and Benson, L.   Spatial coding and tactual stimulation. Journal of Experimental Psychology,   1969, 81, 216-222.

Behar, I., and Bevan, W.   The perceived duration of auditory and visual intervals: cross-modal comparison and interaction. American Journal of Psychology,   1961, 74,   17-26.

Broca, P.I.   Remarques sur la siege de la faculte du langage articule, sources d'une observation d'aphemie (perte de la parole).   Bulletins de la Societe Anatomique de Paris Tome VI, 1861, 36, 330-357.

Conrad, R.   Order error in immediate recall of sequences. Journal of Verbal Learning and Verbal Behaviour, 1965, 4, 161-169.

Ellis, N. R.   Memory processes in retardates and normals. In N. R. Ellis (Ed.) Research in Mental Retardation 4. New York: McGraw-Hill, 1963.

Ellis, N. R.   The stimulus trace and behavioural inadequacy. In N. R. Ellis (Ed.) Handbook of Mental Deficiency. New York: McGraw-Hill, 1963.

Frankenhauser, M.   Estimation of Time, an Experimental Study. Stockholm: Almguist & Wiksell, 1959.

Hermelin, B. and O'Connor, N.   Psychological Experiments with Autistic Children.   Oxford: Pergamon Press, 1970.

Hirsh, I. J., Bilger, R. C. and Deatherage, B. H.   The effects of auditory and visual background on apparent duration. American Journal of Psychology,   1956, 69,   561-574.

Lashley, K. S.   Brain Mechanism and Intelligence.   University of Chicago, Illinois, 1929.

Luria, A. R.   The Role of Speech in the Regulation of Normal and Abnormal Behaviour (J. Tizard, Ed.) London: Pergamon Press, 1961.

O'Connor, N. and Hermelin, B.   Speech and Thought in Severe Subnormality.   London: Pergamon   Press,   1963.

Savin, H. B.   On the successive perception of simultaneous stimuli. Perception and Psychophysics,   1967,   95,   285-289.

Shepherd, R. N. & Metzler, J.   Mental retardation of three-dimensional objects. Science, 1971, 171, 701-703.

Wing, L.   The handicaps of autistic children - a comparative study. Journal of Child Psychology and Psychiatry, 1969, 10, 1-9.

THE NATURE OF INTELLIGENCE

H. J. Eysenck

Institute of Psychiatry

London, England

The theory of intelligence goes back a long way. Plato and Aristotle already separated out cognitive performance from emotional and conative behaviours, and Cicero used the term intelligentia very much in its modern meaning. Spencer revived the term, and together with Sir Francis Galton gave it wide acceptance among educated people in the 19th century. Spearman's notion of general intelligence or g was essentially based on these foundations, adding only a testable deduction, which in modern terms we would phrase as follows: different measures of intelligence, suitably chosen and applied to random samples of the population, should intercorrelate in such a manner as to produce a matrix of rank 1. In this context "suitably chosen" simply means that the tests should not show undue similarity, but constitute an approximation to a random sample of all possible tests of cognitive ability.

To this psychological and statistical definition of intelligence, Sir Francis Galton added the notion that intelligence was inherited, a notion already prominent in the writings of Plato, but now made testable by Galton's use of familial correlations and twin studies.

A third line of approach was that of the physiologist, where the clinical work of Hughlings Jackson, the experimental investigations of Sherrington and the microscopic studies of the brain carried out by Campbell, Brodman and other did much to confirm Spence's theory of a "hierarchy of neuro-functions," with the basis of type of activity developing by fairly definite stages into higher and more specialised forms. Thus in the adult human brain marked differences in the architecture of different areas and of different cell-layers are perceptible under the microscope, specialisations which appear and develop progressively during the early months of infant life. The brain, so it was found, always acts as a whole; its activity, as Sherrington pointed out, is "patterned, not indifferently diffuse," and the patterning itself always "involves and implies integration." Lashley contributed, from his massive research activity, the concept of 'mass action" of the brain, a mass action theoretically identified with intelligence by several writers.

Most of this work was concerned with intelligence as an intra-species concept, but there were also writers concerned with the evolutionary approach and inter-species comparisons. The early work of Lartet (1968) and Marsh (1874) resulted in concentration on what Jerison (1973) calls the "principle of proper mass"; "the mass of neural tissue controlling a particular function is appropriate to the amount of information processing involved in performing the function." As he points out, this implies that in comparison among species the importance of a function in the life of each species will be reflected by the absolute amount of neural tissue for that function in each species, a principle which gave rise to the detailed study of brain size, both in relation to body size and also as an independent measure of mental capacity of different species, evolving through the last 50 million years or so.

These notions, theories and findings gave rise to the testing movement, beginning with Binet, and going on through Stern, Burt, Terman, Thorndike, Thurstone and Thomson to present-day figures like Cattell and Guildford. The practical success of IQ tests, first demonstrated in the American Army tested during the First World War, and later in consolidated in educational practice, tended to "freeze" the form of IQ testing, with the single addition of the separate measurement of group factors, or "primary factors," as Thurstone called them--verbal ability, numerical ability, visual-spatial ability, perceptual ability, memory, divergent as opposed to convergent ability, etc. Thurstone's early attempts to disprove the existence of g and reduce all mental measurement to primary factors, was abortive, as he himself later admitted; by only working with highly intelligent students he reduced the range of intelligence so much that general ability factors were difficult to find. When he and Thelma Thurstone extended their work to random samples, they soon found that correlations between primaries themselves fell into the pattern predicted by Spearman, giving a matrix of approximately rank 1 (Eysenck, 1979). Criticisms of the theory of intelligence, and of intelligence testing, have become prominent in recent years, but many of them rest on misunderstandings that can easily be cleared up. Thus it is often asked: "How do you know that IQ tests measure intelligence?" The answer expected is of course some actual demonstration of the correspondence between IQ tests and some undoubted measure of intelligence, but this is a quite unreasonable and unscientific expectation. Intelligence is not a thing, existing in outer space, which would make it possible to demonstrate isomorphism; intelligence is a concept, like mass, or velocity, or electric resistance, and as such is part of a nomological network of facts and hypotheses; it is meaningless to ask whether such a concept "exists" in the sense that real object exist--although even there philosophers might ask some searching questions about the meaning of "existence."

It is curious that on the theoretical side psychologists have shown themselves largely disinterested; with occasional exceptions, not usually very serious ones, psychologists have refrained from formulating testable theories about the nature of intelligence, i.e., theories which would bind together the different types of tests used for the measurement of IQ, and predict the g loadings of different types of test. The major exception to this rule is of course Spearman (1927, 1923) whose laws of neogenesis are too well known to require restatement here. These laws are of course too general to be as useful as they might be, although they have proved effective in that some of the best culture-fair tests, such as Raven's Matrices, were explicitly constructed in line with them, and at the suggestion of Spearman himself. Quite recently Sternberg (1977) has produced a componential analysis of human abilities which is explicitly based on Spearman's laws, but breaks them up into much more specific ponents. This is an important and interesting attempt at theory-making, giving rise to testable deductions, many of which have in fact been tested, and it is to be hoped that others will follow his example and improve the existing model until it is able to take into account even greater numbers of typical IQ test paradigms than it does at present.

When it is said that "intelligence is what intelligence test measure," this is not, as is often assumed, either a tautology, or a joke, or an excuse for the psychologist's inability to find a better definition. Bridgeman (1936) argued for the usefulness of operational definitions in physics, and it is difficult to find any reason why operational definitions should be forbidden to the psychologist. The layman does not usually understand quite what is implicit in such an operational definition; he believes that the psychologist arbitrarily selects, on an almost random basis, tests of one kind or another, and then simply defines intelligence in terms of these tests. But as we have seen, this is quite unreasonable. Starting with the theory of intelligence as an all-pervasive force in creating individual differences in cognitive functioning, the psychologist goes on to predict the existence of certain very unlikely patterns of intercorrelations; his proof for the meaningfulness of the theory is the actual discovery of such patterns of intercorrelations. These then define the choice of tests, in the sense that "good" intelligence tests have high loadings on the general factor, and "bad" tests have low loadings. Thus the selection of tests is largely objective, and the very notion of a "good test" contains within it the whole theoretical approach leading to the findings of matrices of low rank among intercorrelations between cognitive tests.

Should we be ashamed of not having a universally agreed theory of intelligence? The expectation that such a theory should exist or that the measurement of intelligence is meaningless unless and until such a theory is forthcoming, is itself evidence of a profound misunderstanding of the scientific method, or the development of scientific theories. Scientists work with a concept of gravitation, but there is no widely accepted theory of gravitation, although 300 years have elapsed since Newton first propounded his theory of "action at a distance." His theory is still with us, and is at present being revived; but there are also two other theories, Einstein's field theory, and the particle interaction theory of gravitons, based on Planck's quantum mechanics. The fact that there are in existence three entirely different theories, none of which is amenable to direct proof, has not led physicists to dismiss the concept of gravitation as meaningless, and it is difficult to see why psychologists should be expected to be more successful than physicists in providing a universally agreed theory, based on cast-iron empirical proof.

What is more worrying, perhaps, is that theorists still exist who not only doubt the existence of g, but who formulate theories expressly excluding it. A good example here is the work of Guilford, whose structure-of-intellect model contains some 120 different abilities, made up of all possible combinations of five types of mental operations, four types of contents, and six types of products. Each ability is defined by its particular position on each of the three dimensions and it is not assumed that abilities sharing positions with respect to two dimensions, but differing in a third, are necessarily more closely related than abilities sharing only a single dimension. Guilford rejects Thurstone's development of oblique rotation, i.e., of correlated factors, and thus would make it impossible for us to derive from his factors any higher order concept of general intelligence.

Guilford's conception stands or falls with his denial of the existence of a "positive manifold," i.e., the universally found tendency that correlations between cognitive tests are uniformly positive. Guilford has pointed out that out of 48,140 correlation coefficients between tests observed in his own work, 8,677 fell in the interval between -10 and +10, and therefore for 24% of the correlations found in his numerous studies the null hypothesis could not be rejected, i.e., they were compatible with the view that the true correlation was zero.

He goes on to argue that data such as these do not support the view of the existence of a single pervasive general factor of intellectual ability.

Guilford's argument is quite unacceptable. In the first place, even in his own work, 76% of the correlations found between his test of allegedly independent abilities are positive and high enough to reject the null hypothesis; such a finding is certainly not compatible with Guilford's view that measures of intellectual abilities are unrelated except insofar as they are measures of the same ability. In the second place it is quite impossible to accept his figure of 24% of correlations being essentially zero. There are three reasons for this doubt.

In the first place, many of the populations studied by Guilford were highly selected for intelligence, e.g., airforce cadets in an officer's training programme. This inevitably reduces the range of ability in the sample, and consequently also the correlations to be found. Restriction of range is a very powerful factor in reducing correlations that are significant and positive in the general population to a level of the insignificance in samples showing this restriction of range.

In the second place, many of the tests used by Guilford have had relatively low reliabilities, occasionally with values of below 0.50. This means of course that a large proportion of the total variance in these tests is error variance, and consequently that these tests cannot correlate highly with other tests, as they measure whatever it is they measure so unreliably.

The third criticism would be that at least some of the tests Guilford has used are of doubtful relevance to the concept of intelligence as a general cognitive ability. Areas covered by behavioural content for instance deal with sensitivity to psychological states and feelings, and these are likely to be related rather to personality particularly neuroticism, than to intelligence. Some at least of the low or zero correlations found by Guilford may be due to the inappropriate choice of tests.

Simply removing all tests with reliabilities lower than 0.6 from the calculations reduces the number of correlations not statistically significant down to below 2%, and in some of Guilford's tables to below 1%. Thus the true number of apparently insignificant correlations is vanishingly small even in Guilford's own work. Furthermore, it has been shown that when tests of general intelligence have been used, they correlate positively and significantly with all the other variables in the batteries in question. When we add that many of Guilford's factors are unreplicable, even in his own work, we must conclude with Horn and Knapp (1973) that Guilford's model-of-intellect is not acceptable, and does not present any real alternative to Spearman's concept of g.

Much the same must be said of Piaget's theories, which have sometimes been held to be antagonistic to orthodox IQ testing, and to give a different, and better, idea of cognitive developments. It is possible to use scores on Piaget-type tests and problems as proper mental tests, and correlate them with existing IQ tests, and also to intercorrelate them with each other, and when this is done it is found that they behave very much as do other types of IQ test items, neither better nor worse than the average good IQ test item. This is not the use intended for his tests by Piaget, of course, but it is notable that results from his own type of approach do not contradict the general rule of statistical relationships deduced from Spearman's theory.

General intelligence was from the beginning regarded as a largely inherited quality, although of course some degree of environmental determination was never denied by Galton and his followers. This view too has come under criticism in recent years, although these criticisms are largely made in ignorance of the methods of analysis, and the models of inheritance, used by modern behavioural geneticists. There are of course many different ways of assessing the relative contributions of nature and nurture, and the important and interesting thing is that these give very similar estimates of heritability. We have studies of identical twins brought up in isolation; we have studies of monozygotic and dizygotic twins, comparing their degree of resemblance; we have familial studies, relating similarity in IQ to degree of consanguinity; we have studies of regression to the mean; we have studies of adopted children, to see whether these resemble their true parents or their adoptive parents more; and we have many different types of environmental studies, such as correlations between environmental factors and IQ, or the study of orphanage children who are provided very similar environments, but whose IQ variance does not seem to be diminished because of this lack of environmental heterogeneity.

Results from studies such as these have to be integrated with a general model elaborated by geneticists which attempts to include all the various sources of variance which determine the phenotype. In addition to additive genetic variance we also have such factors as assortative mating, which is quite prominent in regard to intelligence, dominance, which also provides important non-additive genetic variance, and similar factors. On the environmental side we have the differentiation between within-family and between-family environmental additive variance, and we have at least two sources of interaction between genetic and environmental factors. Thus the model claims to be a comprehensive one, unlike the usual sociological types of models which only pay attention to environmental factors, and completely disregard genetic ones (Eysenck, 1979).

The Coleman report is an excellent example of this environmental bias. Coleman carried out his famous analysis of educational
effects on the basis of a model which completely neglected genetic
factors, and came to the conclusion that the school made little or
no contribution to differences in scholastic achievement. This
conclusion is dependent on the assumptions made; when realistic
estimates of genetic variance are introduced, we find that the
effect of the school becomes as strong as the effect of the home
environment. Thus do wrong assumptions vitiate important social
conclusions. Relatively specialised methods are used to provide
evidence for different aspects of this model. Dominance, for
instance, can be studied by looking at "inbreeding depression,"
i.e., the lower levels of IQ achieved by the children of consang-
uinous matings, as for instance matings of cousins. Inbreeding
depression is a direct consequence of directional dominance, and
the results show that high intelligence is in fact dominant over
low intelligence.

It is interesting that Jensen has used this phenomenon in a
very suggestive manner to demonstrate the existence of g. He
argued that if g was dominant, and if inbreeding depression
demonstrated this dominance, then the degree of inbreeding depression would be a function of the g loading of each of the tests in
the Wechsler battery. He therefore compared the g loadings of the
Wechsler tests with the degree of inbreeding depression observed,
and found a very highly significant relationship. This would be
completely unexpected if some such model as Guilford's were accepted, and thus adds another argument against the spreading of the
g variance amongst a number of factors.

The general finding from all these different types of investigations is that the heritability of intelligence is somewhere in the
neighbourhood of 80%. Leaving out Burt's data, regarding the admis-
sability of which there has recently been some argument, a reanalysis
of all the available data disclosed a heritability of 70%, which, when
corrected for attenuation, rose to the figure of 79.5%; this may be
contrasted with a figure of 80% given by Burt's data taken by themselves (Eysenck, 1979). It is of course important to recognize the
limitations of such figures. They are population statistics, i.e.,
they do not refer to the degree of genetic and environmental determination for any particular individual, and they apply to a particular
group, at a particular time, and cannot be generalised to other
groups or other times. The considerable degree of equalisation of
educational opportunities that has taken place in the last 30 years
would almost certainly have the effect of increasing the genetic
effects, and reducing environmental ones, and if the process continues then we may expect a somewhat higher heritability of g in
100 years' time than that which obtains now.

Neither would it be correct to regard genetic factors as producing a permanently "fixed" level of ability. What is found applies to a given environment, and profound changes in that environment may lead to profound changes in the development and distribution of intelligence. If it is true that glutamic acid can raise the IQ of dull children by something like 10 points, while leaving that of bright children or average children unaffected, then we could alter the heritability and even the mean value of IQ in a given population by administering this drug to all dull children (Eysenck, 1973). However, it should not be assumed that such alterations in the environment as would make a profound change in our statistics of heritability would be easy to produce, or even possible; while we must recognise the restrictive nature of our findings, nevertheless the possibility of profound changes must be demonstrated in practice before their reality can be admitted. Simply to press for greater equality in education, in salaries, and in similar matters would not greatly alter the observed differences in IQ, as the experiment on orphanage children demonstrates. Those who believe in the possibility of manipulating intelligence by manipulating environmental variables bear the onus of proof, and so far that proof has not been forthcoming.

So far I have laid particular emphasis on what one might call the internal proof for the existence of a meaningful concept of intelligence; there is of course also an external proof of validity, which depends on demonstrating that IQ tests are predictive in certain areas where one would normally expect intelligence to be prominent. These areas are essentially education, work, and achievement. I have surveyed the results of such studies elsewhere (Eysenck, 1979), and the results certainly are in line with expectation in all these fields. Occupations where the man-in-the-street would expect intelligence to be required show on the whole higher average levels of intelligence amongst those in these occupations than would be found in other areas where low intelligence would be expected; doctors, professors and accountants have mean IQs a great deal higher than do dustmen, unskilled labourers and farm workers. In education, there is considerable correlation between achievement and IQ, both in schools and at university. Intelligence tests have proved their value in officer selection, in selection for the civil service, and in relation to other methods of selection. Interestingly enough there is also evidence of heteroscedasticity when IQ values are measured against achievement; this is expected because intelligence is a necessary but not a sufficient determinant of achievement, so that people who are high on achievement are nearly always high on IQ, but people high on IQ may be low in achievement. This failure may be due to personality defects; thus in the famous Terman studies of genius. Those children who later on turned out to be failures had been rated as being emotionally unstable, neurotic, etc., at the time of the first testing.

We have a meaningful, well authenticated psychological concept, intelligence;   we have ample evidence that this concept describes adequately individual differences within (Eysenck, 1979) and between species (Jerison, 1973), and we know that these individual differences are largely produced by genetic factors.   We must now go one step further and ask ourselves a question which is crucial for the biological approach to which we are committed: Can we formulate a physiological theory which can account for the major psychological and genetic facts, and which can produce measuring instruments capable, on the biological side, of repro- ducing the results which IQ tests can produce on the psychological side?   This is a tall order, but I do not think that we can rest content until and unless some such isomorphism has been established. Fortunately a beginning at least has been made in this direction, and although what I have to say now is obviously highly specula- tive, I believe that it is essentially in the right direction, and it also seems to be the case that there is some impressive evidence in favour of the theory in question.

In presenting this theory, which owes its formulation to two of my colleagues, I shall follow closely their own development of it (Hendrickson, 1972, 1973;   Hendrickson and Hendrickson, 1978). Inevitably the statement here will be too brief and dogmatic to be satisfactory, but it may give some idea of the sort of reasoning involved, and the sort of data to be looked at.   Essentially, the theory is concerned with the transmission of information in the cortex, the hypothesis being that (1) correct (error-free) trans- mission is the essential basis of intelligent behaviour, and (2) degree of error-free transmission can be measured in terms of certain characteristics of the averaged evoked potential (A.E.P.). Historically the measurement preceded the theory (Chalke and Ertl, 1965;   Ertl, 1969;   D. E. Hendrickson, 1972;   Plum, 1969; Shucard and Horn, 1972;   Weinberg, 1969), but in presenting the theory I will discuss it in advance of the factual findings.

How can information be processed through the cortex, bear- ing in mind the all-or-none principle of neural transmission? Hendrickson, an expert in computer technology, used his profes- sional knowledge to suggest that information was transmitted through certain characteristics of the so-called spike or pulse trains in the axons of nerve cells.   There are two major such characteristics.   (1) Such spike trains have exactly 22 pulses, giving 21 intervals;   Figure 1 shows a "long" and a "short" pulse train;   it is of course known that the <u>intensity</u> of the stimulus is directly related to the firing rate of the neuron.   Trains such as these can be recorded from individual axons (single units), and Hendrickson's hypothesis states that <u>all information is contained in the pattern of the 21 intervals between the 22 spikes in the pulse train.</u>   (There are also to be observed many isolated pulses that can be seen from time to time in single unit recordings;   these

are probably random events which convey no information, and are
ignored by the more distal neurons the pulses eventually reach.)
(2) The second characteristic of the spike trains or pulse trains
is that the series of time intervals between pulses is selected from
a set of only four possible intervals. Brink (1951) gives a diagram
showing in histogram form a series of pulse train intervals;
there are clearly four groups, centering on 6, 12, 18 and 24
milliseconds. The spacing of these intervals constitutes the code
that is used in the transmission of information through the brain,
each of the 21 intervals being able to assume one of the four
lengths.

So far the theory has dealt with transduction and transmission
of information; how does the brain receive and decode this
information and how does it deal with it? According to the theory,
events at the synapse explain this next step. At the synapse,
neural stimulation causes Ach to leave the synaptic vesicle and
enter the synaptic cleft. On the other side of the cleft, Ach
causes sodium (Na) to pour into the postsynaptic neuron. This
sodium ion carries a positive charge, and is attracted by the
negative charge on an RNA molecule (template) attached to a
microtubule (Mt). This RNA molecule is comprised of four possible
nucleotide bases, and the interaction between the RNA and the

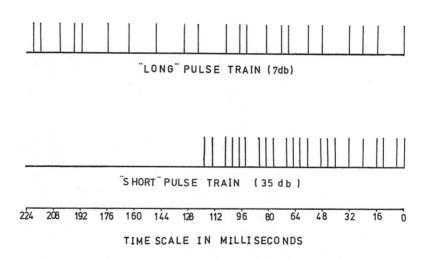

"LONG" PULSE TRAIN (7db)

"SHORT" PULSE TRAIN (35 db)

| 224 | 208 | 192 | 176 | 160 | 144 | 128 | 112 | 96 | 80 | 64 | 48 | 32 | 16 | 0 |

TIME SCALE IN MILLISECONDS

Figure 1. Pulse trains illustrating length of pulse train as a
function of intensity of stimulation.

sodium determines the further transmission of whatever message was originally encoded in the spike train. There is some evidence that human RNA has 21 nucleotide bases in its sequence, giving a very neat (and certainly not coincidental) correspondence between the size of the molecule and the number of intervals between the 22 pulses in the spike train. Hendrickson gives further details about the "recognition" process and the learning process involved in this molecular change, but this is not the place to go into detail.

However, it is important to realize that transmission and interchange of information are statistical, not deterministic events. They are affected by many different happenings taking place in the body, including for instance changes in temperature--the hydrogen bonds of the RNA are very sensitive to temperature, losing their strength in high temperatures (fever) and over-responding to the pulse trains. Our reaction times are quicker, our mental clock ticks faster in real time so that more time seems to pass, and so on. The opposite hapens in hypothermia; we cease to react to stimuli we would normally react to.

Intelligence, in Hendrickson's theory, is the summation of all of the factors which can affect the synaptic recognition process. When the process is working well, we have a very high probability of recognizing what we should and of ignoring meaningless pulse trains; this corresponds to, or is basic to, high intelligence. When the process is influenced by too many extraneous variables, or when there are faults and errors implicit in it, we have low intelligence. The theory bears some relation to Thomson's (1939) famous "number of bonds" theory, which he offered as an alternative to Spearman's neogenetic formulation; instead of "number of bonds" we now have "probability of recognition of pulse trains," with the bonds being substituted by the correctly identified pulse trains.

Before we get to the actual thinking process which underlies our conception of intelligence, we need to realize that single pulse trains are rare (as in simple reaction times), and that much more usually whole series of pulse trains are chained together, increasing dramatically the probability of error (mis-recognition). Hendrickson has given some quantitative estimates of the probabilities of such breakdowns of recognition, linking these with IQ estimates. Errors in the transmission process require more frequent repetition of the message, in order to produce recognition and learning, and hence lead to slower learning in dull as compared with bright subjects. The more complex the message, the more likely is a breakdown in the recognition sequence, or the learning process; this agrees well with the fact that the more complex a mental task is, the more does it require high IQ in order to solve it.

Perhaps the most frequent source of error in the process
under consideration is the failure of the spike train to preserve
intervalintegrity, i.e., failure of the axons to keep the pulses
moving down them at a constant speed. Hendrickson has shown
by computer simulation how such failures would cumulate in pulse
chains. He programmed the computer to generate a pulse train.
Each pulse was set as a predetermined interval, to which was
added a controlled and random amount of error. As soon as the
first pulse interval was generated by the computer, it started the
clock for the second pulse interval. This also had a preset
interval and some random error. However, the actual point at
which the second pulse occurs is a function not only of its own
random error, but also the random error of the first pulse interval. In other words, as we generate interval after interval in
the pulse train, the errors are cumulative.

This line of argument leads us directly to the averaged
evoked potential (A.E.P.). This is a measure of the wave activity
observed in the EEG consequent upon presentation of an auditory
or a visual stimulus, averaged over several trials to increase the
signal/noise ratio. Typical AEPs are presented in Figure 2,
taken from 10 bright and 10 dull subjects whose IQs were determined
on the WISC. Researchers have usually taken the latency of
consecutive waves to correlate with IQ, typical results showing
higher correlations of later waves rather than earlier ones, and
with correlations usually in the 30s or 40s at best. Hendrickson's
(1972) own research also demonstrated significant correlations
between the amplitude of the AEP waves and IQ, as determined
by the AH4 test, with correlations for both latency and amplitude
slightly higher for the verbal than for the spatial tests. Amplitude and latency were not correlated, giving a multiple correlation
with IQ of about .60.

The theory discussed above leads us to a more meaningful
measure of AEP intelligence than simple latency or amplitude,
although correlated with both. The computer simulation study
showed degeneration of the pulse train with cumulating errors;
this leads to the disappearance of components, as pulses that are
close together merge into each other. Consider the A.E.P. as a
direct picture of such pulse trains; in a person characterized by
low IQ (greater error frequency) the major components of the
waves should gradually merge and disappear, leaving a less
differentiated record. This is precisely what we see in Figure 2,
comparing the results from the less bright with those of the
bright subjects. We now have to go further into the record to
obtain the same number of components as the "noise" level increases;
thus the latency scores are an artifact, rather than a measure of
some "speed of response." This hypothesis also explains why it
is the later waves which give the higher correlations with IQ;
the degenerative effects are cumulative.

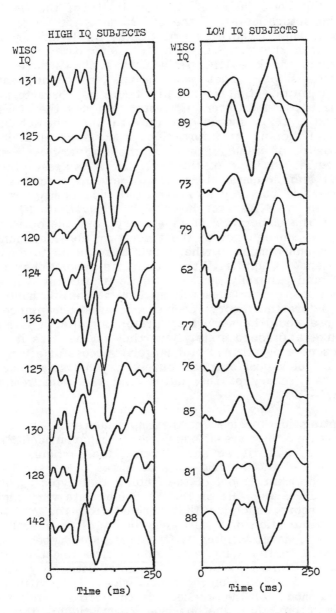

Figure 2. Evoked potential wave-form for 10 high and 10 low
          IQ subjects. (From Ertl and Schafer, 1969.)

This type of consideration immediately leads us to the suggestion that the appropriate score is in fact neither latency nor
amplitude, but some index of the complexity of the wave form,
such as the actual length of the line forming the envelope of the
wave. Using this measure on Ertl's data, E. Hendrickson found a
correlation with the W.I.S.C. of .77; using data of her own, on
subjects given the W.A.I.S. preliminary analyses have produced
similar relationships. These correlations are getting into the
range of magnitude that is usually taken as characteristic of
correlations between different IQ tests; we may therefore perhaps
say that the A.E.P., scored according to the Hendrickson theory
of mental functioning or information processing, is at least as
good a measure of intelligence as is the ordinary IQ test, and
probably better in view of the fact that it is less influenced
by cultural and educational factors of an environmental kind.
The argument is partly postdictive, but also partly predictive;
the use of the A.E.P. was shown to be relevant to IQ measurement before the elaboration of the theory, but the optimization of
scoring was a consequence of the theoretical considerations outlined
above. Obviously much further work is required to develop the
theory, extend its applicability, verify its predictions in several
directions, and generally demonstrate its usefulness. It is almost
certain that many anomalies will appear which will have to be
eliminated before the theory receives universal acceptance, and no
good purpose would be served by pretending that it is already in
anything like a finished state. Nevertheless, even as it stands it
does represent a determined and largely successful effort to bring
together the biological and the purely psychological sides in a
comprehensive theory of the nature and the measurement of
intelligence.

Accepting for the moment the empirical results reported, we
see at once that they are of considerable importance for a theory
of intelligence, even if we should reject the particular biological
theory advocated by Hendrickson, or agree to regard it as still
unproven. The main import of the finding that typical multi-faceted
IQ tests, such as the AH4 or the WISC correlate very highly with
a biological score, such as that derived from the A.E.P., is
surely the vindication of Spearman's theory of a general factor of
intelligence, g, as underlying all the variegated types of cognitive
tasks included in these IQ tests, and a firm rejection of such
theories as Guilford's, which would distribute the g variance
among unrelated group factors or primaries. It is difficult to see
how such a model of the intellect as Guilford's could possibly
predict, or account for, the observed correlations; these are not
only compatible with Spearman's or Thomson's model, but can be
directly predicted from it. The theoretical link provided by
Hendrickson between IQ measure and A.E.P. may or may not be
along the right lines; the simple empirical findings are sufficient

to rule any theory not including a g factor out of court.

It is possible to take this line of argument a step further. The different tests included in the WISC have different g loadings; if the A.E.P. is a good measure of g then and only then would we expect the correlations of the different WISC tests to be proportional to their g loadings. Elaine Hendrickson carried out this computation, and the correlation between g loadings and A.E.P. scores for the 10 tests turned out to be .697, which is highly significant statistically. Such a finding too is incompatible with any theory which rejects the concept of intelligence (g), and relies instead on groups of unrelated factors. The finding does not exclude the presence of additional cognitive factors, related to test content (verbal, numerical, perceptual) or to mental processes involved (memory, convergent, divergent), but it makes these distinctly less important than g itself.

Processing of information, as emphasized by Jerison (1973) in relation to the evolution of the brain, and by Hendrickson in the physiological model discussed above, is closely related to learning, i.e., the modification of synaptic transmitters; is learning meaningfully related to intelligence? Many early attempts to do so resulted in apparent failure because of the lack of correlation between different learning experiments (Eysenck, 1979). This failure was partly due to the low cognitive content of many of the activities involved. Learning to play tennis, or billiards, or football, are examples; so are abilities to learn to drive a motor car, to make love, or to sit on top of a pole for four weeks in order to be mentioned in the Guinness Book of Records. When we insist on the cognitive content of the task to be learned, we find that intelligence is highly correlated with such tasks, depending on the degree of complexity shown in the task. Such a relation is apparent in the theory proposed by Gagne (1968), in which he tried to construct a generalised learning hierarchy in terms of different levels of complexity, a hierarchy which has some interesting resemblances to Piaget's levels of development. He lists in order: stimulus-response, motor chaining, verbal chaining, multiple discrimination, concepts, principles and problem solving. Alvord (1969), in his research on transfer in mental hierarchy, has shown that measures of general intelligence become increasingly predictive of performance at each successively higher level in the learning hierarchy, and similar findings have been reported by Fox and Taylor (1967) and by Jensen (1970); all these studies are in agreement with the notion that the more complex the learning task, the greater the IQ required for its accomplishment. A summary of all this work, and the conclusions it gives rise to are given in Eysenck (1979).

We have so far laid emphasis on the meaningfulness of the concept of intelligence, as measured by IQ tests and as mirrored in biological measures such as the evoked potential. It is this meaningfulness, or proven theoretical usefulness in explanation and prediction, that is important in a theoretical concept; as mentioned before, the notion of "existence" is philosophically meaningless in relation to concepts, although it may be usefully employed as an alternative expression for meaningfulness in this context. Granted this meaningfulness, it may nevertheless be possible to break up the concept of IQ in various ways, just as the concept of the atom is still useful in physics, but has lost its meaning as an elementary particle which could not be further subdivided, and has instead given rise to a whole host of over a hundred different and more elementary particles. One such subdivision is that made by Cattell between fluid and crystallized ability. The Galton-Spearman notion of g is probably to be identified with fluid ability, crystallized ability being the result of applying this fluid ability to the learning of specific responses. The term crystallized ability is probably badly chosen, in the sense that we are dealing here not with an ability, but rather with an achievement. A good vocabulary represents an achievement; it is hardly to be called an ability, although it is a good measure of Cattell's crystallized ability.

Levels of development, whether those recognised by Piaget or by Gagne seem to incorporate a definite break, categorized by White (1965) in terms of an associative and a cognitive level. Rather similar to this distinction is that made by Jensen between level 1 ability and level 2 ability. "Level 1 ability is essentially the capacity to receive or register stimuli, to store them, and to later recognize or recall the material with a high degree of fidelity... it is characterised especially by the lack of any need of elaboration, transformation, or manipulation of the input in order to arrive at the output. The input need not be referred to other past learning in order to issue effective output." Originally Jensen called this "the basic learning ability.

Level 2, on the other hand, is at the high complexity end of the Gagne scale of learning. "Level 2 ability...is characterised by transformation and manipulation of the stimulus prior to making the response. It is the set of mechanisms which make generalisation beyond primary stimulus generalisation possible. Semantic generalisation and concept formation depend upon Level 2 ability; then coding and decoding of stimuli in terms of past experience, relating new learning to old learning, transfer in terms of concepts and principles, are all examples of level 2. Spearman's (1927) characterisation of g as the "deduction of relations and correlates" corresponds to level 2." This is an important and meaningful distinction, although whether it is truly a qualitative one, or simply a quantitative one making a break between tests with low g loadings and tests with high g loadings is a question that is still unanswered.

A third attempt to break down the IQ into constituent parts has been attempted by Eysenck (1979), who has criticised the usual run of factor analytic studies in terms of the scores chosen. As he points out, a given score on an IQ test can be arrived at in many different ways by many different people, and may therefore reflect different combinations of putative elements. He has suggested that the fundamental unit of analysis should be the item, not the test score, and furthermore that much information is thrown away by simply regarding an item as correctly or incorrectly solved, rather than measuring the latency of the solution. Furneaux (1973) and White (1973) have collaborated in an attempt to produce a model based on such more fundamental measures, and it has been shown that when this is done three uncorrelated and fundamental abilities seem to be involved in producing the total IQ score. These are mental speed, persistence of effort, and error checking, producing individual differences in the latency of correct and incorrect responses, the latency of giving up on items the individual feels he cannot solve correctly, and the number of erroneous solutions. A mathematical model has been constructed, incorporating these measures as well as the difficulty levels of the items involved, but this is not the place to go into details regarding this model (White, 1973). It seems almost axiomatic that from the applied point of view three independent factors repreoducing perfectly the single IQ test score must make prediction more accurate than this undifferentiated score, but direct evidence is still sparse. It is possible that some of the factors involved may be personality rather than cognitive factors, and this possibility is strengthened by the finding that different types of neurosis can be differentiated from normality in terms of these three variables (Brierly, 1961). On the theoretical side this approach, although dating back over 25 years, has not been discussed widely enough by psychologists in this field to make it possible to pronounce on its value.

We may summarise very briefly the main conclusions of this attempt to review the evidence respecting the nature of intelligence. It is found that the concept is theoretically meaningful, that it can be used to generate testable hypotheses, and that these hypotheses have on the whole been borne out by empirical studies. Intelligence as so conceived is related to learning, particularly of complex material, and it is determined to a large extent by genetic factors, including non-additive genetic factors such as assortative mating and dominance. The concept is meaningful in an evolutionary context, brain structures subserving it having been developing over the past 50 million years or so. The concept can be identified fairly closely with specific theories of neurological and physiological functioning, particularly with the evoked potential, and the processing of information through the cortex; theories exist which would unify the psychological and the physiological aspects of intelligence. The concept has internal and external validity,

and it seems justifiable to conclude that it constitutes a true
scientific paradigm in the Kuhnian sense.

References

Alvord, H. Quoted in Eysenck, 1979.

Bridgeman, P. W. The Nature of Physical Theory. Princeton:
University Press, 1936.

Brierley, H. The speed and accuracy characteristics of neurotics.
British Journal of Psychology, 1961, 52, 273-280.

Brink, F. Excitation and conduction in the neuron. In S. S.
Stevens (Ed.), Handbook of Experimental Psychology. New
York: Wiley, 1951, 50-93.

Chalke, F. C. R., and Ertl, J. P. Evoked potentials and intelligence. Life Sciences, 1965, 4, 1319-1322.

Ertl, J., and Schafer, E. W. P. Brain response correlates of
psychometric intelligence. Nature, 1969, 223, 421-422.

Eysenck, H. J. The Measurement of Intelligence. Lancaster:
Medical and Technical Publishers, 1973.

Eysenck, H. J. The Nature and Measurement of Intelligence.
London: Springer, 1979.

Fox, W. L. and Taylor, J. E. Adaptation of training to individual differences. Paper presented to the NATO Conference
on "Manpower Research in the Defence Context," London, 1967.

Furneaux, W. D. Intellectual abilities and problem-solving behaviour. In H. J. Eysenck (Ed.), The Measurement of Intelligence. Lancaster: Medical and Technical Publishers, 1973.

Gagné, R. N. Contributions of learning to human development.
Psychological Review, 1968, 75, 177-191.

Hendrickson, A. E. An integrated molar/molecular model of the
brain. Psychological Reports, 1972, 30, 343-368.

Hendrickson, A. E. Translations of auditory stimuli into neural
codes. Psychological Reports, 1973, 32, 315-321.

Hendrickson, A. E. and Hendrickson, D. E. The biological basis
and measurement of intelligence. Paper read at XIXth International Congress of Applied Psychology in Munich, August
4th, 1978.

Hendrickson, D. E. An examination of individual differences in
cortical evoked resonses. London: Unpublished Ph.D. thesis,
University of London, 1972.

Horn, J. and Knapp, J. R. On the subjective character of the
empirical base of Guilford's structure-of-intellect model.
Psychological Bulletin, 1973, 80, 33-43.

Jensen, A. Learning ability, intelligence, and educability. In V.
Alle, (Ed.), Psychological Factors in Poverty. Chicago:
Markham, 1970.

Jerison, H. J. Evolution of the Brain and Intelligence. London:
Academic Press, 1973.

Plum, A. Visual evoked responses: their relationship to intelligence. Florida: Unpublished Ph.D. thesis, University of
Florida, 1969.

Shucard, D. W. and Horn, J. L.   Evoked cortical potentials and
     measurement of human abilities.   Journal of Comparative and
     Physiological Psychology, 1972, 78, 59–68.
Spearman, C.   The Nature of "Intelligence" and the Principles of
     Cognition.   London: Macmillan, 1923.
Spearman, C.   The Abilities of Man.   London: Macmillan, 1927.
Sternberg, R. J.   Intelligence, Information Processing, and Ana-
     logical Reasoning.   London: Wiley, 1977.
Thomson, G. H.   The Factorial Analysis of Human Ability.
     London: University of London Press, 1939.
Weinberg, H.   Correlation of frequency spectra of averaged visual
     evoked potentials with verbal intelligence.   Nature, 1969, 224,
     813–815.
White, P. O.   Individual differences in speed accuracy and persis-
     tence: A mathematical model for problem solving.   In H. J.
     Eysenck (Ed.), The Measurement of Intelligence.   Lancaster:
     Medical and Technical Publishers, 1973.
White, S. H.   Evidence for a hierarchical arrangement of learning
     processes.   In L. P. Lipsitt and C. C. Spiker (Ed.), Advan-
     ces in Child Development and Behavior, Vol. 2.   New York:
     Academic Press, 1965.

THE PRIMARY MENTAL ABILITY

Lloyd G. Humphreys

University of Illinois

Champaign, Illinois, U.S.A.

Abstract

My intent in this paper is to bring up to date a discussion of correlational models in intelligence which I started in the American Psychologist in 1962 to provide data concerning the importance of general intelligence, and in the end to renew my support for the approach to definition and theory of Godfrey Thomson (1919).

Factor and Other Models

In the earlier paper I discussed the hierarchical model of intelligence and especially the model espoused by Vernon (1950). Since that time, Cattell (1963) and Horn (1968) have written extensively concerning a variation of the hierarchical model which is somewhat similar to the one of Vernon. The Cattell and Horn approach is noteworthy, however, in that the model is incomplete. Their higher order factors of fluid, crystallized, and visualization abilities, among others, are themselves intercorrelated, but the general factor is missing. I still find the hierarchical model attractive, and I still do hierarchical factoring. During the years, however, my skepticism concerning the meaning and importance of lower order factors has increased.

In 1962 I also made favorable comments concerning the possibilities of Guttman's facet analysis (1944) for an understanding of human abilities, but did not carry the matter very far. I now return to that discussion.

87

Facet Analysis of Item Types

Guilford's structure of intellect model (1967) assumes three dimensions or facets consisting of 6, 5, and 4 elements, respectively, among cognitive tests. Let us take these as dimensions of item types rather than primary mental abilities. Looked at in this way Guilford had far from exhausted the possible facets in tests. For example, items differ in the extent to which they are speeded. Let us arbitrarily assign three elements: highly, moderate, and unspeeded. The examinee's set toward guessing can be manipulated by instructions. At least two elements are required. Items differ with respect to their level of difficulty if given unspeeded. Three elements should suffice: so simple that only highly speeded conditions produce errors, moderately difficult but solvable given enough time, and difficult enough to administer under speeded conditions. The decision to score number right or number wrong adds another facet. Finally, let us add sensory modality which provides a wide gamut of item types by only considering visual and auditory elements. I avoid for purposes of this paper the possibility that auditory presentation might add elements elsewhere in the set. I also omit elements that could be added to Guilford's content facet in the form of different kinds of information.

The Cartesian product of my facets results in 8,640 item types, although some of the cells cannot be filled in any realistic manner. Sensory modality accounts for some of these vacant cells, the combination of speed and item difficulty accounts for others and error scores are at times determined by the number right. Nevertheless, I have now defined many more tests than occur in Guilford's model, but I hasten to add that I do not consider them the equivalent of the chemical elements, and I do not call them "primary mental abilities."

I do make some psychological assumptions about responses to these many different kinds of items. To this extent each item type could be said to measure a different ability. If two tests have everything in common except for different elements on a single facet, given a large enough sample the correlation between the two will be distinguishably smaller than the maximum value allowed by the respective reliabilities. In other words when one element of a facet is substituted for another, the rank order of examinees true scores will change somewhat, but the facet analysis does not allow one to predict the amount of change. Shifting elements on one facet may also produce more change in individual differences than shifting elements on some other facet. We know that a shift in content from words to numbers, for example, is quite potent.

Order in the Correlations

Let us now consider what the table of intercorrelations of these tests would be like in a wide range of talent. Tests that had all elements in common except one would generally have the highest correlations. Tests that had fewest elements in common would generally have the lowest correlations. Elements on certain of the more potent facets might produce some degree of clustering, but the clustering produced by other potent facets would cut across other clusters orthogonally. The most obvious impression concerning these correlations would be their almost continuous gradation in size. Furthermore, the gradations in size would proceed in several directions from any one starting point. There would be no obvious order in the matrix.

A second important observation would be the virtual absence of zero or negative correlations. No matter how little two tests had in common in terms of the facet analysis, the correlations would be virtually all positive in direction. The only exceptions would be a very small number of negative correlations associated with the correct answers to the simplest items on highly speeded tests given under the condition that did not sufficiently discourage guessing. Error scores on these same tests after reflection, however, would be positively correlated with both the right and the reflected wrong scores on the remaining tests. These inferences are of course extrapolations but from a rather wide range of empirical observations.

If number of attempts were substituted for number right, the size and number of negative correlations would increase. This indicates that mere speed of response undisciplined by the need to produce correct answers may belong in a different domain than the cognitive. Other data indicate that tests of this type are substantially affected by temporary psychological dispositions and bodily states and that high reliability at one point in time gives way to low to moderate stability of scores over time.

The Common Factor Model

Let me now make some predictions about the factors in this massive matrix. In one sense each test already defines a factor. The ultimate factor in the Guilford sense is defined by two or more parallel forms of the same test. There will be many factors remaining in the matrix, however, and the uniqueness component of each test will be nonzero. No matter how large the sample or how reliable the tests, there will be no obvious breaks in the size of the Eigenvalues as a function of the ordinal number of the factor. A simple structure is as important in the decision concerning the number of factors as it is in guiding the rotations.

In the universe of tests defined by the facets and their elements, one would not be able to define in an objective manner rotated first order factors that would replicate existing so-called primaries. These latter would be fractionated into several smaller factors at best and at worst the variance of a "primary" would be scattered among very diverse factors. Rotations directed visually and judgmentally could of course produce factors that would resemble the Thurstone primaries (1938), but it is easy to capitalize on the very large number of rotational choices when there are many tests and many factors. I am also very certain that objectively rotated factors could readily be interpreted by almost any factor analyst, or for that matter any psychologist, but my confidence in this possibility is based upon the almost limitless capacity of pschychologists to interpret any relationship after the fact.

Without a dependable first order rotation of factors there would be no dependable way to define higher order factors. The variances associated with Vernon's verbal-educational and practical-mechanical are in the matrix just as are the variances associated with the Cattell and Horn fluid, crystallized, and visualization abilities, but these would shade over into other factors in an almost continuous manner. It is also unlikely that anything resembling these so-called second order factors would appear in the second order in my large matrix.

This discussion leads inevitably to the conclusion that nice, clear first and second order factors reflect mainly our habits of test construction and our selection of the tests to factor. I thought that I had laid to rest in 1962 the belief that certain factors were intrinsically first order factors, that other factors were intrinsically second order factors, and that only first order factors were primary. Whether a factor will appear in a given order depends upon the density of sampling from the universe of tests. In one battery a factor can appear in the first order, in another battery in a second or higher order. Cattell at first had a simple second order structure: fluid, crystallized, and speed abilities. Now the number of second order factors has grown substantially. New factors are not discovered. Rather they are invented, albeit by a complex, indirect process. If Horn or Cattell were to obtain a correlation matrix designed to define all of their so-called second order factors, it is highly probable that a third order analysis would reduce the number of major group factors to a more manageable number, possibly fluid, crystallized, and speed of response. The general factor would now be in the fourth order in their data.

The Schmid-Leiman (1957) transformation of oblique factors in several orders into a single order of orthogonal factors defined by the original variables shows very clearly that the only difference between a first order or so-called primary factor and a higher order factor lies in the number of variables which define it. Breadth is the key concept, not superordination, yet factor theorists continue to discuss factors in two orders as if they belonged to different species of abilities and as if their factors had completely independent existences.

The Nature of Tests

In reaching my conclusions concerning the inadequacy of the common factor model to describe for psychological purposes the intercorrelations of tests I have not overlooked Thurston's box problem (1947) or other factoring of the dimensions of physical objects. These analogies are not convincing because there are fundamental differences between physical measures and psychological tests. As I have just described, tests can be invented in almost limitless numbers. They also contain multiple items which can vary widely in their level of intercorrelations. The characteristics of a total score which is a linear combination of numerous items are completely determined by the characteristics of those items, and psychologists select the items to be included in psychological tests. The definition of a test is much more arbitrary than is the definition of a measure of length. The primary difference between a test and a physical measure is represented by the test's characteristic of homogeneity, which is absent from physical measures.

High homogeneity of items has been considered a desirable goal by most test constructors, but there is little basis for this. High item intercorrelations are not synonomous with psychological unidimensionality. If each item is complex and to the same extent, the test will be homogeneous, but not unidimensional. One can argue convincingly that factors in Guilford's model are inextricably complex psychologically since each factor is a combination of content, operation, and product. Test constructors need to think in terms of an appropriate level of homogeneity for the measurement of the psychological attribute they are interested in. A factor does not ipso facto represent a useful psychological attribute, and a claim that a test is factorially pure represents little in the way of recommendation.

A test of general intelligence constitutes an excellent example of this reasoning. The items in a standard intelligence test may define numerous common factors, but if the number of these factors is large and the contribution to total variance of each is small the test may still be considered relatively homogeneous with respect to the latent attribute it is designed to measure; i.e., intelligence. It is perhaps unfortunate with respect to a statistical definition of homogeneity that intelligence items have to have content and require operations which result in products. Perturbations are produced by these sources of variance, but their effects can be kept small. The presence of such perturbations constitute the very essence, however, of behavioral measures of latent traits.

Alternatives to Common Factor Analysis

Since I have rejected the common factor model as developed by Thurstone (1938) and others for this universe of tests, or from a random sample of tests from that universe, alternative model and procedural suggestions are in order. The model which best describes the intercorrelations requires a combination of Spearman's hierarchical order (1904, 1927) and Guttman's order models (1944). One can find both simplex and circumplex orders in complex relationships to each other in the universe of tests I have defined. But there will also be evidence, no matter how criss-crossed with other orders, of Spearman's hierarchical order. Some of the tests defined by the facet analysis are more heavily loaded on the general factor than others. I wonder if perhaps a tree model of the sort being used in scaling might be applicable.

My second suggestion is one of strategy and is the precise duplicate of a suggestion made in 1962. This is to develop tests no narrower than those for the "main effects." A test of a main effect is restricted to items homogeneous for a single element of one facet, but heterogeneous for all other elements of the other facets. This strategy would be parimonious with respect to the number of tests needed. Unless certain combinations of elements combined in a nonlinear fashion, partial correlations could be used to estimate individual scores in any one cell of the "space" defined by the facet analysis. In a constant amount of testing time this could also be done more reliably than by constructing a separate test for each element. Also, I would see very little to be gained by subjecting the intercorrelations of the tests of the main effects to traditional methods of common factor analysis.

A third suggestion might be labeled one of tactics. Since I am interested in the general factor in intelligence primarily, I can obtain a reasonable estimate by means of higher order factoring of existing cognitive tests. Habits of test construction allow one to find structure that would otherwise be obscured.

The Basis for a General Factor

Since designating certain measures defined by my facets and their elements as noncognitive--not quite as arbitrary as it sounds --leaves no negative correlations in the master matrix of cognitive measures, a firm basis is provided for a general factor. The pervasiveness of positive correlations among item types which have correct answers and for which there is presssure to obtain correct answers, suggests that the general factor in human abilities reflects a good deal more than habits of test construction. Lower order factors may represent little more than convenient descriptive dimensions, but the general factor may be psychologically more important, may be more than merely descriptive. A number of years ago I tried without much success to factor a matrix in such a way that the broadest factors would be the first ones extracted and the narrowest ones would be last. In contrast, standard methods extract the least important factors first, which the investigator mistakenly calls primary, and the most important ones last. Someone better equipped than I should try again. A successful method could serve to revolutionize psychological thinking in several areas beyond the intellectual domain. Over-interpretation of first order factors is endemic in psychological research. The solution is not to substitute the first principal factor, except under exceptional circumstances and with acknowledgment of its approximate nature.

A reasonable conclusion for this section is that the general factor among cognitive tests is a candidate for the designation primary. It is still necessary, however, to look beyond the intercorrelations of tests for evidence concerning its importance.

The Importance of the General Factor

There are several sources of evidence concerning the importance of the general factor in human affairs. With respect to some of the evidence, all of us are so close to it that its importance is neglected. Other evidence stems from the research of that very small number of psychologists who do research in something approaching the full range of human talent.

Up the Educational Ladder

If pressed, most college teachers would admit that their students, no matter how dull they seem at times, are actually a superior group compared to the general population, but the amount of selection on broad measures of intelligence is not generally known with any accuracy. There is actually a dearth of studies that estimate the quantity of selection as students ascend the educational ladder. A mere recital of the hurdles along the way, with each being selective, indicates qualitatively the amount that occurs. Staying in public school, high school graduation, application to an institution of higher education, acceptance by the institution, completion of the undergraduate degree, application to professional or graduate school, acceptance by the institution, completion of the professional or graduate training, application for a postprofessional or graduate position, acceptance into the profession, staying in the profession, all of these involve selection. Some selection is imposed by the student or his family, some by the institution. Both types are on the general factor in intelligence.

It is not possible to quantify accurately the amount of selection involved by looking up the scores on intelligence tests taken when postdoctoral persons were in primary school. The intercorrelations of scores on intelligence tests follow the simplex pattern during development. Thus relative position on the general factor is not constant throughout the maturational period. There is undoubtedly some small amount of change in the rank order of individuals after age 18, but during the first 18 years change is relatively large. In the six years between the fifth and eleventh grades the correlation between two composite measures of the general factor, each measure designed to be as nearly identical to the other as possible, is .862 for more than 1400 white boys and girls in a national sample. Reliability estimates are .937 and .947, and the estimate of common true score variance is 84% (Humphreys and Parsons, 1979). Change is more rapid than this in the earlier years.

With adequate testing instruments and with overlapping samples it is possible to estimate accurately the selection that does occur. When I headed the USAF personnel research facility in the fifties, we discovered that the scale used for qualification as an Air Force officer in the stanine range from 2 through 9 covered only the highest 30% of enlisted personnnel. The lowest 4% of the standardization group of officer candidates represented approximately 70% of the enlisted group. Furthermore, most of the officer scale, the portion in particular that distinguished between minimally acceptable and unacceptable, was crowded within the highest 10% of the enlisted group. Practically 100% of officer candidates in our military academies were in the upper 10% range.

It was also possible to draw some tentative conclusions about civilian institutions of higher education on the basis of data from the reserve officer's training program (AFROTC). Our most selective private institutions were slightly more selective than the military academies. Public institutions were lower and more variable, but the officer quality stanine still provided an adequate scale for the lowest of these.

In the light of the amount of selection that does take place up the educational ladder, critics of intelligence tests have over-interpreted the small correlations obtained between measures of intelligence and criterion measures among samples of college graduate or holders of graduate degrees. The effect of selection, or of many successive selections, on correlations can approach the effect of holding constant in a partial correlation a variable (general intelligence) having high communality with the other variables

(college aptitude tests, or college grades, and professional achievement). Depending on the pattern and size of the correlations,
partialling out general intelligence can change a large positive correlation to a negative one.

Who Goes Where to High School

It is well known that there are ability differences among
school means for cognitive variables. The residential patterns
associated with social class are also well known and are generally
considered to be responsible for the cognitive differences among
schools. However, social class and general intelligence are correlated so that a more analytical look at school differences is in
order.

We requested from the Talent Data Bank the intercorrelations
of means for 83 cognitive measures, a composite measure of socio-
economic status of individual students, and 21 school variables on
10th grade boys and girls (Humphreys, Parsons, & Park, 1979).
Complete data including means and variances as well as intercorrelations were available for 713 and 678 schools for males and
females respectively. The amount of selection can be assessed by
the ratios of the standard deviations of school means to those for
individuals in the schools. These ratios, incidentally, are approx-
imations to the etas for the regressions of tests on schools.
When squared, these ratios are estimates of common variance.
For the individual Project Talent tests the median ratio is above
.50 while the ratio for the SES index is above .60, which suggests
more selection on SES than on individual abilities. Most of the
Project Talent tests had very modest reliabilities, however, because tests had to be kept short to conform to limited testing
time. Thus the size of these ratios is reduced by errors of
measurement. In contrast the SES index, a composite of 9 types
of verifiable information, was undoubtedly highly reliable. In
contrast the median ratio for the 24 linear composites is well
above that for SES, and the highest ratios (above, 70) are found
for reliable composites which would be highly correlated with a
standard measure of intelligence.

We also factored tests and demographic measures. There is
a large general factor on which tests that are known to be good
measures of "g" have loadings from .9 to .95 for both boys and
girls. For example, General Vocabulary and Reading Comprehension define the upper level. The three highly speeded clerical
type tests referred to earlier, and which in this research were
scored by number right only, are the only ones which do not
have appreciable loadings on the general factor in either sex.
Hunting and fishing information for the girls are also not loaded
appreciably on the general factor. The socio-economic index for
the student's families has a general factor loading in the seventies.

The general factor loading for this index is about at the mean of the cognitive tests; i. e., selection on socio-economic factors appears to be indirect. Of the school variables rate of college going has the highest general factor loading.

Size of the General Factor

One measure of the importance of the general factor is its contribution to total variance. In the research of Atkin et al. (1977) both first-order oblique factors and hierarchical orthogonal factors were reported. The contrast in the size of the loadings of the group factors in the two rotations is dramatic. The group factors almost disappear after the general factor has been extracted, but in the first-order oblique solution the various factors are very well defined. Loadings of the defining variables are high and there are substantial numbers of variables in the hyperplanes.

These authors obtained two different hierarchical solutions with somewhat different characteristics. The first was the result of the second-order factoring of Binormamin rotations of first-order factors. The second, which spread the total variance somewhat more evenly over the general and group factors, was a Procrustes rotation with targets consisting entirely of either unities or zeros. In the first the general factor accounts for 83% of the common factor variance, in the second 69%.

Predictive Validities

When a statistically naive person, who unfortunately is frequently a psychologist, looks at the correlations between tests measuring various components of general intelligence and socially important criteria, the impression gained is one of great variability. That impression is largely, though not entirely, due to the prevalence of small samples in validation research. Another source of variability is variation in the range of talent from one population sampled to another. A third source is associated with differences in the amount of measurement error from test to test and from criterion to criterion. A fourth potential source which is of interest in the present discussion is the composition of the tests in common factor terms.

The extent to which different common factors contribute to variation in validity coefficients from test to test and from criterion to criterion depends on several parameters of the situation. It is more difficult to establish differential validity in a wide range of talent, and when the individuals in the population are relatively young, have little occupational and only secondary school educational experience. As the age, education, and occupational experience in the population increases and the range of talent decreases, the possibility of establishing differential validity increases.

With respect to the younger and less experienced population of military enlisted personnel, after almost seven years of trying to achieve a useful degree of differentiation in the early and middle fifties, I determined that it was possible to distinguish between mechanical and clerical criteria with two broad clusters of tests, but that finer discrimination was highly problematic. The broad clusters of tests are correlated, all load on the general factor, and the latter still accounts for a major portion of the valid variance of each cluster. I have also had occasion recently to review current military personnel research reports and have not been able to observe any advance in that regard. Differential classification of pilots and navigators in W. W. II, although made easier by the restriction of range of talent on the geneal factor, .was based on similar clusters of cognitive tests.

These broad factors in cognitive tests do not conform to the usual Thurstone primaries. Neither do they conform to the broader factors of Cattell and Horn. They do approximate the Vernon model. One of the occupational clusters contains "dirty hands" mechanical occupations. The tests having the most differential validity for this cluster include all forms of mechanical information and comprehension. The other cluster is represented by clerical, white collar occupations, and the related tests are speeded clerical checking, speeded numerical operations, and mathematical information. Spatial visualization, general vocabulary, and arithmetic reasoning are in the center of the space defined by the two broad factors with the first named being closer to the mechanical tests and the last to the numerical and mathematical tests. Vocabulary is closer to the mechanical cluster than the Vernon model suggests. Unfortunately past and present military tests do not include a recognized measure of the construct of fluid intelligence, but it is quite clear that crystallized intelligence is split down the middle in these data. My guess is that fluid intelligence would fall in the middle between the two broad factors and would be related about equally to the two clusters of military occupations.

The limited differential information for purposes of guidance or classification furnished by military tests would almost certainly be duplicated with civilian tests and civilian occupations in a similar population if adequate data were available. I am not thereby claiming support from these predictive validities for two broad traditionally defined aptitudes over and beyond general intelligence. It seems much more plausible to me that we have here again a transfer of training phenomenon. The two broad military factors are defined by variables that reflect a very common split in the secondary curriculum which in turn produces differential exposure of high school students to information and skill training.

There is, of course, additional differential exposure that is
extracurricular.

Other Evidence Briefly Noted

An indirect indication of the importance of the general factor
is the ease with which a good measure can be developed from
seemingly very different content. In a culture in which almost
100% of the children are in school for the first 6 to 8 years, a
composite of achievement tests late in that period will correlate
about as high with the Stanford-Binet as does the Wechsler.
Without near universal education this would not be true. It is
also possible to reproduce these findings with a test composed of
many types of nonacademic information. Project Talent, for
example, included a wide range of information tests. It is possible
to obtain a composite from these tests, after excluding the ones
that overlap most with standard academic achievement tests, that
is an excellent measure of the general factor. A third measure as
highly correlated with a standard test of intelligence as the latter
is correlated with a second standard test, is a composite formed
of Piagetian tasks. We have a manuscript in press (Humphreys &
Parsons, 1979) in which the correlations of a Piagetian composite
with a Wechsler and academic achievement composite is .88 in a
wide range of talent.

Another indirect indicant appears in teaching methods research. This is the relative size of the contribution to total
variance of the dependent variable of the independent variable or
variables and individual differences in general intelligence. It is
no wonder that experimental psychologists prefer to report their
research findings in the form of t and F-ratios rather than correlation coefficients. A related research finding is the small contribution to variance of differences in outcomes associated with different
institutions (public schools, colleges) when there is adequate
control for the quality of the incoming students. The analysis of
covariance does not provide completely adequate control under the
best of circumstances for differences among intact groups. It is
especially inadequate when children are changing appreciably and,
in so far as we can determine, without regard to the treatment
differences imposed on the groups.

The Nature of General Intelligence

In this section I shall discuss some research, theory, and
speculation concerning the nature of the construct of general
intelligence. The first section contains some research findings
concerning a possible genetic component to the variance of scores
on intelligence tests. My approach here is quite narrow; I do not
attempt to review the very voluminous literature on this subject.
Next I relate the construct of general intelligence as it has devel-

oped in the psychometric tradition to the approach of experimental cognitive psychologists. Then I conclude the paper with a brief characterization of the psychometric construct.

Genetics and the General Factor

Several years ago I made use of the ratio of cross-twin to within-twin correlations obtained from the Project Talent data bank to investigate whether different types of cognitive tests showed evidence for differential degrees of heritability (Humphreys, 1974). The means of these ratios do not differ for information tests and for noninformation tests, or for standard intelligence test subtests and for subtests not commonly found in intelligence tests.

Two other methods of analyzing the within-twin and cross-twin correlations led to the same conclusion: namely, there was no evidence for differences in heritability from one type of test to another within the rather wide limits of the tests studied. I interpreted these findings as indicating that the genetic contribution to these cognitive tests, whatever its amount, was restricted to the general factor.

The General Factor and Process Research

Cognitive experimental psychologists have been proceeding rapidly with research on intellectual processes in recent years. Some have been relating their research quite directly to general intelligence or to its components. The work of Hunt and his associates (1976) exemplifies this approach.

While this research is extremely interesting and gives promise that it will eventually shed considerable light on our understanding of both process and the present construct of general intelligence, there is reason to believe that measures of these processes may eventually merely supplement the information provided by a standard test of intelligence. The supplementary information may be very useful, but it will not supplant intelligence tests.

I believe that no one would presently claim that this research has reached a point when it can be applied usefully. Good measures of cognitive processes of the sort studied by Hunt and others will require psychometric as well as experimental analysis. It is highly probable that an indicant of process obtained from a single experimental paradigm with a particular set of content carries a large nonerror specific in addition to the variance of the process being studied. That is, a useful measure of process will require multiple "items" just as a reliable and valid test requires multiple items. Only by zeroing in on a particular process from several methods and types of content can a valid measure of the latent trait be developed.

Given a measure of process having adequate psychometric characteristics, however, the correlation between the test of process and the test of general intelligence will be substantially less than unity. Also, the test of intelligence will have higher correlations with many socially important criteria than the measure of process. I believe, and have stated elsewhere, that intelligence is the resultant of the processes of acquiring, storing, retrieving, combining, comparing, and using in new contexts information and skills. (Guilford's operations are, in this context, acquired skills rather than basic processes). General intelligence is, therefore, the resultant of the fundamental processes cognitive psychologists are studying. Although the latter are more fundamental, they can still be less valid for socially important criteria. The test of general intelligence samples a very large repertoire of information and skills which transfer to further education and to occupations. In part the predictive validities of a test of general intelligence are transfer phenomena. A person's current level of proficiency in a wide ranging intellectual repertoire furnishes two kinds of information: about the effectiveness in the past of the intellectual processes that produced the repertoire and the availability of the elements in that repertoire for transfer to new learning situations.

Another insight into process has been opened up by a non-experimental method of analysis: cross-lagged correlation analysis. Atkin et al. (1977) found a highly significant difference between the cross correlations for a psychometric measure of aural comprehension and a composite of 15 other cognitive measures of reading, achievement, and information when the two are separated by as much as six years. The direction of the difference is that individual differences in aural comprehension anticipate individual differences in the intellectual composite. Humphreys and Parsons (1979) have shown that the lag between aural comprehension and the general factor is probably about three years between the 5th and 11th grades in public school.

The orally administered test includes content similar to that in a measure of reading comprehension; presumably there must be a difference in some fundamental process which allows individual differences in one to anticipate individual differences in the other. At the present time one can only speculate about possible processes.

Interpretation of the Construct

General intelligence is a phenotypic construct of considerable importance in human affairs. I can characterize it no better than I did in a recent paper (Humphreys, 1979) a characterization which follows the Godfrey Thomson tradition (1919) which allows the acceptance of a general factor without requiring an entity within the organism. To the extent that there is a genetic contribution to individual differences in general intelligence that contribution is polygenic. Environmental contributions are also multiple. To coin a term we might call these contributions polyenvironmental. Similarly, the biological substrate for general intelligence is poly-neural, and the behavioral observations which define the phenotypic construct are polybehavioral.

"This intepretation of general intelligence is very similar to descriptions of fluid ability. The recommended measures of fluid ability, however, are not the only nor possibly even the best measures of general intelligence. Intelligence is too fluid to be tied to a particular subset of cognitive tests, and there is a fluid (general) component in the variance of the most crystallized information or achievement test."

References

Atkin R., Bray, R., Davison, M. Herzberger, S., Humphreys, L, and Selzer, U., Ability factor differentiation, grades 5 - 11. Applied Psychological Measurement, 1977, 1, 65-66.

Atkin, R., Bray, R., Davison, M., Herzberger, S., Humphreys, L, and Selzer, U. Cross lagged panel analysis of 16 cognitive measures. Journal for Research in Child Development, 1977, 48, 944-952.

Cattell, R. Theory of fluid and crystallized intelligence: A critical experiment. Journal of Educational Psychology, 1963, 54, 1-22.

Guilford, J. The nature of intelligence. New York: McGraw-Hill, 1967.

Guttman, L. A new approach to factor analysis: the radex. In Lazarsfeld, P. (Ed.), Mathematical Thinking in the Social Sciences. Glencoe, Ill.: Free Press, 1954.

Horn, J. Organization of abilities and the development of intelligence. Psychological Review, 1968, 75, 242-259.

Humphreys, L. The organization of human abilities. American Psychologist, 1962, 17, 475-483.

Humphreys, L. The misleading distinction between aptitude and achievement tests. In D. R. Green (Ed.), The aptitude achievement distinction, Monterey, California: CTB/McGraw Hill, 1974, 262-274.

Humphreys, L. The construct of general intelligence. Intelligence, 1979, 3, 105-20.

Humphreys, L. and Parsons, C. A simplex process model for describing differences between cross-lagged correlations. Psychological Bulletin, 1979, 86, 325-334.

Humphreys, L. and Parsons, C. Piagetian tasks measure intelligence and intelligence tests assess cognitive development: A reanalysis. Intelligence, 1979, in press.

Humphreys, L. and Parsons, C., and Park, R. Dimensions involved in differences among school means of cognitive measures. Journal of Educational Measurement, 1979, 16, 63-76.

Hunt, E. Varieties of cognitive power. In L. B. Resnick (Ed.), The nature of intelligence. Hillsdale, NJ: Lawrence Erlbaum, 1976.

Schmid, J. and Leiman, J. The development of hierarchical factor solutions. Psychometrika, 1957, 22, 53-61.

Spearman, C. The theory of two factors. Psychological Review, 1914, 21, 101-115.

Spearman, C. The abilities of man. New York: MacMillan, 1927.

Thomson, G. On the cause of hierarchical order among correlation coefficients. Proceedings of Royal Society, A, 1919, 95.

Thurstone, L. Primary mental abilities. Psychometric Monograph, 1938, (No. 1).

Thurstone, L. Multiple factor analysis. Chicago: University of Chicago Press, 1947, XIX, 515.

Vernon, P. The structure of human abilities. New York: Wiley, 1950.

Footnote

This research was supported by a grant from the National Institute of Mental Health, MH 23612-06, Studies of Intellectual Development and Organization. Requests for reprints should be sent to: Dr. Lloyd G. Humphreys, 425 Psychology Building, University of Illinois, Champaign, IL 61820, U.S.A.

GENETIC DIFFERENCES IN "g" AND REAL LIFE

Sandra Scarr

Yale University

New Haven, Connecticut, U.S.A.

Despite the author's substantial agreement with the spirit of presentations by Professors Eysenck and Humphreys, arguments about the extremity of their views and their certainty are presented. In the author's research on twins and adopted children, there are important age differences in the effects of home environments and important differences in the effect of environments on tests of academic achievement, aptitude, and IQ. The confident claims that there is one important general ability and one figure for the heritability of intelligence are disputed.

So overwhelming is the agreement between Professors Eysenck and Humphreys that I feel moved to pick at both of them around the edges of their arguments and to propose modifications on their certainties.

They are in concert by saying:

(1) "g" is the major intellectual dimension, the major portion or variability in intelligence;
(2) differences in "g" are highly heritable and biologically based, a revival of the theory of Godfrey Thomson;
(3) the speed of neural transmission is a key to understanding individual differences in intelligence; and
(4) genetic variance in all kinds of cognitive tests and tasks overwhelms any measurable or measured environmental variance.

There are those who could and would dispute each point. Unfortunately, I am in modest agreement with their views, but I will strive to describe my discomfort with the extremity of their views and their certainty in the face of inconsistent evidence. Particularly, I will discuss what I believe to be important age differences in the effects of family environments on children's intellectual development--that is, younger children are far more affected than older children by the intellectual climate of their homes--and the important differences in the effects of home environment on measures of IQ, academic aptitude and school achievement. Although this conference centers attention on intelligence, I hope that we agree that one's intellectual achievements include what one can do with "g" in socially meaningful contexts, including school.

First, let me address the issue of IQ heritability, which Prof. Eysenck proposes confidently is .80. My data on adoptive and biological families and on twins support the conclusion that about half of the individual variability in IQ test scores and other cognitive tests is due to genetic differences. In the course of showing you some of these data, I will also illustrate the inconsistencies in age effects and in parent-child versus sibling resemblance. Second, I will turn to the effects of measured family background in the adoptive and biological families and show that family environments have more effect on intelligence manifested in school achievement than on IQ tests. Although, as Prof. Humphreys said, there may not be differences in the heritability of IQ, aptitude, or of achievement scores, there is evidence for greater effects of home environment on differences in school achievement than IQ.

Three Studies

First, let me turn to issues of the magnitude of genetic variance in intelligence, as measured by individually-administered IQ tests. Three of my studies are relevant--the transracial adoption study, the adolescent adoption study, both carried out with Prof. Richard A. Weinberg of the University of Minnesota, and the Philadelphia Twin Study, a second research on 400 pairs of twins, to be published soon (Scarr, in press, 1980).

The resemblance of genetically-related and unrelated persons in the same families is a particularly interesting test of the effects of family environments, because those who are genetically unrelated resemble each other only because they are reared in the same household. The correlations between unrelated siblings reflect the impact of differences among environments of the adoptive families. The comparison of the correlations of biological and adoptive siblings yields an estimate of the magnitude of genetic differences in the population from which the families are sampled.

Adoptive families are not representative of the general population, of course, because they are selected by agencies for their virtues. In the transracial adoption study, the same families provide their own biological controls, but they do not represent the range of environments in the general population.

The transracial adoption study was carried out from 1974-1976 in Minnesota to test the hypothesis that black and interracial children reared by white families (in the culture of the tests and of the schools) would perform on IQ tests as well as other adopted children (Scarr and Weinberg, 1976). For the present purposes, the parent-child and sibling resemblances of genetically-related and unrelated members of these families is salient.

In the transracial families were 143 biological children, 111 children adopted in the first year of life (called the Early Adoptees) and 65 children adopted after 12 months of age--up to 10 years at the time of adoption. Most of the later adoptees were in fact placed with the adoptive families before four years of age, but they were not the usually-studied adopted children who have spent all of their lives past the first few months with one adoptive family. As we described in an earlier paper (Scarr and Weinberg, 1976), the later adoptees have checkered pre-adoptive histories.

The second adoption study included 115 adoptive families with adolescents who were adopted in the first year of life, and 120 biological families with their own adolescent offspring. In this study, separate samples of adoptive and biological families were necessary, because these were white, Minnesota families who had adopted white infants, usually for reasons of infertility. The samples of biological and adoptive families are very comparable, however, in socio-economic status, as reported in Scarr and Weinberg (1978).

The third study, of identical and fraternal twins, included about 400 pairs of 10 to 16 year old, same-sex twins, 175 black pairs and 225 white pairs. The twins' families varied widely in socio-economic status and were very representative of the distribution of whites and blacks in the Philadelphia SMSA.

With these three studies, then, I hope to illustrate that genetic differences do contribute to intellectual differences among people in all segments of the population, but that the magnitude of genetic effects seems to vary among age groups and that environmental differences among families are more important for school achievement than for IQ test scores.

Biases in Comparisons of Twins and Siblings

Critics of twin and adoption studies often claim that one cannot make genetic inferences from the comparison of identical with fraternal twin correlations or comparisons of biological with adopted sibling

correlations or comparisons of biological with adopted sibling correlations because perceived and expected similarities are greater for
identical than fraternal twins and for biological than adopted siblings.
Moreover, identical treatment because of the strikingly similar appearance of many identical twin pairs has been claimed by Kamin (1974)
to be sufficient to explain their greater cognitive similarity than
fraternal pairs. The parallel argument against family studies is
that adopted children know that they are not genetically related to
their parents or siblings, and therefore may not expect to be like
them; biological offspring may be expected to resemble their family
members If such biases exist, then one ought not to conclude that
the greater similarity of identical than fraternal twins or of biological
than adopted relatives is due to their greater genetic similarity.

In the comparison of twins, many people do not realize that
twins themselves, their parents, and others are often wrong about
whether the twins are identical or fraternal (Carter-Saltzman and
Scarr, 1977). In two studies (Scarr, 1965; Scarr and Carter-Saltzman,
in press) we have shown that cognitive similarities between co-twins
are related to their actual zygocity and not to the zygocity they or
others believe them to be. Other investigators (Plomin, Willerman
and Loehlin, 1976; Lytton, 1977) have used other strategies with
results that lead to the same conclusion. Thus, the greater perceived similarities in the appearance of identical twins do not seem
to be related to their greater cognitive similarities.

In adoptive families, all members know that the children are
genetically unrelated to the parents and to each other. No one is
confused, as in the case of twins' zygocity. To test for possible
biases in the perceptions of adoptive and biological families, we
asked the adolescents and their parents to rate their similarity to
other family members (parents to their adolescents and adolescents
to their parents) on six scales, one of which was intelligence (Scarr,
Scarf, and Weinberg, Note 1). I am relieved to report that although,
on the average, biological family members think they are more similar
to each other than do members of adoptive families, neither group
is accurate about their IQ resemblance to relatives. That is, differences in WAIS scores between family members are not related to
their self-perceptions of similarity. Thus, the fact that biological
relatives tend to believe that they are more similar than adoptive
family members does not bias the comparison of IQ correlations in
biological and adoptive families.

Parent-Child IQ Resemblance

To turn to the results of the first adoption study, Table 1
shows the correlations of the parents and children in the transracial
adoption study. The adoptive families had adopted at least one
black child, but there were also other adopted children and many
biological offspring of these same parents. The children ranged in

Table 1.   Comparisions of Biological and Unrelated Parent-Child IQ
Correlations in 101 Transracial Adoptive Families

Parents-Biological Children

	N (pairs)	r
Adoptive mother—own child	141	.34
**Natural mother—adopted child	135	.33
Adoptive father—own child	142	.39
**Natural father—adopted child	46	.43

Parents-Unrelated Children

	N (pairs)	r
Adoptive mother—adopted child	174	.21 (.23)*
**Natural mother—own child of adoptive family	217	.15
Adoptive father—adopted child	170	.27 (.15)*
**Natural father—own child of adoptive family	86	.19

* Early Adopted Only (N=111)

** Educational level, not IQ scores

age from four to about 18. Because of the age range, children from four to seven years were given the Stanford-Binet, children from eight to 16 the WISC, and older children and all parents the WAIS. The adopted children averaged age seven, and the natural children about ten.

Table 1 shows the parent-child IQ correlations for all of the adopted children in the transracial adoptive families, regardless of when they were adopted. The total sample of adopted children is just as similar to their adopted parents as the early adopted group! The mid-parent child correlation for all adoptees is .29, and for the early adoptees, .20. Mothers and all adopted children are equally similar, and fathers more similar than they are to the early adopted children.

Table 1 also shows the correlations between all adopted children's IQ scores and their natural parents' educational levels. Because we did not have IQ assessments of the natural parents, education is used here as a proxy. Despite this limitation, the correlations of natural parents' education with their adopted-away offspring's IQ scores are as high as the IQ correlations of biological parent-child pairs and exceed those of the adopted parent-child IQ scores. The mid-natural parent-child correlation of .43 is significantly greater than the mid-adopted parent-child r of .29.

Because the adoptive parents are quite bright, their scores had considerably restricted variance. In Table 1 the correlations between parents and their natural and adopted children are not corrected for restriction of range in the parents' IQ scores. When corrected, the correlations of biological offspring with their parents rise to .49 and .54 and the mid-parent (the average of the two parents) is .66. Adopted child-parent IQ resemblance rises to .36 (Scarr and Weinberg, 1977). When the IQ scores of the parents are corrected for restriction of range, the magnitude of the resemblance between biological parents and children reared together exceed that of the natural parents' educational level and the IQ scores of the adopted-away offspring, but the latter are still higher than the correlations of corrected IQ score correlations for the adoptive parents and adopted children.

The correlations between natural parents of adopted children and the biological children of the same families is an estimate of the effects of selective placement. If agencies match educational and social class characteristics of the natural mothers with similar adoptive parents, then the resemblance between adoptive parents and children is enhanced by the genetic, intellectual resemblance of natural and adoptive parents. Selective placement also enhances the correlation between natural parents and their adopted-away offspring, because the adoptive parents carry out the genotype-environment correlation that would have characterized the natural parent-child pairs, had the children been retained by their natural parents. Thus, neither the adoptive parent-child correlations nor the natural parent-adopted child correlations deserve to be as high as they are. In another paper (Scarr and Weinberg, 1977), we adopted the solution proposed by Willerman et al. (1977), to subtract half of the selective placement coefficient of .17 from both the natural parent-adopted child correlation and half from the adoptive parent-adopted child correlation. There are other corrections that could be justified by the data set, but I will leave the "ultimate" solution(s) to biometricians. My simple figuring of these data yields "heritabilities" of .4 to .7.

Sibling Correlations

In Table 2, the sibling correlations reveal a strikingly different picture. Young siblings are quite similar to each other, whether genetically related or not! The IQ correlations of the adopted sibs, genetically unrelated to each other, are as high as those of the biological sibs reared together. Children reared in the same family environments and who are still under the major influence of their parents score at similar levels on IQ tests. The IQ correlations of the adopted sibs result in small part from their correlations in background, such as their natural mothers' educational levels (.16) and age at placement in the adoptive home (.37), which is in turn related to the present intellectual functioning of the children--the earlier the placement the higher the IQ score. Age of placement is itself correlated with many other background characteristics of the child and is a complex variable (Scarr and Weinberg, 1976). It seems that some families accepted older adoptees and others didn't, and that the families differed on the average in the rearing environments that they provide. But note that the correlation among the early adopted siblings is fully .39! Even among the families who had early adoptees, differences in family environments and selective placement account for an unexpectedly large resemblance between unrelated children.

The major point for this symposium is that the "heritabilities" calculated from the sibling data are drastically different from those calculated from the parent-child data. We have explained our interpretation of this result elsewhere (Scarr and Weinberg, 1977, 1979). The point to Professors Eysenck and Humphreys is that h^2 is not uniformly .80. As Christopher Jencks pointed out in his earlier book (1972) the correlations of unrelated young siblings reared together do not fit any biometrical model, because they are too high. This study only makes the picture worse.

Twin Correlations

The second study of young adolescent twins reveals a variety of "heritabilities" for several cognitive tests in black and white

Table 2. Sibling Correlations: Natural and All Adopted Children
          of Adoptive Families

Natural Sibs	N (Pairs)	r
All IQ scores	107	.42
Stanford-Binet	10	.50
WISC + WAIS	63	.54
Natural Sib-Adopted Sib		
All IQ scores	230	.25
Stanford-Binet	57	.23
WISC + WAIS	63	.20
Natural Sib-Early Adopted Sib (All IQ scores)	34	.30
All Adopted Sibs		
All IQ scores	140	.44
Stanford-Binet	36	.31
WISC + WAIS	50	.64
Early Adopted Sibs (All IQ scores)	53	.39

populations. It is not possible in this brief presentation to describe
the measures in full (see Scarr, 1979, in press). In Table 3 are
the MZ (identical) and DZ (fraternal) twin correlations for black
and white samples on five tests: Raven Standard Progressive
Matrices (1958, Sets A-D), the Columbia Test of Mental Maturity
(1959), the Peabody Picture Vocabulary Test (Dunn, 1959), Benton's
Revised Figural Memory Test (Benton, 1963), and a Paired-Associate
task devised by Stevenson, Hale, Klein, and Miller (1968). The
last is a largely rote or Level 1 task, whereas the others are com-
parably cognitive and correlated with each other about .5 in both
racial groups. The scores on each test were corrected for age,
which would naturally inflate twin correlations because twins are

always exactly the same age. Age correction reduced all of the
twin correlations. Although the internal consistency of all of these
measures is over .85 in all groups (over .95 for most), the twin
correlations are not as high as many other studies of other age
groups and other samples reported. The reason, I think, is that
the ages 10 to 16 years are a period of very rapid intellectual change.
Note particularly how low the DZ twin correlations for the white
group are, lower than any of the sibling correlations in the study
of younger, biological siblings, or even adopted siblings. In any
case the "heritabilities" are not uniformly .8.

The Adolescent Adoptees

This study was conceived to assess the cumulative impact of
differences in family environments on children's development at the
end of the childrearing period (Scarr and Weinberg, 1978). All of
the adoptees were placed in their families in the first year of life,
the median being two months of age. At the time of the study they
were 16 to 22 years of age. We administered the short form of the
WAIS to both parents and to two adolescents in most of the 115
adoptive families. A comparison group of 120 biological families had
children of the same ages. Both samples of families were of similar
socioeconomic status, from working to upper middle class, and of
similar IQ levels, except that the adopted children scored about 6
points lower than the biological children of similar parents.

Table 4 gives the parent-child and sibling correlations for the
WAIS IQ and the four subtests on which it is based. The parent-
child IQ correlations in the biological families are what we were led
to expect from our earlier study and others--around .4 when uncorrected for the restriction of range in the parents' scores. The
adoptive parent-child correlations, however, are lower than those of
the younger adopted children and their parents. And the IQ correlation of adopted children reared together is zero! Unlike the
younger siblings (who, after all, are also of different races), these
white adolescents reared together from infancy do not resemble
their genetically-unrelated siblings at all.

The IQ "heritabilities" from the adolescent study vary from .38
to .61, much like the parent-child data in the study of younger
adoptees, but very unlike that data on younger sibs.

Our interpretation of these results (Scarr and Weinberg, 1978),
is that older adolescents are largely liberated from their families'
influences and have made choices and pursued courses that are in
keeping with their own talents and interests. Thus, the unrelated
sibs have grown less and less alike. This hypothesis cannot be
tested fully without longitudinal data on adopted siblings; to date
all of the other adoption studies sampled much younger children, at
the average age of 7 or 8. I can think of no other explanation for

Table 3. Comparisons of MZ and DZ Correlations and Heritabilies for Normalized Standard Scores on Five Cognitive Measures by Race

Black

Test	MZ (65)	DZ (95)	t	MZ–DZ
Raven	.63	.36	2.07*	.27
Columbia	.46	.25	1.51	.21
Peabody	.66	.52	1.37	.14
Benton Error	.61	.31	2.49**	.30
P-A Task	.65	.40	1.66*	.25

White

Test	MZ (121)	DZ (91)	t	MZ–DZ
Raven	.59	.15	3.65***	.44
Columbia	.39	.11	2.25*	.28
Peabody	.64	.40	2.44**	.24
Benton Error	.57	.22	3.05**	.35
P-A Task	.56	.49	.64	.07

* $p < .05$

** $p < .01$

*** $p < .001$

Table 4. Correlations Among Family Members in Adoptive and Biologically-Related Families (Pearson Coefficients on Standardized Scores by Family Member and Family Type) for Intelligence Test Scales

Child Score	Reliability (*)	Biological (120 families)				Adoptive (104 families)			
		MO	FA	CH	MP	MO	FA	CH	MP
Total WAIS IQ	(.97)	.41	.40	.35	.52	.09	.16	-.03	.14
Subtests									
Arithmetic	(.79)	.24	.30	.24	.36	-.03	.07	-.03	-.01
Vocabulary	(.94)	.33	.39	.22	.43	.23	.24	.11	.26
Block Design	(.86)	.29	.32	.25	.40	.13	.02	.09	.14
Picture Arrangement	(.66)	.19	.06	.16	.11	-.01	-.04	.04	-.03

__ = biological $>$ adoptive correlation, $p < .05$

Sample Sizes: Pairs of Family Members

	Biological				Adoptive			
	MO	FA	CH	MP	MO	FA	CH	MP
Children	270	270	168	268	184	175	84	168

Assortative Mating

	Biological FA-MO	Adoptive FA-MO
WAIS IQ	.24	.31
Arithmetic	.19	-.04
Vocabulary	.32	.42
Block Design	.19	.15
Picture Arrangement	.12	.22
Sample Size	120	103

MO = mother-child; FA = father-child; CH = child-child; MP = midparent-child

*   reliability reported in the WAIS manual for late adolescents

the markedly low correlations between the adopted sibs at the end of the childrearing period, in contrast to the several studies of younger adopted sibs, who are embarrassingly simliar. For none, however, is the heritability of differences in IQ uniformly .8.

Effects of Family Background on IQ, Aptitude and Achievement Scores

For contrast with the material that is forthcoming, let us look first at the effects of family environments on young adoptees' IQ scores differences. Table 5 shows two regression equations, one for the biological children and one for the early adopted children of the transracial adoptive families. The predictive variables are more substantially related to the IQ scores of the biological children, with an R^2 of .30, compared to an R^2 of .156 for the young adoptees. The major difference in the two equations is the predictive value of the parents' IQ scores for the biological children's IQ scores. The IQ scores are correlated, of course, with parental demographic characteristics, whose coefficients are pulled in a negative direction when they co-exist in the equation.

Now, let us look at similar data for the adolescent adoptees and their biological, comparison families. The adolescents' IQ, school aptitude, and achievement test scores were regressed on family demographic characteristics, sibling order, and parental IQ. The adopted adolescents' scores were regressed on those variables plus the natural mothers' age, education, and occupational status. The goal of these analyses was to estimate how much the indexed differences in family environments contribute to individual differences in IQ and school test scores. The contribution of genetic differences to test score differences is grossly underestimated by this procedure, because the only parental scores available are WAIS IQ for the biological parents. There are no comparable data on the natural parents of the adopted children nor are there school test scores on any of the parents. Nonetheless, it is interesting to examine the pattern of R^2's obtained from the regression of the IQ, aptitude, and achievement scores on social and genetic background. Table 6 gives a summary of the regression analyses. (Detailed versions of the regressions are given in Scarr, Note 2).

Let us concentrate on the adoptive families first. Because the parents in this case provide only the social environment, it is possible to estimate the effects of differences in these environments, which range socioeconomically from working to upper-middle class. The R^2 values, shrunken from each equation, give the estimated percentages of variance in test scores accounted for by socioeconomic differences between families--that is, those social environmental features that siblings share--and by environmental differences between siblings within the same families, which are indexed here by sibling order (in biological families this would be called birth order).

Table 5. Regressions of Child IQ on Family Demographic Characteristics, and Parental IQ in Transracial Adoptive Families with
their Own Children

| | Biological Children (143) | | Early Adopted Children (111) | |
	B	beta	B	beta
Mother's IQ	.474	.32	.141	.13
Father's IQ	.513	.40	-.028	-.02
Father's Education	.682	.14	.389	.09
Mother's Education	-.943	-.15	1.501	.25
Father's Occupation	-.174	-.23	.008	*
Family Income	.445	.06	-.371	-.06
Total R^2	.301		.156	
Shrunken R^2	.269		.116	

* $F < .01$, variable did not enter the equation.

Between Family Effects

The most striking result is that differences in adoptive families'
income, parental education, fathers' occupations, and parents' IQ
scores account for minus one percent of the variance in their adolescents' IQ scores. In fact, the uncorrected R^2 for the regression of
adopted adolescents' IQ scores on their adoptive parents' characteristics is only .02, which shrinks to -.01 with correction. This
means that differences among families' social class and intellectual
environments have virtually no effect on IQ differences among their
children at the end of the child rearing period. By comparison,
the same variables accounted for 11.6 percent of the IQ variance
among the younger adopted children.

The same regression equation for the biologically-related adolescents is given at the bottom of Table 6. In contrast to -.01, their
corrected R^2 is .26 for the same measures of between-family differences in social class and parental IQ. This value is identical to the
shrunken R^2 for the younger sample of biological children in the

Table 6. R^2 Estimates of the Effects of Social Environmental and Genetic Differences on IQ, Aptitude, and Achievement Test Scores (Stepwise Regressions)

		WAIS	Aptitude			Achievement		
		IQ	Verbal	Num.	Total	Read	Math	Total
	Adopted Adolescents N =	150	147	128	128	140	128	128
Step	**Social Environmental Indices**							
1.	Between Families[1]	.01	.05	.03	.04	.09	.08	.10
2.	Within Families[2]	.02	.02	.00	.01	.01	.05	.03
	Total Environment	.03	.07	.03	.05	.10	.13	.13
3.	Genetic Indices[3]	.06	.08	.02	.05	.07	.07	.09
	Total R^2	.09	.15	.05	.10	.17	.20	.22
	Biological Adolescents N =	237	231	158	158	195	187	187
	Social Environmental Indices and Genetic Indices							
1.	Between Families[1]	.26	.19	.13	.18	.14	.14	.18
2.	Within Families[2]	.03	.04	.04	.07	.01	.02	.02
	Total R^2	.29	.23	.17	.25	.15	.16	.20

Shrunken R^2's

Notes

1 = parental education, father's occupation, family income, parental WAIS IQ's

2 = sibling order

3 = natural mothers' education, occupation, and age (to correct for young mothers)

transracial adoptive families. In the case of biological children, of course, these differences between families are due to both environmental and genetic differences, the latter being of overwhelming importance in explaining the IQ differences both among younger children and adolescents in these families.

As we move from IQ to school test scores, there are three important trends to notice: first, the effect of differences in social environments between families increases as the tests sample more recently taught material; second, natural mothers' genetic contribution to test score differences is similar and moderate across the various tests; and third, that the contribution of biological parents' IQ scores to their offsprings' test score differences is far less for school aptitude and achievement tests than for IQ tests.

The first point is that the major difference in explained variance between IQ and school achievement test scores is that social class differences--that is, differences among families--account for the majority of the explained variability in achievement scores and virtually none of the IQ differences. It is the social environment differences among the adoptive families, indexed by parental demographic characteristics, that contribute most to school achievement differences among the adopted adolescents. In one sense, then, school achievement tests are more biased against working class environments than are IQ tests!

Natural Mothers' Effects

To test the second point, the effects of genetic differences among the adopted adolescents, the index of genetic differences is admittedly very weak. We have information on only one of the natural parents, and that information is limited to educational and occupational level at the time of the child's birth and age, which was entered into the regression equations to correct for any underestimation of younger mothers' educational and occupational levels. Regardless of the limitations of those variables, one can see from Table 6 that natural mothers' characteristics are substantially related to their offspring's intellectual achievements, even though any variance due to selective placement has been removed by entering social environmental variables into the equations first.

Biological Parents' IQ Effects

On the third point, the predictive power of biological parents' IQ scores, the detailed tables of regression analyses (available in Scarr, Note 2) show that parental IQ's decline from 15 percent of the variance in adolescents' IQ scores (holding everything else in the equation constant) to less than 2 percent of the variance in aptitude and achievement test scores (again holding constant education, income and other variables). Parental IQ is by far the best predictor of IQ differences among biologically-related children, but parental education and family income are as good predictors of school aptitude score differences and better predictors of school achievement scores. This does not mean that the genetic differences are less important for aptitude and achievement scores, as we can note from both the natural mothers' data and from sibling correlations of test scores to be reported. But it does mean that parental IQ differences are more closely related to their offspring's differences in IQ than in school achievements. If we had obtained reading and mathematics achievement scores for the parents, however, it may

Table 7. Sibling Correlations of IQ, Aptitude, and Achievement Test
Scores of Adopted and Biologically-Related Adolescents

	Biological		Adopted		
	N(pairs)	r	N(pairs)	r	$h^2 = 1.6(r_{bio} - r_{adopt})$
WAIS Verbal	168	.23	84	.07	.26
Performance	168	.21	84	.07	.22
IQ	168	.35	84	-.03	.61
Aptitude, Verbal	141	.29	68	.13	.26
Numerical	61	.32	49	.07	.40
Total	61	.32	49	.09	.37
Achievement, Reading	106	.27	73	.11	.26
Math	104	.35	58	-.11	.53
Total	104	.33	58	-.03	.58

well be that the between family genetic differences would remain
relatively constant across the kinds of tests while the impact of social
environments would rise, giving a higher total between-family R^2
for achievement than IQ test scores. From the adopted family results
it is clear that environmental differences among families are a trivial
source of IQ differences and a substantial source of differences in
school test scores.

Sibling Correlations

Another method for checking on the effects of family environment on test scores is to calculate the correlations between pairs of
siblings who are genetically unrelated but who have been reared
together from early infancy, as are our adopted children. Their
sibling correlations are given in Table 7, with the corresponding
biological sibling correlations for comparison.

As one can see, the effects of being reared in the same household, neighborhood, and schools are negligible unless one is genetically related to one's brother or sister. The correlations of the

biological siblings are modest but statistically different from zero.

With the most simple-minded version of the heritability coefficient and an assumption that parental assortative mating is the same for aptitude and achievement as for IQ, we multiply the difference between the biological and adopted siblings' correlations by 1.6. The heritability estimates vary from .22 to .61, with a median of .37. Although these values are not .8, as some would claim, neither are they zero. There seems to be no consistent difference in heritability by the kind of test.

The negligible differences in heritability of IQ, aptitude and achievement scores in this study of late adolescents is congruent with Lloyd Humphreys' findings of equal heritabilities for all cognitive measures in the Project Talent data (Humphreys, 1979) and the Texas Adoption Study result of equal sibling resemblances of IQ and school achievement measures in a sample of younger children (Willerman, Horn and Loehlin, 1977). In other words, there seems to be no greater sibling resemblance for one or another kind of intellectual achievement, when they are all "g" loaded. Humphreys and I agree, however, that some specific skills may have different heritabilities.

More relevant for this discussion of the papers by Professors Humphreys and Eysenck are the findings that the effects of family environments vary with the age of the child and the material sampled on the test. Younger children seem to be far more influenced by differences among families. Children reared in working class families are more disadvantaged in comparison to upper middle class children when the tests sample specifically and recently taught material, that is, by school achievement tests rather than IQ tests. And, finally, I hope you will agree that the evidence from these studies argues for a heritability of intellectual measures in the .4 to .7 range, and not .8.

References

Benton, A. L. The Revised Visual Retention Test, Form C, Dubuque, Iowa: William C. Brown, 1963.

Burgemeister, B. B., Blum, L. H., and Lorge, I. Columbia Mental Maturity Scale. New York: Harcourt, Brace and World, 1959.

Carter-Saltzman, L., and Scarr, S. MZ or DZ? Only your blood grouping laboratory knows for sure. Behavior Genetics, 1977, 7(4), 273-280.

Dunn, L. M. Peabody Picture Vocabulary Test. Circle Pines, MN: American Guidance Service, 1959.

Eysenck, H. J. The nature of intelligence. Paper presented at the NATO International Conference on Intelligence and Learning," University of York, England, July 16, 1979.

Humphreys, L. G. The primary mental ability. Paper presented at the NATO International Conference on Intelligence and Learning, University of York, England, July 16, 1979.

Jencks, C. Inequality: A reassessment of the effects of family and schooling in America. New York: Basic Books, 1972.

Kamin, L. J. The science and politics of IQ. New York: John Wiley, 1974.

Lytton, H. Do parents create, or respond to, differences in twins? Developmental Psychology, 1977, 13(5), 456-459.

Plomin, R., Willerman, L., and Loehlin, J. C. Resemblance in appearance and the equal environments assumption in twin studies of personality traits. Behavior Genetics, 1976, 6, 43-52.

Raven, J. C. Standard Progressive Matrices: Sets A, B, C, D, and E. London: Lewis, 1958.

Scarr, S. Environmental bias in twin studies. Eugenics Quarterly, 1968, 15, 34-40

Scarr, S. IQ: Race, social class and individual differences, new studies of old problems. Hillsdale, NJ: Lawrence Erlbaum Associates, in press, 1980.

Scarr, S., and Carter-Saltzman, L. Twin Method: Defense of a critical assumption. Behavior Genetics, in press, 1979.

Scarr, S., and Weinberg, R. A. IQ test performance of black children adopted by white families. American Psychologist, 1976, 31, 726-739.

Scarr, S., and Weinberg, R. A. Intellectual similarities within families of both adopted and biological children. Intelligence, 1977, 1(2), 170-191.

Scarr, S., and Weinberg, R. A. The influence of "family background" on intellectual attainment. American Sociological Review, 1978, 43, 674-692.

Scarr, S., and Weinberg, R. A. Nature and nurture strike (out) again. Intelligence, 1979, 3, 31-39.

Stevenson, H. W., Hale, G. A., Klein, R. E., and Miller, L. K. Interrelations and correlates in children's learning and problem solving. Monographs of the Society for Research in Child Development, 1968, 33(7), Serial No. 123.

Willerman, L., Horn, J. M., and Loehlin, J. C. The aptitude-achievement test distinction: a study of unrelated children reared together. Behavior Genetics, 1977, 7, 465-470.

Reference Notes

1.   Scarr, S., Scarf, E., and Weinberg, R. A. perceived and actual similarities in biological and adoptive families: Does perceived similarity bias genetic inferences? Unpublished paper, Yale University, 1979.

2.   Scarr, S. Heritability and Educational Policy: Genetic and Environmental Effects on IQ, Aptitude and Achievement. Invited address presented at the annual meeting of the American Psychological Association, New York City, September 1, 1979.

PHYSIOLOGICAL EVIDENCE THAT DEMAND FOR PROCESSING CAPACITY VARIES WITH INTELLIGENCE

Sylvia Ahern and Jackson Beatty

University of California, Los Angeles

Los Angeles, California, U.S.A.

Spearman, in proposing his influential two-factor theory of intelligence, adopted an implicit biological model which was untestable in his time. Spearman suggested both a general factor of intelligence (g), which corresponded to the amount of "general mental energy" available to an individual for information processing, and a set of specific ability factors, which were brain systems or "mental engines" drawing upon the general energy pool (Spearman, 1904). Of this analogy Halstead (1947) later observed:

> "As a simple, deterministic, mechanistic scheme, nor more forthright view of the biological nature of intelligence is to be found in the whole of the literature on the subject. Yet it is chiefly from the biological standpoint that the theory remains in the realm of speculation, for thus far no systematic program of research has appeared for the testing of the biological...implications of this conception. (p.11)"

Spearman's general factor bears a striking similarity to concepts now dominant in cognitive psychology. His idea of "mental energy" is very much like Kahneman's (1973) notion of "mental effort" and Norman and Bobrow's (1975) general processing resource. Furthermore, a reasonable amount of evidence in human neurophysiology suggests that this general resource may be identified with the functioning of the brainstem reticular activating system (Beatty, 1978; Luria, 1973). Because of these developments, we now seem in a position to empirically test this first biological theory of individual differences in general mental ability.

One measure of reticular function that is particularly useful in human neurophysiology is the task-evoked pupillary response (Kahneman and Beatty, 1966; Beatty and Wagoner, 1978; see Lindsley, 1961, or Moruzzi, 1972, for a review of the functional significance of the reticular activating system). A task-evoked pupillary response is a time-locked averaged record of pupillary dilation and constriction occurring during the performance of a mental task. The amplitude of the task-evoked pupillary response functions as a sensitive and accurate measure of the "mental effort," or the demand for processing resources imposed by the task requirements (Beatty, 1978). More complex and demanding cognitive tasks elicit larger task-evoked pupillary responses (Goldwater, 1972), but very little is known about inter-individual differences in demand for cognitive processing resources in cognition.

In the present set of experiments, task-evoked pupillary responses were employed as an index of the demand for processing resources imposed by four different cognitive tasks. In all four experiments, university students of either very high or relatively low intelligence were tested using items of fixed objective difficulty. Pupillary responses during cognitive processing could be related to psychometric intelligence in at least three ways. First, Spearman's hypothesis would predict that persons characterized by high psychometric g would exhibit larger task-evoked pupillary responses when pressed to the limits of their information-processing capacity. This hypothesis was not adequately tested in the present experiments. A second hypothesis suggested by his analogy is that more intelligent individuals have more efficient specific abilities. Therefore, less "mental energy" is necessary to perform a given task in more intelligent individuals. According to this line of reasoning, more intelligent individuals would tend to have more efficient specific abilities and, for that reason, should exhibit smaller task-evoked pupillary responses during successful processing of any given set of items for which they have a particular aptitude. A third opposing hypothesis is suggested by the motivational view of intelligence differences. According to this theory, more intelligent individuals bring more resources to bear on the solution of any problem and this may be reflected in larger task-evoked pupillary responses. Finally, it might be argued that reticular function has nothing to do with psychometric intelligence and therefore the task-evoked pupillary responses should not differ as a function of intelligence. Thus, several hypotheses relating processing resources and intelligence may be considered. These experiments serve as an initial step in evaluating the concept of intelligence as a neuro-physiological construct.

General Method

Twenty four males and 19 female undergraduates (ages: 17-25 years) with combined Verbal and Quantitative scores on the

Scholastic Aptitude Test of either 950 or less (low intelligence group) or 1350 or more (high intelligence group) served as subjects.

Each subject was tested in four cognitive tasks that were adapted for concurrent pupillometric measurement and were modelled after one or more sub-tests in a standard intelligence test. The tests may be briefly described as follows:

1. Mental multiplication. Subjects were required to solve auditorily-presented multiplication problems at three levels of difficulty.

2. Digit span. Subjects were presented with strings of 6 or 13 digits at the rate of 1/sec for immediate recall. The superspan condition was employed as an attempt to assess the task-evoked pupillary response when subjects are pressed to the limits of processing capacity. This portion of the experiment is not discussed here because of space limitions (but see Ahern, 1978).

3. Vocabulary. Subjects were required to judge whether two words had the same meaning. The initial word was drawn from either the easiest or most difficult portions of one of three standard vocabulary subtests.

4. Sentence comprehension. Baddeley's Grammatical Reasoning Test (1968) was employed, in which subjects hear a sentence of the form "A precedes B," which is followed by an exemplar, "A-B" or "B-A." The subject is required to judge if the sentence describes the letter pair. Item difficulty is manipulated by transforming the sentence into the passive, by negating the sentence or both.

Pupillary diameter was measured using a Whittaker 1051 video pupillometer, the output of which was digitized at 50 msec intervals by a general purpose computer controlling the experiment. Individual pupillary responses were stored for each trial in each experiment, examined for recording and eye movement artifacts, and then averaged. Averaged task-evoked pupillary responses were obtained for each subject in each experiment using only error- and artifact-free trials. The amplitude of each task-evoked pupillary response was taken as the average dilation while processing the test information with respect to pretrial pupillary diameter. Task-evoked pupillary responses may range up to .6 mm. in amplitude and are independent of baseline pupillary diameter over a wide range of baseline values.

In addition to the four experimental tasks, the amplitude of the pupillary light and darkness reflexes was measured to assess peripheral differences in pupillary responsivity between groups.

Each subject was also given the following battery of psychometric tests for control purposes: The Eysenck Personality Inventory, the Wesman Personnel Classification Test, and the Spielberger State/ Trait Anxiety Inventory.

Pattern of Results

The results obtained in these experiments were remarkably straightforward. First, in each of the four experiments, the manipulation of task difficulty had its expected effect on performance: the percentage of errors was larger for the more difficult conditions in every task. Furthermore, in the multiplication, vocabulary and sentence comprehension tasks, increasing task difficulty was associated with significantly larger task-evoked pupillary responses (Multiplication, $p < .001$; Vocabulary, $p < .001$; and Sentence comprehension, $p < .0001$, all by analysis of variance). These results assure that the well documented relationship between severity of task requirements and the amplitude of the task-evoked pupillary response are replicated in our data. For the digit span task, a comparison of task-evoked pupillary responses for errorless performance was not possible, due the difficulty of the 13-digit condition. However, the familiar effects of task loading on pupil (Kahneman and Beatty, 1966) were evident in the shape of the response for errorless 6-digit trials: The amplitude of the response increased monotonically with the number of digits presented.

The effects of intelligence were examined by comparing the performance and pupillometric data between experimental groups. In each of the four experiments, the subjects in the high intelligence groups made significantly fewer errors (Multiplication, $p < .0001$; Digit span, $p < .001$; Vocabulary, $p < .01$; and Sentence comprehension, $p < .0001$, all by Mann-Whitney U-tests). Thus the tasks employed were sensitive to the between group differences indexed by the combined SAT sorting variable.

Between group differences in the amplitude of the task-evoked pupillary response also were present for three of the four tasks employed. With the exception of the vocabulary task, in which the pupillary responses were essentially identical in both groups, the response amplitudes were consistently smaller for the more intelligent subjects than for their less intelligent counterparts. The significance of these mean differences were tested by analysis of variance with the following results: Multiplication, $p < .03$; Digit span, $p < .10$; and Sentence comprehension, $p < .02$. Figure 1 presents the task-evoked pupillary responses obtained in the mental multiplication task, for purposes of illustration (See Ahern and Beatty, 1979, for details). We interpret these results as supporting Spearman's conjecture that the secondary abilities or "mental engines" of more intelligent persons are more efficient or

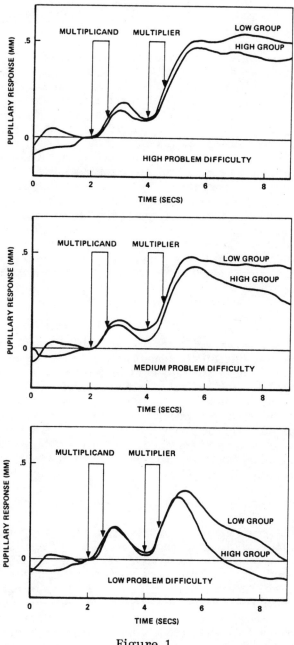

Figure 1.

Averaged task-evoked pupillary responses for correctly solved
problems at three levels of difficulty for subjects in the high
and low groups of psychometrically measured intelligence. At
all difficulty levels, larger pupillary responses are observed
for subjects in the low group. (From Ahern and Beatty, 1979.)

automatic, requiring less "mental energy" or processing capacity for their operation.

The 13-item digit span condition was included to provide an estimate of the task-evoked pupillary responses when the task forced subjects to the limits of their processing resources. In fact, this manipulation was only partially successful. An examination of both the behavioral and pupillometric data suggested that subjects were not attempting to process the entire 13 items, but rather limited themselves to some subset with which they were more able to cope. Nonetheless, these data indicate that the more intelligent subjects may have greater processing resources to employ at the limits of performance. The magnitude of the task-evoked pupillary response tended to be larger for the high intelligence group in the 13 item digit span condition, the only condition in all four experiments where this was the case. This difference itself was not statistically significant. However, the interaction of group and difficulty (6 versus 13 items) for pupillary response amplitude in the digit span task was highly significant (p $<$ .0001). Thus we may conclude that the effects of intelligence on the task-evoked pupillary response are very different for the within capacity and the above capacity conditions.

It could be, however, that individuals differing in psychometrically measured intelligence also differ in autonomic responsivity and that the differences in task-evoked pupillary response observed between groups do not reflect central attentional processes, but merely peripheral autonomic differences. This is not the case: there is virtually no difference in the magnitude of either the autonomically mediated light or dark pupillary response between groups. Therefore, the observed differences in the pupillary response during cognitive processing must be attributable to the operation of brain systems central to the pupillary control nuclei of the autonomic nervous system. This conclusion points either to the reticular core or to the neocortical structures that modulate its activity.

The relation of the task-evoked pupillary response to other, more traditional measures of individual differences is also of interest. Two composite variables representing baseline pupillary diameter and task-evoked pupillary response amplitude were constructed by averaging the values obtained over all four experiments for each subject. In the matrix of correlations between personality and ability variables with reflex and task pupillary variables, only four significant correlations emerged. State anxiety correlated positively with amplitude of the light reflex (.38, p $<$ .01) and no other pupillar variable. More anxious persons, therefore, tend to exhibit larger pupillary constrictions to increases in illumination. No scale of the Eysenck Personality Inventory showed significant correlations with any pupillary measure. For the quantitative scale of the Scholastic Aptitude Test, there was a significant negative correlation (-.35,

p $<.05$) with the mean amplitude of the composite task-evoked pupillary response. Similarly, the quantitative scale of the Wesman Personnel Classification Test also correlated negatively with this pupillometric variable ($-.35$, p$<.05$). Finally, the amplitude of the composite task pupillary variable also was negatively correlated with a reasonable estimator of fluid intelligence, the WAIS digit-back-ward subtest
($-.49$, p $<.001$). Taken together, these correlations are in accord
with the primary between group pupillometric finding, that the
performance of a cognitive task results in smaller task-evoked pupillary responses in more intelligent individuals.

In Summary

Using the task-evoked pupillary response during mental activity as an index of processing capacity utilized in the performance of
a mental task, important differences emerge between groups of university students differing in psychometrically measured intelligence.
For three of four tasks using items of fixed objective difficulty,
individuals in the more intelligent group consistently exhibited smaller
pupillary responses during cognitive processing. This is interpreted
as indicating that more intelligent individuals possess more efficient
specific cognitive structures for information processing. Furthermore,
there was an indication that they may also possess a greater quantity
of processing resources or Spearman's "mental energy" which was
suggested by the reversal of the effects of intelligence on pupillary
response amplitude in information overload. These data provide
clear evidence that physiological differences between individuals of
differing psychometric intelligence emerge during mental activity.

References

Ahern, S.K. Activation and intelligence: Pupillometric correlates
    of individual differences in cognitive abilities. Unpublished
    doctoral dissertation, University of California, Los Angeles,
    1978.
Ahern, S. K., and Beatty, J. Physiological signs of information
    processing vary with intelligence. Science, 1979, 205, 1289-
    1292.
Baddeley, A. D. A three-minute reasoning test based on grammatical
    transformation. Psychonomic Science, 1968, 10, 341-342.
Beatty, J. Pupillary dilation as an index of workload. In Proceed-
    ings of the symposium on man-system interface: Advances in
    workload study. Air Line Pilots Association, Washington, D.C.,
    1978.
Beatty, J., and Wagoner, B. L. Pupillometric signs of brain acti-
    vation vary with level of cognitive processing. Science, 1978,
    199, 1216-1218.
Goldwater, B.C. Psychological significance of pupillary movements.
    Psychological Bulletin, 1972, 77, 340-355.
Halstead, W. C. Brain and intelligence: A quantitative study of

the frontal lobes. Chicago: University of Chicago Press, 1947.
Kahneman, D. Attention and effort. Englewood Cliffs, N.J.: Prentice-Hall, 1973.
Kahneman, D., and Beatty, J. Pupil diameter and load on memory. Science, 1966, 154, 1583-1585.
Lindsley, D. B. Attention, consciousness, sleep and wakefulness. In J. Field (Ed.), Handbook of Physiology (Volume III). Washington, D.C.: American Physiological Society, 1960.
Luria, A. R. The working brain. New York: Basic Books, 1973.
Moruzzi, G. The sleep-waking cycle. Reviews of Physiology: Biochemistry and Experiments Pharmacology. New York: Springer-Verlag, 1972.
Norman, D. A., and Bobrow, D. G. On data-limited and resource-limited processes. Cognitive Psychology, 1975, 7, 44-64.
Spearman, C. General intelligence objectively determined and measured. American Journal of Psychology, 1904, 15, 201-293.

CLOSURE FACTORS: EVIDENCE FOR

DIFFERENT MODES OF PROCESSING

M. J. Ippel and J. M. Bouma

Vrije Universiteit

Amsterdam, The Netherlands

Abstract

The study investigates whether the perceptual factors Closure speed (Cs) and Closure flexibility (Cf) reflect individual differences in mode of information processing. Forty subjects were selected for their factor scores on both factors and placed into four groups: high Cf, low Cf, high Cs and low Cs. Each subject participated in three tachistoscopic tasks: one verbal recall task and two binary classification tasks with visuo-spatial stimulus material. The results were tentatively interpreted in terms of differences in focal attention related to Cf under conditions that favoured analytic processing. In one experiment differences in speed of wholistic processing appeared to be related to Cs.

Cognitive psychologists have paid little attention to the influence of individual characteristics on the mode of processing of particular visuo-spatial stimuli. In the cognitive psychological literature, the primacy of the nature of task and stimulus in the choice of mode of information processing is emphasized (e.g., Kahneman, 1973; Garner, 1974). Recently, however, individual differences in perceptual strategies were experimentally demonstrated (e.g., Hock, Gordon and Marcus, 1974; Cooper, 1976). If people systematically vary in the way in which they process particular stimulus, this might have an influence on the meaning of psychometric tests of perceptual and cognitive abilities. In fact, there are some studies suggesting the possibility of multiple processing on psychometric tests with visuo-spatial material (e.g. French, 1965; Hunt, 1974). An information processing approach to intelligence and its measurement will have to account for both task characteristics and individual characteristics as determinants of the choice of mode of information processing on visuo-spatial tasks. In a recent study we found complex interactions between a subject factor and experimentally manipulated task characteristics of the embedded figures test (Ippel, 1979).

How are we to identify individuals that differ in mode of processing of particular visuo-spatial stimuli? In absence of a cognitive theory that specifies what kind of individuals display what kind of behavior, our starting point was somewhat arbitrary. We assumed that factor analytic studies of certain perceptual and cognitive tests revealed stable sources of variance and covariance that might be of use in the search for individual differences in readiness for certain modes of information processing. In this study the perceptual factors "Closure flexibility" (Cf) and "Closure speed" (Cs) (French, Ekstrom and Price, 1976) were investigated. In several factor analyses a Cs-primary was loaded by a second-order factor that was intuitively interpreted as a "synthetically or configurationally functioning" factor. Cf was repeatedly loaded by an "analytically functioning" second-order factor (e.g., Botzum, 1951; Pemberton, 1952; Messick and French, 1975).

An exploratory device that might be helpful in this connection is the study of lateral asymmetries of the visual system. Usually stimuli are unilaterally presented by tachistoscope to the left and the right hemifield. Visual asymmetry is defined as a performance difference when stimuli are distinctly presented to both hemifields and can result in either right or left hemifield superiority. Since each hemifield is contralaterally connected with a hemisphere, visual asymmetry reflects differences in processing: left hemifield superiority indicates a right hemisphere dominance, etc. There is a growing conviction among neuropsychologists that the two hemispheres differ in their capacity for wholistic and analytic information processing. The question whether both hemispheres may perform both modes, albeit at a different level of competence, or whether the hemispheres are exclusively specialized cannot be resolved at present. Bradshaw, Gates and Patterson (1976), in reviewing some of their experiments, tentatively conclude a quantitative rather than a qualitative difference. Evidence suggests that not the nature of the stimuli, but the mode of processing determines the visual asymmetry: with wholistic processing the right hemisphere is superior and with serial analytic processing the left hemisphere is superior. Hence it seems plausible to hypothesize that if people differ in their mode of processing the same stimuli, they will tend to display differences in visual asymmetry.

The purpose of this study is to investigate whether individual differences alongside the perceptual factors Cf and Cs are related

to patterns of visual asymmetry suggesting a relatively greater
readiness for wholistic processing in high Cs subjects and a
relatively greater readiness for analytic processing in high Cf
subjects.   The task conditions were arranged in such a way that
in one experiment an analytic approach is favourable (the form-
color task).   In another experiment wholistic approach is expected
to be favoured by the task conditions (the dot patterns task).
From the third experiment, a verbal recall task that is mediated
by the left hemisphere, scanning processes may be inferred.

Method

Ninety-two right-handed subjects took 14 perceptual and
reasoning tests including marker-tests of the factors: Closure
flexibility, Closure speed, Space-Visualization, Inductive Reasoning
and Perceptual Speed.   Following a principal component analysis,
three factors were extracted and rotated to simple structure by
means of the oblimin procedure.   Factor I loaded the Cf-markers,
Space-Visualization tests and the Inductive Reasoning tests.   This
factor is interpreted as a rather broad Cf factor.   Factor II
loaded mainly the Perceptual Speed tests, and factor III loaded
exclusively the Cs-markers.   As a consequence of the rather high
intercorrelation between Cf and Cs (.44), it was impossible to
study both organismic factors in one orthogonal analysis of variance design.   So we decided to perform two separate ANOVAs:
one with Cf and one with Cs as organismic factor.   Forty subjects
were selected for their factor scores of the factors I (Cf) and III
(Cs).   Subjects were placed into four groups: a low Cf group
and a high Cf group, a low Cs group and a high Cs group.   The
Cf groups were composed of average Cs scorers and vice versa.
Each group consisted of 10 subjects.   Each subject participated in
three tachistoscopic tasks: two binary classification tasks and a
verbal recall task.   In the binary classification tasks a memory
stimulus was centrally presented.   After an interstimulus interval
it was followed by an unilaterally presented test stimulus.   A
same/different judgment was required.   Stimuli were depicted on
slides and presented by a three-field Scientific Prototype tachistoscope with automatic projection time control and reaction time and
response registration.   The sequence of presentation of these
tasks was randomized across the subjects.   More experimental
details will be published elsewhere.

Differences in pattern recognition

It seems reasonable to assume that in binary classification
tasks preattentive processes will only result in correct responses
if (a) the stimuli can be physically encoded and if (b) the judgment does not require any directed attention.   In other words,
wholistic processing is possible with stimuli that are completely
identical or - under certain conditions - completely different.   In

our first experiment we used dot patterns developed by Garner
and Clement (1963). These patterns consist of configurations of 5
dots placed in an imaginary 3 x 3 matrix in such a way that no
row or column is empty. Garner (1974) formulated a theory of
perception of single stimuli which specifies various subsets of
equivalent dot patterns. After reviewing research with these
stimuli, Garner concluded that - under normal conditions - the
patterns are configurationally or wholistically processed. A configurational quality of these dot patterns, namely the "figural goodness," appears to be inversely related to the size of a subset of
equivalent dot patterns. In our experiment solely dot patterns
from equivalence subsets of size 8 were used. In all there are
seven distinct equivalence subsets of this size. A pair of stimuli
was defined as "same" if and only if the stimuli were completely
identical (not merely equivalent). A "different" pair never consisted of dot patterns from the same subset.

Our data clearly revealed differences in speed of processing
related to Cf and Cs. Surprisingly, these differences are not
likely due to different modes of processing between the low-level
and high-level groups of both organismic factors: the interactions
Cs x hemifield and Cf x Hemifield were not significant. A left
hemisfield superiority is found in the ANOVA of the Cs data.
This pattern of visual asymmetry suggests a wholistic processing
of the dot patterns by the Cs subjects. Although there seems to
be a slight tendency toward a right hemifield superiority within
the high Cf group, this effect did not reach a level of significance.

In order to create an experimental task that favoured analytic
processing, we used visuo-spatial stimuli for which the judgment
"same" was not based on complete physical identity. The second
experiment utilized pairs of two-dimensional stimuli. These dimensions had three levels each: color, with yellow, green, and
brown as levels; form, with circle, triangle, and square as levels.
In this way both dimensions generated a total set of nine stimuli.
A pair of stimuli was defined as "same" if and only if they had
the same color or the same form. A pair of stimuli was defined
as "different" if and only if they matched in neither color nor
form. None of these paired stimuli were completely identical.
Thus for each stimulus half of the remainder of the total set
could be classified as "same" and the other half as "different".

Logically, if subjects use a serial selective mode of processing,
a shorter $\overline{RT}$ for the "same" responses is to be expected. In
order to give a correct "same" response subjects must decide one
dimension to be identical. A correct "different" response, however,
requires subjects to decide two dimensions to be different.
Thus, on the average fewer decisions have to be made for correct
"same" responses. Accordingly as figure 1 shows, in the high Cf
group both "same" and "different" responses tend to show a right

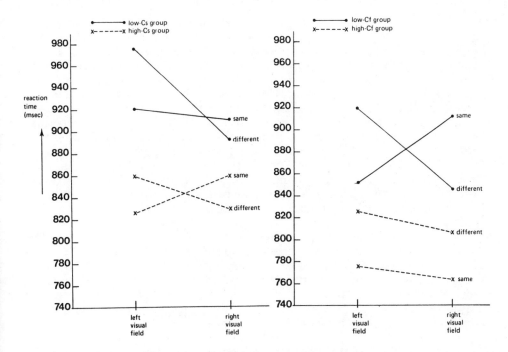

Figure 1. Mean response times in the form-color task of the experimental groups: high and low Cf, high and low Cs in four experimental conditions.

hemifield superiority, with a shorter $\overline{RT}$ for "same" responses in both hemifields. The pattern of "same" responses in the low Cf group is quite dissimilar and more closely resembles the response patterns of the Cs groups. No right hemifield superiority is found, and surprisingly, in the right hemifield the "same" responses tend to be processed more slowly in comparison to "different" responses. The latter result might tentatively be explained by a less selective processing by the low Cf subjects. In that case the "same" pairs constitute an ambiguous stimulus compound: one dimension is identical and one different. This may lead to response interference, an effect that might be weaker in the right hemisphere than in the left hemisphere.

Differences in scanning strategies

Letters, letter sequences and words are better perceived in the right hemifield. It appears that this visual asymmetry in verbal recall is influenced not only by hemispheric dominance in verbal encoding, but also by effects of scanning processes. The two-stage conceptualization model as suggested by White (1976) specifies two different types of scanning. Firstly, a peripheral-to-foveal scanning. This type of scanning occurs relatively early in the iconic memory stage. Since letters scanned first will also be better identified because of relatively strong trace images, this type of scanning will result in a better recall of the leftmost letters with left hemifield presentation, and the rightmost letters with right hemifield presentation. Secondly, a postexposural scanning (Heron, 1957) resulting in a better recall of the leftmost letters within each field of presentation. White suggests that this type of scanning takes place when information is transformed from the iconic to an auditive memory. The postexposural scanning follows the rules of normal reading, i.e. from left to right. White's two-stage model is consistent with Neisser's (1967) distinction between an early preattentive processing of the stimulus information and a later focal processing.

In order to investigate whether Cf and Cs are related to differences in scanning strategies the subjects were asked to participate in a verbal recall task. Three letters (consonants) were projected horizontally on the right or on the left hemifield, and a free recall of the letters was required. For every correctly recalled letter the subject was awarded one point. A total score per subject was computed for each letter position.

The ANOVA with the Cs subjects and that with the Cf subjects both showed a right hemifield superiority. Cf and Cs appeared not to be related to differences in degree of hemisphericity as indicated by overall performance measures. A letter position analysis, however, revealed some interesting recall differences between the high and low Cf group. The high Cf group produced a superior recall of the letters that were closest to the fovea. In the right hemifield the left-hand letters were better recalled than the central and the right ones. This suggests a left-to-right scanning. The low Cf group -and also both Cs groups - showed a better recall of the rightmost letters with right hemifield presentation. According to White's (1976) model this suggests a stronger influence of peripheral-to-foveal scanning. This interpretation is also supported by group differences in recall of the central letters: low Cf subjects showed a relatively large decay in recall in comparison to the outside ones, whereas the high Cf group did not.

With left hemifield presentation the high Cf group showed a superior recall of the rightmost letters and negligible differences between the central and left ones. This recall pattern differs greatly from our expectations; it cannot be explained by the peripheral-to-foveal scanning hypothesis, nor by the left-to-right scanning hypothesis. An interpretation based on the rather well-founded empirical statement that letters scanned first will be better identified (White, 1976) suggests a right-to-left scanning by high Cf subjects. This would indeed be the mostly efficient

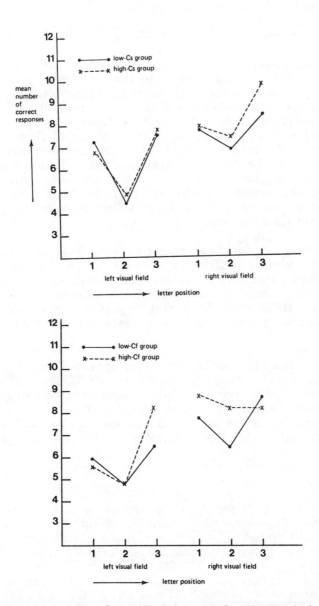

Figure 2. Mean number of recalled letters in the verbal recall task. Experimental groups: high and low Cf, high and low Cs. Experimental factors: field of presentation and letter position.

approach, but it contradicts the natural peripheral-to-foveal law
as well as the learned left-to-right rule.

General Discussion

Our data provided some tentative support for our expectation
of processing differences related to the closure factors. In case
of Cf we found indications of differences in conscious allocation of
attention in two quite different tasks: a binary classification task
and a verbal recall task. Although in these experiments the Cs
groups displayed performance patterns similar to those of the low
Cf subjects, that does not imply that the Cs dimension can be
characterized merely by lack of ability for detailed analysis of
stimulus configurations. Although there are some difficulties in
interpreting the results of the dot patterns task they revealed
positive indications of differences in speed of wholistic processing
related to the Cs dimension.

Our interpretations are tentative especially because of the
serious methodological difficulty created by the highly intercorrelated factors Cf and Cs. We are now analyzing a broader range
of tests in an attempt to isolate the more independent second-order
factors "analytic functioning" and "synthetic functioning," in
order to be able to do some experiments with a more balanced
design.

What is this all worthwhile? The approach reported here is
meant as an initial attempt to identify individuals that differ in
readiness for wholistic and analytic modes of processing of visuo-
spatial stimuli. Knowledge about interaction between this individual characteristic on the one hand and task and stimulus characteristics on the other hand may be of great use in the development
of intelligence tests.

References

Botzum, W. A. A factorial study of the reasoning and closure
    factors. Psychometrika, 1951, 16, 361-386.
Bradshaw, J. L., Gates, A., Patterson, K. Hemispheric differences
    in processing visual stimuli. Quarterly Journal of Experimental
    Psychology, 1976, 28, 667-681.
Cooper, L. A. Individual differences in visual comparison processes.
    Perception and Psychophysics, 1976, 19, 5, 433-444.
French, J. W. The relationship of problem-solving styles to the
    factor composition of tests. Educational and Psychological
    Measurement, 1965, 25, 9-28.
French, J. W., Ekstron, R. B. and Price, L. A. Manual for Kit of
    Reference Tests for Cognitive Factors. Princeton: Educational
    Testing Service, 1976.
Garner, W. R. The processing of information and structure. New

York: John Wiley & Sons, 1974.

Garner, W. R. and Clement, D. E.   Goodness of pattern and pattern uncertainty.   Journal of Verbal Learning and Verbal Behavior, 1963, 2, 446-452.

Heron, W. Perception as a function of retinal locus and attention. American Journal of Psychology, 1957, 70, 38-48.

Hock, H.S., Gordon, G. P., and Marcus, N. Individual differences in the detection of embedded figures. Perception and Psychophysics, 1974, 15, 47-52. Hunt, E. Quote the Raven? Nevermore! In: L. Gregg (Ed.) Knowledge and Cognition. Hillsdale, New Jersey: Lawrence Erlbaum Associates, 1974.

Ippel, M. J. Generalizability of Performance-scores on Embedded Figures Material. Manuscript submitted for publication, 1979.

Kahneman, D. Attention and Effort. Englewood Cliffs, N.J.: Prentice Hall Inc., 1973.

Messick, S. and French, J. W. Dimensions of Cognitive Closure. Multivariate Behavioral Research, 1975, 1, 3-16.

Neisser, U. Cognitive Psychology. New York: Appleton-Century-Crofts, 1967.

Pemberton, C. The closure factors related to other cognitive processes. Psychometrika, 1953, 17, 267-288.

White, M. J. Order of Processing in Visual Perception. Canadian Journal of Psychology, 1976, 30, 140-156.

Footnote

The first author's share in the preparation of this paper has been made possible by a grant from the Dutch Organization for the Advancement of Pure Research (Z.W.O.). A more detailed technical report is in preparation.

TEST STRUCTURE AND COGNITIVE STYLE

J. Weinman, A. Elithorn, and S. Farag

Guys Hospital Medical School

London, England

Abstract

Some studies examining the nature of performance differences on a single cognitive test are reported. Most of the work is concerned with the analysis of response times and with the fragmentation of these into component times for different phases of problem-solving. The results indicate that while overall processing speed is primarily cognitively determined, the way in which time is distributed over different phases of performance is greatly influenced by personality factors. The analysis of responses revealed characteristic strategies and errors associated with level of ability. The importance of task parameters in eliciting these indices of performance differences is discussed.

Introduction

Experimental and psychometric approaches to the study of cognitive functions have tended to develop quite separately over the years. Experimental psychology has largely avoided the problem of individual differences and many psychometric tests do not readily lend themselves to a more experimental approach. In this paper there is an attempt to examine the underlying nature of individual differences in performance on a single psychometric test, the Perceptual Maze Test (P.M.T.) which has been described elsewhere (Elithorn, et al, 1963). This is a binary-structured route finding task which Butcher (1968) commended as a tool for combining experimental and psychometric approaches since it "can be used in the same way as other intelligence tests...but can be readily adapted to study the parameters of problem-solving." In addition it is sensitive to changes

in level of functioning due to such factors as ageing and cerebral
dysfunction.

Previous experimental work with the P.M.T. was primarily concerned with investigating the effects of task parameters on subjective
difficulty (Lee, 1965). The aim of the present investigations has
been to derive evidence of individual cognitive styles and of differences in cognitive strategies as related to different levels of overall
performance. Two distinct ways of achieving these aims have been
investigated:
  (1) the analysis of response times on single items
  (2) The analysis of response pathways for evidence of consistent
      response strategies and error patterns.
The present paper will mainly be concerned with describing
the first of these approaches although a brief outline of the second
approach will be presented towards the end of the paper.

The Analysis of Response Times on Single Items

There has been a tacit agreement that speed of performance is
a prime source of differentiation in cognitive ability. This notion is
implicit in the structure and scoring systems of many cognitive test
procedures and has also received some attention from experimental
psychologists more recently. Eysenck (1967) has also claimed that
something akin to speed of information-processing slopes could be
produced for individuals from their response times on single test
items arranged in order of complexity. He suggested that these
slopes would be parallel but with lower intercepts for more able
subjects. In the first experiment to be reported there is a direct
test of this suggestion and in subsequent studies there is an attempt
to look more closely at temporal differences in specific phases of
maze solving.

Experiment I. Developmental changes in rates of maze solving

This first study was primarily concerned with investigating the
use of "speed of processing" slopes as a way of differentiating
levels of performance on the P.M.T. One version where the items
are arranged in order of difficulty is the recently developed children'
version. It was therefore decided to derive and compare performance
slopes of children of different age levels.

Subjects. The subjects for this study comprised 111 school
children ranging in age from 8 to 17 years. The total sample was
divided into the following five groups in order to compare developmental   changes   in   performance:
(1)   8 - 9 years (N = 30)   (4)   14 - 15 years (N = 17)
(2) 10 - 11 years (N = 22)   (5)   16 - 17 years (N = 20)
(3) 12 - 13 years (N = 22)

Procedure.   Each subject was tested individually with the children's P.M.T., which comprises sixteen items arranged in order of structural and empirical difficulty (see Fig. 2).   All the items were presented with the maximum solution number specified and subjects were instructed to find and draw in the solution path as quickly as possible.   The time taken to complete each item was recorded.

Results.   A clear monotonic increase in the overall pass/fail score was found with increasing age.   In order to derive performance slopes, the mea solution time for each level of complexity in each age group was calculated.   Regression slopes were calculated for mean response speed on item complexity in age group and these are shown in Figure 1.

An analysis of variance for differences between regression slopes was carried out and revealed no significant differences in slope function although it is apparent from Figure 1 that the slopes are not strictly parallel.

Discussion.   At first glance these results appear to provide support for Eysenck's (1967) hypothesis concerning individual differences in cognitive performance.   Moreover they appear to be directly comparable with the findings of Hooving, Morin and Konick (1970) who found developmental increases in speed of memory scanning without any changes in slope function.   However, the present result still raises some additional questions as to the nature of this speed difference.   If these findings were totally compatible with the Hooving et al (1970) study and with those of Hunt, Lunneborg and Lewis (1975), then it would be concluded that all these subjects are solving mazes in an essentially similar fashion which speeds up with increasing age.   While such a conclusion might be valid for memory scanning performance which is a fairly well defined process, there seems little doubt that maze solving must incorporate a number of different processes.   Thus the total maze solution time does not represent the time taken for a unitary activity but is made up of the

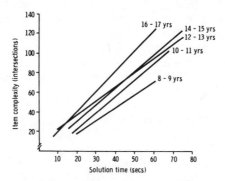

Figure 1. Regression slopes for response speed against item complexity in each age group.

times taken on various, distinct phaes of performance such as search, tracking and checking. The experiments which follow are therefore intended to examine how subjects distribute their time on various phases of maze solving and the extent to which task parameters, ability level and non-cognitive factors may determine this.

Experiment 2. The nature of binary response times

From the discussion of the first study it became clear that there may be difficulties associated with comparing levels of P.M.T. performance on the basis of response times on individual items. One obvious problem is that of differentiating two subjects who obtain the same overall time on an item of a defined level of complexity. While it would be possible to regard two such subjects as identical, there are good grounds for suspecting that the same total response time might be achieved in a number of quite distinct ways since subjects may differ consistently in the way they distribute their time on the various phases of the task. The computer-generated version of the P.M.T. (Jones and Weinman, 1973) offers a good opportunity to test this possibility since the time taken to traverse each binary node on the maze can be automatically recorded. Thus it is possible to see how long is spent on the initial search phase and whether any further searches are subsequently carried out during the tracking phase. Such secondary searches would therefore correspond to what Newell and Simon (1972) have referred to as "subgoal searches." The aim of the second experiment is to examine whether there is a large range in the number of subgoal searches used by individuals and the extent to which these are independent of structural aspects of maze patterns.

Procedure. Twenty-four undergraduate students each attempted six, 16 row computer-generated P.M.T. patterns. These were presented on a visual display and subjects responded using a keyboard for tracking in their response paths. Each pattern had a single solution path and was presented without the maximum solution number. Subjects were instructed to find and track in the optimal solution path and the times for each binary response were recorded. Since all the subjects responded optimally, the binary decision times were therefore for identical paths as each maze was designed with a unique solution path. This allowed an analysis to be made of the contribution of both pattern parameters and individual differences to the overall variance in response times.

Results. From the distribution of single decision times, it could be seen that the initial search times form an almost separate distribution from the tracking response times but that some of the latter overlap and these are considered to be "subgoal" search times. This classification of "subgoal" times is based on an inspection of the data rather than on a formal statistical procedure and may be too conservative an estimate of a subgoal search time. Even so, using this "ad hoc" criterion a range of 0.5 to 4.3 subgoal searches per maze was found in this group of subjects. An Analysis of Variance was carried out on all the times excluding the initial search times. Significant effects were found due to subjects (F = 4.33; $p < 0.01$), to mazes (F = 9.61; $p<0.001$) and to specific nodes (F = 6.49; $p<0.001$). Moreover a very significant interaction between mazes and nodes was found (F = 19.17; $p< 0.001$) but no other interactions were significant.

Discussion. From the distribution of the single decision times, it is immediately clear that the initial search times form an almost entirely separate distribution from the other response times. The small overlap between these two distributions suggested that a number of the tracking times were in fact subgoal search times. Although pattern parameters were found to play a role in determining the latter, a wide range was found in the number of subgoals used. Moreover since this was a pretty homogenous group, who had all taken the same maze paths, these results strongly suggest that different scanning strategies are being adopted. Some subjects appear to search large areas of the maze before responding, whereas others either chose or are forced to sample much less information. These differences may be explained cognitively in terms of a subject's "working memory" capacity (Baddeley and Hitch, 1974) or by a more non-cognitive explanation in terms of differences in "conceptual tempo," as described by Kagan (1967). Some of these possibilities are examined more closely in experiment 4.

Experiment 3. The contribution of search, tracking and checking times to the overall response time

The previous study showed that the total solution time can be separated into a search phase and a tracking phase. In the present study and in the subsequent studies, a third phase of performance can also be distinguished and this comprises the period when the subject is checking the solution prior to its evaluation by the computer. The present study was designed to assess the effects of increasing item complexity on these three phases of performance.

Procedure. Sixteen young adult subjects attempted sets of ten mazes at four levels of complexity (7, 10, 13 and 16 rows), the order of sets being randomised. The time taken for searching, tracking and checking was recorded together with a measure based on the proportion of the total time spent on the search phase (proportional search).

Results. The ANOVA on the log-transformed data showed that search and tracking times increased significantly, and in a fairly linear fashion with increased item complexity. In contrast the checking time and proportional search measure remained quite constant although both of these showed a large significant variance due to individuals.

Discussion. Large individual differences were found in two aspects of performance, namely proportional search and checking, suggesting that qualitative differences exist in the relative amount of time spent searching for and verifying a solution. These differences were found amongst a relatively homogenous group of individuals in terms of ability level and appear to be consistent over a wide range of item complexity although they are more marked on larger items. Taken together with the results from the previous experiment, these results indicate that if the overall response time can be fragmented into component times, then it is possible to detect consistent individual differences in the way subjects distribute their time between the difference stages of maze solving. The next experiment attempts to identify correlates of these differences.

Experiment 4. The nature of individual differences in response speed

The present experiment was designed to identify cognitive and non-cognitive correlates of the various differences in response speed which were found in the three previous studies. Earlier results had also indicated that identifying cognitive style differences may also depend on such task parameters as complexity. It was therefore hypothesized that non-cognitive or stylistic factors will exert more influence on performance on larger patterns attempted without the maximum number. Secondly, it was hypothesized that measures of cognitive ability would be more related to actual speed of performance than to the way in which time is distributed over the various phases of maze solving.

Procedure. Twenty-four undergraduate students all completed the following: (a) P.M.T. Ten mazes were presented at two levels of item complexity (7 and 13 rows) using the computer-automate version. At each level, five mazes were presented with the maximum solution number specified and five without this information. The median solution, search and check times were derived together with the median proportional search for each subset of five mazes.

(b) Eysenck Personality Inventory (E.P.I.). A 64-item personality questionnaire which provides measures of extraversion and neuroticism (Eysenck and Eysenck, 1964).

(c) A.H.5. Test. A group of "high grade" intelligence which provides separate measures of verbal/numerical and visuo-spatial ability (Heim, 1965).

Results. Correlations between the four indices of P.M.T. performance on each set of mazes and the A.H.5 and E.P.I. scores are shown in Table 1. It was found that the A.H.5. scores have a consistent negative correlation with the P.M.T. total time and search time although this only reached statistical significance on the largest patterns presented without the solution number. No particular

TABLE 1

Correlation coefficients between the E.P.I. and A.H.5. scores and the P.M.T. performance indices (where T = Total time; S = Search time; C = Check time; % = proportional search).

	7 Rows (with)				7 Rows (without)				13 Rows (with)				13 Rows (without)			
	T	S	C	%	T	S	C	%	T	S	C	%	T	S	C	%
Extraversion (E.P.I.)	.16	-.19	.31	-.36*	-.12	-.31	.33	-.47**	-.17	-.2	.28	-.32	-.27	-.36*	.31	-.61***
Neuroticism (E.P.I.)	.21	.1	.04	-.15	.21	.34*	-.29	.4*	.22	.18	-.11	.18	.35*	.45**	-.27	.37*
A.H.5.	-.33	-.33	-.14	-.09	-.24	-.12	.16	-.16	-.2	-.16	.07	.01	-.36*	-.22	-.11	-.05

*p < 0.05   **p < 0.025   ***p < 0.01   (for 23 d.f.)

relation between the A.H.5. scores and either the check times or proportional search was found.

The extraversion scores were found to correlate negatively with proportional search, particularly when the maximum solution was not given. A consistent but non-significant positive correlation between extraversion and check time was also obtained, but no clear relation with the total time was found. Neuroticism was found to be associated with slower performance particularly under when the maximum solution was not given.

Discussion. These results show quite strikingly that the personality factors appear to play an important role in determining the distribution of a subject's time over different phases of maze solving. In particular more extraverted subjects spend a relatively short time on the search stage and possibly as a consequence appear to spend more time verifying their responses. Neuroticism is associated with slower search times and all these correlations were found to be more marked when the maximum solution number is not given, particularly on more complex items. In contrast the intelligence test scores were associated with faster overall performance but not with proportional search or check times.

Taking these results together with these from the first three studies, it can now be seen that although it is possible to use overall response time an index of individual difference in P.M.T. performance, a range of performance difference can be observed by carrying out a more detailed chronometric analysis. The following three general conclusions can be drawn:

(a) Personality factors appear to determine how time is distributed over different sequential stages of maze-solving.

(b) Overall performance speed appears to be a function of cognitive factors.

(c) Task parameters play an important role in determining some of these correlations, especially the non-cognitive ones. In this respect the present results are consistent with those of Kagan (1967) who noted that cognitive style differences are best observed on tasks with greater response uncertainty.

The Analysis of Response Pathways

The second approach in investigating individual differences in P.M.T. performance has involved the analysis of response pathways. When the maximum solution number is not presented on the P.M.T. then 2^n different pathways could be taken on any maze (where n = number of rows in the maze lattice). Many of these paths may never be taken and many may be similar in overall outcome but the point is that there is a wide range of potential responses. Thus the analysis of response pathways may provide information about consistent qualitative differences. Summary results will be reported from a study of 817 eleven-year children, who were trisected into low, medium and high scoring groups based on their overall pass/fail score.

Comparisons of the complete pathways of the three groups revealed considerable differences in the routes selected, as can be seen in the two examples shown in Figure 2. From these examples, two characteristics of the less able subjects could be discerned. Firstly they are more inclined to make decisions based on a more restricted look-ahead in that their paths are directed towards areas of the maze which are more immediately attractive rather than towards a more long-term gain. Secondly their paths appear to keep to a straight line more than the high ability solvers.

On larger maze patterns the three groups diverge even more and it becomes difficult to characterize all the differences between them. It was therefore decided to restrict the pathway analyses to

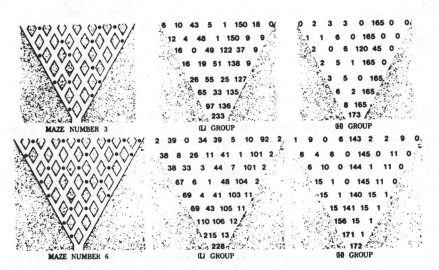

Figure 2. Two mazes from the children's P.M.T. together with the routes chosen by the low scoring (L) and high scoring (H) groups. (The numbers of subjects starting the maze is shown at the vertex and the distribution of their routes can be seen by the numbers shown at each node.)

certain decision junctions, where the solver is faced with a choice of taking or rejecting an immediate gain. This technique, which is described in detail by Lee (1965), can provide an analysis of both immediate gain and straight line response tendencies and errors, by evaluating the outcome of each such decision with respect to its binary alternative. The three ability groups were found to make characteristic types of response and errors. Poorer maze solvers were consistently found to make more straight-line error and responses particular in upper halves of maze patterns, indicating a more limited look-ahead in these subjects.

General Conclusions

These studies have shown that it is feasible to analyze aspects of performance on a single psychometric test in order to understand the nature of differences in performance level. Clear qualitative differences in response strategies have been found to be associated with level of ability and these appear to result in faster overall processing in high ability subjects. Even amongst individuals of similar ability there are substantial stylistic differences in the way time and effort is distributed over the various sequential stages of maze solving. However it appears to be very necessary to be able to manipulate task patterns in a systematic way in order to best observe these differences in cognitive style and a task such as the P.M.T. is particularly appropriate in this respect.

Using these techniques it is therefore possible to collect a large amount of data from a single test which in turn can give greater insight into the nature of individual differences. In our experience this approach offers clear advantages in the clinical setting where cognitive tests are frequently used to quantify changes in overall level of functioning. With a strictly psychometric approach it is rarely possible to understand the underlying nature of such changes, whereas this has been a prime consideration of the work which has been outlined in this paper.

References

Baddeley, A. D. and Hitch, G. J. Working memory. In G. H. Bower (Ed.), The Psychology of Learning and Motivation. New York: Academic Press, 1974.

Butcher, H. J. Human Intelligence: Its nature and assessment. London: Methuen, 1968.

Elithorn, A., Jones, D. and Kerr, M. O. A binary perceptual maze. American Journal of Psychology, 76, 3, 506-508.

Eysenck, H. J. Intelligence assessment: a theoretical and experimental approach. British Journal of Educational Psychology, 37, 81-89.

Heim, A. W. A.H.5. Group Test of High-Grade Intelligence. Windsor: N.F.E.R. Publishing Company Ltd., 1968

Hooving, K. L., Morin, R. F. and Konick, D. S. Recognition reaction time and size of memory: a development study. Psychonomic Science, 21, 247-248.

Hunt, E., Lunneborg, C. and Lewis, J. What does it mean to be high verbal? Cognitive Psychology, 7, 194-227.

INTELLIGENCE AND THE ORIENTING REFLEX

H. D. Kimmel

University of South Florida

Tampa, Florida, U.S.A.

A series of studies is described involving measurement of the orienting reflex in retarded, gifted, and intellectually average children. These studies show that measured IQ is positively correlated with the strength and persistence of orienting reactions. In addition, some evidence is presented to support the conclusion that orienting reactions may be strengthened by conditioning and that this may lead to improved performance in intellectual tasks.

What can we learn about intelligence from a consideration of its relationship with the orienting reflex? Just what is the orienting reflex anyway? And how is it related to intelligence? These are some of the questions I will try to answer.

The orienting reflex refers to an assortment of bodily reactions elicited by novel or unexpected stimuli. These include postural adjustments, such as pricking up the ears in response to an auditory stimulus, autonomic nervous system reactions, such as digital vasoconstriction, as well as EEG desynchronization. The vigor of these components of the orienting reflex is a positive function of the intensity of the eliciting stimulus, and depends upon its novelty or unexpectedness and the time between stimulations. The reflex tends to habituate quite readily with repeated administrations of the stimulus. There is substantial evidence indicating that the elicitation of the orienting reflex is followed immediately by heightened sensitivity to exteroceptive stimulation, manifested in lowered absolute and difference thresholds. This increased sensitivity appears to extend beyond the sensory modality of the eliciting stimulus (Sokolov, 1963).

Broadly speaking, the orienting reflex reflects the heightened attention, that must be maintained to ensure that potentially significant events will not pass unnoticed. These events may be important in identifying sources of possible nourishment or danger. Because most stimuli usually have no significance at all, the ease of habituation of the orienting reflex is an energy conservation mechanism and a safe-guard against a positive feedback spiral. The orienting reflex is evolutionarily recent; yet its role in adaptation is vital. Nevertheless, the reaction itself is better viewed as pre-adaptive rather than adaptive because it does nothing to actually manage the eliciting stimulus. If sensitivity to environmental stimulation is a fundamental factor in its adaptive processing it is reasonable to assume that the orienting reflex is a basic component of intellectual functioning. We began our work on the relationship between intelligence and the orienting reflex with this assumption.

Our research on the relationship between intelligence and the orienting reflex has involved comparisons of autonomic indices of orienting in mentally retarded and gifted children with those of intellectually average children. The initial impetus for this research was the discovery that reactions mediated by the autonomic nervous system are capable of being modified by response-contingent reinforcement rather than being conditionable only classically as had previously been believed. This discovery suggested the possibility that humans' orienting reactions could be strengthened by instrumental conditioning and that this strengthening might result in improved intellectual performance as well. This description of our starting point should make it clear that our research was conceptualized in an environmentalistic frame of reference, although it was also based upon the assumption that intelligence is fundamentally biological.

In our first study (Kimmel, Pendergrass, and Kimmel, 1967) we compared a group of severely retarded children with normal controls in habituation and conditioning of the electrodermal orienting reflex. None of the retarded children had IQ's above 50 while the normal controls' IQ's were between 100 and 120. In the habituation phase of this study visual stimuli of different shapes (square, triangle, and circle) were presented and the child was instructed simply to pay attention and avoid unnecessary movements. Figure 1 shows the average magnitude of the electrodermal orienting reflex in the normal and retarded groups of children during five blocks of 3 habituation trials. The figure shows that the retarded children's reactions reduced in strength much more rapidly than the controls'. Statistical analysis of these data indicated that the interaction between Groups and Blocks of trials was statistically significant, $F (4, 120) = 4.90$, p 0.01).

During conditioning, half of the retarded and half of the normal children were reinforced with candy and "good" each time

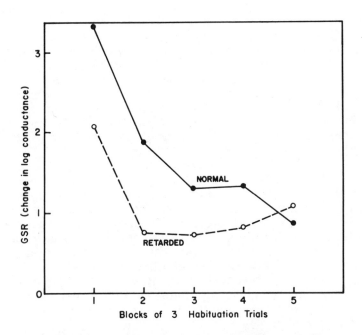

Figure 1. Average magnitude of orienting reflex elicited by visual stimuli during habituation in retarded (N=16) and normal (N=17) groups, in blocks of 3 trials. (Reproduced with permission from Kimmel, Pendergrass, & Kimmel, 1967).

they made an electrodermal reaction to a stimulus, while the other half of each group were reinforced for not reacting to the stimulus. Figure 2 presents the average strength of the orienting reaction in the four subgroups of subjects formed in this way (adjusted for differences in their habituation reactions). As is shown in Figure 2, only the intellectually normal children who were reinforced for nonresponding showed any tendency to change in response strength - and this was a paradoxical increase. Analysis of variance of these data indicated that the overall difference between the retarded and normal groups was significant, $F (1, 28) = 4.75$, $p < 0.05$, as was the 3-way interaction of Groups, Type of reinforcement contingency, and Trial blocks, $F (4, 112) = 2.49$. $p < 0.05$. The triple interaction reflects the fact that the two groups reinforced for nonresponding diverged across trial blocks but the groups reinforced for responding did not.

The children had been tested on the Seguin form board 2 months prior to conditioning and were retested with the Seguin immediately following. Seven of the 8 controls who received response-contingent reinforcement improved on the form board

from pretest to posttest and one got worse, while 6 out of 10
controls who received nonresponse-contingent reinforcement improved and 4 got worse. This difference was not quite statistically significant but may be compared with test-retest data from
10 other normal children who did not receive the conditioning.
Four of these children improved, 5 got worse, and 1 was unchanged. Eleven of the 12 retarded children who received response-
contingent reinforcement improved on the Seguin while 1 got
poorer. Of 11 retarded children who received non-response-
contingent reinforcement, 6 improved on the form board and 5 got
worse. This differential effect in the retarded group was statistically significant, Chi Square = 4.10, $p < 0.05$.

Although there were no consistent differences between the
electrodermal reactions of children reinforced for responding or
nonresponding, the conditioning experiences must have been the
reason for the improvement in form board performance shown by
the children who received response-contingent reinforcement.
This was of course the most interesting result of the study but it
sorely needed verification. The finding that the retarded children's
orienting reactions habituated more quickly than the controls was
a confirmation of previous findings showing weaker and less
persistent orienting reflexes in the retarded (Grings, Lockhart,
and Dameron, 1962).

We conducted a larger study to examine more systematically
the possibility of transfer from conditioning of the orienting
reflex to subsequent intellectual performance (Pendergrass, 1969),
using stimulus change to elicit the orienting reflex. Although
Pendergrass found that it was possible to alter children's preferences for using shape and color concepts in a simple concept
utilization task, there was again very little evidence of conditioned
orienting reflex effects. In the Pendergrass study all of the
children were within the normal to bright normal intelligence
range and the relationship between intelligence and the orienting
reflex was not directly examined.

The methodology developed in the Pendergrass study was
used in a comparison of intellectually gifted children (IQ = 130 -
170) with normal controls (IQ = 90 - 110) (Kimmel and DeBoskey,
1978). Money was substituted for candy as reinforcement and
nonreinforced control groups were also run. Stimulus change was
again used to elicit the orienting reflex and a variation of Pendergrass' concept utilization task was used to examine transfer
effects. In this study a stimulus was presented repeatedly until
the electrodermal reaction reduced to zero. Then either the
shape or the color of the stimulus was changed, dishabituating
the orienting reflex, and the child was reinforced with money and
"good." Analysis of the number of trials needed to reach habituation showed that the gifted children needed an average of 75%

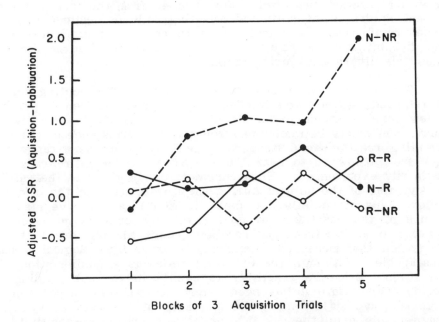

Figure 2. Average adjusted magnitude of orienting reflex elicited
by visual stimuli during 15 conditioning trials in N-R
(normals reinforced for response), N-NR (normals reinforced for nonresponse), R-R (retarded reinforced for
response), and R-NR (retarded reinforced for nonresponse), in blocks of 3 trials, N=8 each. (Reproduced
with permission from Kimmel, Pendergrass, & Kimmel, 1967).

more trials than the normal controls to habituate in the first
stimulus series, a significant difference, $\underline{t}$ = 3.33, p < .05. The
two groups did not differ in habituation rates in subsequent
stimulus series, due to a floor effect. In addition, the gifted
children made significantly larger initial orienting reflexes than
the normals, F (1, 89) = 5.97, p < .05, and the interaction between
Intelligence, Reinforcement, and Trials was also significant, F
(17, 1530) = 2.15, p < .01 in the initial orienting reflex magnitudes.
The triple interaction apparently reflected the fact that money
influenced the strength of the initial orienting reflexes of the
normals throughout session but did not influence the gifted children's orienting reflexes until the later stages.

The dishabituated orienting reflex was followed by reinforcement for half of the children in each group but not in the other
half. The effect of money reinforcement was significant overall,
F (1, 89) = 4.43, p < .01. The triple interaction in the conditioned

orienting reaction magnitude data stemmed from the fact that the normal children's dishabituated orienting reflexes declined without money reinforcement but increased with money reinforcement, while the dishabituated orienting reflexes of the gifted children did not show this divergence during training.

We now have sufficient information about the orienting reflex in retarded, gifted, and intellectually average children to permit a few generalizations to be stated. It is clear that a positive correlation exists between the strength and persistence of children's orienting reactions and measured intelligence, across a range of IQ's from below 50 to near 170 - essentially the entire range ordinarily experienced. When consideration is given to the rather passive role of the subject during the measurement of the orienting reflex ("just sit quietly and pay attention"), and the vegetative nature of the reactions involved (i.e., the child is not even aware that an electrodermal reaction occurs when it does), the conclusion that measured intelligence is basically biological seems inescapable. Although the orienting reflex is a manifestation of the brain's primitive reactivity to events in the surrounding world, even before it has been determined whether these events have adaptive significance, the plasticity of the nervous system comprehends even the modification of this primitive sensitivity, with the possibility that enhanced intellectual performance may result. It is unlikely, for this reason, that research of the type described in this presentation can contribute definitively to the resolution of the nature-nurture question as it is most commonly posed.

References

Grings, W. W., Lockhart, R. A., and Dameron, L. W. Conditioning autonomic responses of mentally subnormal individuals. Psychological Monographs, 1972, 76, Whole No. 558.

Kimmel, H. D. and Deboskey, D. Habituation and conditioning of the orienting reflex in intellectually gifted and average children. Physiological Psychology, 1978, 6, 377-380.

Kimmel, H. D., Pendergrass, V. E., and Kimmel, E. Modifying children's orienting reactions instrumentally. Conditional Reflex, 1967, 2, 227-235.

Pendergrass, V. E. The effect of reinforcement on preferences for simple classification concepts. Conditional Reflex, 1969, 4, 199-211.

Sokolov, E. N. Perception and the conditioned reflex (trans. S. W. Waydenfeld). New York: MacMillan, 1963.

INDIVIDUAL DIFFERENCES IN MEMORY SPAN

William G. Chase
Carnegie-Mellon University, Pittsburgh, Pennsylvania, U.S.A.

Don R. Lyon
University of Dayton, Dayton, Ohio, U.S.A.

K. Anders Ericsson
Carnegie-Mellon University, Pittsburgh, Pennsylvania, U.S.A.

Abstract

One series of experiments examined the correlation between memory span and the speed of symbol manipulation in short-term memory, and another experiment analyzed the effects of extended practice on memory span. In the first study, most of the estimates of processing speed did not correlate with memory span, and it was concluded that short-term memory capacity is not determined by the speed of symbol manipulation in short term memory. In the second study, memory span greatly increased with extended practice, but this increase was due to the acquisition of a mnemonic system. Short-term memory capacity was unaffected by practice.

Individual differences in memory span are interesting from both a psychometric and an information-processing point of view. From a psychometric perspective, memory span is an important item on IQ tests because of the high correlations between memory span and IQ scores. It has been suggested that memory span is a good index of mental retardation and brain damage, but in the normal adult population, it probably is not a very good predictor of high-school or college grades (Matarazzo, 1972). Some people have even gone so far as to suggest that a pure measure of memory span--span ability--is the best culture-free determiner of intelligence (Bachelder & Denny, 1977a,b).

From an information-processing point of view, memory span is the most often used measure of short-term memory capacity, which in turn is one of the most important human limitations in thinking and problem solving (Newell & Simon, 1972). Recent information-processing studies by Cohen and Sandberg (1977) and Lyon (1977) have ruled out any obvious mnemonic coding strategies as causes of individual differences in short-term memory capacity.

It has been suggested by several people in the information-processing literature that memory span is related to the speed of mental processes in short-term memory. For example, Hunt, Frost and Lunneborg (1973), in their attempt to link psychometric and information-processing theories of intelligence, suggested that verbal intelligence is related to the speed of short-term memory processes. Baddeley, Thompson and Buchanan (1975) suggested that the speed of the rehearsal loop determines the memory span, in large part, because verbal items--those based on a phonemic code--tend to decay away within about 2 sec, and the function of rehearsal is to keep them from decaying. From their analysis of reading rates and memory spans, Baddeley et al concluded that people's memory spans are roughly equivalent to the number of words they can read in 2 sec. In a similar analysis, Cavanagh (1972) has suggested that there is a direct relationship between memory span and short-term memory search rates. From his analysis of memory span and scanning rates, Cavanagh concluded that it takes about $\frac{1}{4}$ sec to search short-term memory. The implication is that people's memory search rates are determined by how many items are searched in $\frac{1}{4}$ sec.

In this paper we will summarize work in our laboratory on two questions. First, are individual differences in memory span due to differences in the speed of symbol manipulation in short-term memory? And second, is it possible to increase one's short-term memory capacity with extended practice?

Speed of Symbol Manipulation

To summarize in advance our analysis of the first question, we have found very little evidence to support the idea that memory span is determined by the speed of symbol manipulation in short-term memory, at least in the college student population. We have run a series of experiments designed to establish the correlation between short-term memory processing rates and memory span, and one of the most interesting things we found was that the correlation between memory span and rehearsal rate is an artifact. In two studies, no relation was found between people's memory spans and their rehearsal rates for lists of digits well below memory span (3, 4, and 5 digits), but for lists that approach the memory span (6 digits), the correlation is about .50. This correlation is an artifact because people with low memory spans experience difficulties in remembering as memory load increases, and as a result, their rehearsal rate is slowed. There is no relationship between rehearsal rate and memory span for lists of digits below memory span.

In a larger study of 31 college students, we obtained, in addition to memory spans, reliable estimates of several information processing rates. These estimates included search for the presence of an item in short-term memory (Sternberg, 1966), search for the location of an item in short-term memory (Sternberg, 1967), and metered memory search (Weber & Castleman, 1969) in both short-term and long-term memory. The long-term metered memory search task in this study was alphabet search. In this task, the subject is presented both with a probe and a meter, and he must find the item located n places from the probe, where n is the meter. For example, a letter (H) and a number (3) are presented and the task is to name, as quickly as possible, the letter that appears 3 places later in the alphabet (K). This same procedure was used for short-term metered memory search except that the material to be searched is a random list of digits in short-term memory. In addition to these memory search tasks, we measured the corresponding visual search speeds because we wanted an estimate of processing rates uncontaminated by memory load. Finally, we estimated several components of the rehearsal process, including the time to start rehearsal and the time to execute rehearsal. Start time is the time between onset of a GO signal and rehearsal of the first item, and execution time is the average inter-item time during rehearsal. The correlations between these various processing rates and memory span are shown in Table 1, along with the reliabilities. (Digit span reliability was .96).

Table 1

Processing Speed Reliabilities (odd-even)
and Correlations with Digit Span

	Reliability Coefficient	Correlation with Digit Span
Visual Search for Presence	.90	.23
Visual Search for Location	.74	-0-
Visual Metered Search	.84	-.17
Memory Search for Presence	.95	-.17
Memory Search for Location	.82	-.63**
Memory Metered Search	.87	-.62**
Alphabet Metered Search	.95	-.46**
Rehearsal Start Time	.99	-.59**
Rehearsal Execution Time	.99	-.38*

$p < .05*$
$p < .01**$

None of the visual search speeds correlated with memory span, nor did memory search for presence. The correlation between memory span and rehearsal execution time increased with memory load as before, but even with large memory loads the correlation was only

-41. Finally, the correlation between memory search for location and memory span is due to the same artifact that underlies the correlation between memory span and rehearsal.

There were only three non-artifactual correlations with memory span: metered short-term memory search, metered alphabet search, and rehearsal start time. At this point we can only speculate about the source of these correlations. In the metered short-term memory search task, it is possible that concurrent indexing (counting items until the meter is reached) imposes an additional load on short-term memory. This concurrent memory load could cause people with low memory spans to slow down. The correlations in the other two tasks--alphabet search and rehearsal start time--may indicate that people with low memory spans are also slower at activating information in memory. That is, people with low memory spans seem to be slower at accessing information in long-term memory, in secondary memory,, or in whatever inactive storage systems are used when information is not in short-term memory, but once information is activated, they seem to process it at the same rate as people with high memory spans.

The data in these studies provide very little support for the idea that memory span is determined by the speed of symbol manipulation in short-term memory. If anything, our data suggest that memory span may indirectly affect processing rates. That is, people with low memory spans may experience delays in processing as the memory load increases because they are forced to take extra time to update their short-term memory.

If the speed of symbol manipulation in short-term memory is not the major cause of individual differences in memory span, then what is? A good case can be made that memory span depends upon long-term memory knowledge structures and processes built up with practice (Chi, 1976). In the next section we explore the issue of whether short-term memory capacity can be increased with practice. An illustrative case study shows that digit span can be increased seemingly indefinitely if long-term memory coding structures are built up with practice.

Extended Practice

There are reports in the literature of increases in memory span with substantial amounts of practice (Gates & Taylor, 1925; Martin & Fernberger, 1924). Since memory span is such an essential ingredient both in psychometric theories of intelligence and information processing theories of thinking, it is of some interest to understand the nature of these practice effects. In our laboratory, we practiced one individual for about an hour a day, 3-5 days a week, for a year on the memory span task. In that time, his memory span increased steadily from seven digits to over fifty digits.   How did he do it, and did he increase his short-term memory capacity?

Our analysis (Chase & Ericsson, 1978) indicates that this subject developed an elaborate mnemonic system, based primarily on running times for various races (e.g., 339 = three minutes and thirty-nine seconds, near world-record mile time).   Our analysis further indicated that there was no increase in short-term memory capacity.   The evidence is the following.   First, when the subject groups digits together to form mnemonic codes, his groups are almost always 3- and 4-digit groups, and he has never generated a group larger than five digits.   Second, the subject always maintains the last few digits (4-6 digits) as an uncoded rehearsal group, and he never allows the rehearsal group to exceed six digits.   In fact, a 6-digit rehearsal group invariably is segmented as two groups of three digits.   Third, the subject also hierarchically groups his groups together into supergroups.   After some initial difficulty in remembering 5-group supergroups, the subject generally uses 3-group supergroups and he never allows a supergroup to exceed 4 groups.   Finally, when the subject was switched from digits to letters of the alphabet, there was no transfer, and his memory span dropped back to about six consonants.

The outcome of this study makes it clear that one must distinguish between memory span and short-term memory capacity.   Memory span is limited both by the capacity of short-term memory and by coding processes, and the more elaborate the coding processes, the greater will be the discrepancy between memory span and short-term memory capacity.   It is certainly possible to increase memory span by learning to code information so that it can be retrieved from long-term memory, but it does not seem possible to increase the capacity of short-term memory.   It remains an important question to determine the extent to which the correlation between memory span and IQ is due to short-term memory capacity per se, and the extent to which coding processes are important.

References

Bachelder, B. L. & Denny, M. R.   A theory of intelligence: I. Span and the complexity of stimulus control.   Intelligence, 1977, 1, 127-150. (a)

Bachelder, B. L., & Denny, M. R.   A theory of intelligence: II. The role of span in a variety of intellectual tasks.   Intelligence, 1977, 1, 237-256. (b)

Baddeley, A. D., Thompson, N., & Buchanan, M.   The word length effect and the structure of short-term memory.   Journal of Verbal Learning and Verbal Behavior, 1975, 14, 575-589.

Cavanagh, J. P.   Relation between the immediate memory span and the memory search rate.   Psychological Review, 1972, 79, 525-530.

Chase, W. G., & Ericsson, K. A.   Acquisition of a mnemonic system for digit span. Paper presented at the nineteenth annual meeting of the Psychonomic Society, San Antonio, Texas, November 1978.

Chi, M. T. H. Short-term memory limitations in children: Capacity or processing deficits? Memory & Cognition, 1976, 4, 559-572.

Cohen, R. L., & Sandberg, T. Relation between intelligence and short-term memory. Cognitive Psychology, 1977, 9, 534-554.

Gates, A. I., & Taylor, G. A. An experimental study of the nature of improvement resulting from practice in a mental function. Journal of Educational Psychology, 1925, 16, 583-592.

Hunt, E. B., Frost, N., & Lunneborg, C. E. Individual differences in cognition: A new approach to intelligence. In G. Bower (Ed.), Advances in learning and motivation, Vol. 7, New York: Academic Press, 1973. Pp. 87-122.

Lyon, D. R. Individual differences in immediate serial recall: A matter of mnemonics? Cognitive Psychology, 1977, 9, 403-411.

Martin, P. R., & Fernberger, S. W. Improvement in memory span. American Journal of Psychology, 1929, 41, 91-94.

Matarazzo, J. The measurement and appraisal of adult intelligence. Baltimore, MD: Williams & Wilkins, 1972.

Newell, A., & Simon, H. A. Human problem solving. Englewood Cliffs, NJ: Prentice Hall, 1972.

Sternberg, S. High speed scanning in human memory. Science, 1966, 153, 652-654.

Sternberg, S. Retrieval of contextual information from memory. Psychonomic Science, 1967, 8, 55-56.

Weber, R. J., & Castleman, J. Metered memory search. Psychonomic Science, 1969, 16, 311-312.

Footnote

Preparation of this article was supported by contract NOOO14-79-C-0215 from the Advanced Research Projects Agency. We are indebted to M.T.H. Chi for her helpful comments.

Requests for reprints should be sent to William G. Chase, Department of Psychology, Carnegie-Mellon University, Pittsburgh, Pennsylvania, 15213, U.S.A.

TOWARDS A SYMBIOSIS OF COGNITIVE PSYCHOLOGY
AND PSYCHOMETRICS

Jim Ridgway

University of Lancaster

Lancaster, England

Abstract

Cronbach (1957) highlighted two distinct traditions of scientific psychology, namely the experimental tradition and the correlational tradition. The paper discusses ways in which the two
disciplines can be brought together. Guttman's (1955) facet
analysis is seen as a way of introducing experiments into psychometrics; Newell's (1973) criticisms of cognitive psychology are
reviewed and are seen to be resolvable if more use is made of
psychometric methods. The review draws attention to the dominance of static structural models both in cognition and psychometrics. The fusion of the two disciplines is viewed as a relatively small problem compared to that of accounting for subject
strategies, and for structural changes which occur over time.

The Role of Experiments in Psychometrics

Consider a central problem in psychometrics, namely, how we
discover what our tests measure; until we can produce clear definitions, attempts to link psychometrics and cognitive psychology
are doomed to failure. A number of techniques are used, for instance, inspection of items, correlations with other tests, and
studies of group differences. Guttman (1955) pointed out that all
these exercises are conducted post hoc. Since tests and test
items are constructed by their designers for some specific purpose, it seems reasonable to ask for a clear statement of the
designer's theory of what is being measured, and, more specifically, for the rules of item construction.

Just as the experimental psychologist may study the effect

163

of factors A and B (with levels a_1, a_2, b_1, b_2. etc.) upon behaviours, so the rational features or facets of a test should form part of a psychometrician's hypothesis about the individual differences to be displayed in a test. Just as the experimenter must determine whether or not factors A and B do indeed have effects on behaviour, for the psychometrician it should be an empirical question whether or not the facets he has chosen to study are effective in varying the nature of the individual differences exhibited on the test. Thus validation can be viewed as a search for correct hypotheses about the correspondence between a system of definitions and specifications, and an empirical data structure.

Guttman refers to this approach as "facet design and analysis," and has demonstrated its use both in test construction and in reanalysing existing data to uncover new structures. Several examples of its use in establishing the construct validity of existing tests have been provided by Levy (1973) and by Ridgway (1979a, 1979b).

What sort of a view does it give us of psychometrics? The first thing that we should notice is that it is an exploratory technique, and has no psychological content at all. It should perhaps be viewed as the thinking man's factor analysis. In itself, it offers no theory of behaviour, and no structure of intellect; however it is a powerful tool for uncovering these structures, if they exist. A benefit which should accrue from an emphasis on the specification of rules is that while cognitive psychologists are quite prepared to investigate, and to provide models of process for the ways in which people deal with some well defined rules, they are far more loath to study a rag-bag of rules labelled, say, "verbal ability." Thus by insisting that the rules for constructing tests be defined unambiguously, a facet analysis can be viewed as a technique for presenting the content of psychometrics in a form which is amenable to investigation by cognitive psychologists.

The Role of Individual Differences in Cognitive Psychology

Newell (1973) gave us a critique of cognitive psychology entitled "You can't play twenty questions with nature and win," in which he made three main criticisms of cognitive psychology. First, psychology is based on phenomena, not theory. Second, our approach of testing binary oppositions (e.g. serial versus parallel processing) does not address any of our main goals (e.g., to understand cognition) directly, and rarely results in a resolution of the dichotomy. Third we generate a body of knowledge whose usefulness is severely limited because we have no way of relating different studies, except in an intuitive way. We can answer the first criticism by saying simply that we have to start somewhere. It seems sensible to address problems which we can see have limited scope, if we can solve them. It is to be hoped that we will be able to generalise our models of phenomena at some later date.

The second criticism is the problem of identifiability, in disguise. Given some observed pattern of responding, a number of different models can be proposed to account for the data. When just experimental data are considered (e.g., reaction times, mean number correct) there is often no way of choosing between alternative models. We can suggest that a facet analysis approach to experiments in cognitive psychology could be used to investigate such models. If we construct two tasks in such a way that one model predicts that they must share common processes, and which another model predicts they need not, and examine the correlation between performance on the tasks, we may be able to discriminate between the models on the basis of correlational data, in a way in which estimating parameters, and establishing goodness of fit never can. Examples of this approach in the areas of perception and memory are provided elsewhere (Ridgway, 1979b).

The third criticism proposed by Newell is that we cannot reliably relate the findings of different experiments to each other. We can consider this problem at two levels, namely at the level of an individual experiment (what other experiments are relevant?), and at the level of the whole area of cognitive psychology (how are phenomena related?). The problem arises because no one attempts to establish the key facets of his experimental task; the key facets are "self evident." The cognitive psychologist is as uncritical of his experimental paradigm as the psychometrician is of his test. In order to know which experiments in the literature are relevant to the one in hand, we should simply correlate performances on tasks which we believe to be the same. High correlations support our beliefs; low correlations lead us to search for the source of the differences between the tasks directly. A psychometric approach to the domain of cognitive psychology will enable us to go some way towards dealing with this problem at a more global level. By examining the relationship between individual differences in performance across a wide range of tasks we will be able to group together tasks which are strongly related, and which may well utilise the same underlying cognitive processes.

Towards a Fusion of Psychometrics and Cognitive Psychology?

It is unfortunate that both cognitive psychology and psychometrics are largely based on static models of mental processes. Mental ability can be measured; measurement must be "reliable," and predictive of behaviour over several years. Cognitive psychology advances by discovering "the" model of the boxes in our heads; we must get the number, nature, and interconnections

right. While both of these statements are caricatures, they are
sufficiently close to the truth to be disturbing. One might argue
that a grand synthesis of current cognitive psychology and current
psychometrics, although a neat trick, is relatively unimportant
compared to the problems of producing a unified approach which
can encompass what we know to be fundamental properties of our
cognitive systems, namely, structure, function, and evolution,
or, being, doing, and becoming.

As soon as we allow a ghost into the machine, which seeks to
optimise performance by assigning different aspects of the task to
different parts of the machine, our problems increase dramatically.
We now have the problem of deducing the invariant structure of the
machine (its architecture) and of infering the method used (the soft-
ware) simultaneously. We should look with optimism, therefore, to
the recent wave of studies which have focused on the strategies
which subjects bring to our experiments. Let us hope that we can
relate these studies into our views of individual differences and
of cognition.

The notion that either structures (over the long term) or functions (over the short term) are changing is also one which has received scant attention; we have precious few models of changing
structures or processes, and the whole problem of accounting for
change is one which we must solve before we can claim to have an
adequate explanation of cognition processes.

Let us draw the discussion to a close. We have argued that
psychometrics can benefit from experimental techniques, and have
suggested that several of the problems in cognitive psychology can
be resolved by the application of psychometric techniques. However,
even if the two disciplines can be reconciled, they provide a poor
framework for the explanation of our cognitive processes, because
of the emphasis on steady state processes. In order to provide an
adequate framework, our theories must be able to encompass the
notions of subject strategies, and of structural changes which
occur over time.

References

Cronbach, L. J. The two disciplines of scientific psychology.
    American Psychologist, 1957, 12, 671-683.
Guttman, L. An outline of some new methodology for social re-
    search. Public Opinion Quarterly, 1955, 18, 395-404.
Levy, P. On the relation between test theory and psychology. In
    P. Klein (Ed.), New Approaches in Psychological Measurement.
    London: Wiley, 1973.
Newell, A. You can't play 20 questions with nature and win. In
    W.G. Chase (Ed.), Visual Information Processing. New York:
    Academic Press, 1973.

Ridgway, J.   New lamps for old: what do you do when predictive
    validity declines?   Proceedings of the Occupation Psychology
    Conference, 1979a, 17, (Summary).
Ridgway, J.   Psychometrics and cognitive psychology.   In R. Wood
    (Ed.), Proceedings of the Social Science Research Council
    Seminar   on   Rehabilitating   Psychometrics,   1979b.

DEVELOPMENT AND MODIFIABILITY OF ADULT INTELLECTUAL PERFORMANCE: AN EXAMINATION OF COGNITIVE INTERVENTION IN LATER ADULTHOOD

Sherry L. Willis, Paul B. Baltes, and Steven W. Cornelius

The Pennsylvania State University

University Park, Pennsylvania, U.S.A.

Introduction

Traditionally, differential and cognitive approaches have emphasized different dimensions of adult intelligence. Differential psychology has sought to represent intellectual functioning in terms of structural models of human abilities (Cattell, 1971; Guilford, 1967). Much of the emphasis in this approach has been on idividual differences in intellectual ability. In contrast, cognitive psychology has focused on identifying the cognitive processes and strategies involved in intellectual functioning (Newell and Simon, 1972; Sternberg, 1977). It has been suggested that cognitive psychology provides a more dynamic approach to the study of intelligence in that the focus is on the processing of information, whereas psychometric ability factors represent static products of cognition.

In one sense, however, both approaches have tended to assume a somewhat static view of adult intelligence. That is, much theory and research associated with each position has involved assumptions regarding stability in adult intellectual performance. Thus, the focus in both approaches has been primarily on the normative or average level of intellectual functioning rather than on an examination of the full range of intraindividual variability in adult intellectual performance (Baltes and Willis, 1977, Willis and Baltes, 1980). However, it will be suggested in this paper that there may be considerable plasticity in intellectual performance, particularly in later adulthood; thus, potential as well as average levels of functioning must be examined.

Several trends have contributed to such assumptions regarding stability in adult intelligence. In differential psychology the notion

169

regarding the static nature of intelligence (Baltes and Willis, 1979;
Brown and French, 1979). Within cognitive psychology the importance of a predictive vs. diagnostic (learning) approach to intellectual assessment is gaining attention (Brown and French, 1979;
Resnick, 1979). The traditional emphasis on prediction appeared
to involve a static perspective of intelligence, such that the individual's current level of functioning (based on prior learning and
assessed by standard intelligence tests) was considered to provide
an accurate reflection of future learning potential. In contrast,
those advocating a diagnostic approach suggest that current level
of functioning may not provide an accurate prediction of the individual's potential zone of intellectual development, if prior learning
opportunity has been limited (e.g., environmental deficits, learning
disability). In this case, a learning or diagnostic approach involving
an examination of the range of plasticity in intellectual functioning
within a short-term experimental, assessment or interventive context
would be useful. Such an approach emphasizes intraindividual variability rather than a normative (average) level of intellectual functioning. A learning or diagnostic approach has been most forcefully
articulated (within cognitive psychology) by those working in the
area of mental retardation (Brown and French, 1979). In addition,
these researchers are engaged in a series of training studies examining the range of modifiability of intellectual performance in learning disabled and retarded populations (Belmont and Butterfield,
1977; Brown, 1978).

Similar concerns regarding intellectual variability within a
psychometric or differential approach to intelligence have been
associated most notably with the recent revival of a life-span
perspective. Within a life-span approach, developmental change
and plasticity are examined across the total life span rather than
primarily in childhood or adolescence. Two lines of recent research
have examined individual variability in intellectual functioning in
adulthood. The first and more extensive line of research, illustrated primarily by the work of Schaie (1979), has focused on the
use of cohort-sequential methodology in the longitudinal study of
adult intelligence. In contrast to cross-sectional findings suggesting
a peak in intellectual functioning in childhood or adolescence, longitudinal research suggests continued intellectual development for
some abilities into young adulthood, such that in current cohorts
of healthy, well-educated adults a peak in intellectual functioning
may not be reached until early middle age. Moreover, much less
pervasive decline in old age has been reported than for cross-
sectional samples. In addition, comparisons of earlier and later
adult cohorts at the same chronological age indicate that more recent
cohorts performed at a higher level for some abilities than did
earlier cohorts at the same age. Such cohort-differences research
suggest that the lower level of intellectual performance of current
older adult cohorts may be partially attributable to cohort-related
obsolescence as a function of socio-cultural change. Thus, the cur-

of stability appears to have been closely related to assumptions regarding the nature of ability factors. Those taking a casual, rather than descriptive, view of the nature of factors have tended to ascribe trait-like characteristics to such ability factors. Cattell (1971) has referred to factors as "source traits," and Guilford (1967) has described a factor as "an underlying latent variable along which individuals differ" (p. 41). Based on a biological perspective of traits as enduring characteristics (e.g., eye color, race) of the individual, there was the tendency to make similar trait-like assumptions regarding ability factors, such that consider-able stability in intellectual performance was expected.

Within cognitive psychology, stability notions have been related to the concern with identifying a set of elementary information proces-es (Newell and Simon, 1972; Sternberg, 1977). These processes were considered elementary in the sense that within a given theory they were the fundamental units of analysis. The elemental nature of these processes appears to have led to assumptions regarding their stability. Moreover, some have suggested that information processes may be a direct reflection of neural efficiency in function-ing, again implying the elemental, stable character of such proces-ses (Jensen, 1978; Ertl, 1971).

In addition, both differential and cognitive approaches have placed heavy emphasis on predictability (Anastasi, 1976; Sternberg, 1977). Within the psychometric approach, the concern was on devel-opment of measures which could predict individual differences in performance in academic or occupational settings, whereas in cog-nitive psychology the goal was to design models of sufficient generality to predict or simulate the manner in which information was processed across a variety of content and task domains. To achieve such predictive power, models were developed which focused on normative or average levels of intellectual functioning and assumed considerable stability in intellectual performance.

Finally, stability assumptions regarding adult intelligence have resulted, in part, from the traditional emphasis within developmen-tal psychology on the earlier portion of the life span (Labouvie and Chandler, 1978; Baltes and Willis, 1979). That is, many models of adult intelligence have evolved from child-oriented theories of in-telligence, such that intelligence was seen as developing in child-hood and adolescence, followed by a period of considerable stability through most of adulthood and a sharp decline in old age. Thus, most developmental change in intelligence was assumed to occur in childhood with relatively little important developmental variability through the remainder of the life span.

However, within both differential and cognitive psychology there appears to be a movement toward reexamination of a normative or average approach to intellectual functioning and of assumptions rent elderly may be at a disadvantage in many academic-related contexts, such as testing situations. As a function of such obsolescence, older adults' average level of intellectual performance as assessed in standardized testing contexts may not provide an accurate reflection of their potential zone of intellectual functioning. In this case, a learning approach may be useful in examining the range of plasticity (variability) in older adults' intellectual performance.

An Examination of Intellectual Plasticity (Variability) in Later Adulthood

In this paper two studies will be reported briefly which are part of an ongoing research program aimed at examining the modifiability of intellectual performance in later adulthood through a cognitive training paradigm. A series of short-term longitudinal training studies focusing on several abilities representing fluid intelligence are being conducted. Within the Cattell-Horn theory of fluid-crystallized intelligence, fluid intelligence is conceived as one of two general dimensions of intelligence, involving stable trait-like properties and exhibiting a normative pattern of decline in later adulthood (Horn and Cattell, 1967; Cattell, 1971). Our training research seeks to examine the range of variability which can be experimentally produced for component abilities representing such a trait-like dimension of intelligence and, thus, to assess the modifiability of normative decline in fluid intellectual performance in the elderly.

In the first study to be reported, the range of variability in intellectual performance as a function of practice (retest) effects was examined. Such a study explored intellectual modifiability under minimal intervention conditions; subjects participated in multiple retest sessions with no instruction on cognitive strategies and no feedback regarding correctness of response. In the second study, subjects received training on cognitive strategies required in solution of the target fluid ability tasks. Training effectiveness was assessed with regard to both durability (maintenance) of training effects and transfer to a theory-based pattern of ability measures.

Research on retest-practice effects. Thirty older subjects ($\overline{X}$ age = 69.2 years, SD = 5.18) participated in eight one-hour retest sessions (Hofland, Willis, and Baltes, Note 1). At each retest session, subjects were administered under standard testing conditions two measures, representing the two fluid abilities of Figural Relations and Induction respectively. The Culture Fair test (Scale 2, Power Matrices Scale 3; Cattell and Cattell, 1957) was identified from previous research (Cattell, 1971) to represent the Figural Relations ability; the Induction ability was marked by an Induction Composite test including Letter Sets (Ekstrom, French, Harman, and Derman, 1976) Number Series and Letter Series (Thur-

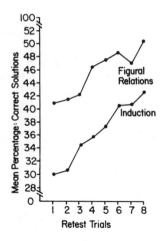

Figure 1

stone, 1962) tests. No external feedback regarding correctness of responses was given during the retest sessions.

The mean percentage of correct solutions for each measure was computed for each of the eight retest sessions and is shown graphically in Figure 1. A one-factor analysis of variance with repeated measurement across the eight trials was performed on the raw scores for each of the two retest measures. Significant performance gains (p < .001) were found across the eight trials for each of the two measures (Figural Relations: $F = 16.81$, $df = 7,203$; Induction: $F = 26.42$, $df = 1.29$). Total improvement in mean scores on both measures was roughly equivalent to one standard deviation. With regard to the performance pattern across the eight sessions, subjects exhibited small, steady gains between consecutive trials. Separate trend analyses for the two measures indicated that only a linear component was significant (p < .001). No apparent performance asymptote was reached.

Training research. Modifiability of fluid intellectual performance in the elderly has also been examined as a function of a series of short-term longitudinal training studies each focusing on one target fluid ability. In one such study (Willis, Blieszner, and Baltes, Note 2) involving the target ability of Figural Relations, training effectiveness was assessed by comparing posttest performance of randomly assigned experimental and control groups (Total N = 58, X age = 69.8, SD = 5.7). Experimental subjects participated in five one-hour training sessions focusing on cognitive strategies identified in task analyses to be involved in solution of Figural Relation-type problems. The two criteria for assessing training effectiveness were durability (Maintenance) of training effects over three posttest occasions

(1 week, 1 month, 6 months) and transfer (generalizability) of training across a broad battery of seven fluid and crystallized measures.
With regard to training transfer, a hierarchical theory-based pattern
of trasnsfer was predicted with the largest training effects occurring
for the three near transfer measures representing the target fluid
ability: ADEPT Figural Relations (Plemons et al., 1978), Culture
Fair (Cattell and Cattell, 1957), Raven (Raven, 1962). Less or no
training effects were predicted for two levels of far transfer, involving far fluid transfer to the fluid ability of Induction and far nonfluid transfer to Crystallized Intelligence and Perceptual Speed.
Induction was represented by two measures: ADEPT Induction
(Blieszner, Willis, and Baltes, Note 3) and Induction Composite
(Ekstrom et al., 1976; Thurstone, 1962) tests.   Crystallized Intelligence was marked by a Vocabulary measures (Ekstrom et al., 1976)
and Perceptual Speed by the Identical Pictures test (Ekstrom
et al., 1976).

The entire data matrix (across treatments and occasions) for
each of the seven posttest measures was standardized using the control group's score on that measure at Posttest 1 as the standardization base with a mean of 50 and standard deviation of 10.   This
standardization procedure was employed to provide a common baseline
of performance on each measure to which all other data points for
that measure could be compared and to eliminate scale level differences between measures, thus facilitating comparison of transfer
effects across measures.   A graphic summary of the training and
control groups' standardized mean scores for the seven transfer
measures averaged across the three posttest occasions, is shown in
Figure 2.   Mean scores of the training group were larger than the
control's scores for all seven measures at each of the three posttests.   The pattern of training transfer is represented by the relative difference between the standardized mean scores for the training and control groups for each measure.   Note that the difference
between mean scores for training and control groups appears larger
for the three near, Figural Relations, measures than for the four
far (fluid and nonfluid) measures.

An overall analysis as a general assessment of training effects
was performed across all measures and occasions, using standardized scores.   That is, a 2 (Treatment: Training, Control) x 3
(Occasion: Posttests 1, 2, 3) x 7 (Measures) analysis of covariance with repeated measures was conducted using the pretest score
on the ADEPT Figural Relations test as the covariate.   There was
no significant difference beten training and control groups at pretest.   This analysis resulted in a significant Treatment main effect
$(F [1, 54] = 11.81, p < .001)$, and a significant treatment x Meas-
sure interaction $(F [6,336] = 2.25, p < .05)$ suggesting a differential treatment effects across the seven transfer measures as predicted.   A significant Occasion main effect $(F [2,112] = 12,00, p < .001$
was obtained and interpreted as suggesting retest effects common to

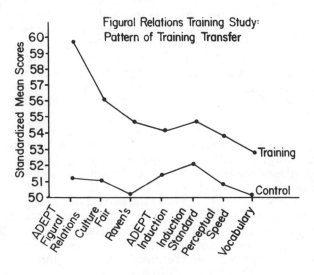

Figure 2

both training and control groups. A significant Measure main effect
(F [6,336] = 3.43, p < .05) occurred as a function of differential
training and retest effects by measure, given the standardization
procedure.

Follow-up analyses via the Tukey WSD conducted separately
by measure indicated that training and control groups differed significantly on each of the three near tranfer measures across posttests: ADEPT Figural Relations (p = .000), Culture Fair (p = .008),
Raven's (p = .018). No significant differences between training and
control were found for the four far transfer measures separately:
ADEPT Induction (p = .151), Induction Composite (p = .16), Vocabulary (p = .138) and Perceptual Speed (p = .122). However, increasing the statistical power by using a repeated measures analysis
of covariance on just the four far transfer measures resulted in a
significant Treatment main effect (F [1,54] = 4.15, p = .047) for the
four far transfer measures.

Discussion

Training research in later adulthood. Findings from both the
retest and training studies suggest considerable variability in intraindividual intellectual performance in later adulthood. In the retest
study significant performance increments were found for each of two
measures, representing Figural Relations and Induction abilities.
Such retest effects occurred under a minimal interventive practice

condition in which subjects received no training or feedback, thus, suggesting subjects possessed or were able to generate on their own cognitive strategies and/or test-taking skills useful in improving their performance. In the Figural Relations training study a pattern of differential training transfer was found with significant training and transfer effects being established and maintained for the three near transfer measures. Such training effects for the three measures represent a broad continuum of training transfer within the target ability. Moreover, these training effects were maintained over a six-month period.

Data from the training study also suggests that transfer effects extended, although to a lesser degree, beyond the target ability. Th training group's scores on all four far transfer measures at all post-test occasions were larger than those for the control. In our view, such an effect on far transfer measures is less likely to result from ability-specific improvement. Rather it may reflect generalized, non-ability-specific transfer attributable to situational or ability-extraneou factors (e.g., increases motivation, anxiety reduction) which were accrued as a function of the training treatment but are not intrinsic to performance on the target ability per se. Such non-ability-specifi transfer would affect performance on a wide variety of ability meas-ures and would show a general effect across the far transfer meas-ures as was found. The likelihood of non-ability-specific transfer occurring may be greater for educationally and/or test-disadvantaged populations, such as the elderly. Considerable retest effects were also found in the training study. They were differentiated from ability-specific training effects as being general such that retest effects occurred for both experimental and control groups and did not follow the predicted pattern of differential transfer.

Such training research would appear to have important implica-tions for theories of adult intelligence. Most current models of adult intelligence, both within the psychometric and cognitive approach, focus on the normative or average pattern of intellectual aging and do not address the potential for plasticity in intellectual functioning in middle and later adulthood. While most intelligence models in childhood and young adulthood have also focused on normative pat-terns of development, cognitive training research has examined the range of modifiability of intellectual performance during these age periods. This training research has contributed to more compre-hensive models of intellectual development early in the life span. Such training research is needed to supplement current theories of normative adult intellectual development. It is suggested that comprehensive theories of intelligence including both potential and normative dimensions of functioning may be particularly important in adulthood, in light of recent cohort research examining the po-tential impact of socio-cultural change on adult intelligence.

Reference Notes

1. Hofland, B. F., Willis, S. L., and Baltes, P. B. Fluid intelligence performed in the elderly: Retesting and intraindividual variability. Unpublished manuscript, College of Human Development, The Pennsylvania State University, 1979.

2. Willis, S. L., Blieszner, R., and Baltes, P. B. Training research in aging: Modification of intellectual performance on a fluid ability component. Unpublished manuscript, College of Human Development, The Pennsylvania State University, 1980.

3. Blieszner, R., Willis, S. L., and Baltes, P. B. Training research on induction in aging: A short-term longitudinal study. Unpublished manuscript, College of Human Development, The Pennsylvania State University, 1980.

References

Anastasi, A. Psychological testing (4th ed.). New York: Macmillan, 1976. Baltes, P. B. and Willis, S. L. Life-span developmental psychology, cognitive functioning and social policy. In M. W. Riley (Ed.), Aging from birth to death. Washington, D. C.: American Association for the Advancement of Science, 1979.

Baltes, P.B., and Willis, S. L. Toward psychological theories of aging and development. In J. E. Birren and K. W. Schaie (Eds.), Handbook of the psychology of aging. New York: Van Nostrand-Reinhold, 1977.

Belmont, J. M., and Butterfield, E. C. The instructional approach to developmental cognitive research. In R. V. Kail and J. W. Haten (Eds.), Perspectives on the development of memory and cognition. Hillsdale, NJ: Erlbaum, 1977.

Brown, A. L. and French, L. A. The zone of potential development: Implications for intelligence testing in the year 2000. In R. J. Sternberg and D. K. Detterman (Eds.). Human intelligence. Norwood, NJ: Ablex, 1979.

Brown, A. L. Knowning when, where and how to remember: A problem of metacognition. In R. Glaser (Ed.). Advances in instructional psychology. Hillsdale, NJ: Erlbaum, 1978.

Cattell, R. B., and Cattell, A. K. S. Test of "g": Culture Fair (Scale 3, Form A., 1963 Edition; Form B, 1961 Edition, Second). Champaign, IL: Institute for Personality and Ability Testing, 1961, 1963.

Cattell, R. B. Abilities: Structure, Growth, and action. New York: Houghton-Mifflin, 1971.

Ekstrom, R. B., French, J. W., Harman, H., and Derman, D. Kit of factor-referenced cognitive tests, 1976 Revision. Princeton, NJ: ETS, 1976.

Ertl, J. P. Fourier analysis of evoked potentials and human intelli-

gence. Nature, 1971, 230, 515-516.
Guilford, J. P. The nature of human intelligence. New York:
     McGraw Hill, 1967.
Horn, J. L., and Cattell, R. B. Age differences in fluid and crys-
     tallized intelligence. Acta Psychologica, 1967, 26, 107-129.
Jensen, A. R. "g"--outmoded theory or unconquered frontier?
     Paper presented at the meeting of the American Psychological
     Association, Toronto, 1978.
Labouvie-Vief, G., and Chandler, M. Cognitive development and
     life-span development theories: Idealistic vs. contextual per-
     spectives. In P. B. Baltes (Ed.), Life-span development and
     behavior (Vol. 2). New York: Academic Press, 1978.
Newell, A., and Simon, H. Human problem solving. Englewood
     Cliffs, NJ: Prentice-Hall, 1972.
Plemons, J. K., Willis, S. L., and Baltes, P. B. Modifiability of
     fluid intelligence in aging: A short-term longitudinal training
     approach. Journal of Gerontology, 1978, 33, 224-231.
Raven, J. C. Advanced progressive matrices, Set II, 1962 Revision.
     London: H. K. Lewis and Co., Ltd., 1962.
Resnick, L. B. The future of IQ Testing in education. In R. J.
     Sternberg and D. K. Detterman (Eds.). Human intelligence.
     Norwood, NJ: Ablex, 1979.
Schaie, K. W. The primary mental abilities in adulthood: An explor
     ation in the development of psychometric intelligence. In P. B.
     Baltes and O. G. Brim, Jr. (Eds.), Life-span development and
     behavior (Vol. 2). New York: Academic Press, 1979.
Sternberg, R. J. Intelligence, information processing, and analogi-
     cal reasoning: The componential analysis of human abilities.
     Hillsdale, NJ: Erlbaum, 1977.
Thurstone, T. G. Primary mental abilities, Grades 9-12, 1962
     Revision. Chicago: Science Research Associates, 1962.
Willis, S. L., and Baltes, P. B. Intelligence and cognitive ability.
     Chapter prepared for Symposium on Contemporary Issues in
     Cognitive Psychology of Aging, American Psychological Associa-
     tion, New York, 1979.

Footnote

     The Adult Development and Enrichment Project (ADEPT), sup-
ported by a grant from the National Institute of Aging (#5-ROI-
AG004403) to co-investigators Paul B. Baltes and Sherry L. Willis,
involves a research program to examine the effects of cognitive
training programs on the intellectual performance of older adults.
Thanks are due to several research assistants of the project
(Rosemary Bliezner, Majorie Lachman, Brian Hofland, Vincent
Morello, Gail Peck, Manfred Schmitt), its field and training staff
(Carolyn Nesselroade, Myrtle Williams) and John R. Nesselroade
and Paul A. Games, statistical consultants.

THE RELATIONSHIP BETWEEN MEMORY SPAN

AND PROCESSING SPEED

Roderick Nicolson

University of Sheffield

Sheffield, England

Abstract

For any individual, processing speed, as reflected by read-ing rate, varies for words of different lengths, and the rate of increase of memory span as a function of reading rate yields an index of memory capacity. A study of memory span in 8, 10 and 12 year old children, using these direct measures of processing efficiency and memory capacity, indicated that the developmental increase in memory span is attributable wholly to the increase in mean reading rate. For all age groups, a subject's memory span for a given set of words was roughly equal to how many of the words the subject could read in two seconds. Furthermore, for any given reading rate, the memory span was independent of age.

Digit span tests have long been included in tests of general intelligence and typically correlate at around 0.50 to 0.60. Bill Chase's article in this volume gives an excellent introduction to the literature, and a brief overview should suffice here. I shall use memory span (MS) to refer to the mean correct recall in any episodic memory task involving immediate serial recall of a series of stimuli, a wider definition than that of digit span. The general findings are that MS correlates fairly highly with intelligence, though not as highly as does backward MS (e.g., Matarazzo, 1972); that it does not correlate highly with any other tests of episodic memory (Underwood et al, 1978); that it is higher for high verbal than for low verbal subjects (Hunt et al, 1975); and,

within subjects, that it is higher for short words than long words
(Baddeley et al, 1975), and for more frequent than less frequent
words (Watkins, 1977). Turning now to developmental studies, all
that is clear is that MS does increase with age. Recent studies
designed to assess the contribution of strategic factors such as
rehearsal, chunking and retrieval strategies (e.g., Chi, 1977;
Huttenlocher & Burke, 1976; Lyon, 1977; Samuel, 1978), have led
to negative findings, indicating that these strategic factors can
be, at best, only a partial explanation of the increase in MS with
age. This leaves two alternative null hypotheses for the residual
increase, namely "structural" and "process" explanations, which
attribute the effect to a developmental increase in memory capacity
and processing efficiency respectively. Capacity explanations are
justifiably unpopular, since the concept of capacity is almost
impossible to define or measure (see e.g., Allport, in press),
but, unfortunately the concept of processing efficiency is little
better. Without the means for measuring directly both memory
capacity and processing efficiency, we cannot evaluate their
relative contributions to the developmental increase in MS.

Fortunately, Baddeley, Thomson, and Buchanan (1975) introduced a technique which elicits such direct measurements. In a
range of experiments with adult subjects, they first demonstrated
the word length effect on MS, that is that, other things being
equal, one can remember more short words than long words.
Next, they investigated the relationship between MS and reading
rate (RR). They constructed five pools of 10 equi-syllabic words
matched across pools for frequency and semantic category, with
the number of syllables increasing from one to five across the
pools. For each word pool they measured mean MS (number
correct in serial order following visual presentation of five words
from the pool at a two second rate), and mean RR (calculated
from the time taken to read aloud a list of 50 words taken from
the pool). As one might expect, both MS and RR suffered a
highly significant decline as the number of syllables increased,
and this was reflected by an overall correlation of 0.69 between
MS and RR. The most interesting finding was that MS varied
linearly as a function of RR, that is that the five pairs of (MS,
RR) points, one pair for each number of syllables, lay on a
straight line,

$$MS = k \cdot RR + c$$

where k, the slope, which has the dimensions of time, was 1.87,
and c, the intercept, was close to zero. In words, regardless of
the number of syllables, the subject was able to recall as many
words as he could read in 1.87 seconds. Baddeley et al interpreted
this in terms of a 1.87 seconds' capacity articulatory rehearsal
loop, a concept taken from Baddeley and Hitch's (1974) working
memory system.

The value of this technique may now be apparent. Reading aloud involves use of all the routine input, lexical access and output processes, and it is reasonable, therefore, to interpret RR as an index of processing speed. Note that this makes explicit the requirement that processing speed depends on factors such as word length. We may now interpret Baddeley et al's results as

MS = "capacity" x "processing speed" + constant,

where MS and processing speed vary as a function of the number of syllables, and the capacity is inferred from the slope of this relationship.

This gives the rationale for the following investigation of the relative contribution of capacity and processing efficiency to the development of MS. We used three groups of 10 children with mean ages 8.1, 10.2, 12.1 years, and the procedure was a close replication of Baddeley et al (1975, Expt. 6) except that we omitted the five syllable word pool, which was too hard.

Overall, the pattern of results was strikingly similar to that of Baddeley et al. Analysis of variance indicated that age and number of syllables had significant main effects on both MS ($F(2,27)$) = 3.95, p <.05; $F(3,81)$ = 38.99, p < .0001) and RR ($F(2,27)$ = 3.58, p < .05; $F(3,81)$ = 152.73, p < .0001). Overall MS and RR were lower than for adults, but improved with age, with the youngest children performing significantly the worst. For each age group both MS and RR decreased as the number of syllables increased and the within-group correlations between MS and RR were 0.71, 0.51, 0.66 for age 8, 10, 12. When the within-group mean MS and RR for each number of syllables was plotted, the relationship was linear for all three age groups (all correlations were greater than 0.98), see figure 1, in which the adult data are

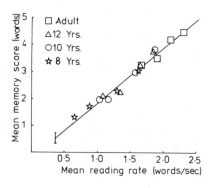

Figure 1

taken from Baddeley et al. It is clear that all four groups are well-fitted by a straight line through the origin. The best fit line is MS = 2.08RR − 0.24, with overall correlation 0.996. The within-group best fit slopes (which we interpret as mean capacity) are in ascending age order 1.83, 2.31, 2.34 and 1.87 respectively. There were no significant differences between groups for the individual slopes or intercepts, and 28 of the 30 subjects were reasonably well fit individually by a linear function (correlation above 0.50).

In the above analysis, mean MS and mean RR were calculated for each number of syllables. In figure 2, the data are collapsed over syllables, and MS is plotted directly as a function of RR. It is clear that there is no difference between the ages in the mean MS for each of the five categories of RR. In other words, for any given reading rate, MS is independent of age.

Conclusions

We have shown that children's MS is affected by RR in qualitatively the same way as adults'; that the relationship is linear, and so the interpretation of its slope as "capacity" is possible; that, despite the significant increase in MS and RR with age, there is no significant age-related change in the slope or the intercept of the MS-RR line; and, finally, that for a given RR, MS is independent of age.

These results provide strong support for the hypothesis that, both within subjects and between age groups, changes in MS are directly attributable to changes in RR, and thus, that the increase in mean RR (processing speed) with age is a sufficient explanation of the increase in mean MS with age.

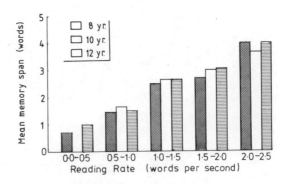

Figure 2

Finally, to return to the relationship between MS and intelligence, we have attributed changes in group MS with age to processing efficiency rather than memory capacity. It should be stressed that this is a group effect rather than an individual effect, and it is very likely that large differences in MS between individuals may be attributable to capacity, not processing, differences. The technique described here in which, for each subject, MS and RR are manipulated by use of words of different lengths, provides a means of investigating this question.

References

Allport, A. Attention and performance. In G. Claxton (Ed.) New directions in Cognitive Psychology. London: Routledge & Kegan Paul (in press).

Baddeley, A. D. and Hitch, G. Working Memory. In G. H. Bower (Ed.), The Psychology of Learning and Motivation. New York: Academic Press, 1974, Vol. 8.

Baddeley, A. D., Thomson, N. and Buchanan, M. Word length and the structure of short term memory. Journal of Verbal Learning and Verbal Behaviour, 1975, 14, 575-589.

Chase, W. G. Individual differences in memory span. This volume.

Chi, M. T. H. Age differences in memory span. Journal of Experimental Child Psychology, 1977, 23, 266-281.

Hunt, E., Lunneborg, C. and Lewis, J. What does it mean to be high verbal? Cognitive Psychology, 1975, 7, 194-227.

Huttenlocher, J. and Burke, D. Why does memory increase with age? Cognitive Psychology, 1976, 8, 1-31.

Lyon, D. R. Individual differences in serial recall: A matter of mnemonics? Cognitive Psychology, 1977, 9, 403-411.

Matarazzo, J. D. Wechsler's measurement and appraisal of adult intelligence. (5th edition). Baltimore, MD: Williams and Wilkins, 1972.

Samuel, A.G. Organizational vs. retrieval factors in the development of digit span. Journal of Experimental Child Psychology, 1978, 26, 308-319.

Underwood, B.J., Bourch, R.F., and Malmi, R.A. Composition of episodic memory. Journal of Experimental Psychology: General, 1978, 107, 393-419.

Watkins, M. J. The intricacy of memory span. Memory and Cognition, 1977, 5, 529-534.

COGNITIVE MECHANISMS AND TRAINING

Magali Bovet

Geneva University

Geneva, Switzerland

Piaget's main concern is to discover how knowledge is formed (= epistemological interest). He uses two methods: historico-critical (or the history of science) and developmental psychology (or the study of the formation of knowledge from birth to adolescence). For his epistemological purposes, the two methods are complementary: for the purposes of our discussion, we shall limit ourselves to the second method.

However, in order to understand the context in which a discussion of Piagetian-type learning is to be situated, it would seem important to first describe Piaget's interest in epistemology and the conceptions which result from it.

Piaget has always been interested in the biological processes, since he considers these to be the basis of all cognitive mechanisms. His point of view is not, however, reductionist, for he sees a continuity between the two sorts of processes and uses functional and structural analogies as a cognitive heuristic. According to Piaget, the same regulatory processes[1] (regulations and equilibrations) are involved in biology and cognition. Piaget's constructivist conception is based on processes of this type, whereas apriorists and empiricists[2] do not seem to see the utility of such mechanisms. In fact, it is these processes which enable Piaget to develop his idea that cognitive growth is an active process and to explain the spontaneous curiosity of the child without external reinforcement. These regulations and equilibrations lead the child via the process of "empirical abstraction," on the one hand, to find a new equilibrium each time that his actions come up against obstacles in the environment; internal perturbations, on the other hand, are overcome by "reflexive abstraction" (Piaget, 1977) (i.e., they are understood and not simply neglected). By means of these two processes, the child reaches a better equilibrium (i.e., augmentative equilibration) (Piaget, 1975).

Piaget's conception also differs from that of apriorists or empiricists in that, for him, cognitive development results from an interaction between the subject (knower) and the object (of knowledge). The fundamental processes characterising this interaction are assimilation (or the modification of objects to conform to the actions of the subject) and accommodation (or the complementary adaptation of the actions of the subject to objects). These processes function at all levels of development, whether the acts involved be reflex actions, practical actions, representation/conceptual actions, or abstract mental actions.

Finally, as is well known, Piaget has tried to analyze what is common to the different types of behaviour which succeed each other in the course of development, and what underlies them. Using algebraic models, he distinguishes different types of structures based on a logical analysis of cognitive behaviour. Generally speaking, he believes that actions are gradually organised into systems of operations, i.e., interiorised and reversible actions which form a grouping characterised by the logic of class relations; these classes and relations are then combined to form the group of formal operations. An elementary system can be observed at the sensori-motor level, where the actions of the baby, by means of coordinations and differentiations, are organised into acts of practical intelligence. The baby thus forms a practical group of displacements where time and space are structured in such a way that the object acquires a permanent status (object permanence). With the advent of representation or the semiotic function, the actions of the subject become organised into logical structures (seriation, inclusion, etc.) and operatory systems; at the same time, the child constructs invariants such as the conservation of number, matter, weight, length, etc. The operations evolve until they finally constitute formal or hypothetico-deductive structures. These behavioural structures are hierarchically organised and give rise to different levels of cognitive development, to wit, three major stages in cognitive development: the sensori-motor stage which goes from birth to the advent of representation; the concrete operational stage; and the formal operational stage, which is attained during adolescence. Piaget insists on the sequential nature of these stages, on the presence of an overall structure which determines all new behaviour at each stage (not only the dominating properties), and on the fact that all structures of a lower level are integrated into more powerful structures.

As far as learning is concerned, the most important require-ment of the stage conception is temporal succession. However, the structures which define these stages form systems which are broken down into the course of development into substructures or partial structures; these are then integrated into broader systems. In addition, the formation of structures of a similar logical level does not necessarily happen synchronously in different epistemo-logical domains (e.g., logico-mathematical and physical-causal) or for different psychological contents. Piaget himself discusses the problem of horizontal "decalages," i.e., time gaps which occur within the overall structural system--in particular at the concrete operational level--(cf. Piaget, 1941, 1966 and currently at the CIEG)[3]. In a similar manner, certain overlaps may occur between the major structures--i.e., sensori-motor, concrete and formal--at the upper limits of one structure and the lower limits of the next. This is neither contradictory to the structuralist conception nor to that of stages, as long as stages are neither reversed nor skipped.

We shall now attack the problem of learning and the research that has been done in this field. The theory of Piaget bears obvious implications for such research. In fact, from an epistemo-logical point of view, it is conceivable that the rhythm of develop-ment may be considerably modified by learning experiments based on the interactionist principle, on the one hand--by increasing the role of external intervention, i.e., by manipulating reality in front of the assimilating subject until he accommodates his schemes --and by playing on the constructivist aspect on the other hand-- i.e., by encouraging the subject to assimilate more, by trying to spark off the integration and coordination of the action schemes with each other and with reality. If the theory of equilibration is taken as fundamental, it should be possible to stimulate progress by creating a disequilibrium in the subject's structural system, thereby producing new restructuring. This would result from the resolution of the cognitive conflict aroused by the experimenter. Finally, from a methodological point of view, the learning exercises would be mainly based on the clinical method (better expressed in terms of critical exploration). This consists of dialogues between the experimenter and the child, where the arguments of the child are confronted with those of the experimenter in order to obtain a certain coherence in the subject's position. This method should also stimulate the "reflexive abstraction" capacities of the child.

In the case of considerable progress being obtained (both in time, i.e., acceleration, and in extent, i.e., generalisation), we might be accused of providing support for the empiricist approach. If, on the other hand, no progress is observed, the apriorists' conception might seem to be more appropriate. However, we feel

that Piaget's position has a reply to both outcomes for, as we
have seen, the cognitive development of the epistemological subject
evolves within the limits of an important structural process, profoundly anchored in interactionism and constructivism. The
function and the capacity of cognitive development is to produce
more powerful logical structures than the present ones, and to
increase their number thanks to the progressive equilibration
process. Finally, as in embryology (see Waddington's notion of
"competence"), Piaget believes in the existence of optimal "time
zones" for assimilation (or "reflexive abstraction"), beyond which
no acceleration can be obtained.

It would seem therefore that Piaget's epistemology, theory
and methodology protect him from the criticisms of other learning
researchers who would like to disprove his conception of cognitive
development or show its weaknesses.

In any case, rather than extremes (acceleration of several
years or absence of all progress), it is more probable that we
shall observe medium improvements which would constitute real
structural elaborations; and rather than invalidating Piaget's
theory, these would add to its flexibility. In addition, the learning
experiments could lead to a better understanding of the Piagetian
model and the processes of cognitive development in general.

In the last ten to twelve years, a great number of learning
experiments related to Piaget's theory have been carried out, especially in anglo-saxon countries.[4] Projects in the first two categories were aimed at accelerating cognitive development (I).
Some (Type I) were carried out within a Piagetian framework,
their aim being to improve construction of knowledge. Several
projects (Type II) were principally aimed at proving the fallacy of
Piaget's model--especially the structuralist part of his theory (and
the notion of stages which underlies it) by accelerating the rhythm
of development; any means were considered good as long as they
were efficient in the short term. Other researchers (Type III)
wished mainly to verify or obtain a better understanding of the
theory, e.g., the dynamic aspects of development, the problem of
decalages, the interconnections between different types of structures
(logico-mathematical, physical, etc.). Finally, the purpose of
some research projects was mainly educational (Type IV), but
these will not be referred to here.

It is the controversy resulting from the first three types of
experiment which will be the subject of our discussion.

NB: We shall limit ourselves to the research done on the
concrete operational period, since this is the period which has
given rise to most discussions. However, recent experiments
have also produced interesting results on the passage to the
formal stage.

Strauss and Brainerd, in a series of three articles, discuss the significance of a collection of data for Piaget's theory (covering the period 1965-1973 approximately).

Strauss's line of research seems to correspond to types I and III outlined above. In 1972, he published a review of a series of learning experiments which had been designed to attain cognitive transformations of a structural nature. Strauss describes various types of learning situations using different methods. He compares the results obtained and gives various possible interpretations of the differences found. This also enabled him to investigate the sequentiality of stages. As regards the numerous authors cited in reference to each type of experiment, we refer you to the articles in question.

Strauss distinguishes two principle categories of training. The basic principle involved in the first category is the creation of a disequilibrium in the structuration process in order to bring about a reequilibration at a higher level. This disequilibrium can be induced by an external source ("adaptational disequilibrium"); the disequilibrium is then between information from the environment and a cognitive structure. "Organisational disequilibrium," on the other hand, is internal and involves a cognitive conflict between cognitive structures. The principle guiding the second category of training, which centers around mental operations, is the induction of operations at a higher level. This was done either by presenting situations involving addition and subtraction or by concentrating training on the notion of reversibility. Other training situations were designed to spark off operational coordinations and integrations.

As regards the first training category, Strauss cites in particular the learning models of Bruner and of Geneva. He compares the two theories: in Bruner's, the only form of conflict is between products of the iconic and symbolic modes of representation; it therefore remains external to the structure. The Piagetian conflict, on the other hand, is between the structure and retroactive feedback.[5]

In a final discussion, Strauss summarizes the results obtained with these different types of training: various degrees of progress are accessible involving either a genuine structural transformation or mainly what Strauss calls structural elaboration, in the sense that a same structure is applied to new notions. As a whole, the difficulty seems to lie in the methodology used in these studies, which is sometimes questionable; further research is needed.

Strauss specifies that to him the analysis of the results seems to be consistant with Piaget's organismic developmental stage hypothesis.

Brainerd (1973) strongly contests Strauss's conclusions. His main criticism is that Strauss, in his review, selected only those experiments which confirmed his point of view, whereas Brainerd, when reviewing other experiments, found many data to the contrary. He mentions the existence of a great number of analyses of relatively recent literature (see his article) and expresses his surprise at the fact that Strauss finds anything new to say on the subject. At the same time, however, he admits that: "It is the first time in recent memory that anyone has suggested that there is some well-established branch of the developmental literature that provides consistent support for the stage hypothesis." (p. 349). Brainerd concludes his own review by enumerating seven points by which he refutes Strauss's 1972 work.

Strauss, in reply (1974), shows after a detailed re-analysis of Brainerd's criticism, that this author was incorrect in his counter arguments: "Thus, my original assessment of the training literature seems to me to be substantiated" (p. 181). Such a statement means, in fact, that, to date, most of the Piagetian learning research does not invalidate the theory as far as the succession of stages and structures is concerned.

Brainerd (Brainerd and Siegel, 1978) re-enters the discussion in a chapter entitled "Learning Research and Piagetian Theory." His main purpose is to attack Piaget's approach to learning which is based on "spontaneous development" as opposed to "laboratory learning." He criticises the Geneva conception of "self-discovery methods" because they are much less successful than methods based on other techniques such as "tutorial training," which produce cognitive improvement of a more substantial nature. Brainerd concludes that Piaget's concept of training is vapid and gives us to understand that this is true of the whole of Piaget's theory of cognitive development. We shall try to reply briefly to Brainerd's criticisms.

It is not at all clear whether or not Brainerd, when he speaks of the Geneva conception of learning, refers to the research carried out at the CIEG[6] in 1958 (Piaget and collaborators, 1959). The epistemological question which led to this research was whether it is possible to construct logical structures on the basis of empirical learning laws (more or less limited to the reading of experience). (NB. This is a very partial conception of the variety of methods used by the empiricists.)

A series of experiments were carried out by Greco, Morf, Wohwill, Smedslund and others. Piaget concluded from their experiments that the formation of logical structures could not be accounted for by learning laws alone, but that the equilibration principle, which involves the constructive activities of the child, is an indispensable complement. Brainerd makes no mention in his chapter of the equilibration principle, which for Piaget is a guideline in his conception of cognitive development, and therefore of his approach to learning. Brainerd once again reduces Piaget's view to one of "laws of spontaneous development," "everyday experience," "natural development" and concludes that Piaget is basically Rousseauian; this would seem to indicate an important lacuna in his understanding and knowledge of the work of Piaget.

On the other hand, Brainerd refers explicitly to the more recent learning experiments carried out in Geneva (Inhelder et al., 1974) when he speaks of the "self-discovery method." He completely misunderstands the meaning of this method, which only has a sense if used in the constructivist context of Piaget's theory. Brainerd's only concern is the efficiency of the various tutorial methods which he recommends in his chapter: whether a procedure works and how well it works. He seems not in the least interested in explaining why this is so. Such an attitude seems to us to constitute a crucial scientific divergence between Brainerd's approach and that of other researchers.

Having read this chapter, it does not seem surprising to us that Brainerd and Strauss should have had such a heated debate about their respective reviews of the learning literature. Their interpretation of the same facts could not possibly coincide owing to the fact that their outlook on the Piagetian view of cognitive development is fundamentally different from an epistemological, a theoretical and a methodological point of view.

Let us now consider the experiments done by Bryant (1974) on number. Bryant has developed learning experiments based on the number conservation experiment of Piaget (Piaget et al., 1941); he is very critical of Piaget's experiment and carries out numerous controls. Bryant's research seems to fall into the category of Type II described at the beginning of this paper, since he tries to bring children of the age of three years to the conservation of number and he theoretically contests Piaget's conception of conservation problems in general.

Bryant gives three possible explanations for the non-conservation responses obtained in the classical experimental situation described by Piaget: one of two collections A and B (constructed in one-to-one correspondence) is modified (B becomes B') and the child is asked to say whether or not there are still as many tiddlywinks in A and B'.

1.   Failure may be due to a lack of memory. Bryant criticizes Piaget's experimental situation because there is no way in which one can verify whether failure is due to the "forgetting"

of the initial correspondence between A and B, when B is
changed into B'. If this were the case, the child would be able
to make a correct inference between B and B' and thus reply
correctly.

2.   Failure could be due to the fact that the child is incapable
      of making transitive inferences A - B'. But Bryant has shown
      in another series of experiments (cf. his chap. 3) that very
      young children know how to make transitive inferences, so it
      is not this problem, in his opinion, which is the stumbling block
      in conservation.

3.   The most important problem, according to Bryant, is that
      Piaget's conservation situation involves a conflict between two
      incompatible judgments. The child is in the presence of two
      cues: the one to one correspondence which leads to judgments
      of equality, and the difference in length which, after trans-
      formation, leads to a response based on the inequality of the
      two rows. The child does not know which of these two cues
      to choose. It is this conflict which in Bryant's opinion prevents
      the child from replying correctly, and this independently of his
      understanding or not the principle of invariance.

NB.   What is the meaning of "understanding" if the child can still
      hesitate? This is not at all clear (see our discussion later on).

Bryant therefore makes the hypothesis that if the conflict is
removed, the child will reply in terms of invariance: "The
solution consists of showing the child that one of the two
judgments which he makes in the conservation situation is soundly
based and the other is not."

We shall now describe an experiment done by Bryant, where
the conflict is eliminated. This experiment was carried out on
children of 3 to 6 years of age, and there were three types of
situations; in each one, one row contained 20 dots, the other 19
(see Figure 1).

The children were asked to make a judgment concerning the
equality/inequality of the number of dots in the three situations.
As could be expected, responses given to A were always correct;
those to B were below chance level; those to C were at chance
level, and this from three years on.

Bryant then carried out two conservation experiments: a)
he modifies A into a configuration of type B; b) he modifies A
into a configuration of type C.

As the children reply correctly to A and incorrectly to B,
the transformation A into B arouses a conflict, and results in

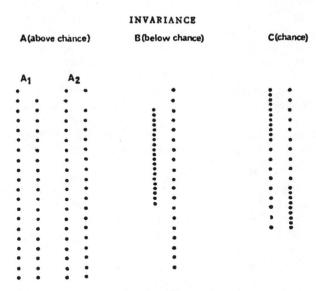

Figure 1.

responses below chance level. On the contrary, since in C the children reply at random, the transformation A into C arouses no conflict; the children therefore base their replies on A and give the correct reply. According to Bryant, this experiment shows that if one eliminates the conflict between the judgments previous to the transformation and after the transformation, the invariance principle is acquired at three years of age.

We shall not go into details about other experiments carried out by Bryant, where he trains the child to understand that the length to cue is incorrect. The aim is the same as in the previous experiment: invalidate the cue which gave rise to the wrong response by reinforcing the other. He concludes "once again we find that training a child that length is an incorrect cue will improve his performance in a conservation type task" (p. 145).

We greatly doubt whether such an improvement in performance has anything to do with an improvement in the child's understanding of the problem. In Piaget's view, the development of intelligence does not correspond to an accumulation of improvements in performance. I think that at this level, the root of the disagreement is of a theoretical nature.

Here, we would like to suggest that the significance of the transformation in the conservation experiments has been completely misunderstood by Bryant and that, for the very reason that he has misunderstood the theory that underlies the experiment. In fact, a conservation experiment, whichever it may be, only has a meaning if it is inserted in a system of operations forming a logical structure. In addition, the development of thought is considered to be an active construction; the child, by coordination and interiorization of his actions, gradually overcomes the difficulties involved in a problem such as invariance. It is not the experimenter's role to simplify the problem, thereby making it trivial; nor is it his role to reduce a logical problem to one of perceptual evaluation.

Now, this is just what Bryant does when he removes all possibility of confusion by eliminating the conflict between the initial and final states of the transformation: the child may give the "right" reply, but nothing proves that he has solved or even tackled the genuine problem of the transformation underlying the concept of invariance.

According to Piaget, when two quantities are equalized (or made different), then one of them (or both) modified so that the shape or spatial arrangement changes (but with no addition or subtraction), the child comes to understand that the transformation is logically annulable by an inverse transformation which restores the initial situation, and that the modifications of the states are completely compensated.

In the case of number, invariance implies a mental return to the one-to-one correspondence and an understanding of the compensation between the length of one of the rows and the density of the elements in the other. NB. It is clear that the cognitive construction which enables the child to find a logical solution to this problem involves gradual developmental process. Piaget (1968) has shown the existence of an intermediate state where the child's reasoning is semi-logical; during this period, the child is only capable of mental empirical return ("renversabilité") and not reversibility. He is therefore incapable of conservation reasoning.

It seems to us that any intervention which tries to bypass an active construction reduces epistemologically the very significance of cognitive development as conceived and analysed by Piaget.

Many errors of this kind occur in various conceptions of intellectual learning. Either the "disturbing" cues are eliminated (Braine, 1959; Bruner, 1964, 1966; Bever and Mehler, 1967), etc., or the child is "taught" or is "shown" (Smedslund, 1958). In

short, an attempt is made to simplify the experimental situation
by eliminating the difficulties. But this desire for the problem to
be solved at an earlier age leads to the problem being oversimplified
to the point of losing all meaning. One of the explicit principles
of Piaget's theory suggests, on the contrary, that it is by introducing "disturbing" elements into the child's present level of
understanding that he may be prompted to seek and thereby to
construct his knowledge. As says Piaget: "Intelligence is structured by functioning."

Bryant's experiments seem to show that the greatest disparity
between Piaget's conception and his own lies in the fact that, for
Bryant, intelligence is drawn from perception whereas for Piaget
this is not the case. For Piaget (1961), perceptual activities are
partly characterized by the global structures of the Gestalt theory
which are non-conserving, non-additive and non-developmental.
Intellectual structures, on the other hand, are reversible, additive
and result from a constructive process (see Piaget, 1947).

We have already raised the question as to what, for Bryant,
is the significance of the invariance principle discovered by the
child thanks to the conflict created in Piaget's experimental situations (point 3 of Bryant).

In fact, cognitive development with regard to this problem is
characterized by a primitive reaction where the child is convinced
that the number has changed: either because of the length, or
because of the density. It is only later that the child begins to
hesitate--not between "is there more here or there?" but between
change and invariance. A partial understanding of reversibility
enables him to reason in terms of conservation; this judgment is
based on the initial one-to-one correspondence and the possibility
of mentally reestablishing it. However, the perceptual configuration
makes the child think as well in terms of a change. Finally,
perceptible states and transformation are coordinated in a logical
understanding: the child   grasps simultaneously the reversible
principle of the operation which is at the base of the rearrangement
and the fact that the perceptual inequalities cancel each other
out. At this point, invariance seems a logical necessity and there
is no longer any hesitation in the child's judgment. It seems to
us that the result of the transformation of A into C in Bryant's
experiment, which leads the child to reply correctly, proves that
the child has not understood the invariance. The only thing
Bryant has managed to do is to increase the number of correct
responses above "chance level," but the responses themselves
have nothing to do with the conservation problem as such.

We shall now speak of an experiment which falls into category

III described above. This experiment tries to clarify the reasons
for the efficiency of learning situations based on Piaget's theory.
The learning of logical notions was studied using a method of
conflict (Lefebvre and Pinard, 1972); a second part of the study
tries to specify the conditions in which such a method can be
most effective.

In their first approach, Lefebvre and Pinard invent conflictual
learning exercises for the conservation of liquids. They begin
with a preparatory phase where, by a judicious choice of glass
recipients of different diameters into which they pour measured
quantities of liquid (sometimes one unit, sometimes two), they try
to destroy the more or less general conviction of the preoperatory
child that the higher the level, the greater the quantity of liquid.
They then carry out two types of conflictual exercises:

a)   a first series deals with compensation. Two identical glasses
      A and B contain different quantities of liquid. In one problem,
      the liquid in B is then emptied into a wider glass (W); the
      level goes down in W to match that of A. In a second problem,
      the contents of A are poured into a narrower glass (N). The
      level in N rises to match that of B. In both problems, the
      child must, to be correct, recognize the inequality of amount.

b)   another series of exercises centers the conflict around
      addition and subtraction. On the one hand (addition exercise)
      equal quantities of liquid are poured into identical glasses A
      and B. A is then poured into a larger diameter glass (L),
      already partially filled with liquid so that the levels are the
      same in B and L. On the other hand (subtraction exercise),
      only part of A is poured into a narrower glass (N), so that
      the levels are the same in N and B. Again, in both problems,
      the child must recognize unequal quantities despite equal levels.

In short, all of these transfers surprise the child: different
quantities suddenly reach the same level but without any liquid
having been added or taken away; or they reach levels which are
the inverse of the quantities; or, despite addition or removal of
some liquid, initially equal quantities maintain the same level.

The results show a significant progress in several items of
the test; the operatory nature of these acquisitions is discussed
by the authors. In addition, during learning, reactions show the
effectiveness of the conflictual dynamics; thanks to them, the
transformation is gradually mastered and the different dimensions
involved compensated.

In a second approach, the authors try to specify the conditions
which are a necessary preliminary to any conflict being felt by
the subject and to its ultimate resolution. The aim therefore is to

determine, in a concrete manner, the initial level which is necessary
if the child is to benefit at a maximum from the conflictual situations, and to improve the method for evaluating progress.   They
thus tackle one of the important questions which bothered Strauss
and Brainerd in their discussion of the literature.   Inhelder et
al. (1974) also mention this problem, but did not deal with it in
detail. Lefebvre and Pinard also refer to Piaget's research on the
preoperatory period which we mentioned in connection with Bryant's
experiments (Piaget, 1968).

In their research, the authors mention three conditions for
any serious prognosis:

a)   The child must possess the functional preoperatory schemes
      which enable him to create functional links of dependence
      between events

b)   he must be fairly consistent in his reactions, i.e., he must
      not contradict himself from one moment to the next (successive
      consistency) and should be able to confer a single meaning
      to a concept (simultaneous consistency)

c)   the child must be able to accept the facts - i.e., acknowledge
      the facts without distorting them

The main hypothesis is that the conflictual exercises involved
in their learning procedure (Lefevre and Pinard, 1972) will spark
off progress.   The extent of this progress will be relative to the
level of the subjects in the preliminary test items (3 categories of
evaluation).   The results confirm this hypothesis in the sense
that performance at the preliminary tests provides a good basis
for prediction of the effectiveness of the exercises (at post-test).
These two experiments provide significant information for the
hypothesis of equilibration, i.e., that cognitive development
consists of processes of internal compensation for perturbations.
These perturbations can either be felt spontaneously by the subject,
or can be sparked off by an outside element.   The compensations
for the perturbations do not simply restore a former state of equilibrium;   they lead the subject to a higher state of equilibrium.

In this sense therefore it would seem that the creation of
learning exercises which favor cognitive conflicts can help us to
gain a better understanding of the development of knowledge.   In
all learning, however, one has to take into account the
integrative capacities of the child--not in the sense of his
producing an accommodation to what the experimenter proposes
him, but by providing him with the possibility of active
constructions.

Lefebvre and Pinard's point of view is very similar to that of

Strauss's in that he also deals with disequilibrium methods, methods which are linked to regulatory and equilibration processes in the theory of Piaget.

At the close of this discussion, we shall leave Piaget's exclusively[8] intellectual theory and examine the conception of Kuhn (1978)[8], who would like to see a synthesis between two approaches which have been rather artificially dissociated in developmental psychology. In this paper, the author asks whether the mechanisms of social development and those of cognitive development cannot be dealt with in a common manner, rather than applying a systematic mechanistic model to social development and an organismic paradigm to cognitive development. While social learning theory as yet possesses no theoretical structures which could account for cognitive processes, so cognitive developmental theory makes no mention of the interaction between the developing individual and the historical and cultural conditions in which this development takes place.

Kuhn defines the respective parameters of mechanistic and organismic paradigms, thus showing their fundamental opposition:

--overt behaviour versus processes internal to the organism

--discrete and autonomous behavioural elements in contrast to elements which are organized into structures

--behavioural elements which are under stimulus control versus an organism activating interactions with the environment in order to construct its own psychological structure and knowledge of the world.

It is possible to conceive of a model which would incorporate these two behavioural aspects and at the same time unify the theoreti concepts which belong to each one.

We would like to mention two research approaches which, without providing a solution to this problem, are likely to help to bring ideas closer together.

In Geneva, Doise (1978) has investigated cognitive development in psychosociological terms; his hypothesis was that social interactio contributes to intellectual progress and that this, in return, provide: the child with more elaborate instruments for social interaction, whicl in turn allow new cognitive constructions, etc. (This conception is 1 in contradiction with Piaget's (See Piaget 1931, 1951, 1967) but accen tuates the role of social intervention in socio-cognitive interaction.) Doise and his team have carried out a series of learning experiments based on different Piagetian notions. By varying the type of social interaction, he studies the degree of progress made by the child.

Either several children of different cognitive levels are put together and asked to solve a problem, or one or two adults intervene. This intervention can take two forms: either the adult proposes a "progressive" model or he proposes an alternative but of the same level as the solution given by the child. Thus, Doise has created situations of cognitive conflict where the progress observed is explained by the necessity to coordinate different points of view. He thus obtained undeniable progress in several cognitive domains. In addition, cognitive "decalages" between groups belonging to different social classes were reduced by socio-cognitive interactions of this type.

Such an approach constitutes a first attempt at bridging the gap that Kuhn mentions between social development conceived as a mechanistic model and the development of knowledge according to Piaget's model. But several problems still remain. Doise's approach involves a structural conception of psychosocial interactions. As Kuhn mentions in the conclusion to her article, one aspect in particular is still neglected and this is the dynamics involved both in social interaction and cognitive development, i.e., an "energetic" or affective parameter. Kuhn had hoped that the study of moral development might constitute a meeting point between the mechanistic and organismic conceptions, but up to now research in this field has not produced the results that were hoped for.

As a last point, we would like to mention the Piagetian cross-cultural comparative research on cognitive development - not to be confused with cross-cultural studies based on IQ, intelligence tests, etc. This research has now been carried out over a number of years. Piaget believes that this type of study is useful and even necessary (Piaget, 1966) if one wants to verify the universality of the main mechanisms of cognitive development. This research has mostly centred on the concrete operational period (see a review of the literature by Dasen, 1972 (Bovet, 1968, 1975) and more recently on the sensori-motor period (Dasen, Inhelder et al., 1978).

This kind of research seems to partly fill another of the lacunae mentioned by Kuhn, i.e., the fact that developmental psychology does not consider the cultural conditions in which this development takes place. Thus, the inter-individual cognitive research of Doise and the Piagetian cross-cultural studies both attempt to go beyond the strictly organismic paradigm of Piaget's theory of cognitive development.

References

Bovet, M.C. "Etudes interculturelles du developpement intellectuel et processus d'apprentissage." Revue Suisse de Psychologie pure et appliquee, 1968, 27, no. 3/4, p. 189-200.

Bovet, M. C. "Etude interculturelle de processus de raissonnement: notions de quantite et relations spatio-temporelles." These presentee a la FAPSE de l'Universite de Geneve, Section de psychologie, 1971, Editions medecine et Hygiene, Geneve, 1975.

Braine, M. S. "The ontogeny of certain logical operations: Piaget's formulation examined by non-verbal methods." Psychological Monograph 1959, 73 no. 5, p. 1-43.

Brainerd, C. "Neo-Piagetian training experiments revisited: Is there any support for the cognitive-developmental stage hypothesis?" Cognition 2(3) p. 349-370, 1973.

Brainerd, C. "Learning Research and Piagetian Theory" in L. Siegel and C. Brainerd (Eds.), Alternatives to Piaget, New York: Academic Press, 1978.

Bruner, J. "The Course of Cognitive Growth." American Psychologis vol. 19, no. 1, 1-15, 1964.

Bruner, J. "On cognitive growth" II - 30-67, in J. Bruner, R. Olver, and P. Greenfield (Eds.), Studies in cognitive growth, New York, 1966.

Bryant, P. E. and Trabasso, T. "Transitive inferences and memory in young children." Nature, 232, 456-458, 1971.

Bryant, P. E. and Trabasso, T. "The understanding of invariance by very young children. Canadian Journal of Psychology, 26, 78-95, 1972.

Bryant, P. E. "Perception and understanding in young children: An Experimental Approach," Methuen and Co., Ltd., 1974.

Dasen, P., Inhelder, B., Lavallee, M., Retschitzki, J. "Naissance de l'intelligence chez l'enfant baoule de Cote d'Ivoire." Hans Huber, Berne, Stuttgart, Vienne, 1978.

Dasen, P. "Cross-cultural Piagetian research: A summary." Journal of Cross-cultural Psychology, vol. 3, no. 1, 23-39, 1972.

Doise, W., Deschamps, J. C., and Mugny, G. "Psychologie sociale experimentale." Paris: Armand Collin, 1978.

Gruber, H., Voneche, J. "The essential Piaget: An interpretative Reference and Guide." London: Routledge & Kegan Paul, 1977.

Inhelder, B., Sinclair, H., Bovet, M. "Learning and the Developmer of Cognition. Cambridge, MA: Harvard University Press, 1974.

Kuhn, D. "Inducing development experimentally: Comments on a research paradigm," Developmental Psychology, 10, p. 590-600, 1974.

Lefebvre, M. "Apprentissage de la conservation des quantites par une methode de conflit cognitif." Canadian Journal of Behavioural Science, 1972, 4, 1-12.

Lefebvre, M., and Pinard, A. "Influence du niveau initial de sensibilite au conflit sur l'apprentissage de la conservation des quantites par une methode de conflict cognitif." Canadian

Journal of Behavioural Science/Revue Canadian Science Comp., 6 (4), 398-413, 1974.

Mehler, J., and Bever, T. "Cognitive Capacity of Very Young Children." Science, Vol. 158, no. 3797, p. 141-142, 1967.

Piaget, J. "The moral judgment of the child," New York: Harcourt, 1932.

Piaget, J. "Le mecanism du developpement mental et les lois du groupement des operations." Archives de psychologie, Geneve, 28, 1941.

Piaget, J. "Psychology of Intelligence." London: Routledge and Kegan Paul, 1947.

Piaget, J. "La pensee egocentrique et la pensee sociocentrique." Cahiers internationaux de Sociologie, Vol. X, Paris, 1951.

Piaget, J., and Szeminska, A. "The Child Conception of Number." London: Routledge and Kegan Paul, 1952.

Piaget, J., and Coll. Apprentissage et Connaissance. EEG, Vol. VII, VIII, IX and X. Paris: PUF, 1959.

Piaget, J. "Preface a Laurendeau, M., et Pinard, A. "Les premieres Notion spatiale de l'enfant: examen des hypotheses de Jean Piaget." Delachaux & Niestle, 1966.

Piaget, J., Grize, J. B., Szeminska, A., and Vinh-Bang. "Epistemologie et psychologie de la fonction." EEG Vol. XXIIII, Paris: PUF, 1968.

Piaget, J. "Etudes Sociologiques." Geneve: Droz, 1967.

Piaget, J. "Quantification, conservation and nativism": Quantitative evaluations of children aged two to three years. Science, 1968, 162, p. 976-981.

Piaget, J. "The mechanisms of perception.". London: Routledge and Kegan Paul, 1969.

Piaget, J. "Biology and Knowledge." Chicago: University of Chicago Press, 1971.

Piaget, J. "L'equilibration des structures cognitives--probleme central du developpement," Paris: PUF, - EEG, XXXIII, 1975.

Piaget, J., et coll. "Recherches sur l'Abstraction Reflechissante." EEG, Vol. XXXIV et XXXV, Paris: PUF, 1977.

Smedslund, J. "Apprentissage des notions de la Conservation et de la Transivite du poids," p. 85-124 in L'apprentissage des structures logiques." Vinh-Bang et al., EEG, IX, Paris: PUF, 1977.

Strauss, S. "Inducing cognitive development and learning: A review of short-term training experiments. I. The organismic developmental approach..." Cognition I (4), 329-357, 1972.

Waddington. "The Nature of Life." London: Allen & Unwin, 1961.

Footnotes

1.   Piaget bases himself on mechanisms alluded to by Waddington

(1961) who calls these "creodes" (organic pathways); they are regulated by homeorhesis (dynamic regulations which, in the case of deviations, bring the organism back into the correct pathways) and homeostasis (regulations which maintain the organism around a certain point of equilibrium).

2. Apriorists believe that knowledge and cognition are innately given as an a priori; empiricists believe knowledge and cognition are learned through empirical experience.

3. International Center for Genetic Epistemology

4. These are generally called training experiments in those countries.

5. NB. We would remark in passing that the Geneva approach is not restricted to this type of conflict alone (see Inhelder et al 1974).

6. International Center for Genetic Epistemology.

7. Our translation.

8. This author has also dealt with the question of learning, first of all experimentally (Kuhn, 1972), then from a more theoretical point of view (Kuhn, 1974); she makes a critical analysis of the literature and presents an original research model. In addition, she has written a very pertinent critical review of Inhelder et al.'s book on learning (Kuhn, 1975).

TRAINING AND LOGIC: COMMENT ON MAGALI BOVET'S PAPER

P. E. Bryant

Oxford University

Oxford, England

Magali Bovet's main point is that people take to training experiments for a variety of reasons and with a number of different aims. She argues that we ought to classify these experiments into different types and she surely is right.

But I wonder about her classification which seems to me to categorise experiments into those done within the Piagetian framework and the rest--those, in other words, for the Geneva view and those against. I worry about it for a number of reasons. It implies a kind of polarisation which certainly adds a little excitement to the subject but which is terribly misleading. The people who run the projects which Magali Bovet describes as "principally aimed at proving the fallacy of Piaget's model" are not on the whole so negative. They too have hypotheses to test even though these are different from Piaget's, and they invariably acknowledge the importance of his discoveries even when they offer a different interpretation.

Nor, I think, is it true that such people misunderstand Piaget's theory. Brainerd's description (1978a) of the assumption behind the Geneva research which Magali Bovet so dislikes seems to me to be accurate and fair, even though the term "self-discovery," which he uses, is rather misleading. The suggestion that he does not really appreciate the disequilibrium-equilibrium model is surely wrong. It is a model described clearly and absolutely correctly in his recent book on Piaget (Brainerd, 1978b).

So I want to suggest another way of classifying training experiments which avoids, I hope, this sort of pitched battle and which would allow people who agree and who disagree with Piaget's

theory to use his work constructively. My classification stems
from a simple assumption which I have about the nature of developmental psychology.

There are two basic questions in developmental psychology.
They are often confused with each other but really quite separate.
The first concerns what children are like at particular ages. How
does a child of three, say, behave and what underlies his behaviour? What will be the difference between the things he does and
the things he understands now and in a year's time? Two years
time? The first question then asks what children are like and in
what ways they change as they grow older. The second question
is the causal one. Given that children do change as they grow
up what exactly makes these changes happen? Learning, language,
disequilibrium, maturation, or something else?

Most developmental psychologists have theories about both
questions. Piaget not the least. His ideas about the first question
are well known. The young child is said to be surprisingly
alogical at first with virtually no understanding of even the
simplest aspects of space or time or quantity. Cognitive development is the gradual acquisition of the correct ideas about these
things, at first in terms of practical routines of behaviour, then
in the form of internal representations which the child is able to
organise and re-organise for himself. The story is complex and
detailed.

Not so Piaget's ideas about the second question, the causal
question. They centre, as Magali Bovet has shown, around the
idea of equilibrium and disequilibrium, and apply equally to the
four month old baby learning what his hand does and to the
fifteen year old trying to design a scientific experiment.

Now the training experiments. Just as there are two developmental questions, so there are two kinds of training experiment.
One is designed to test hypotheses about what children at a
particular age or stage are like and can do--the first question.
If the child is like this he should be able to learn under condition
A but not under condition B. Perhaps the best example of this
genre is Gelman's (1969) highly successful conservation training
experiment. She trained children to solve the conservation problem by using an oddity learning set procedure. Briefly, her
training task involved teaching children to spot which of two
quantities was the odd one out (i.e. more or less in quantity than
the other two which were equal) and to discard perceptual criteria
like length when making this judgment. She was trying to test
her ideas that children usually fail the conservation task because
they are attending to the wrong cues. I think that her results
strongly supported her hypothesis, but my point here is that this
training experiment is not a test of a causal hypothesis. It is
trying to show the reason for a typical childish error and no

more. In principle I can see no difficulties at all in this kind of use of training experiments.

But we do meet some pretty serious problems with the second main kind of training experiment. The purpose of this second type is to test the other question, the causal question. The rationale is obvious. You are interested in a developmental change. You think that it is factor X which brings this change about. So you run a training experiment to test this hypothesis. Let's say that the developmental change is from being wrong in a conservation task to being right. What is done is to take a group of non-conservers, to give some a concentrated dose of factor X and others (the control group) exactly the same set of experiences too but with the vital factor X missing. If the first lot begin to conserve and the controls do not you can argue that your hypothesis is supported. Factor X is the thing. But is this support overwhelming?

Unhappily the answer is usually "no" and it is to the distressing problems of training experiments as tests of causal hypotheses that I now wish to turn, for they touch closely on Magali Bovet's paper. There seem to be three main hazards. The first is in getting the controls right. To ensure that it is factor X, the control group should be given exactly the same experience, the same verbal and nonverbal encouragement and so on but with factor X removed. By this standard I am afraid that all the Geneva training research which I know of fails dismally. Sometimes there are no controls at all. When there are these are inadequate. The well known liquid conservation training experiment by Inhelder, Sinclair and Bovet (1974) is an example. The experimental group were shown two identical containers side by side with the same amount of liquid as each other. Then the liquid from both was released into two different shaped containers, so that the level was higher in one than in the other. Then again it was released, again into two identical containers. The purpose of this procedure was to promote conflict between the child's expectations of what would happen and his perception of what did happen. Conflict leads to disequilibirum, disequilibrium to developmental change. But, though no doubt this procedure did involve conflict, it also involved an awful lot of other things as well, such as seeing the transformations, possible attentional changes, or both. Did the controls sort this out? Well, the control group did nothing. They should have been given the same kind of experiences with the same equipment, but with the conflict removed, but this did not happen.

The same weaknesses are to be found in the Lefebre and Pinard studies (1972) extolled by Magali Bovet. Their experimental groups were given training (which involved conflict) in compensation or in addition and subtraction or in both: their control groups were taught about (of all things) how to make causal judgments. The experimental groups therefore were given a great deal of experience with the kind of material and judgment involved in conservation tasks. The control groups were not. What right have the experimenters to say that it was conflict which improved the performance of the experimental group? But at least this problem about controls, though often unsolved, is soluble.

This may not be so of the next problem, which is the inherent artificiality of training experiments. Suppose you do demonstrate in your experiment that it really was factor X which did the trick. This does not mean that it is factor X which normally causes the development in real life. Conservation is actually a very good case in point. People sometimes write as though there is one successful way to train conservation. Nonsense. There are many ways. Now it may be that despite their apparent heterogeneity all these methods have some as yet undetected factor in common, but I doubt it. It seems much more likely to me that at least some of the successful techniques may have absolutely nothing to do with the real causes.

My view is that the way round this problem is to combine the training experiment with other measures and particularly with correlations. The strengths and weaknesses of correlations and of training experiments are actually complementary. Together they make an impressive team. Suppose you find a correlation between two developmental factors, X and Y, let's say between the ability to produce some part of speech and the ability to do well in the conservation task. The advantage of this correlation is that, provided your research is well done you really have established a connexion between the two at any rate in time. But the weakness is that you cannot say whether A causes B, or B A. Suppose now that you then do a series of training experiments in which you train some children on A to see if it affects B, others on B and see if it affects A. Let's say that you find training on A improves B, but training on B does not affect A. You can tell from this experiment, provided the controls are right, that A did cause B in the experiment but of course the training experiment itself does not tell you if A is connected to B in real life. But the correlation tells you that. The correlation establishes the existence of the connexion, the training experiment gives you its causal direction.

Sinclair-deZwart (1967) has come nearer to this design than anyone else working in logical development. She established a correlation between the spontaneous production of comparative words ("larger," "smaller"), and conservation in one study, seriation in another. Then in both experiments she trained the children to produce the appropriate words spontaneously. Finally

she tested them on conservation or on seriation. The training
did not seem to improve conservation, but had a considerable
effect on seriation. She concluded from the conservation study
that she had shown that language acquisition does not affect
logical development--much. They had the right words, but still
they did not conserve. But she was reluctant to draw the opposite
conclusion from the opposite result in the seriation experiment
(which seems to be far less known in, to use Magali Bovet's
memorable and Gaullist term, the Anglo-Saxon world). The
opposite conclusion, that language does affect logic, would have
been less sympathetic to her colleague Piaget.

That Sinclair thought to combine correlations and training
experience is truly impressive. But there is something missing.
Take the better known negative result in the conservation experiment. Sinclair found a correlation between language and conservation, and then trained language to see if it would affect conservation. She should also have trained conservation to see if it
affected language, and her hypothesis surely would have predicted
a positive effect. As it is her reliance on a negative result to
support one causal hypothesis and to dismiss another is a clear
example of a permissive use of the null hypothesis.

The third problem of the training experiment as a test of
causal hypotheses is a psychological one. People find that younger
children perform one way, older children another, in a particular
experiment and then they go hell for leather to find out what
causes the change without bothering to think what the change
actually is. Conservation is a good example. In dozens of experiments the paradigm has taken over. "What turns a non-conserver
into a conserver?" is the question--with hardly a thought for
what being a non-conserver or even a conserver actually means.

Yet we should not forget that the conservation experiment
was set up to test something in children and it is now, to say the
least, very debatable whether it really does test what it intended
to test. The purpose of the conservation experiment is to test
children's understanding of the principle of invariance. A child
who fails is thought not to have grasped the principle.

Is this true? Do non-conservers really think that spreading
a row of beads increases its number, or pouring some water into
a narrower container thereby giving it a higher level, increases
its volume? I myself have always doubted whether five and six
year old children are that naive, and I used to doubt it for
reasons to do with the structure of the task--reasons which
Magali Bovet has summarised so clearly. But now I have had to
change my mind. I still think that quite young children understand
invariance, but I have had to re-think my ideas about the reasons
for their errors in the conservation task.

I have had to do this because of experiments in which the structure of the conservation task is kept intact, and yet children who make mistakes in the usual form of the task begin to get the whole thing right. Jim McGarrigle and Margaret Donaldson's (1976) experiment seems to me to be the best example. They gave four and five year old children number and length conservation tasks under two conditions. One was the traditional procedure with the experimenter asking the questions, transforming the quantities and so on. In the other condition one change was made. After the child had made his first judgment a teddy bear emerged, misbehaved and in the ensuing chaos as if by accident changed the appearance of the counters (number) or the pieces of string (length). After the miscreant was put away the child was asked the conservation question. This rather cloying routine had a thumping effect. Few of the children were right in the usual task; very many indeed were in the teddy bear version.

I think that there are two things to be said about this result. The first is that it suggests very strongly that many children who fail in the conservation task nevertheless do understand invariance. As such it is in line with many other recent, and now not so recent, experiments which seem to show that Piaget's procedures add up to a massive underestimate of the logical abilities of four, five, six and seven year old children. There are experiments which point the same way with class inclusion (Donaldson, 1978), transitive inferences (Bryant and Trabasso, 1971) and perspective taking (Borke, 1978). None of these studies disputes Piaget's results, but all seem to show that in other tasks which equally test the logical abilities which interest Piaget, young children--duffers in Piaget's tasks--now do very well. Why?

The second thing to note about McGarrigle and Donaldson's experiment is that it maintains the basic structure of the Piagetian task (something which is not on the whole true of the other experiments which I have just mentioned). The two quantities, the transformation, the two questions--one before one after the transformation--all the ingredients were still there. Only the character who pushes the things around was changed. Why did it make such a difference?

It is an awkward question, not only for Piagetians but also for people like myself (1974) who have argued that the children fail because of various faults in the experiment's design. But McGarrigle and Donaldson kept the design unchanged. That is why I do not wish to defend some of my own views which Magali Bovet questioned. They must be wrong.

But how do we analyse this experiment? One possibility is

to say that it means that the whole conservation experiment is a ghastly, trivial misunderstanding and that children are simply playing the wrong game with the adult, but the right one with the teddy bear. To take this view would be to write off the whole conservation enterprise--a staggering achievement. But I think that that would be defeatist. The conservation may still be more important than that.

Let us take another tack. Suppose we accept a distinction between (1) the possession of a logical mechanism--in other words the basic ability to make a logical move--and (2) knowing exactly when to make this logical move. It is not a bad distinction and must in a way be true. We all know that there are occasions when we could have made the right inference but did not.

How else is it possible for Hercule Poirot (a noted Anglo-Saxon) but not us to work out who did it? Piaget's theory is about the first of these two things, the possession of logical mechanisms. When children make mistakes in his tasks he argues that they lack the basic underlying logical structures (give or take a bit of horizontal decalage). But the other experiments which I have just mentioned argue against this and suggest the second alternative very strongly; children fail in one version of the task but not in another and their success in one task indicates the possession of the logical mechanism, while their failure in another suggests that they do not always deploy this mechanism appropriately.

Of course there are other ways of explaining their success in one version of a logical test and failure in another. Information processing is a popular one. But that would be difficult to apply to McGarrigle and Donaldson. So let us consider the possibility that children sometimes fail in logical tasks because they do not know that they must now make a logical move which they can in principle make. What is the evidence on this point?

Well, we (Bryant and Kopytynska, 1976) have some evidence of the Hercule Poirot syndrome in 5 and 6 year old children. We have shown that children, who do not use an intervening measure to compare the height of two brick towers, nevertheless do measure when they have to compare the depth of two holes in wooden boxes. They cannot see those holes, and it is perfectly clear to them that they cannot compare them directly. They know now that they do not know and that they need to make a direct move to fill the gap.

I should like to suggest that this kind of analysis could be applied to the conservation experiment. David Elkind (1968) pointed out some time ago that the conservation task demands an inference. If for example it is the liquid task, the liquid in the two containers A and B is first judged to be equal; then the liquid in one container (B) is tipped into another (B_1) and the child then has to compare A and B_1. Since a direct comparison between A and B would be most unreliable the correct thing to do is to work out that because A=B, and B and B_1 are the same (the invariance principle) A must equal B. This means that the child has to do at least three things. He must recognise that a direct comparison in the second display between A and B_1 is most unreliable, he must realise that B and B_1 are the same, and he must use this knowledge in a transitive inference: $A = B$, $B = B_1$, $A = B_1$.

Now if we apply this analysis to the McGarrigle and Donaldson experiment we have to conclude that the child manages to do all three things in the successful teddy bear condition. What then goes wrong when the adult carries out the transformation? I can offer one speculation. It is that the adult unwittingly makes the child think that an inference is unnecessary and that a direct comparison between A and B_1 is perfectly all right. Here he is--the grown-up--solemnly pouring the liquid from B to B_1, and making its level higher. Clever fellows, these grown-ups: so maybe the level is important enough to be used in a direct comparison after all. But a teddy bear--that's quite a different matter.

This is mere hypothesis, but I produce it as a witness to my belief that the conservation failure is not a trivial phenomenon. It may tell us a great deal about the way children decide whether or not to make a logical move, which in principle is well within their capacity. And surely the question of how children decide when to use their own logical capacities is at least as important, theoretically and educationally, as what capacities they have.

Among other things it forces us to look again at the training experiment. The argument between Magali Bovet and Brainerd is about the acquisition of the principle of invariance. Perhaps we should stop thinking about this for a while, and instead use training experiments to find out how children who at first use the principle only in some circumstances eventually apply it to other situations as well. It is not the usual question, but it could be the right one.

References

Borke, H. (1978) Piaget's view of social interaction and the theoretical construction of empathy. In (Eds) Siegel, L. and Brainerd, C. Alternatives to Piaget. New York: Academic Press.

Brainerd, C. (1978a) Learning research and Piagetian theory. In (Eds) Siegel, L. and Brainerd, C. Alternatives to Piaget. New York: Academic Press.

Brainerd, C. (1978b) Piaget's Theory of Intelligence. New York:
     Prentice Hall.
Bryant, P. E. (1974) Perception and Understanding in Young
     Children. London: Methuen.
Bryant, P. E. and Trabasso, T. (1971) Transitive inferences and
     memory in young children. Nature, 232, 456-458.
Bryant, P. E. and Kopytynska, H. (1976) Spontaneous measure-
     ment by young children. Nature, 260, 77.
Donaldson, M. (1978) Children's Minds. London: Open Books.
Elkind, D. (1967) Piaget's conservation problems. Child Develop-
     ment, 38, 15-27.
Gelman, R. (1969) Conservation acquisition: a problem of learning
     to attend to relevant attributes. Journal of Experimental
     Child Psychology, 7, 167-187.
Inhelder, B., Sinclair, H. and Bovet, M. (1974) Learning and the
     Development of Cognition. London: Routledge and Kegan Paul.
Lefebre, M. and Pinard, A. (1972) Apprentissage de la conservation
     des quantites par une methode de conflit cognitif. Canadian
     Journal of Behavioral Science, 4, 1-12.
McGarrigle, J. and Donaldson, M. (1978) Conservation accidents.
     Cognition.
Sinclair-de Zwart, H. (1967) Acquisition du language et Developpe-
     ment de la Pensee. Paris: Dunod.

In this process the existing constructions in the child's mind are challenged; uneasiness, tension, awareness results and the potential for change is then present. This kind of inquiry is a necessary but not sufficient means for modifying an individual's constructions. The argument is that cognitive growth proceeds through mental activity and the potential for activity can be activated through the inquiry process. If teachers proceed with an active dialectical-inquiry strategy as the preferred course students may not only seek understanding in terms of their own constructs, but also listen with more active and challenging minds to anyone. If they do, they should tend to be more critical evaluators of all information and less passive receivers of knowledge as truth.

Concomitant to the cognitive aspects of the inquiry are affective states; e.g. comfort, pride, interest, fear, etc. As Piaget (1967) says, the cognitive and the affective are both sides of the same coin. They are fused into organic unity. Questions can be asked in a benign way or in an imperious way, as if overtly demanding a response; it can be a putdown or a seemingly true, sincere request for information. Thus, while the cognitive consequences of the questions are to activate thought, the affective ones can have an impact that may be counter-productive or joyful. Distancing behaviors when presented in the form of an inquiry, are only effective if comprehended by the respondent. Thus, to anticipate positive outcomes from such interactions without considering the status of the respondent is to overlook the interdependence of inquiry. The language, the structure and the tempo, along with the message, are all necessary features for inquiry to be effective (Sigel, 1978).

The content of the inquiry orients the individual to cognitive and affective features in the interaction. From the cognitive perspective an inquiry focuses the individual on time/space dimensions, subject matter, and processes. Cognitively, the individual is being asked to evaluate a situation. Examples of such demands follow: inference, e.g. How will Mary feel if she is not invited to your party; causality, e.g. What makes a sailboat move?; justification, e.g., How can you explain the decrease in oil reserves in the United States. Sigel has identified about 40 types of inquiries involving cognitive processes, such as classification, relations, cause-effect, and the like.

In either physical or social problem-solving, the child and the teacher begin with incomplete knowledge; that is, the teacher does not know what the child knows and the child probably does not have all the information necessary to solve the problem, and if he/she does, he may not be aware of it. Inquiry helps to: 1) elicit what knowledge exists in the child, 2) get at bits of knowledge the child may not see as related or relevant, 3) provides a basis for the child growing what he/she does not know, and 4) tells the teacher what the child does not know or needs to know. The degree to which this interchange enhances the child's movement toward problem-solving and, in fact, thinking, will be dependent on subsequent steps the teacher and child take to complete the knowledge base. Inquiry, then contributes to the child's aware- ness of his/her knowledge, and the gaps in his/her knowledge. It is also an opportunity to objectify what he/she does and does not know. This movement toward objectification is a step in the direction of obtaining consensual knowledge about events. Obtaining knowledge is but one step in the entire process of coming to know something. The level of the young child's knowledge is limited to his/her capability to assimilate and concomitantly accom- modate to this new information. To assume that the child will "know" an event, that is, to understand the operations involved as well as the implications, would be presumptuous. The child's knowledge level is best described in terms of a spiral where each level of knowledge is constructed and integrated with subsequent integration proceeding as the child's competence to abstract and interrelate proceeds. This is analogous to Piaget's notions of equilibration.

In this kind of inquiring relationship the child and the teacher think together. They are engaged in becoming aware of the gaps in each others' knowledge. Filling these gaps or dis- crepancies is a step in the process of coming to know something. Coming to know something in this context is the first step in problem-solving. This orientation, when internalized by teachers or students can help them challenge and critically evaluate existing knowledge. If the teacher uses this approach, the probability is that the child will internalize the strategy and as a result not only use it but develop a listening capability that is tied to internal questioning of what is being offered as complete knowledge. It is calculated to create a constant uneasiness with knowledge as it now exists. Like an artist the thinker becomes a detached observer of society. Detached in the sense of not being bound by society's committed thinking on problems. In a manner similar to the young James Joyce in Portrait of the Artist as a Young Man, one must experience an extended and laborious apprenticeship of inquiry before one achieves any degree of certitude of understanding which Joyce tells us through Stephen Dedalus is the greatest gift one can offer his generation: "no one served the generation into which he had been born so well as he who offered it, whether in his art or in his life, the gift of certitude" (p. 264). Before attaining any degree of "certitude of understanding" he tells us he had "a sense of fear of the unknown ... a fear of symbols and portents." From these fears, and doubts, he emerges stronger and surer as to how to forge in the smithy of his consciousness his own concepts of reality as he sees, hears, and feels it, "his own consciousness...was ebbing from his brain and trickling into the very words themselves in wayward rhythms" (p. 68). He tells us he did not fully understand these words at first, but as he pursued (inquired into) their meaning "through them he had glimpses of the real world about him" (p. 108).

This kind of development requires a personal environment that is characterized by genuine inquiry, warmth, and understanding. The genuineness of the inquiry enterprise is influenced by the motivational and affective features of the environment. In addition to a warm, understanding atmosphere, it is critical that teachers or parents continue the dialogue with children, posing alternatives and discrepancies which make continual demands on the child to think further. Like James Joyce the child gets "glimpses of reality" which is another way of saying he begins to become conscious of knowing that he knows. His consciousness, as Teilhard tells us, creates a new world putting him/her to some degree in control of his/her development. With this kind of control there is growth, as exemplified in Joyce, from fear of the unknown to his proclamation as a college graduate that

> I do not fear to be alone or to be spurned for another or to leave whatever I have to leave. And I am not afraid to make a mistake, a life-long mistake and perhaps as long as eternity, too. (p. 55).

There is here the consciousness that he is now the one in control and he must come to grips with how and in what way he will represent reality. In similar manner we believe the distancing and inquiry strategies, whether used at home or in classroom situations can contribute to the child's growing awareness that he/she can be in control. To date the Sigel data seems to support the theory.

References

Erickson, E.H. Identity youth and crisis. New York: W.W. Norton and Company, 1968.

Joyce, J. A portrait of the artist as a young man. New York: Vision Press, 1916.

Piaget, J. "Piaget's theory." In P. Mussen (Ed.), Carmichael's handbook of child psychology. New York: Wiley, 1970.

Russell, Bertrand. An outline of philosophy. New York: Meridian Books, 1960.

Sigel, I. and Saunders, R. An inquiry into inquiry: question asking as an instructional model. In L.G. Katz (ed.) Current topics in early childhood education (vol. 2). Norwood, NJ: Ablex Publishing Corp., 1979.

Sigel, I.E. The distancing hypothesis: A causal hypothesis for the acquisition of representational thought. In M. R. Jones (Ed.), The effects of early experience. Miami, FL: University of Miami Press, 1970.

Sigel, I.E. Social experience in the development of representational thought: distancing theory. Presidential address presented at the meeting of the Jean Piaget Society, Philadelphia, PA: May, 1978.

Teilhard deChardin, Pierre. The future of man. New York: Harper & Row, 1960.

Werner, H. Comparative psychology of mental development. New York: Science editions, 1948.

KNOWLEDGE DEVELOPMENT AND MEMORY PERFORMANCE

Michelene T. H. Chi

University of Pittsburgh

Pittsburgh, Pennsylvania, U.S.A.

Abstract

It is commonly accepted that memory development is accompanied by the acquisition of strategies such as rehearsal. This paper argues for focusing on children's content knowledge base as a locus of development of strategic knowledge. The paper cites some direct and indirect evidence in favor of the view that cognitive development is largely the increment of content knowledge, both declarative and procedural, and further suggests that strategies might be generalized forms of specific content-related procedural knowledge.

To understand learning, one must make a detailed examination of the structure and development of children's knowledge bases. The intention of this paper is to propose that the structure and growth of a child's knowledge base are important components in the study of learning. The paper begins with a definition of the knowledge base, followed by theoretical and empirical rationale for focusing on the knowledge base, and closes with an illustration of the interaction of the use of processing strategies with the structure, content, and representation of a child's knowledge in memory tasks.

Knowledge Base

It is trivial to assert that a child's knowledge base grows with age. To be more specific, it is this growth that accounts for learning and improved memory performance. But it is not trivial to describe the structure of a child's knowledge base at each stage of development, or to explain how this structure accounts for learning and memory performance. The latter is the goal of this research.

For pragmatic reasons, a distinction will be made between three types of knowledge: procedural, declarative, and strategic. Procedural knowledge can be characterized as knowledge of rules; knowing how to multiply two digit numbers, for example. Declarative knowledge may be viewed as lexical knowledge or the knowledge of facts. For example, factual knowledge about animals can be thought of as declarative knowledge. The game of chess provides an excellent illustration of the dfiferences between procedural and declarative knowledge. Knowledge about the chess pieces, games and players corresponds to declarative knowledge, while knowledge about which move to make corresponds to procedural knowledge. Both procedural and declarative knowledge are domain-specific. In this paper, they will be referred to as content knowledge.

In contrast, strategic knowledge may be viewed as knowledge of heuristic rules that are presumably applicable across several domains. For example, the process of rehearsal may be seen as a heuristic rule, and it can be used with digits, letters, or words, etc.

Although the distinction among procedural, declarative, and strategic knowledge may be artificial in the sense that a single formalism such as a production system may be able to capture all three types of knowledge, it provides a useful framework for the discussion of developmental research at the present time.

Developmental researchers in the past have centered their attention primarily on the acquisition, production, and mediation of strategies as a major component of cognitive development, because the evidence has consistently shown that the use of strategies increases with age, and that the increasing use of these strategies is accompanied by an improvement in memory performance. Developmentalists now are faced with the problem of accounting for the acquisition of these strategies. It is proposed here that the increasing use of strategies may be the result of a complex set of processes involving the acquisition and perfection of the strategies themselves, coupled with the development of content knowledge to which these strategies are to be applied. Hence, one initial research goal is to explore the extent to which the richness, structure, and representation of content knowledge affect and influence the use of processing strategies. Before doing so, both the theoretical and empirical rationale for focusing on content knowledge are discussed.

Theoretical Rationale for the Study of Content Knowledge

The prevailing assumption of a major aspect of developmental research is that adults possess a small set of strategies. In memory tasks, for example, a set of strategies might include rehearsal, recoding, grouping, labeling, imaging, elaboration, and so on. Development is thus seen as the acquisition of a limited set of strategies that have been identified in the adult literature as essential to the successful performance of a task. In order to understand how these strategies are acquired with development, however, one may need to examine how the development of content knowledge can facilitate the acquisition of strategic knowledge.

There are basically two theoretical positions that can be taken. The weaker position is to accept the prevailing hypothesis, but with the stipulation that beyond strategic development, memory development is also accompanied by the development of the content knowledge. Hence, whenever the use of deliberate processing strategies cannot account for all the age differences in memory performance, any remaining variance can perhaps be explained by differences in content knowledge. A stronger position is to state that development is the growth of content knowledge, both procedural and declarative, and that strategies are initially domain-specific procedural knowledge that eventually become more generalizable. This view necessitates studying the representation and nature of the content knowledge that children possess, and how domain-specific procedural knowledge might evolve into general strategies.

To summarize, the weaker hypothesis states that development is mainly the acquisition of strategic knowledge, with incremental content knowledge contributing only to a small portion of performance improvement. The stronger hypothesis assumes that development is mainly the increment of more content knowledge, both declarative and procedural. The greater use of strategies with increasing age is a byproduct of greater content knowledge, in the sense that strategies are a generalized form of specific procedural knowledge.

Either hypothesis is consistent with the observation that there is a correlation between age, content knowledge in general, strategy usage, and performance, as shown in Matrix 1 of Figure 1. What Matrix 1 shows is that memory performance generally improves with age, and it also improves with strategy usage and greater general knowledge. Hence, it seems difficult to attribute all performance deficits to processing deficits when performance is also correlated with knowledge deficits. The goal is thus to assess the extent of the knowledge effects.

Empirical Support for the Study of Content Knowledge

Theoretical arguments have been made for the study of content knowledge. Is there any empirical evidence to further

Matrix 1

Age	Content Knowledge	Strategic Knowledge	Memory Performance
Children	Less	Less	Less
Adults	More	More	More

Matrix 2

Age	Content Knowledge	Strategic Knowledge	Memory Performance
Same	Same	Less	Less
Same	Same	More (Training)	More

Matrix 3

Age	Content Knowledge	Strategic Knowledge	Memory Performance
Children	More	Less	More
Adults	Less	More	Less

Matrix 4

Age	Content Knowledge	Strategic Knowledge	Memory Performance
Same Child	More	Same	More
Same Child	Less	Same	Less

Figure 1. The Relationship among age, knowledge, strategy usage, and performance outcome of designs used in developmental research.

suggest such an investigation? Although not explicitly designed to test this hypothesis, several studies have produced results which can be interpreted as support for the weaker hypothesis.

One domain of empirical support arises from training studies that attempt to improve children's memory performance. A limitation is often found in these training studies in their ability to elevate young children's performance to the level of adults or older children. For example, training a rehearsal strategy can generally elevate children's memory performance so that their recall is superior to those of other children of the same age who did not get such training (see Matrix 2, Figure 1). However, training the use of a strategy often cannot elevate recall to the level of older children (Belmont & Butterfield, 1971); some other factor, such as the knowledge base, may be limiting performance.

The limitation of strategy training shows up in another way. When children of all age groups are trained to use a strategy such as grouping, the recall level of all age groups improves, which means that the initial age differences still remain, and must be explained by some other factor (Huttenlocher & Burke, 1976). The same observation also holds for individual differences within an age group. That is, if all the individuals are provided with the same training, whether they need it or not, the initial individual differences will remain after training (Lyon, 1978).

A third limitation of training studies is that they often fail to generalize (Brown, 1974). That is, if children are trained to use rehearsal processes with digits, they may not necessarily be able to generalize the application of such a strategy to words. The failure of generalization can be interpreted in at least three ways: (a) the definition of a strategy as being general is faulty (i.e., strategy usage is necessarily tied to content domain, which supports the stronger hypothesis); (b) training was ineffective in some way, or (c) the role of a strategy in affecting performance is not as powerful as one might think. However interpreted, lack of generalization suggests that an examination of content knowledge is crucial.

Finally, if adults are inhibited from using strategies that have been identified a priori as critical to the performance of a given task, the level of performance of the adults does not drop to the level of the child (Chi, 1977). This again suggests that strategy usage is not entirely responsible for the observed age differences in recall.

Although training studies as a set are difficult to interpret when they fail, the studies cited above collectively point to the possibility that the weaker hypothesis is supported. That is, it appears that beyond deliberate strategy usage, a portion of age

differences in memory performance can be attributed to some
other factor, such as knowledge differences.

In order to seek evidence in support of the stronger hypothesis, a situation analogous to Matrix 3 of Figure 1 can be created,
where the correlation between age and knowledge is disrupted by
manipulating knowledge independently of age. In a study using
this design (Chi, 1978), adults with limited knowledge of chess
were unable to memorize as many chess pieces as 10-year-old
children who had some knowledge of chess. The adults also took
longer (required a greater number of trials) to memorize the
entire chessboard positions than children. For this same group
of subjects, children could memorize fewer digits on a given trial,
and required a greater number of trials to learn 10 digits than
adults. For the first time, it has been shown that age need not
correlate with memory performance when it does not correlate with
knowledge. For the same group of subjects, the strategic knowledge necessary to perform in a memory task presumably did not
change when the stimulus material was changed from digits to
chess. What did change was the amount of content knowledge.
The reversal in the outcome of the performance measures (comparing Matrix 1 and 3) suggests that children who possess more
knowledge in a content domain can overcome whatever limitation
is imposed by more limited strategic knowledge.

Although it is not clear from the chess study whether children's superior performance arises from more developed declarative
or procedural chess knowledge, either assumption is consistent
with the stronger hypothesis, if we want to maintain a distinction
between procedural and strategic knowledge. That is, if we
assume that better memory performance on chess arises from
greater chess-related procedural knowledge, then it suggests that
domain-specific procedural knowledge may serve the function that
strategies serve in mediating performance. Hence, it may only be
fruitful to study domain-specific procedural knowledge.

Another source of data which also supports the stronger
hypothesis comes from Myers and Perlmutter's (1978) research on
2- to 5-year-olds. They found that memory performance in that
age range improved, but they observed no evidence of an increase
in the application of processing strategies. These results tend to
put more emphasis on general knowledge growth as a major focus
for development in that age range, although other less straightforward interpretations are possible.

A final piece of evidence in support of the stronger hypothesis comes from a study in which a situation analogous to Matrix 4
(Figure 1) is created. The approach here was to study intensively
an individual child so that age and general strategic knowledge
are constant, but to vary how much the child knows about a

particular domain of knowledge (Chi, 1979). The subject in this case study was a four-year-old child who is an expert on the topic of dinosaurs. It was possible to partition the child's repertoire of 40 dinosaurs into two sets: One with which he was very familiar and another with which he was less familiar. Using a link-node semantic network structure, the representation of the greater-knowledge set of 20 dinosaurs was shown to be much denser and more complexly organized than the representation of the lesser-knowledge set of 20 dinosaurs. In comparing memory performance on the two sets of dinosaurs, it was not surprising to find that the child's recall, retention, and clustering performance was superior in the more knowledgable set. Hence, the design of this study is essentially the counterpart of a training study. In the one case (Matrix 4), content knowledge was manipulated, and in the other case (Matrix 2), strategic knowledge was manipulated. Both types of manipulations produced superior memory performance under conditions where there was more knowledge, suggesting that both types of knowledge--strategic and content--have powerful influences on memory performance.

Interaction of Content Knowledge and Processing Strategies

Up to this point, the research goal has been to seek evidence of the importance of content knowledge on memory development. Since both content and strategic knowledge have been shown to be important, one needs to examine the interaction of the two.

A study is currently in progress where we describe a five-year-old girl's representation of her 22 classmates. We found that her basic representation was organized according to the seating arrangement of her class, taking the form of a spatial hierarchical structure, in which the 22 children were divided into four sections, with five to six children attached to each section. Associated with each child is additional information, such as the sex, race, and grade levels of the child. In other words, the 22 classmates were not organized hierarchically according to dimensions such as the sex of the child. We know this because when we asked her to recall all the boys' names (or girls' names), she did so by using the spatial seating arrangement.

When we obtained a "stable" representation, (stable means that the same representation was manifested using multiple procedures), we explored how well she could use a retrieval strategy, in this case, recalling the names in alphabetical order. The child easily learned to apply such a strategy when the knowledge was very stable and overlearned, even though the strategy was fairly new to her repertoire. However, she had difficulty applying the same strategy to a learned set of names of people she did not know. Hence, it appears that when and how well a strategy can be used depends a great deal on the structure of the content knowledge to which it is applied. When the content knowledge is overlearned and highly familiar (and perhaps has real-world semantic reference), a young child has no difficulty adopting and using newly acquired strategies. However, when the content knowledge is novel and unfamiliar, the child has greater difficulty. Such preliminary results begin to suggest that powerful strategic heuristics may be acquired only after the content knowledge is fully developed.

In conclusion, the conceptual approach to development proposed here makes a deliberate distinction between strategic and content knowledge. These strategies have been implicitly defined as task-specific but not content-specific. At the end, we alluded to the possibility that these task-specific strategies may be more content-related than had been presupposed.

It would be unwise to conclude without remarking that there are other kinds of strategies that were not considered in this paper. These are non-task- and non-content-specific strategies, commonly known as metastrategies. A metastrategy might be knowing when or in what situation to apply a strategy. These metastrategies seem to be broader and even more general than those that have been dealt with here. The obvious question is to ask in what ways metastrategies are related to content knowledge. We of course would predict that metastrategies cannot develop for any useful purposes without the concurrent development of content knowledge. This is somewhat substantiated by the inconsistent findings regarding the benefits of training meta-strategies for performance (Brown, 1978). Hence, it still seems a worthy goal to pursue the study of the significance of content knowledge, and how it interacts with strategies and metastrategies.

References

Belmont, J. M., and Butterfield, E. C. What the development of short-term memory is. Human Development, 1971, 14, 236-248.

Brown, A. L. The role of strategic behavior in retardate memory. In N. R. Ellis (Ed.), International review of research in mental retardation (Vol. 1). New York: Academic Press, 1974.

Brown, A. L. Knowing when, where, and how to remember: A prob of metacognition. In R. Glaser (Ed.), Advances in instructional psychology (Vol. 1). Hillsdale, NJ: Lawrence Erlbaum Associates, 1978.

Chi, M. T. H. Age differences in memory span. Journal of Experimental Child Psychology, 1977, 23, 266-281.

Chi, M. T. H. Knowledge structures and memory development. In R Siegler (Ed.), Children's thinking: What develops? Hillsdale, NJ: Lawrence Erlbaum Associates, 1978.

Chi, M.T.H. Exploring a child's knowledge of dinosaurs: A case

study. Paper presented at the Society for Research in Child
Development, March 1979.

Huttenlocher, J., and Burke, D. Why does memory span increase
with   age?   Cognitive Psychology,   1976,   8,   1-31.

Lyon, D. Sources of individual differences in digit span size.
Cognitive Psychology, 1977, 9, 403-411.

Myers, N. A., and Perlmutter, M. Memory in the years from two
to five. In P. A. Ornstein (Ed.), Memory development in
children. Hillsdale, NJ: Lawrence Erlbaum Associates, 1978.

REASONING AND PROBLEM SOLVING IN YOUNG CHILDREN

Marion Blank
Rutgers Medical School
Rutgers, New Jersey, U.S.A.

Susan A. Rose
Albert Einstein College of Medicine
New York, New York, U.S.A.

Laura J. Berlin
Albert Einstein College of Medicine
New York, New York, U.S.A.

The reasoning skills of preschool children were examined through three types of problems: Prediction, Explanations, and Explanations of Predictions. The 3-year-olds successfully answered half of the Prediction problems; the 4-year-olds two-thirds of the Prediction problems and better than a third of the other two groups; the 5-year-olds showed high levels of success on all three groups. The results indicate that preschoolers have a capacity for reasoning that has often not been sufficiently appreciated.

Psychological research is notable for the diametrically opposed positions that often surround significant issues. One such area is the interpretation placed on the mental life of preschoolers. On the one hand, observations of naturalistic behavior suggest that they possess a host of complex mental abilities (Issacs, 1945; Maratsos, 1973; Rees, 1978). This view is captured in Tolstoi's observation that: "From myself as a five-year-old to myself as I now am there is only one step. The distance between myself as an infant and myself at five years is tremendous" (cited in Chukovsky, 1968, p. 14). On the other hand, experimental work with young children has shown them to deal poorly with such valued spheres as concepts, inferencing and problem solving (Kendler & Kendler, 1962 and Farnham-Diggory & Gregg, 1975). The negative results have been particularly characteristic of work conducted within the Piagetian tradition wherein children under 7 years of age are typically characterized in terms of weakness (Piaget, 1959, 1962). Their thinking is termed "egocentric, prelogical, affective, un-

231

differentiated, precausal, personal, vague and unanalyzed."

Recently investigators have attempted to resolve the discrepancy between naturalistic and laboratory behavior (see Donaldson, 1978; Karmiloff-Smith, 1978; Rose & Blank, 1974). Relatively little consideration has been given, however, to the fact that the Piagetian problems characteristically demand that the child simultaneously deal both with multiple concepts (e.g., concepts of sameness, co-occurring variations in height and width, etc.) and complex reasoning skills (e.g., if a change was made, was it significant?, how can one justify the basis of the inference?, etc.).

It could be argued that several behaviors must co-occur before a child is judged as having attained a particular stage. This does not mean, however, that the various constructs cannot or ought not to be studied independently. In order to explore this issue, we chose to examine children's ability to reason about experiences in their environment when the problems were not simultaneously burdened by the presence of complex concepts. (Conceptual complexity here refers to ideas which have no perceptual referents). Three sets of processes were selected with 6 problems designed for each process. The processes were: 1. prediction (what do you think will happen if...); 2. explanation of an observation (why do you think that...); and 3. explanation of a prediction or inference (what do you think will... then why do you think that...).

An example of a Prediction task is the following: the child observes objects being placed on and taken off a balance scale and is then asked what would happen if an additional object were to be put on one side of the scale.

An example of an Explanation problem is: a child is shown a boot, near it are a piece of rubber and a piece of paper. The adult says, "Boots are made of rubber like this (pointing) and not paper like this (pointing). Why do you think that boots are made of rubber and not paper?"

An example of an Explanation of a Prediction is: a child is shown a yellow rectangular sponge. Below it are a yellow paper triangle and a yellow sponge triangle. The adult says, "If the sponge were made of this (pointing to the paper triangle) and not this (pointing to the sponge triangle) would it still be a sponge?" After the child response, the adult asks, "Why?"

Subjects

The subjects were 72 children who ranged in age from 36 to 71 months. All the children were white, came from middle class backgrounds and attended private nursery schools in the suburban New York area. There were 12 boys and 12 girls within each 12 month age range; i.e., 3, 4, and 5 years.

Results

Each child received a percentage score based upon the number of problems he/she answered correctly relative to the number of problems administered. Table 1 presents the mean percentage scores. (A more extended discussion of the scoring procedures is available in Blank, Rose & Berlin, 1978).

None of the differences between the sexes was significant. Because the data were not normally distributed, the results were analyzed through a series of nonparametric measures. Three Kruskal-Wallis analysis of variance tests were carried out to determine the effects of age, with one test being used for each type of problem. On all three measures, significant age effects were found with progressive improvement shown as the children moved up the age span. (For Prediction H = 14.9, df/2, p < .01,

Table 1
Mean Percentage of Problems Correct

Age	Sex	Type of Problem		
		Prediction	Explanation	Explanation of Prediction
3	Male	57	19	24
	Female	39	21	18
4	Male	71	40	51
	Female	65	49	47
5	Male	83	76	69
	Female	75	83	65

for Explanation H = 36.6, df/2, p   .001 and for Explanation of
Predictions $\underline{H}$ = $\overline{21}$.3, df/2, $\underline{p}$ <.0$\overline{0}$1).   Overall, by five years, the
children responded appropriately on the great majority of problems
(on no problem was there fewer than 50% correct), while for the
three year olds, only the prediction problems yielded results that
were close to 50% correct.

Three Friedman two-way analyses of variance were carried
out to assess the effect of the type of problem posed.   There
were significant differences among the problems at all three age
groups (for the 3, 4, and 5 years olds respectively, Xr^2 = 20.3,
df = 2, $\underline{p}$ < .01, Xr^2 = 11.3, $\underline{df}$ = 2, $\underline{p}$ < .01 and $\overline{X}r^2$ = 16.6,
$\overline{df}$ = 2, $\underline{p}$ < .001. Predictions were easier than either of the other
two types of problems at 3 and 4 years while only Explanations about
Predictions were noticeably more difficult for the 5-year-olds.   The
majority of tasks were handled successfully by 4-year-olds; the
5-year-olds displayed well over seventy-five percent correct performance.   The consistency and extent of appropriate responses
obtained suggest a level of cognitive ability often deemed to be
beyond the capabilities of children under 6 years.

Discussion

Ever since Piaget began writing about the mental life of the
young child, his views have been challenged by such leading figures
as Buhler (1921), Isaacs (1945), and Vygotsky (1962) who argued
that Piaget had either misinterpreted or underestimated the preschooler.   In almost all cases, Piaget did not deny the validity of
their evidence, but rather argued with their interpretations.   Thus,
what Vygotsky (1962) saw as externalized, self-directed language,
Piaget saw as egocentric speech and what Isaacs (1945) saw as
logical thinking, Piaget saw as transductive reasoning.

In light of this history, it is reasonable to assume that
Piaget would not find the results presented here either surprising
or discomforting.   Disagreement would arise, however, in interpreting these behaviors as true (meaning "logical") reasoning, rather
than as instances of "intuitive" or "transductive reasoning."   At
first glance, the differing interpretations might seem to be only
a matter of semantics.   As Piaget (1959) states "But who would not
see that the two explanations come to the same thing?" (p.274).

If, as a focus on semantics implies, the terms pre-causal or
pre-conceptual thinking were simply labels, then they would pose
little difficulty.   But these terms convey a range of judgmental,
albeit implicit associations in which the preschooler is viewed
almost solely in terms of weakness.   The following quotations are
illustrative of the position taken by Piaget (1962, p. 241):

It is clear ... that distortion of reality is a direct
result of the first deductive constructions" (p. 233).
"Between the ages of four and seven, we find only few
intuitions capable of articulation...but without
generalization or reversibility.

The general view of the preschooler that emerges from a Piagetian
interpretation is thus one of weakness and limitation.

Discomfort must arise when it is recognized that the major
theory of intellectual development currently available sees a
critical period of rapid change as one marked mainly by limitations.   This focus on limitations has come to be recognized by
Piaget's followers and attempts are being made to place the
preschooler in a better light.   The "errors of the nonconservers"
for example are seen to "represent powerful heuristics" rather
than "merely shortcomings to be surmounted later." Within this
framework "attention to spatial cues" may thus be seen not as a
limitation bust as representing the child's "endeavor to gain
predictive control over his environment" (Karmiloff-Smith, 1978,
p. 189).

This reinterpretation, while significant, still fails to account
for inportant and positive developments that we know are
occuring.   For example, it would not lead one to anticipate the
extensiveness of the problem solving behavior observed in the
present research.   Success among the four-year-olds was common
and among the five-year-olds was almost uniformly the rule, rather
than the exception.

The precise nature of the child's learning remains to be
determined.   If we are to advance in this area, we must begin to
delineate, with much greater precision than has heretofore been
available, the forms of reasoning and concept formation that may
exist in the young child.   Only in this manner can we begin to
gain an insight into the rapid strides that the preschool age child
so dramatically displays in a host of complex cognitive areas.

References

Blank, M., Rose, S.A. & Berlin, L.J.   The language of learning:
     The preschool years.   New York: Grune & Stratton, 1978.
Buhler, K.   Die geistige entivicklung des kindes, Jena, 1921.
Chukovsky, K.   From two to five.   (Translated & Edited by M.
     Morton), Berkeley: University of California Press, 1963.
Donaldson, M. Children's minds.   Great Britain: William Collins
     Sons & Co., 1978.
Farnham-Diggory, S. & Gregg, L.W.   Color, form and function as
     dimensions of natural classifications: Developmental changes
     in eye movements, reaction time and response strategies.

Child Development, 1975, 46, 101–114.

Isaacs, S. Intellectual growth in young children. London: George Routledge & Sons, 1945.

Karmiloff-Smith, A. On stage: The importance of being a nonconserver. Behavioral and Brain Sciences, 1978, 2, 188–190.

Kendler, H.H. & Kendler, T.S. Vertical and horizontal processes in problem solving. Psychological Review, 1962, 69, 1–16.

Maratsos, M. Nonegocentric communication abilities in preschool children. Child Development, 1973, 44, 697–700.

Piaget, J. Language and thought of the child. London: Routledge, 1926/1959.

Piaget, J. Play, dreams and imitation in childhood. New York; Norton, 1962.

Rees, N. Pragmatics of language: Application to normal and disordered language development. In: R.L. Schiefelbusch (Ed.), Bases of Language Intervention, Baltimore: University Park Press, 1978.

Rose, S.A. & Blank, M. The potency of context in children's cognition: An illustration through conservation. Child Development, 1974, 45. 499–502.

Vygotsky, L.S. Thought and Language. Cambridge, MA.: MIT Press, 1962.

Footnote

This research was supported by funds from the William T. Grant Foundation. The address of the senior author is Department of Psychiatry, College of Medicine and Dentistry of New Jersey-Rutgers Medical School, Piscataway, New Jersey, 18854.

LOGICAL COMPETENCE IN INFANCY:

OBJECT PERCEPT OR OBJECT CONCEPT?

George Butterworth

The University

Southampton, England

Abstract

Three experiments are summarized which test Piaget's explanation for errors in infant manual search. In these studies, all
possible combinations of 3 spatial location cues were changed
independently between trials at A and at B: a) position defined
with respect to the infant (left or right), b) position defined
with respect to the cover occluding the object (blue or white), c)
position defined with respect to the background on which the
object stood.

It was found that patterns of search depend on changes in
background and cover cues between trials at A and at B. Furthermore, with constant background and a change in cover infants
search correctly, i.e., identify the object over a change in its
position. It is concluded that spatio-temporal criteria for identity
which are inherent in perception guide search.

Introduction

A major assumption of Piaget's theory of sensori-motor development is that the infant does not directly perceive the objective
properties of reality, a world that is spatially structured and that
contains objects which are permanent and retain their identity
through time. Perception is subordinate to and progressively
structured by the infant's instrumental actions in a series of
stages where particular motor strategies mediate the infants'
commerce with objects.

Particularly important evidence for Piaget's view is an error that occurs between the ages of about 8 and 11 months in the infants' manual search for hidden objects. This is known as the stage IV or AB error. Although infants are perfectly capable of retrieving an object hidden at an initial location, A, they will often continue to search at A when they see the self same object being hidden at a new location B.

Piaget maintains that such errors indicate the infant is unable to perceive the object to retain its identity over a change in position. Instead, the infant merely repeats the initially successful action as a "magical" procedure to restore the object to immediate experience. The object is understood to exist and to retain its identity only at the initial location, defined in relation to the infant's successful action. Since the child actually saw the object move from A to B, and it is logically impossible for the object at B to be at its initial location, this is definitive proof for Piaget that perception must be subordinate to the infant's motor strategies.

The problem with Piaget's procedure for testing the infant's perception of object identity is that it confounds several types of position change. Piaget may hide an object to the left of an infant under a cushion. Then when the infant has retrieved the object successfully, it is hidden to the right, perhaps under Piaget's beret. Not only does the object undergo a change in its position as defined with respect to the infant (i.e. to left or right, its egocentric position) but also it changes its position with respect to the cover (an allocentric position) and perhaps with respect to other spatial reference points in the background. Thus, any errors the infant makes may result from a change in egocentrically defined position as Piaget maintains or a change in allocentric position cues, or both. In fact, previous research of my own with A and B locations arranged to left and right of the infant, with identical covers shows that babies search consistently either at the initial location A or at the final location B. There is little evidence that babies will search consistently at the wrong location as Piaget maintains. Instead, the old and new locations seem to be equiprobable after the object is moved and infants will search consistently at one place or the other. I have suggested that this pattern of errors might be explained by a conflict between an egocentric spatial reference system defined by the infant's own body (i.e. to left or right of the midline) and an allocentric system given by spatial cues in the immediate visual field (Butterworth 1977, 1978).

The aim of the present investigation is to examine in more detail the contribution of different kinds of spatial cue to error in Piaget's stage IV task. In carrying out these experiments we discovered that infants can actually search correctly for an object

seen to change its location when the immediate visual field was
spatially structured in particular ways. Thus, there is nothing
inevitable about errors in manual search and the infant seems able
to identify an object over a change in position at least when the
necessary spatial cues are provided. This competence seems
based on the infant's perception of the relation between object
and the spatial properties of the surround, since the same infants
show the typical divided patterns of search when retested under
conditions where the critical spatial cues are not available. Hence
it may be better to characterize search at stage IV as based on a
complete object percept in which the rules for object identity are
inherent in perception than on an incomplete object concept that
presupposes structuring of perception by beliefs. Even without
any beliefs or self conscious rules about permanence and identity,
processes inherent in perception may be sufficient to provide the
infant with veridical information about the objective properties of
the environment.

Procedure

In the experiments to be reported, the procedure was always
constant. Infants were seated in a chair opposite a small platform
used for hiding the objects. There were 24 infants in each
experimental group, comprising 8 babies in each age group, 8
months, 9 months and 10 months. All the infants retrieved an
object (a bunch of keys or a toy car) once from the initial location (A). Then the object was hidden at a new location (B) in
such a way that changes in three spatial location codes were made
independently between the trial at A and trials at B. The codes
were defined in the following way:
    (i) Position defined with respect to the infant i.e. to the
    left or right, the absolute position in space.
    (ii) Position defined with respect to a distinctive blue or
    white cover.
    (iii) Position defined with respect to a distinctive background,
    black or green on which the object rested (the surface of
    the platform used in all the studies).

After the infant had retrieved the object from A once, the apparatus
was drawn out of reach and the object was hidden at a new location
defined with respect to the infant, the cover or the background (in
some conditions the cloth background could be flipped over so that
that portion of the platform that had been green on A trials was
now black and vice versa). Following a three second delay, the
table was pushed back into reach and the infant was allowed to
search. To establish the persistence of error, infants were
tested five times at the new location.

When the experiment was complete, all infants were retested
on a standard stage IV task, in which A and B locations were two

identical white covers arranged to left and right of the infant on
a green background. In the retest, infants searched once at A
and 3 times at B. Thus is was possible to conclude that performance in the experiment proper was a function of the spatial conditions of the task.

Experiment 1

The first experiment was designed to study the effect of
position cues given by the covers and cues given in the background on error. There were four conditions and 96 infants took
part, 24 in each of 4 conditions. Since we were trying to establish whether infants can search correctly, a very stringent criterion for error was adopted. Any move by the infant toward the
incorrect location was deemed an error, even if the baby corrected
himself subsequently.

In the first experiment different combinations of change in
covers and background cues between A and B trials were tested.
Condition I: Two different covers arranged to left and right of
the infant on two different backgrounds. Condition II: Identical
covers on different backgrounds. Condition III: Identical covers
on a homogeneous green background. Condition IV: Different
covers on a homogeneous green background.

The results are presented extremely schematically to save
time. The critical trial to demonstrate competence in search is the
first trial at B and this is adopted here as criterion. Accurate
search is inferred when the number of infants making an error is
significantly less than would be expected by chance. Other
criteria are possible, e.g., comparison with performance on the
first A trial, or with the control condition where nothing changes.
These results are also available but since they do not alter the
major conclusions, data for the first B trial will be reported here
(see Fig. 1).

The main result of this study was to show that cover cues
and background cues were not equivalent in their effects. Infants
searched correctly when different covers were arranged to left
and right on a constant background (condition IV) but showed
the typical divided pattern of search when identical covers were
arranged to left and right on different backgrounds. (condition
II). Different covers on different backgrounds or identical covers
on a constant background (conditions I and III) also showed the
divided pattern of search. When the infants were retested with
identical covers on a constant background, they all showed the
typical divided pattern of search. Thus performance in the
experiment proper was determined by the spatial condition of the
task.

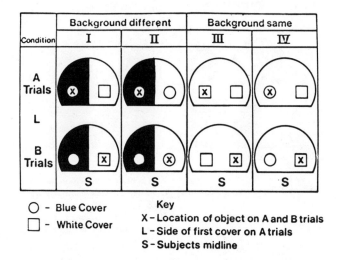

Figure 1. Cover cues and background cues in infant manual ς ›arch.

Experiment 2

In a second experiment, the effect of a change in background cues alone was examined. We had already established that infants will search correctly for an object hidden under a distinctive cover on a constant background, wherever the object was located with respect to the infant in an earlier series of experiments. So in this study, the object was always hidden under the same cover, a condition known to lead to successful search, but its position relative to infant and background was changed. The design is shown in Figure 2. A new group of 96 infants was tested.

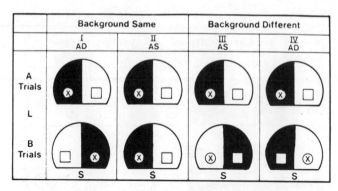

	Background Same		Background Different	
	I AD	II AS	III AS	IV AD
A Trials	⊗ ▢	⊗ ▢	⊗ ▢	⊗ ▢
L				
B Trials	▢ ⊗	⊗ ▢	⊗ ▢	▢ ⊗
	S	S	S	S

○ Blue cover
▢ White cover

X   Location of object on A & B trials
L   Side of first cover on A trials
S   Subjects midline
A.S. Absolute position of object same
A.D. Absolute position of object
      different

Number of infants making an error on first B trial

	I	II	III	IV
	5/24 (statistically correct)	3/24 (statistically correct)	9/24	13/24
Retest	12/24	11/24	11/24	9/24

Figure 2.

The major result was to demonstrate that infants will search correctly, so long as the object is hidden under the same cover on a constant background, regardless of its position with respect to the infant (conditions I and II). If the relation between cover and background changed, even though the cover was constant, the divided pattern of search reappeared. It is of particular interest to note that infants showed the divided pattern of search in condition III, where the object was at the same location with respect to the infant and cover but the background changed. Even though the infants had the opportunity to retrieve the object by making a perseverative response to the same location, the divided pattern nevertheless reappeared. On retesting under the standard conditions, all the infants showed the divided pattern of search.

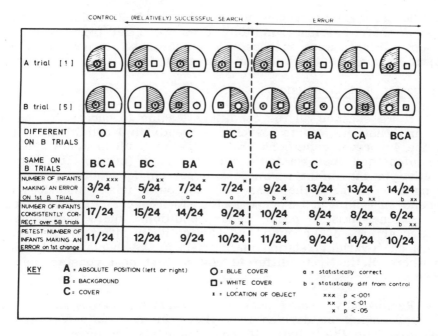

Figure 3. Spatial factors of determining search. (Effects of change
in position defined by background cover or absolute location.)

Combined study, experiment 3

If studies 1 and 2 are combined, only three more groups are
required to test all possible combinations of simultaneous change
in cover, background and absolute position cues. Therefore, the
three extra groups of infants were tested and the results for all
combinations of conditions are shown in Figure 3.

Figure 3 shows the data for all possible combinations of
change in cover, background and absolute location cues arranged
in order of difficulty. A stringent criterion for correct search is
also included, the number of infants in each condition who searched
correctly over all 5 B trials.

The results fall into two groups, three conditions in which
search was relatively successful both by comparison with performance expected by chance and performance in the no change control
condition, and four conditions in which performance did not differ
from chance and also differed significantly (on the first trial at
B) from the control condition. It is not simply the case that
infants can cope with a change in one location code but not with
two since either change can lead to successful search or error.

On the criterion of successful search over 5 B trials, only the first of the comparison conditions does not differ from the control group. Infants were successful when the absolute position of the object changed but cover and background remained the same.

Discussion and Conclusion

If we examine all the conditions leading to successful search we find that the infants searched correctly in 4 conditions. In 2 of the 4 correct conditions the display on B trials would arise if the infant had been rotated around the table (or the table rotated relative to the infant). Successful search does not depend on being able to make the same response on B trials, nor does it depend on making a response to the same cover. Rather, successful search seems to occur under conditions where the whole spatial array, background and distinctively different covers, bears an invariant relation on B trials to the display on A trials or where distinctively different covers rest on a constant background.

Reciprocally, where search is divided between A and B, the B array cannot be derived from the A array by a movement of the infant or the table. It would of course require further experiments to establish whether movement of the infant or the array are equivalent to the transformations leading to successful search in the present study.

In conclusion, infants can identify an object over a change in its spatial location under certain spatial conditions and this competence seems based on the spatial relation between an object and its surround. The surface on which the object rests seems to act as a stable context. Where this context is structured with landmarks in relation to which the infant can keep track of the object's movements, the infant can identify the object correctly. Where these conditions do not apply, the outcome is a conflict between A and B. Although the object is known in relation to a perceived context, it can be identified when it moves if the context contains sufficient spatial structure. In much the same way, a landmark allows a map reader to relate his own movements to the physical environment. So even if the infant is completely lacking in conceptual rules or beliefs about object permanence and identity, he can and does rely on processes in immediate perception to connect the separate places at which the object disappeared . through the invariant spatial context. Hence it may be more appropriate to consider performance on Piaget's stage IV task to be based on a complete object percept that necessarily depends on the spatial structure of the visual field, than on an incomplete object concept.

References

Butterworth, G.E.   Object disappearance and error in Piaget's
    stage IV task.   Journal of Experimental Child Psychology,
    1977, 23, 391-401.
Butterworth, G.E.   Thought and Things: Piaget's Theory. In
    Burton A. and Radford J. (Eds.) Perspectives on Thinking.
    London.   Methuen 1978.
Piaget, J.   The Construction of Reality in the Child.   New York,
    Basic Books, 1954.

Footnote

This research was carried out with the aid of a project grant
from the Medical Research Council of Great Britain.   Author's address: Department of Psychology, The University, Southampton
S09 5NH, England.

PIAGETIAN PERSPECTIVE IN DRAW-A-HOUSE TREE TASK:

A LONGITUDINAL STUDY OF THE DRAWINGS OF RURAL CHILDREN

Violet Kalyan-Masih

University of Nebraska

Lincoln, Nebraska, U.S.A.

Abstract

Drawings of "a house with a tree behind it" were analyzed for developmental changes for 3-5, 6-8, and 9-11 year old rural children. Relationship with Piagetian tasks, Peabody IQ, and WISC-R and correspondence with the Luquet-Piaget sequence of graphic representation were also investigated. Significant developmental changes in House-Tree task with repeated test effects controlled were noted. The House-Tree corresponded significantly with selected Piagetian tasks and WISC-R Block Design at all ages. Regression equations were computed for prediction. The Luquet-Piaget sequence was inferred with additional intervening strategies between intellectual and visual realism. The House-Tree task has potential for assessing cognitive development of younger children. Its use with older children of restricted mental functioning needs to be explored.

Graphic representation or drawing is one of the five semiotic functions (symbolic play, deferred imitation, drawing, mental imagery, and verbal evocation) of the preoperational period according to Piaget and Inhelder (1969). While discussing the evolution of graphic representation among children and endorsing the Luquet stages and interpretations of children's drawings these authors further state: "Thus we see that the evolution of drawing is inseparable from the whole structuration of space, according to the different stages of this development. It is not surprising, then, that the child's drawing serves as a test of his intellectual development." (p. 68).

Statement of the Problem

A three-year longitudinal study[1] was undertaken to investigate the mental and social development of rural children. In Nebraska, this study also included the investigation of graphic representation. More specifically, the main objectives of this part of the study were to determine: 1) if the drawings of "a house with a tree behind it" would reflect developmental changes for 3-5, 6-8, and 9-11 year old children; 2) if these changes would correspond with their cognitive performance; and 3) if these drawings would reflect the Luquet-Piaget sequence in general.

Sample

Multistage area sampling techniques[2] meeting the NC-124[3] guidelines of "representative randomness," stratified by defined criteria of ruralness, farm-derived income, family-intactness, and appropriate age were used for selecting the sample. Forty 3-year, 41 6-year, and 40 9-year old children from rural Nebraska were tested in 1976, 1977, and 1978. Control cohorts were added in 1977 and 1978. Total sample consisted of 121 3-, 6-, and 9-year old children in 1976; 173 4-, 7-, and 10-year old children in 1977; and 224 5-, 8-, and 11-year old children in 1978.

Instruments and Data Collection

The relevant assessment measures are:

1. House-Tree Task (HT). Children were requested to draw the picture of a house with a tree behind it and encouraged to tell about their pictures. These drawings were later scored on a revised scale of 1-11 points.

2. The Nebraska Wisconsin Cognitive Assessment Battery[4] (NEWCAB), derived from the Piagetian theory, was used for collecting cognitive data.

3. Two standard measures - Peabody Picture Vocabulary Test (PPVT) and Wechsler Intelligence Scale for Children (WISC-R) were also administered in the standard manner.

Findings and Implications

A. Quantitative Analysis

For the first objective, ANOVA and Scheffé tests were used for comparing the mean performances of different sub-groups for significance. The three-year longitudinal sample at each age level made significant mean gains on HT from 1976 to 1977 to 1978. The two-year longitudinal sample at each age level also made significant mean gains from 1977 to 1978. To check for repeated test effects, the mean performances of cohorts and their control counterparts were compared at each age level. No significant mean differences were noted. Therefore, practice effects might be ruled out in favor of significant intra subject improvement across time. Also noted were increases in mean scores from younger to older children in an ordered direction suggesting inter-subject progression across age levels.

The second objective was investigated in three ways:

(1) Analyses of zero-order correlation coefficients showed that HT correlated significantly with several Piagetian tasks at each age level. Some empirical relationship between HT and selected Piagetian tasks is, therefore, postulated. Such a relationship was confirmed in an earlier study (Kalyan-Masih, 1976). HT correlated significantly with PPVTIQ at ages 3, 5, 6, and 9. The HT also correlated significantly with WISC Picture Completion Scale at ages 7, 8, and 9. However, HT correlated positively and significantly with WISC Block Design at ages 6, 7, 8, 9, 10, and 11, suggesting that HT and WISC Block Design may possibly overlap in assessing similar abilities at ages 6 through 11. In an earlier study significant and positive relationships between HT and Stanford Binet were noted, suggesting overlap, in spite of different theoretical formulations (Kalyan-Masih, 1976).

(2) Multiple regression analyses were performed to explain variance in the criterion variable (HT) accounted for by the predictor variables (Piagetian tasks). Using the forward stepwise multiple regression analysis procedure several multiple regression equations, with regression coefficients being significant, were computed at each age level. Alternatively, the simple regression analyses were performed for predicting a selected Piagetian task from HT. Several simple regression equations, with regression coefficients being significant were computed at each age level.

(3) Means expressed as a percentage of the maximum score for each task were plotted for each age level across the three years in several line graphs. These lines were neither coincident nor perfectly parallel, but showed an upward trend from 1976--1977--1978, suggesting some correspondence in performance between HT and Piagetian tasks across time (Fig. 1).

B.   Qualitative Analysis

For the third objective, a qualitative analysis of 518 drawings of "a house with a tree behind it" was done. The Luquet-Piaget sequence of graphic representation was inferred with several intriguing strategies between intellectual and visual realism.

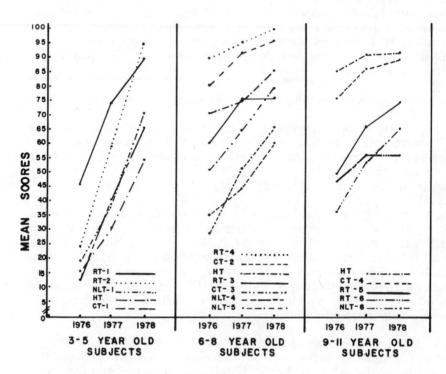

Figure 1

The following summarizes the qualitative analysis of these data:

3-5 years:
1.   Scribbling - Sensory motor pleasurable activity with little or no
        representation.
2.   Fortuitous Realism - "Front-behind" relationship is completely
        ignored. Interest is centered on the discovery that lines/
        dots can represent something.
3.   Failed Realism - The details are juxtaposed or drawn appropri-
        ately. Attention is focused on drawing a tree and a house
        rather than a tree behind the house.

6-8 years:
4.   Intellectual Realism - First time the front-behind relationship
        is handled.
        Tree is drawn inside the house.
    a.   Transitional - Tree is placed outside the house. The

tree drawn inside looks funny to the child, so he toys with the idea of placing the tree somewhere outside-- beside, in front, or on the roof.

b.   Compromise - Four contrived situations when the tree may reasonably be considered behind the house:
--Drawing the house on one side of the paper and the tree on the reverse side.
--Drawing the tree first and then superimposing a house on it.
--Drawing a far away tree on a hill.
--Drawing a tree which is seen through a large open window.

c.   Partial - Partially hidden and partially visible tree trunk behind the side wall or above the roof.

9-10 years:
5.   Visual Realism - Tree top is seen above the roof of the house.

These findings suggested that the graphic representation im- proved with age and that this improvement was associated with their cognitive performance during the preoperational period. The Luquet-Piaget sequence was inferred with additional intervening strategies between intellectual and visual realism.

The HT task is simple and economical. It uses minimum lang- uage which may be of advantage when assessing younger children between the ages of three and seven. After age 8, HT begins to lose its discriminating effectiveness because of the ceiling effect imposed by the score range. However, its usefulness for older children functioning within a restricted mental range needs to be further explored. Its potential for preschool assessment needs to be utilized.

Footnotes

1.   This study used the same sample and the same cognitive data as utilized for the NC-124: A Life Span Analysis of Rural Children's Mental and Social Development. This is a regional project in which Illinois, Indiana, Iowa, Kansas, Michigan, Nebraska, Wisconsin, and Missouri are participating.

2.   H. Whitt, Director Bureau of Sociological Research, Univers- ity of Nebraska-Lincoln, drew the Nebraska sample in accord- ance with the NC-124 guidelines.

3.   The Screening Form was developed under the supervision of S. Clark, D. Pease, and S. Crase, Iowa State University, Ames, Iowa.

4.   The NEWCAB was developed for the NC-124 under the super- vision of V. Kalyan-Masih, University of Nebraska-Lincoln and W. Marshall, University of Wisconsin-Madison.

References

Kalyan-Masih, V. Graphic representation: from intellectual realism
    to visual realism in draw-a-house-tree task. Child Development,
    1976, 47, 1026-1031.
Piaget, J. and Inhelder, B. The Psychology of the Child. New
    York: Basic Books, 1969.

METACOGNITION AND INTELLIGENCE THEORY

John G. Borkowski   and   John C. Cavanaugh

University of Notre Dame   University of Minnesota

Notre Dame, IN., U.S.A.   Minneapolis, MN., U.S.A.

Abstract

In complex memory tasks, states of awareness about memory processes (metamemory) are related to strategic behavior. Two studies are reported in which metamemory is measured independently of strategies and recall; "connections" among the factors show the predictive value of metamemory in tests of strategy transfer.

Introduction

In a recent theoretical paper, Campione and Brown (1978) presented a reformulation of intelligence theory. The model suggests two fundamental levels: an architectural system that features efficiency in coding and decoding and an executive system which has control processes (e.g., rules and strategies), a knowledge base, and Piagetian schemes. In the present paper, we address the issue of whether metacognition should be included in Campione and Brown's version of intelligence theory. First, we explore the explanatory merit of metacognition from a theoretical perspective; then we review data on a specific type of metacognition --metamemory.

Introspective knowledge about cognitive systems defines metacognition. Its function is to aid decisions about how best to deploy cognitive resources as individuals face complex, novel problems. Metamemory represents a special type of metacognition; it refers to self-knowledge about the memory system's operation (Brown, 1978; Flavell & Wellman, 1977). Knowledge in metamemory includes information about the various strategies employed during learning and/or retrieval of information and the interface of these strategies with other forms of background knowledge and with self-knowledge about processing abilities.

I. Metamemory Validation

Should the construct metamemory, or the more general term metacognition, be included in theories of intelligence and in corresponding assessment batteries? In order to answer this question, two issues need consideration: Does metamemory make theoretical sense? Does it have construct validity?

Metamemory has theoretical import insofar as it proves useful in explaining variations in strategy selection, implementation, modification, and invention. Flavell and Wellman (1977) have suggested that an awareness of the person, task, and strategy variables operating in specific memory or problem solving situations might be critical for the successful implementation of memory strategies. Subsequently, Flavell (1978) noted that an important concept in the explanation of production deficiencies could be the notion of metamemory. That is, a child's failure to use available strategies might be due to lack of appreciation for a strategy's utility. We propose that in many complex, novel memory tasks there is a causal relationship between metamemory, strategy use, and recall accuracy. It should be noted that Sternberg (1979) has recently implicated analogous concepts--metacomponents--as explanations of reasoning proficiency.

With respect to the validation issue, previous research searching for metamemory connections has focused on recall span prediction, memory monitoring, and recall readiness tasks. With such tasks it is difficult to obtain objective, independent measures for all three components of the connection--metamemory, strategies, and recall (cf. Borkowski and Cavanaugh, 1979). We now report two studies that surmount some of the problems in logic and measurement inherent in past metamemory research. Independent assessments of knowledge about memory, memory strategies, and recall performance characterize each study. The context of strategy transfer serves as the focus of the search for metamemory-strategy-recall connections.

II. Metamemory and Encoding Strategies

A.   Metamemory and the transfer of a cumulative-cluster strategy. Cavanaugh and Borkowski (in press) assessed the relationship between a task-specific index of metamemory and the maintenance of a trained cumulative-clustering rehearsal strategy. Third graders participated in five sessions; Sessions 1 and 5 assessed metamemory through a modified version of the Study Plan subtest from the Kreutzer, Leonard, and Flavell (1975) questionnaire. Sessions 2 and 3 consisted of training in the use of a cumulative-clustering rehearsal strategy. A list of words comprising a number of semantic categories was presented in blocked form and children were told to rehearse items in each category cumulatively until a change in categories occurred; then the cumulative rehearsal process began anew. Session 4 tested the maintenance and extention of the strategy to a new list, countries blocked by continents.

Children's responses to the metamemory questions were quantified according to the Kreutzer et al. (1975) guidelines. Results indicated a modest but significant correlation between strategy form and metamemory pretest scores at transfer for strategy-instructed children. Furthermore, all 18 of the instructed children who successfully maintained the strategy adequately described it three weeks after the transfer session and then correctly rearranged a random-order list to fit the requirements for using the cumulative-clustering strategy; only one of the nine instructed children who did not maintain the strategy correctly rearranged the list. Apparently, level of metamemory predicted strategy use at maintenance which, in turn, altered subsequent metamemorial-based action. Bidirectionality appeared to define the relationship between strategy use and metamemory (cf. Brown, 1978).

B. Metamemory and an interrogative strategy. On the basis of these initial findings, we extend our research to investigate metamemory-strategy-recall connections with EMR children (Kendall, Borkowski, & Cavanaugh, in press). We hypothesized that meta-memorial knowledge should predict the maintenance and generalization of an acquired interrogative strategy with paired-associate (PA) tasks. Children learned pairs of unrelated items by posing questions about them, then answering these inquiries with semantic elaborations relating each item's main attributes. For example, if the to-be-learned pair was nurse-toaster, the child might say: "Why is the nurse holding the toaster?" Then a relationship was formed: "The nurse is holding the toaster so she can make toast for the sick people." Two groups of EMR children (MA = 6 and 8) participated in pre- and post-test metamemory assessments (the Story-List, Study Plan, and Preparation-Object subtests from the Kreutzer et al. questionnaire), four training sessions in which a four part self-instructional study strategy was taught, a long-term test for retention of the pairs learned during the final training session, a strategy maintenance test with a new PA task (and new experimenter), and a strategy generalization test to lists of word triplets. Metamemory data was quantified as in the Cavanaugh and Borkowski (in press) study. The index of strategy use was based on probe tests at maintenance and generalization which assessed the extent of elaboration for each pair immediately after the recall trial. The most important results were the significant correlations relating quality of elaborations at generalization to metamemory pretest $(r = .50)$ and to metamemory posttest $(r =$

.46). Metamemory was related to performance and strategy use
during strategy maintenance and generalization but not during
strategy acquisition.

III. Feedback and Metamemory

Feedback refers to information supplied to an individual concerning accuracy of performance or the efficacy of a strategy.
In cognitive instructional research the purpose of feedback is to
increase the likelihood of strategy utilization during maintenance
and generalization. Kennedy and Miller (1976) reported that
verbal feedback following training of a rehearsal strategy significantly improved maintenance. Borkowski, Levers, and Gruenenfelder (1976) found that a brief film depicting the correct use of
an active mediational strategy preceding training enhanced strategy
maintenance for first-grade children. Cavanaugh and Borkowski
(in press) showed that feedback concerning a strategy's efficacy,
administered following a maintenance task, significantly improved
task-specific metamemory. Finally, Asarnow (1976) included
feedback in a self-instruction training package designed to implement
a repetitive rehearsal strategy; impressive strategy maintenance
was achieved and production deficiencies eliminated. We suggest
that the major role of feedback in instructional research is to
enhance metamemorial knowledge about a strategy's utility.
Feedback heightens metamemory by emphasizing the match between
task demands, strategic actions, and successful performance, including the experience of doing well on the task.

IV. Summary

Research on metamemory-strategy-recall connections indicates
a modest relationship between metamemory and encoding strategies
across ability groups with different memory tasks and strategies.
Metamemory-strategy connections are strengthened by feedback
procedures following strategy training and are more likely discovered when an acquired strategy is applied to a new problem.

Research is needed to develop metamemory tests that possess
greater reliability and more acceptable psychometric properties;
presumably such tests will rely less heavily on verbal questioning
and more on behavioral observations of children performing meta-
memorial actions. For example, Best and Ornstein (1979) measured
children's knowledge of acquired organizational strategies by asking
them to tell a younger child how to perform a memory task. Behavio
ally-based indices of metamemory may be more reliable and valid in-
dicators of knowledge about memory processes than verbal questionin
techniques.

These conclusions have several theoretical implications. For
memory theory, the fact that strategies are predictable on the basis

of amount and type of metamemorial knowledge needs to be recognized. Metamemory, as one component of metacognition, has an important, perhaps causal, relationship to memory processes (cf. Sternberg, 1979). As such, metacognition stands as a potential conceptual candidate for inclusion in a general theory of intelligence and its accompanying assessment batteries.

V. References

Asarnow, J. R.   Verbal rehearsal and serial recall: The mediational training of kindergarten children.   Unpublished master's thesis, University of Waterloo, 1976.

Best, D. L., & Ornstein, P. A.   Children's generation and communication of organizational strategies.   Paper presented at the Biennial Meeting of the Society for Research in Child Development, San Francisco, March 1979.

Borkowski, J. G., & Cavanaugh, J. C.   Maintenance and generalization of skills and strategies by the retarded.   In N. R. Ellis (Ed.), Handbook of mental deficiency: Psychological theory and research (2nd ed.).   Hillsdale, NJ: Erlbaum, 1979.

Borkowski, J. G., Levers, S. R., & Gruenenfelder, T. M.   Transfer of mediational strategies in children: The role of activity and awareness during strategy acquisition.   Child Development, 1976, 47, 779-786.

Brown, A. L.   Knowing when, and how to remember: A problem of metacognition.   In R. Glaser (Ed.), Advances in instructional psychology (Vol. 1).   Hillsdale, NJ: Erlbaum, 1978.

Campione, J. C., & Brown, A. L.   Toward a theory of intelligence: Contributions from research with retarded children.   Intelligence, 1978, 2, 279-304.

Cavanaugh, J. C., & Borkowski, J. G.   The metamemory-memory "connection":   Effects of strategy training and maintenance.   Journal of General Psychology, in press.

Flavell, J. H.   Metacognitive development.   In J. M. Scandura & C. J. Brainerd (Eds.), Structural/process theories of complex human behavior.   Alphen a. d. Rijn, The Netherlands: Sijthoff & Noordhoff, 1978.

Flavell, J. H., & Wellman, H. M.   Metamemory.   In R. V. Kail, Jr. & J. W. Hagen (Eds.), Perspectives on the development of memory and cognition.   Hillsdale, NJ: Erlbaum, 1977.

Kendall, C. R., Borkowski, J. G., & Cavanaugh, J. C.   Metamemory and the transfer of an interrogative strategy by EMR children.   Intelligence, in press.

Kennedy, B. A., & Miller, D.   Persistent use of verbal rehearsal as a function of information about its value.   Child Development, 1976, 47, 566-569.

Kessel, F. S.   Meta-this, meta-that, meta-the-other.   Paper presented at the Biennial Meeting of the Society of Research in Child Development, San Francisco, March, 1979.

Kreutzer, M. A., Leonard, C., & Flavell, J. H.   An interview

study of children's knowledge about memory. Monographs of
the Society of Research in Child Development, 1975, 40, (1,
Serial No. 159).
Sternberg, R. J.   The nature of mental abilities.   American Psychologist, 1979, 34, 214-230.

ADAPTATION TO EQUILIBRATION: A MORE COMPLEX

MODEL OF THE APPLICATIONS OF PIAGET'S THEORY

TO EARLY CHILDHOOD EDUCATION

Christine Chaillé

University of Connecticut

Storrs, Connecticut, U.S.A.

Abstract

Three levels of educational applications of Piagetian theory are delineated in this paper--the level of the learning experiences of the child; the level of the teacher's views of teaching and learning; and the level of the transmission of culture through the educational process. Each of these levels has been approached with a simple interactionist model of equilibration. Although this model has been a fruitful one and has generated considerable thinking and research, it may rest on oversimple epistemological assumptions. The problems that need to be addressed are restated given the more complex model of equilibration set forth by Piaget (1977).

Concerns of Piaget's work are centered around epistemological questions, involving the development and refinement of a formal theory that focuses on, among other issues, the construction of knowledge in children. Applying Piagetian theory to early education is thus contingent upon interpretation of the theory, and issues related to education arise from different sources than issues related to either genetic epistemology or developmental psychology. More critically, perhaps, attempts to apply theories of psychology to education are clouded by the oftentimes unstated, even unconscious assumptions that underly our already existing educational theories and practices. These assumptions can restrict or distort the understanding of a theory such as Piaget's that may be postulating different, or conflicting, assumptions about the nature of knowledge and the principles of growth and change.

Such distortion may have contributed to the fact that educational applications have been for the most part based on an overly simple equilibration model. In this model, interaction is conceptualized as occurring between subject and object--in educational terms, between the child and the environmental input. I will argue that a more complex model of equilibration should be applied to educational issues at this micro-level of analysis. Moreover, a more complex equilibration model will open up other levels of analysis, other areas of educational implications in which Piagetian theory should make important contributions.

What are the levels at which one can examine the educational applications of a theory? On the first, or "micro-level," analysis can be made of the processes of learning and teaching. On a second level, one can analyze the teacher's framework--the assumptions about teaching and learning that guide the teacher's actions (teaching style and method) and the choice of material to teach (curriculum selection). On a third level, one can analyze the educational process in its broadest sense, the processes by which a culture is perpetuated through educational and social institutions, or what has been termed "cultural reproduction" (Bourdieu & Passeron, 1977). For the most part, Piagetian theory has provoked response from educational research and practice only on the first, micro-level of analysis.

The Simple Equilibration Model at the Micro-Level

There have been numerous and diverse attempts to apply Piaget's theory to early childhood education. But what has been the theory that has been applied? Generally, the problem has been to determine how a child assimilates a new concept to given operational structures. Thus, it is appropriate in this approach to break down the material presented to, or encountered by, the child; to analyze the concept or the concrete materials. This would include examining the appropriateness of materials that are verbally presented versus concretely presented; the amount or type of teacher direction in a "lesson"; interest elicited by particular concrete manipulatives; the degree of perseverence shown by a child in solving a problem; and the amount and type of peer interaction stimulated by a situation. All of these examples involve analysis of the type and appropriateness of the environmental input. On the other hand, analysis can be made of the child's contribution to the teaching-learning situation, e.g., a description or diagnosis of the child's operational stage, symbolic representation capabilities, or perception of a particular concept, analyzing the type and appropriateness of the child's input in an interaction. Taken together, this encompasses the now-classic "match" (Hunt, 1961) between the child and the environment.

Applications to education which have resulted from this

approach have ranged from entire preschool programs (such as
Weikart's Cognitively Oriented Curriculum, described in Weikart,
Rogers, and Adcock, 1971) to what are essentially descriptions of
activities (such as the work of Kamii and DeVries (1976, 1978) on
number and physical knowledge). Two major problems have
resulted in difficulties in the adaptation of Piagetian approaches
to early childhood education programs: determination of operational
levels of children, and ascertaining appropriate environmental
modifications.

First, placement of a child at a given operational level is
difficult. This is partly due to the phenomenon of decalege (cf.
Pinard & Laurendeau, 1969), that theoretically predictable lack of
consistency across tasks. It is also increasingly recognized that
such factors as motivation, interest, and values play important
roles in performance, raising the recurrent issue of the competence/
performance distinction that applies no less to Piagetian "testing"
than to psychometrics. The more the determination of the operational level of a child becomes situation- and concept-specific, the
less importance will be placed in the notion of determining the
child's specific operational level. Concern becomes directed
toward the particular concept or situation, and the real life
functioning of children.

Second, the range of environmental modifications is not clear
and/or not feasible in the realities of many preschool classrooms.
That is, once a child's "level" is determined, the appropriate environmental response, whether it be the type of materials presented
to the child, the type of teacher-initiated dialogue, or the manipulation of possibilities for social interaction, is far from clearly
dictated by either Piagetian theory or the current state of educational research. In fact, most Piagetian "prescriptions" for teaching emphasize the need for flexibility and responsivity to the cues
given by the particular situation (cf. Kamii & DeVries, 1978;
Forman & Kuschner, 1977), an "attitude in teaching" similar to
and based on the principles underlying the clinical method of
Piagetian research.

What is particularly interesting about this conclusion is the
question of why educators are looking to cognitive theory to
guide practice. Description of how learning takes place--which is
where the concept of a match can be most useful--does not necessarily imply prescription. The problem is that prescription has
become as much a part of "teaching" as experimental control is a
part of behaviorist psychology. Both require, and both assume
the possibility and desirability of, control. The need to bring
these definitions of teaching to light becomes essential in a more
complex interactionist model of equilibration.

A More Complex Model of Equilibration

There is nothing new about the model of equilibration that is described here. Increasing dissemination of Genevan research and theory, and constant reformulations and articulations of the theory by Piaget, have occurred too quickly for their assimilation by those who would infer educational implications.

In the equilibration model presented in Piaget (1977), the interaction described above--an interaction between subject and object--represents a causal interaction of the most simple and elementary type. In addition to interactions describing causal actions, elementary interactions can describe logico-mathematical actions (cf. Kamii & DeVries, 1978). Even at this "simple" level, interaction in neither case is really between subject and object. Rather it is between the object as assimilated (the scheme) and the accommodation to the object. The object as assimilated Piaget terms the "observable" which is "anything that can be recorded thorugh a simple factual (or empirical) observation...In this wide sense, regular relationships or functions between two observable features are themselves also observable features" (Piaget, 1976, p. 345). Even at this level of interactions (Type I), it is not a question of a match between empirical reality and operational level. At the next level of interactions (Type II)--those involving inferential coordiations, or coordinations of Type I interactions--it is deduction rather than observation that is "acted on." In both types of interactions, active construction rather than copying of an empirical object results in constant new interactions precisely because of the imperfect match between the scheme as assimilated and the accommodation. An example can be seen in the small (and large) gaps between the infant's grasping scheme and the object the infant is trying to grasp. These disturbances, whether actual or "virtual" (inferred), necessitate reconstruction. Reference to equilibration rather than equilibrium (which may only be theoretically possible) underscores the functional dynamics of this process, as compared with the usual connotations of structuralism with stasis and stages (Chaillé, 1978, 1979).

The reinterpretation of the Piagetian equilibration model involves a significant turning away from a focus on operational stages, and a renewed focus on functional dynamics of all kinds and at all levels, including problem-solving, object exploration, early symbolization, language acquisition, and peer interactions. These topics have been the subject of recent and ongoing Genevan research (cf. Karmiloff-Smith & Inhelder, 1975). Attention is now being focused on a more detailed functional analysis of the variables described as environmental inputs and as aspects of the child's operational level. Examination of the complex interactions involved in the active construction of various kinds of knowledge requires more than an interactionist model of learning; it involves the acceptance and understanding of more profound epistemological assumptions.

The Level of Transmission

Relatively few have addressed the issue of whether or not the views of curriculum that can be based on Piagetian equilibration theory can be adopted by teachers or teachers in training who may be approaching education with a different set of epistemological assumptions (some exceptions are Duckworth, 1972, and Sigel, 1978). What can happen, in fact, is the systematic distortion of Piagetian "curriculum" ideas to conform to the assumption that teaching involves the transmission of knowledge from teacher to child, with the child in a relatively passive role vis-a-vis the learning process. This distortion can occur when an individual teacher is learning the Piagetian model or when an institution is adapting aspects of Piagetian theory. At the preschool level, these problems of a mismatch between assumptions and activities can be seen in the diverse views among early childhood educators on the specific values of children's play, a traditionally "encouraged" form of activity that is seen as essential in an early childhood program yet for many different reasons depending on the orientation of the teacher or researcher (cf. Chaillé & Young, in press). A more complex model of interaction implies the need to directly confront the nature of these assumptions and the specific ways they can be translated into educational practice.

Cultural Reproduction

Assumptions about the nature of teaching and learning, which we are saying need to be examined more closely than they have been examined in the past, are embodied in the institutions in which teaching and learning occur. Recent work in the sociology of education (cf. Bourdieu & Passeron, 1977), presents models for considering educational curriculum and methods as problemmatic. This raises the possibility of parallels between the processes of learning at the micro-level, processes of teaching at the level of transmission, and processes of reproduction at the level of educational institutions. There are some interesting similarities between Piagetian structuralism and Bourdieu's theories of cultural reproduction that should be explored with such parallels in mind.

Expanding views of development are opening up new directions for educational theory and practice, and it is time for a reexamination of the theoretical models as they are applied to early education.

References

Bourdieu, P. and Passeron, J. Reproduction in education, society and culture. Beverly Hills, California: Sage Productions, 1977.

Chaillé, C. The child's conceptions of play, pretending, and toys: Sequences and structural parallels. Human Development, 1978, 21, 201-210.

Chaillé, C. Adaptation and equilibration: An essay review of Jean Piaget, The Development of Thought and The Grasp of Consciousness. Harvard Educational Review, 1979, 49, 101-106.

Chaillé, C. and Young, P. Some issues linking research on children's play and education: Are they "only playing"? International Journal of Early Childhood, in press.

Duckworth, E. The having of wonderful ideas. Harvard Educational Review, 1972, 42, 217-231.

Forman, G., and Kuschner, D. The child's construction of knowledge: Piaget for teaching children. Belmont, California: Wadsworth Publishing Company, 1977.

Greenfield, P. M. Structural parallels between language and action in development. In A. Lock (Ed.), Action, Gesture and Symbol: The Emergence of Language. New York: Academic Press, 1978. Hunt, J. McV. Intelligence and Experience. New York: Ronald Press, 1961.

Kamii, C., and DeVries, R. Piaget, Children, and Number. National Association for the Education of Young Children, 1976.

Kamii, C., and DeVries, R. Physical Knowledge in Preschool Education: Implications of Piaget's Theory. New York: Prentice Hall, 1978.

Karmiloff-Smith, A., and Inhelder, B. "If you want to get ahead, get a theory." Cognition, 1975, 3, 195-212.

Piaget, J. The Grasp of Consciousness: Action and Concept in the Young Child. Cambridge, Mass.: Harvard University Press, 1976. Piaget, J. The Development of Thought: Equilibration of Cognitive Structures. New York: The Viking Press, 1977.

Pinard, A., and Laurendeau, M. "State" in Piaget's cognitive-developmental theory: Exegenesis of a concept. In Elkind, D., and Flavell, J. (Eds.), Studies in Cognitive Development. New York: Oxford University Press, 1969.

Sigel, I. E. Constructivism and teacher education. The Elementary School Journal, 1978, 78, 333-338.

Weikart, D. P., Rogers, L., Adcock, C., and McClelland, D. The Cognitively Oriented Curriculum. ERIC, National Association for the Education of Young Children, 1971.

A MODEL OF COGNITIVE DEVELOPMENT

J. A. Keats

The University of Newcastle

Australia

The fact that the intelligence quotient in either of its common forms is an inadequate representation of cognitive development has been widely recognized for many decades. Most longitudinal studies over wide age ranges indicate that with at least a considerable proportion of subjects there are systematic trends in IQ changes. One of the recent studies, McCall (1973), classified subjects in terms of types of changes and was able to relate type of change to type of child raising practice in a number of cases. Such studies indicate that a one parameter representation of cognitive development is inadequate.

Problems associated with the definition of a cognitive growth curve are described by Bayley (1955) but these problems are largely removed by tests constructed to meet the requirements of the ability model proposed by Rasch (1960). The British Ability Scales (1978) for example meet these requirements. The model proposed here assumes the availability of ability scores in this sense for subjects at a number of different age levels. The ability of subject i at time since birth, t_j, is represented by A_{ij}. The two parameters associated with each subjects are denoted by c_i and d_i and the equation relating these parameters to ability at a given age is assumed to be:

$$A_{ij} = \frac{t_j}{c_i t_j + d_i}$$

This expression differs somewhat from that suggested by Halford and Keats (1978), and explored further by Keats (1978) but the new form leads to independent estimates of the parameters.

The following properties of this cognitive growth curve may be derived:

1) Ability A_{ij} approaches, but never reaches an asymptotic value of $1/c_i$.

2) The subject reaches half this asymptotic value at an age of $t_j = d_i/c_i$ in whatever units age is measured, that is, months or years, etc.

3) If a group curve is defined in terms of the harmonic mean, $H(A.j)$, of the subjects' abilities at each of a number of age levels t_j then:

$$H(A.j) = \frac{t_j}{\bar{c}t_j + \bar{d}}$$

Thus the group curve has the same mathematical form as the individual curves and approaches an asymptote of $1/\bar{c}$. The semi-asymptotic value of $\frac{1}{2\bar{c}}$ is reached when

$$t_j = \bar{d}/\bar{c}$$

4) From the group curve it is possible to define a mental age t_k corresponding to the chronological age t_j for subjects with A_{ij} values of less than $1/\bar{c}$. For such subjects a ratio IQ may be defined as:

$$IQ_R = \frac{100 \cdot t_k}{t_j} \qquad \frac{100\bar{d}}{(c_i - \bar{c})t_j + d_i}$$

This derivation reveals clearly the weaknesses of this type of IQ, that is the lack of definition when $A_{ij} \geq 1/\frac{1}{c}$ and the fact that for subjects for whom $\frac{1}{c} = \frac{1}{\bar{c}}$ the IQ_R is stable, but at a value unrelated to the adult asymptotic level in that it depends solely on d_i, the _rate_ of development.

5)   An alternative IQ measure, which may be shown to approximate the deviation IQ, can be defined as:

$$IQ_D = \frac{100 \cdot A_{ij}}{H(A.j)} = \frac{100(\bar{c}t_j + \bar{d})}{c_i t_j + d_i}$$

which may be rewritten

$$1/IQ_D = c_i + \frac{(\bar{c}d_i - \bar{d}c_i)}{\bar{c}t_j + \bar{d}}$$

from which it can be shown that IQ_D approaches $1/c_i$, the asymptotic value as t_j becomes arbitrarily large. Furthermore for subjects who reach their own semi-asymptotic value, $1/2c_i$, at the same age as the group curve, $H(A.j)$ reaches its semi-asymptotic value, $\frac{1}{2\bar{c}}$, $IQ_D = \frac{1}{\bar{c}_i}$ at all age levels. For subjects of this kind, IQ_D will be constant apart from random fluctuations attributable to errors of measurement. The findings of McCall et al. (1973), reveal that approximately 40 percent of subjects are of this kind.

One of the consistent findings of developmental studies is that the correlation between a child's IQ and his mother's IQ increases with age. Munsinger (1975) notes that this finding also holds for children who have been with foster parents from an early age. Unfortunately the available data on this point are not very extensive but those reported by Skodak and Skeels (1949) are consistent with the conclusion that c_i, the parameter related to the asymptotic ability level has a substantial genetic component whereas, d_i the parameter determining rate of development does not. Apparent anomalies in these data noted by Munsinger (1975) are explicable in terms of the difficulties in using the ratio IQ already noted.

Spada and Kluwe (1977) examine the problem of relating psychometric models to Piaget's (1947) theory of cognitive development. Their data suggest that a strictly deterministic model is not appropriate. They then examine a probabilistic model related to the Rasch model which yields a better representation, but still does not suggest that stage-wise development will occur. They then propose a more complex form which could represent stage-wise development, but do not develop it to a testable stage.

One of the problems with the usual psychometric models of cognitive development is that, unlike the model proposed here, they do not include time or age as a variable. Stage-wise development can be accounted for in the present model by assuming that for some tasks a minimum ability level (A_o) must be reached before the subject has any possibility of giving the correct response with adequate explanation as required for Piaget's tasks. However for $A_{ij} > A_o$ the subject will have a probability greater than zero of giving the correct response.

It has been shown that $1/A_{ij}$ has additive properties with respect to $1/t_j$. For this reason it will be assumed that the probability of a correct response will depend on $1/A_o - 1/A_{ij}$ rather than $A_{ij} - A_o$. Using a form of the Rasch model one obtains:

$$P_{ijk} = \frac{1/A_o - 1/A_{ij}}{1/A_o - 1/A_{ij} + e_k}$$

where P_{ijk} is the probability of the individual (i) at time (t_j) will give the correct response to an item of difficulty e_k.

Then:

$$P_{ijk} = \frac{d_i(t_j - t_{io})}{d_i(t_j - t_{io}) + e_k \cdot t_j \cdot t_{io}}$$

where t_{io} is the age at which individual (i) reaches the ability level A_o.

The interesting feature of this expression is that it depends only on the developmental parameter (d_i) and not on the asymptote parameter (c_i). For this reason it would be expected that items which show stage-wise effects will be much more susceptible to environmental conditions. Piaget has often insisted that the cognitive development he is describing depends on assimilating and accommodating to the environment. It would thus be expected that the present formulation would be consistent with this type of development. Empirically it has been found very difficult to devise items which test the operations described by Piaget and satisfy the usual psychometric properties required to give reliable tests. This phenomenon of relative instability across tasks would be expected if environmental influences are important. Even though these operations have been developed through exchanges with the environment, it would appear that a certain minimum ability is required before the individual can benefit from these exchanges. Thus according to Inhelder (1968) certain types of mental defectives do not completely master any concrete operational tasks despite many years of interacting with the environment.

The model of cognitive development proposed here appears to be the first to separate the developmental aspects of cognition from the asymptotic level approached in adulthood. Its theoretical usefulness will depend on the results of further research to investigate the interpretations placed on the two proposed parameters. Whether or not the estimation of a second parameter is of sufficient practical significance to justify the extra effort in applied areas also remains to be established.

References

Bayley, N. On the growth of intelligence. American Psychologist, 1955, 10, 805-818.

British Ability Scales. Windsor: NFER Publishing Co. Limited, 1978.

Halford, G. S., and Keats, J. A. An integration. In J. A. Keats, K. F. Collis and G. S. Halford (Eds.), Cognitive Development: Research based on a neo-Piagetian approach. Chichester: John Wiley and Sons, 1978.

Inhelder, B. The diagnosis of reasoning in the mentally retarded. (translation from French, 1942), 1968.

Keats, J. A. A proposed form for a developmental function. In J. A. Keats, K. F. Collis and G. S. Halford (Eds.), Cognitive development: Research based on a neo-Piagetian approach. Chichester: John Wiley and Sons, 1978.

McCall, R. B., Appelbaum, M. I., and Hogarty, P. S. Developmental changes in mental performance. Monographs of the Society for Research in Child Development, 1973, 38 (3, Serial No. 150).

Munsinger, H. The adopted child's IQ: A critical review. Psychological Bulletin, 1975, 82, 623-659.

Piaget, J. The psychology of intelligence. London: Routledge and Kegan Paul, 1947.

Rasch, G. Probabilistic models for some intelligence and attainment tests. Copenhagen: Danmarks Paedogoïsch Institut, 1960.

Skodak, M., and Skeels, H. M. A final follow-up study of one hund adopted children. The Journal of Genetic Psychology, 1949, 75, 85-125.

Spada, H., and Kluwe, R. Psychometric models of intellectual develc ment and their reference to the theory of Piaget. Draft paper, Institute for Science Education, University of Kiel, 1977.

THE USE OF A PIAGETIAN ANALYSIS OF INFANT
DEVELOPMENT TO PREDICT COGNITIVE AND
LANGUAGE DEVELOPMENT AT TWO YEARS

Linda S. Siegel

McMaster University

Ontario, Canada

Abstract

The ability of a Piagetian based infant test, the Uzgiris-Hunt, to predict subsequent language and cognitive development and to detect infants at risk for developmental problems was assessed. The infants were administered the Uzgiris-Hunt scale at 4, 8, 12 and 18 months, and the Bayley Scales of Infant Development and the Reynell Developmental Language Scales at two years. The Uzgiris-Hunt scale and most of its subtests, were significantly correlated with cognitive and language development at two years. Object relations items and the understanding of means-end relationships were predictive of language development. The analysis of infant cognitive abilities within a Piagetian framework appears to be a promising method for assessing early development.

It has been suggested that the testing of infant abilities may be of greater value if more specific cognitive functions are measured (e.g., Honzik, 1976). The Uzgiris-Hunt test (1975) measures the development of various Piagetian concepts such as object permanence and the understanding of means-end relationships. As most infant tests do not measure these functions in detail, the present study used the Uzgiris-Hunt to assess specific cognitive abilities.

It has been postulated that certain aspects of early cognitive development are related to the acquisition of language. According to Moore and Meltzoff (1978), the understanding of object permanence

271

and identity is a critical aspect of language development. When children become aware that objects can retain their identity in the face of transformation, they have acquired the basis by which they can attach labels to objects. Another aspect of language is its function as a communicative activity. If, as Bates, Camaioni and Voltera (1976) have suggested, the child must understand the role of language in influencing others and the significance of intentionality in communication, then the understanding of how to manipulate and control the environment should be important to language development. Gestural imitation has also been assumed to be relevant to language acquisition (Morehead & Morehead, 1974). The Uzgiris-Hunt test was designed to assess these and other aspects of early cognitive development and to relate them to subsequent language development.

Method

Subjects

The subjects, 148 infants from Hamilton of Ontario and surrounding area (100 kilometer radius) enrolled in a prospective study of preterm (birthweight under 1500 grams) and fullterm infants. The sample is described in detail in Siegel, Saigal, Rosenbaum, Morton, Young, Berenbaum, and Stoskopf (1979).

Procedure

The children were administered an adaptation of the Uzgiris-Hunt scale at 4, 8, 12 and 18 months, the Reynell Developmental Language scale, a standardized test measuring language expression and comprehension, and the Bayley Scales of Infant Development at 24 months.

Uzgiris-Hunt Scale - These are tests of cognitive capacities of infants based on Piagetian theory. The following scales were used: (a) Schemes - a test of the type of variety of activities that a child exhibits with familiar objects (e.g., doll, car); (b) Visual Pursuit and Object Permanence - test of the child's ability to visually and/or manually search for objects that are hidden; (c) Means - the extent that a child tries to influence and problem solve in the environment by, for example, using tools such as a stick to reach an object beyond his or her immediate reach; (d) Concepts of Space - the capacity of the child to understand and use containers, recognize obstacles; (e) Gestural Imitation-- the ability of the child to imitate familiar (e.g., stirring a spoon in a cup) and unfamiliar (e.g., scratching a surface) gestures; (f) Vocal Imitation--the ability of the child to imitate familiar and unfamiliar sounds and words; and (g) Causality--the ability of the child to understand and try to activate some environmental event (e.g., pulling a string to make a music box work).

Results

The correlations between the Uzgiris-Hunt scales and the 24 month Bayley scores are shown in Table 1. As can be seen in Table 1, many subscales of the Uzgiris-Hunt correlate highly with the Bayley, particularly the Means, Schema, Object, and Space subtests. At 18 months the correlations are lower, probably because many of the infants are performing at ceiling level on the tests.

Table 1. Correlations between Uzgiris-Hunt Scales in infancy and Bayley Scores at 2 years

Schema	Visual Pursuit & Object Permanence	Means	Space	Gestural Imitation	Vocal Imitation	Total
		4 Months				
.42***	.34***	.49***	.46***	.20	.32***	.49***
		8 Months				
.33***	.25**	.42***	.38***	.29**	.23*	.40***
		12 Months				
.27**	.37***	.30**	.43***	.40***	.11	.51***
		18 Months				
.20	.37***	.15	.23*	.09	.14	.41***

***p < .001
**p < .01
*p < .05

Table 2 shows the correlations between Piaget scales and the Reynell Language scores at 24 months. Certain subscales (e.g., Means, Space, Schemas) are predictive of language development. As with the correlations between the Uzgiris-Hunt and the Bayley, the lack of correlations of 18 month scores are probably a reflection of ceiling effects.

Table 2. Correlations between Uzgiris-Hunt Scale in Infancy and Reynell Language Scale at Two Years

Uzgiris-Hunt

	Schemas	Visual Pursuit & Object Permanence	Mean	Space	Gestural Imitation	Vocal Imitation	Total Score
Reynell			4 Months				
Comprehension	.33***	.26*	.40***	.34***	.22*	.22*	.38***
Expression	.37***	.07	.35***	.37***	.17	.26***	.37***
			8 Months				
Comprehension	.27***	.27**	.42***	.35***	.21*	.35***	.39***
Expression	.19	.14	.40***	.27**	.17	.27**	.29**
			12 Months				
Comprehension	.22*	.39***	.41***	.22*	.33**	-.01	.45***
Expression	.41***	.30**	.40***	.18	.31**	.05	.45***
			18 Months				
Comprehension	.32***	.39***	.14	.18	.23*	.15	.37***
Expression	.31**	.37***	.23*	.30**	.31**	.13	.43***

*** $p < .001$

** $p < .01$

* $p < .05$

The Uzgiris-Hunt scales are differentiated between the infants who were delayed (Bayley MDI   85) at 2 years and those who were not. The Uzgiris-Hunt total scores differentiated between the delayed and the non-delayed at each age, 4, 8, 12, and 18 months. The following subscales differentiated between the delayed and non-delayed groups: 4 months - schemas, object permanence, means, space, vocal and gestural imitation;   8 and 12 months - schemas, object, means, space, gestural imitation;   18 months - object, means, causality.

Discussion

The Uzgiris-Hunt scale and a number of its subtests predicted cognitive development at 2 years as measured by the Bayley. These scales can be viewed as tests of problem solving ability and the significant correlations at different ages indicate certain continuities in mental development.

The Uzgiris-Hunt also predicted language development. The object concept items are predictive of language development indicating that the rudimentary symbolic functions involved in searching for a vanished object may be precursors of language development, as predicted by Meltzoff and Moore (1978). The means subtest was also correlated with language development; this subtest involves an understanding of the relationships with the environment, and the abilities tested may be precursors of the skills involved in understanding the communicative functions of language. Gestural and vocal imitation were, in come cases, significantly correlated with language development but these correlations were of a lower magnitude than the means and object relations subtests.

The analysis of infant cognitive abilities using a Piagetian framework appears to be a useful one for predicting normal and atypical development.

References

Bates, E., Camaioni, L. & Voltera, V. Sensorimotor performatives. In E. Bates Language and context: The acquisition of pragmatics. New York: Academic Press, 1976.
Honzik, M.P. Value and limitation of infant tests: An overview. In M. Lewis (Ed.) Origins of intelligence: Infancy and early childhood. New York: Plenum Press, 1976.
Moore, M.K. & Meltzoff, A.N. Object permanence, imitation, and language development in infancy: Toward a neo-Piagetian perspective on communicative and cognitive development. In F.D. Minifie and L.L. Lloyd (eds.), Communicative and cognitive abilities: Early behavioral assessment. Baltimore: University Park Press, 1978.

Morehead, D.M. & Morehead, A. From signal to sign: A Pia-
    getian view of thought and language during the first
    two years. In R.L. Schiefelbusch and L.L. Lloyd (Eds.),
    Language Perspective: Acquisition, retardation, and inter-
    vention. Baltimore: University Park Press, 1974.
Siegel, L.S., Saigal, S., Rosenbaum, P., Morton, R.A., Young,
    A., Berenbaum, S. and Stoskopf, B. Predictors of develop-
    ment in preterm and fullterm infants: A model for detecting
    the "at risk" child. Submitted for publication.

TESTING PROCESS THEORIES OF INTELLIGENCE

Earl C. Butterfield

University of Kansas

Kansas City, Kansas, U.S.A.

I. Introduction

A dauntingly complex but necessary research strategy follows from two simple beliefs about intelligence. The first belief is that intelligence develops: behavior becomes increasingly complex and abstractly organized with age. The second belief is that individual differences in intelligence are general: people who perform intelligently in one situation are more likely than people who don't to perform intelligently in another situation. Given that there are specialized forms of knowledge and specialized modes of thought, it is still true that to be termed intelligent a person must behave in generally effective ways. Despite their simplicity, these two beliefs are universally accepted. The developmental character of intelligence is accepted by process and structural theorists alike; it is accepted by continuity and noncontinuity theorists, by those who do and those who do not subscribe to stage theories, as well as by those who accept the antitheoretical view that intelligence is only what IQ tests measure. The belief that intellectual differences are general can be seen in the functionalist argument that intelligence is adaptability, since adaptability amounts to performing well in diverse situations. It can be seen in the Piagetian argument that an instructional experiment cannot be claimed to have influenced intelligence unless it has changed a wide range of uninstructed behaviors as well as the instructed ones. It can be seen in any standardized test of intelligence, since even the factorially purest tests yield composite IQ or MA scores. The research implications of these two beliefs fall on all who would test theories of intelligence.

My purpose in this paper is to translate the implications of these two beliefs into a research strategy for validating process theories of intelligence. I realize that it may not be possible to maintain completely the distinction between process and structural theories, since a key distinction between process and structure is that the former varies across time and the latter does not. Whether a factor is observed to change can depend crucially upon the rate at which its behavior is sampled, so that a slowly changing process can appear stable, like a structure, if too little time passes between observations. I realize too that most theories of intelligence are reasonably considered a mix of process and structural concepts, and there are probably no purely structural theories. Such considerations notwithstanding, I will be concerned in this paper with theories or aspects of theories that concern processes. By process, I mean a factor whose manner of change is specified in theory and is manipulable. A factor that cannot in theory be experimentally manipulated is termed structural. There are accepted ways of studying structural concepts, as by showing invariance from one setting or person to another of a parameter specified in a mathematical model. But to use such an approach with any precision requires control of relevant process variables, which will not be possible until all aspects of the strategy required to study intellectual processes have been implemented programmatically. For this reason, a clarification of how to study intellectual processes should strengthen research approaches to both process and structural aspects of intelligence.

II.   Research Implications of Intellectual Development

In cognitive theory, behavior is distinguished from processes that are said to underlie it. The theoretical goal is to explain behavior by reference to processes. Therefore, testing cognitive theory requires the use of research designs and dependent measures that allow separate inferences about process and performance. To assure that performance has been explained, it is also necessary to show relationships between performance measures, on the one hand, and process measures and manipulations, on the other hand.

The belief that intelligence develops is based on the observation that as children age their behavior becomes more complex and abstractly organized. The generic hypothesis of developmental cognitive theory is that at least some of the processes that underlie performance also become more complex and abstractly organized with age. The research strategy required to determine whether changes in underlying processes explain intellectual development must allow for the possibility that only some processes develop, and it must make provision for determining which processes do and which do not change with age.

The factor of cognitive development and the process/performance distinction require the use of the entire strategy outlined in Table 1 to validate a process theory of intelligence. The strategy begins with three preliminary steps, the first two of which are judgmental. Step 1 is to choose an important intellective domain of investigation. As in all judgmental matters, importance lies in the mind of the investigator, but there are consensual constraints. Since Galton's time, few have judged the study of sensory thresholds or simple reaction time as importantly intellective. Matters having to do with language, world knowledge, or memory are far more likely to be agreed upon nowadays as importantly central to intelligence. Having selected a domain of investigation, one must settle on some criterion task(s). Most investigators settle on one, though there is a trend in cognitive research toward the use of multiple performance measures. This stems in part from an increased recognition of the importance of establishing the generality of one's cognitive analyses, and more will be said of this later in this paper. The third step is to establish that performance on one's criterion measure(s) changes with age. This is a simple correlational matter. Having taken care of these preliminaries, the research strategy begins in earnest at Step 4 (see Table 1) with a process analysis of performance done within narrowly defined age groups. It continues, in Step 5, with demonstrations that the processes identified in Step 4 are age related. It moves, in Steps 6 and 7, to instructional experimentation designed to make the process theory meet the logical requirements of manipulative experimentation. The paragraphs in the next section of this paper offer reasons for including Steps 4 through 7 in the strategy. After that comes consideration of the implications following from the generality of individual differences.

A.   Analyze Processes Within Age Groups

Even though a goal of process theories of intelligence must be to explain intellectual development, Table 1 calls first for analyses performed within narrowly defined age groups. The purpose is to give validity, independent of age, to each process that accounts for any performance variance. Without such validity, no clear conclusions can be drawn from establishing process/age relations, which is called for in Step 5. Since age cannot be accelerated, reversed, or otherwise manipulated, some way must be provided to determine whether any process correlate of age arises from some unidentified confound of age. One such provision is to establish the validity of each process within narrowly defined age groups, before correlating it with age. Another provision is to produce and validate a process theory that accounts for all of the within-age variance in performance on the criterion tasks used to study intellectual development. Having such a complete account allows, during Step 5, determinations of which processes do not develop. A complete process theory of intelligence will include concepts that are not developmental as well as those that are. Moreover, until a process theory accounts for all of the variance in its target performance measure(s), any other incomplete process account can be claimed to be more basic, and someone will always accept the claim (Chase, 1973; Newell, 1973). As long as any appreciable variance remains unaccounted, there will be irrelevant disputes about which processes explain most elegantly or parsimoniously particular sorts of performance. Only an exhaustive account of variance cannot be challenged capriciously. Given one exhaustive account of variance, debate ends. Given two or more exhaustive accounts, disciplined considerations of elegance, parsimony, generality, and personal preference become relevant.

The goal of accounting completely for performance variance within ages is necessary to validate fully a process theory, but it cannot be held as prerequisite to moving to Step 5 of the validation strategy. If it were a prerequisite, developmental studies could not yet be performed. Including this goal is intended to remind investigators that, until it is reached, strong interpretations of developmental studies are not possible.

B. Correlate Process Measures With Age

The purpose of Step 5 is to determine whether a process changes with age. It also provides a test of the developmental completeness of a process theory. If the analysis upon which a theory builds is developmentally complete, then it will be possible to reduce performance/age correlations to zero by partialling out indices of processes that develop.

Step 5 also provides information necessary to respond effectively to a question that inevitably arises in response to studies of the sort outlined in Steps 6 and 7. Such instructional studies are generally reported by investigators with a behavioral rather than a cognitive orientation. Usually, such studies are not preceded by developmental process studies of the sort outlined in Step 5. Rather, the behavioral analyst takes raising or lowering criterion performance as his goal, and he modifies his instructional approach intuitively until he accomplishes his goal. Having done so, a cognitivist will almost invariably ask whether his instructional routines mimic or can be taken as a model of the normal course of development. A thoughtful behaviorist will say that his instructions stand as a possible model of how development might normally proceed, but he will confess that he cannot assert that it is a model of how development does proceed. Then, it will often happen that the behaviorist's work will be dismissed by the cognitivist as developmentally irrelevant, particularly if the cognitive critic can think of some developmental fact to suggest that the model implicit in the behaviorist's instruction might not be a good one for normal development. Step 5 provides data to justify the assertion that the processes instructed in Steps 6 and 7 do in fact change in the normal course of development. Thus, if behaviorists who have used instruction in generalized imitation as a prerequisite to teaching language to severely retarded children had shown first that generalized imitation precedes language development and accounts for normal children's language acquisition, their work would have been less readily dismissed by cognitivists as irrelevant to normal development.

Step 5 is stated in terms of chronological age, but mental age can be a more appropriate index of developmental level. The strategy allows the use of MA or IQ as well as CA. In fact, the strategy in Table 1 is applicable to any sort of comparative research. Thus, the study of differences among cultures would begin, in Step 4, with analyses performed separately within different cultures, and it would proceed, in Step 5, to comparisons among cultures. A more general expression of the strategy can be found in Butterfield (1978).

C. Eliminate Age Differences With Process Instruction

Cognitive theory in general is vulnerable to the criticism that its empirical bases are weak. It can fairly be said that the ties between the concepts of basic cognitive science and its data are tenuous (Anderson, 1976; Schank, 1976; Townsend, 1972, 1974). Developmental cognitive theory is only slightly less immune to this criticism than basic cognitive theory (Butterfield and Dickerson, 1976; Butterfield, 1978). Some argue that it may be impossible with empirical methods alone to affirm any theory satisfactorily (Lachman, et al., 1979; Weizenbaum, 1976). Nevertheless, the premise of Steps 6 and 7 is that applying the logic of manipulative experimentation to process explanations will greatly strengthen the ties between cognitive theory and data. In the first place, process instruction that affects performance shows most directly that the process is real. Perhaps more importantly, applying the full instructional logic provides the strongest possible basis for claims about the normal course of cognitive development.

The logic of Steps 6 and 7 is that instructed processes can be invoked as explanations of age or other group differences only if identical instructions are applied to various (age) groups, and then only if the instructions leave the groups performing at identical levels. The effect of the instructions can be to raise the performance of the younger group (Step 6) or to lower the performance of the older group (Step 7). However, if after instruction there remain reliable differences between the age groups, then the processes affected by the instructions may not be responsible for differences between the ages under uninstructed conditions. In Step 6, where instructions are intended to improve poor performance, the notion is that older groups who naturally perform better are already using the instructed processes, but the younger groups who perform poorly are not. Therefore, the more accurate group should benefit little or none from the instructions, but the less accurate group should benefit greatly. Conversely, in Step 7, where the instructions are intended to eliminate the processing thought to account for adults' accurate performance, the inaccurate children should be impaired relatively little, since they are presumably not using the target processes anyway. If the goal is to account for why young children perform inaccurately, then the instructional approach requires that older people be instructed along with the younger ones.

In its most definitive form, which is admittedly not yet attainable, the instructional experiment leaves the performance of either the older (Step 6) or younger (Step 7) group unchanged, and the performance of all groups identical. Implementing such an experiment would require a complete process understanding of the development of some intellectual performance, as well as accurate age norms of when the relevant processes have developed as completely as they will without special tuition. Given that there is no process analysis that will account completely for any cognitive performance, producing identical group performances is improbable: older groups will likely perform better than younger ones even after instruction, unless ceiling or floor effects are encountered. Moreover, there is ample evidence that fully mature individuals do not process optimally, so that older groups will almost always benefit from process instructions that are not carefully constrained by a knowledge of how far development carries people toward optimal processing. As long as the older group benefits, the process account of development is incomplete, even if the process analysis of within-age performance is complete. For these reasons, there must be a constant interplay and recursiveness between the various steps in the research strategy, and rules to guide this interplay are given in connection with Steps 6 and 7 (see Table 1).

In order to make process instruction experiments fully interpretable it is necessary to take inobtrusive measures of the instructed processes. The goal of such experiments is to change performance by manipulating processes, and, especially when process analyses are incomplete, it is entirely possible to influence process without having a marked influence on performance. It is necessary to determine when a failure to change performance markedly results from a failure to change the target process, and making that determination requires the use of inobtrusive process measures during instruction. I noted above that instructions designed to improve the performance of younger people will often improve that of older people too. When that happens, inobtrusive process measures taken prior to instruction are needed to determine whether the older people who benefited did so because they were processing relatively youthfully before instruction. Whenever the effects of instructions are assessed with a posttest, process

Table 1

HOW TO VALIDATE A PROCESS EXPLANATION
OF COGNITIVE DEVELOPMENT

Step 1. CHOOSE AN IMPORTANT COGNITIVE DOMAIN

Step 2. SELECT CRITERION TASK(S) THAT FAIRLY REPRESENT
PERFORMANCE IN YOUR CHOSEN DOMAIN

Step 3. SHOW THAT PERFORMANCE ON YOUR CRITERION TASK(S)
CORRELATES WITH AGE

So far, the work will have been judgmental (Steps 1 & 2) and descriptive (Step 3). Steps 4 through 7 are efforts after explanation.

Step 4. PERFORM A PROCESS ANALYSIS OF PERFORMANCE ON
YOUR TASK(S), WITHIN AGES

    A. Make process measurements that correlate with performance.

    B. Show correlations between independent measures of each
process.

    C. Manipulate each process.

    D. Show that each process manipulation changes performance.

    E. Determine by multiple correlation whether the validated pro-
cesses combine to account for all variance in criterion task
performance. If they do not, more process analysis will be
needed (Steps 4-A through 4-D).

Step 5. SHOW THAT PROCESSES UNDERLYING PERFORMANCE
CHANGE WITH AGE

    A. Demonstrate correlations between age and each process
measure

    B. Using performance measures, demonstrate interactions
between age and process manipulations. Collect concurrent
process measurements.

    C. Determine by partial correlation whether those processes
which correlate with age reduce the age/performance correlation to zero. If they do not, more process analysis
will be needed (Step 4).

Step 6. TEACH CHILDREN TO PROCESS AS ADULTS, THEREBY
RAISING THEIR PERFORMANCE TO THE LEVEL OF SIMILARLY
INSTRUCTED ADULTS.

If instructed children's performance falls short of instructed
adults', check concurrently collected process measurements to
see that instructions actually induced children to process as
adults.

    A. If instructions failed to induce adult processing, revise
them and try again.

    B. If instructions did induce adult processing, retreat to
Steps 4 & 5 for further process analysis.

    C. If children's and adults' instructed performance are equal,
but the instructions raised adult performance too, use concurrently collected process measures to see that adults who
contributed to the increase were using childish processing.

Step 7. TEACH ADULTS TO PROCESS AS CHILDREN, THEREBY

LOWERING THEIR PERFORMANCE TO THE LEVEL OF SIMILARLY
INSTRUCTED CHILDREN.

  If the instructed adults' performance lies above instructed
  children's, use concurrently collected process measurements
  to see that instructions actually induced adults to process
  childishly.
  A. If instructions failed to induce childish processing, revise
      them and try again.
  B. If instructions did induce childish processing, retreat to
      Steps 4 & 5 for further process analysis.
  C. If children's and adults' performance are equal, and the
      instructions lowered children's performance, use concurrent-
      ly collected process measures to see that children who con-
      tributed to lowering were using relatively mature processing.

measures must be taken during posttest, to assure that the
subjects continued to use the instructed processes following the
termination of instruction.

  The number of intellective tasks for which it is presently
possible to secure inobtrusive process measures is small. Cognitive
scientists have invested heavily in inferential procedures and
lightly in developing relatively direct measures of cognitive processes (cf., Belmont and Butterfield, 1977). Until this lamented
trend (Newell, 1973) is reversed, satisfactorily complete instructional tests of developmental cognitive theory will be few indeed.
Moreover, the few tests will be performed with criterion procedures that have been around for a long time, because it is only
well studied tasks for which underlying processes have been
identified and the necessary range of inobtrusive measures has
been developed. There have been marked changes in the kinds
of criterion performance that cognitive theorists study, so that
any investigator who tries seriously to follow the strategy outlined
in Table 1 will be criticized as old fashioned and outdated with
respect to his performance measures. My best advice is to turn
the other cheek and persist, because I see no way other than
the strategy in Table 1 to produce valid developmental cognitive
theory.

III. Research Implications of General Individual Differences

  The fact that individual differences in intellectual performances are general across tasks adds other steps to the research
strategy required to validate process analyses of cognition. The
only sort of generality established by any of the steps outlined in
Table 1 is generality across independent measures of the same
processes within the same task. This is not the sort of generality
that psychometricians have in mind when they speak of intelligence
as a general factor. They have in mind performance differences
that cut across tasks whose solutions are presumed to rely upon

substantially different processes.

Within cognitive theory there is a distinction between subordinate processes, which operate on environmental input or representations of it, and superordinate processes, which operate on subordinate processes. Subordinate processes include, among many others, recognition (matching a representation of incoming information to a representation from long-term memory), labelling (applying a name drawn from long-term memory to a representation of incoming information), rehearsal (repeated covert verbalization of a label or group of labels), and elaboration (retrieving from long-term memory the diverse sorts of information connoted by a label). A major goal of the process analyses called for in Table 1 is to identify the subordinate processes required for accurate performance of particular cognitive performances. A premise of cognitive theory is that different performances rely on different combinations of a limited set of subordinate processes. Each subordinate process has some range of problems to which it applies. The wider that range, the more general the subordinate process.

The role of superordinate processes is to select and coordinate the subordinate processes required to solve any particular performance problem. Superordinate mechanisms have been called by various names, such as metaplan (Miller, Galanter, and Pribram, 1960), self-instruction (Reitman, 1970), and the executive (Anderson and Bower, 1973; Greeno and Bjork, 1973; Neisser, 1967). By whatever name, superordinate processes are in theory completely general, since they are responsible for the selection of subordinate processes for the solution of every information processing problem encountered by any person. Table 2 outlines how to test the generality of both subordinate and superordinate processes.

Table 2 is constructed as a continuation of Table 1, since it is concerned with establishing the generality of analyses performed as outlined in Table 1. Thus, Table 2 begins with Step 8, which calls for a decision whether each process is subordinate or superordinate, and there are four questions to guide this decision.

A. Is the Process Subordinate of Superordinate?

A process theorist who asks about the generality of his analyses must begin by determining whether he is asking about subordinate or superordinate processes. Different research approaches are required to test the generality of the two, because superordinate processes exist at a much higher level of inference than subordinate processes.

1. Does process select, coordinate or modify other processes?

Table 2
HOW TO CONTINUE A PROCESS VALIDATION
SO AS TO ESTABLISH GENERALITY

Step 8. DECIDE WHETHER A PROCESS EXPLANATION CONCERNS
SUBORDINATE OR SUPERORDINATE PROCESSES, by answering
these questions:
    A. Does the process select, coordinate or modify other processe
    B. Does the process operate between trials?
    C. For people who fail to select, coordinate or modify other
        processes, does brief instruction induce them to do so
        and improve their performance?
    D. Do the performance gains derived from brief instruction
        fail to endure or generalize?
If the answer to all four of these questions is YES, the process
is superordinate. Go to Step 11.
If the answer to all four is NO, the process is subordinate.
Go to Step 9.
If the answers are mixed, or if the questions are not yet
answerable, it is premature to test generality. Instead,
you should perform more experiments like those outlined
in Step 4 or described in text.

Step 9. INDEPENDENTLY ANALYZE VARIOUS TASKS TO DETERMIN
WHETHER THEY SHARE SUBORDINATE PROCESS(ES). For eash tas
determine whether
    A. Analogous process measures correlate with performance.
    B. Comparable manipulations change use of process.
    C. Comparable manipulations influence performance.
Compare analyses of various tasks to see whether they share
subordinate process.

Step 10. RELATE PROCESS USE ON ONE TASK TO PROCESS USE C
OTHERS, by
    A. Correlating measures of subordinate processing across
        tasks for heterogeneous group of subjects.
    B. Using one task, instruct deficient subjects in the use of th
        subordinate process, and test for transfer of the process
        to other tasks requiring use of the process.

Step 11. USING TWO OR MORE TASKS THAT SHARE NO SUBORDIN
ATE PROCESSES, AND WORKING WITHIN HETEROGENEOUS GROUPS
    A. Correlate, across tasks, quality of subordinate process
        selection.
    B. Correlate, cross tasks, rates of effective subordinate
        process selection.

Step 12. USING TWO OR MORE TASKS THAT SHARE NO SUBORDIN
ATE PROCESSES, WORK WITH INACCURATE SUBJECTS AND GRADE
SEQUENCES OF SUBORDINATE PROCESS INSTRUCTION
    A. To determine the completeness of instruction required to
        secure proficient performance on one task and to correlate
        an index of that completeness with indices of quality of
        strategy selection derived from uninstructed performance

on another task.

B.   To assess the extent to which instruction in superordinate processes reduces the completeness of subordinate instruction required to secure accurate processing.

The first question designed to decide whether a process of interest is subordinate or superordinate asks whether it selects, coordinates, or modifies other processes (Step 8, Table 2). The alternative possibility is that the process operates on environmental information or its transformations. An equally satisfactory way to define one's method of approach to a problem type or whether it is intended to pose the question is to ask whether the process is intended to yield an answer to an instance of the type. If the process is intended to define or change an approach, it is superordinate. If it is intended to yield an answer, it is subordinate. In either form, this question is about one's theory of what is required to arrive at a solution to his criterion problem. Answering this first question requires examination of one's theory. The second question is the empirical analogy of the first.

2.   Does process operate between trials?

Superordinate processes serve to match subordinate processing abilities to problem demands. Even though the chief source of input for that matching process must be the subject's experience during experimental trials, they must use the time between trials to revise their understanding of the requirements of a problem in view of their accumulated experience with it, to evaluate the effectiveness of the approach, and to set new goals for the next trial. Therefore, an empirical test of whether superordinate processes are at work is to determine whether there are systematic trial-by-trial changes in the deployment of subordinate processes.

Superordinate processing should be indicated by several sorts of trial-by-trial changes in subordinate processing. When faced with a novel and reasonably complex problem, a person should require direct experience with it to determine its information processing requirements. Early trials should be more informative than later ones, so there should be greater changes in people's strategies across early trials, which is to say, they should be strategically more consistent on later trials. Given enough experience, people should arrive at a strategy which they use on subsequent trials, but their fashioning of this strategy should be the gradual result of accumulated experience with the task, so that growth in the degree of similarity to their final strategy should be seen in early trials. Assuming that people have comparable information processing mechanisms the degree of similarity or concordance among them in strategy should increase across trials. Assuming a problem for which there is an optimal approach, people should come, across trials, to approach that optimum more closely. If superordinate processes are related to performance, the foregoing expectations about consistency, gradual approximation to one's own final strategy, strategic concordance among people and approximation to an optimal strategy should be more pronounced for accurate than for inaccurate problem solvers. Testing for such trials effects requires direct measures of subordinate processes employed on each trial of a problem.

3. Do simple instructions rapidly induce effective subordinate processing?

The third question in Step 8 of Table 2 is the typical production deficiency question. It is answered by an instructional experiment of the sort described above in connection with Step 4-D of Table 1. That is, experiments from which children are inferred to be production deficient rely on instruction of subordinate mediational processes (Flavell, 1970). Investigators first determine the subordinate processes required for good performance on a particular task; then they instruct children to use them (e.g., Brown, Campione, Bray, and Wilcox, 1973; Butterfield, Wambold, and Belmont, 1973; Moely, Olson, Halwes, and Flavell, 1969). Even though the instructions are designed to influence subordinate processing, investigators have emphasized superordinate immaturities to explain why children benefit from instruction, which is to say, why they are production deficient.

The reason for emphasizing superordinate explanations, such as metamemory (Flavell and Wellman, 1977; Brown, 1975, 1978) and executive functions (Butterfield and Belmont, 1977), is that the performance gains resulting from subordinate instruction are swift and dramatic. Investigators have found it unreasonable to suppose that such simple and effective instructions teach children the specific processes upon which the instructions focus. It has seemed more reasonable to suppose that such simple and effective instructions teach children the specific processes upon which the instructions focus. It has seemed more reasonable to suppose that the investigator is selecting, through his instructions, which subordinate processes the child will use. The failure of the child to select effective subordinate processes for himsel is viewed as a failure of superordinate processes. Question 3 under Step 8 incorporates this logic. It says that whenever simple subordinate process instruction results in swift performance gains, the problem is one of superordinate processing.

4. Do instructed performance gains fail to endure or generali:

The fourth question designed to decide whether a process is superordinate (Step 8, Table 2) is based on an extension of the logic underlying the production deficiency instruction. Transfer tests are given only to people who require instruction on a training task.   The fact that training is successful, as it must be before the investigator tests for transfer, says that the people who are tested for transfer never did lack the appropriate subordinate processes.   They simply failed to invoke them without training. This follows from the fact that no instructional experiment in the cognitive literature can fairly be represented as an effort to impart subordinate processes. Cognitive instructional experiments can only be represented as ways of telling people to do what they already know how to do.   It follows that the trained subjects' failure to invoke the trained processes on their own was in the superordinate business of assessing the cognitive requirements of the training task.   Since training and transfer tasks come from the same class of cognitive problem, the superordinate matter of assessing cognitive requirements will be no less important for the transfer test than it was for the pretest that indicated a need for training.   In view of the child's superordinate failure on the pretest, the best prediction is that the child will fail similarly on the transfer test, because it cannot be reasonably supposed that subordinate process instruction will have improved superordinate processing.   It follows that failures of successful subordinate process instruction to transfer, or even to endure, are evidence that the processing problem has occurred at the superordinate level.

5.   Conclusion.

Any problem for which it is possible to say that superordinate processes contribute to successful solutions will be a problem for which it has been well established that particular subordinate processes are required.   Without good measures of the requisite subordinate processing it will be impossible to assess trial-by-trial changes in strategy, it will be impossible to advance a compelling theory to examine for the presence of superordinate processes, and it will be impossible to design effective process instructions whose transfer can be tested.   If the answer to all of the questions under Step D is Yes, there will remain the choice of whether to   focus   on   subordinate   or   superordinate   processes.

B.   To Determine the Generality of Subordinate Processes....

1.   Process analyze various tasks.

In principal, all that is required to demonstrate the generality of a subordinate process is to show that it contributes to accurate performance on more than one task.   For each task tested, one should determine whether measures of the target process correlate with performance, whether comparable manipulations change use of the process, and whether the comparable manipulations influence performance.   These requirements are specified in Step 9 (Table

2), and are the same as Steps 4-A, 4-C, and 4-D (Table 1). To establish the degree of generality of a subordinate process, one needs to determine the number and range of tasks for which the process contributes to accurate performance.

In practice, developmental cognitive psychologists have not adopted the foregoing approach to establishing generality of subordinate processes. They have striven instead to establish generality by correlating the use of subordinate processes across tasks (cf., Butterfield and Dickerson, 1976) or, more frequently, by showing that instruction in the use of a subordinate for one problem induces its use for another (cf., Brown and Campione, 1978; Borkowski and Cavanaugh, in press). These are approaches whose methodological demands can seldom be fully satisfied, so they are risky ways to seek evidence of generality. They are listed in Table 2 under Step 10 more as a way of setting goals for investigators than as currently required approaches.

2. Correlate process use across tasks.

Step 10-A calls for correlating measures of a process across problems. Its use implies that determining that a process is used for the solution of more than one problem (Step 9) is insufficient to establish process generality. Step 10 calls additionally for a demonstration that there are stable differences across tasks in people's use of the process. It is Underwood's (1975) individual differences test turned to testing generality. It is a risky test because there are no completely analyzed cognitive performances. Therefore, failure to obtain a correlation across problems can easily result from a failure to use some process other than the one whose generality is being tested. For example, a child might appreciate the value of rehearsal and use it when the names to be rehearsed are supplied, as in aural presentation of words in a subject-paced recall task. The same child might fail to appreciate the need to generate labels, as when pictures are presented in a subject-paced memory experiment. Such a child would rehearse in an aural task, but not in a pictorial one, and would contribute error variance to a study of cross-task generality. But that error variance should be attributed to a failure to label, a process whose generality is not at question, not to a lack of generality of rehearsal. The investigator's chore when seeking cross-problem correlations of process use it to insure that his subjects use all processes other than one he is focusing on. This is the only condition under which a failure to use the target process is interpretable as a failure of generality. The lack of complete process analysis for any cognitive task makes it impossible to verify that such a condition has been met. An investigator might nevertheless proceed, on the gamble that his test will not be destroyed by subjects' failures to use nontarget processes. His gamble might establish cross-task generality, but failure cannot

be taken as evidence against generality. Unless the investigator
has used either concurrent measurement or direct instruction to
insure that pertinent nontarget processes were used, he can take
a failure to observe generality across tasks only as an indication
that more process analysis is required to make his failure interpretable.

3. Show transfer of process training from one task to
another.

The reason most often given in the literature for studying
transfer of training of subordinate processes is to establish that
having a process has changed some "real," "true," or "genuine"
aspect of cognition (Borkowski and Wanschura, 1974; Kuhn, 1974;
Denney, 1963). The idea is that any cognitive process worthy of
the name is a general one. Most efforts to secure transfer have
failed, and the reason seems to be that investigators have not
appreciated the importance of analyzing the roles of processes
other than the ones whose generality is being tested, before
undertaking transfer studies.

It would be lovely if informed guessing or loose reasoning
could provide the process analysis required for tests of transfer.
Unfortunately, the task analytic requirements are much too specific
and detailed. The investigator who would demonstrate transfer
must thoroughly understand both his training and his transfer
tasks. By definition, training and transfer tasks are similar;
both require processes taught during training. However, they
are not identical. Performing the transfer task must also require
processes not taught during training. If it did not, the test
would be for durability rather than for transfer. Since the tasks
are not identical, both must be analyzed to demonstrate that they
require the instructed processes. But certain knowledge that the
two tasks require shared processes does not guarantee that
failing the transfer task results from not transfering the trained
processes. The child might well understand that the transfer
requires use of his newly learned processes, and he may use
them but fail the transfer test for not engaging the untrained
processes it requires. Without knowing precisely where each
subject's performance breaks down, an investigator cannot interpret a failure to obtain transfer. No investigator has known
these things about his transfer test, because no performance
studied in instructional tests of transfer has been well analyzed.
As with correlational studies of processes across problems, an
investigator might choose to perform a transfer test knowing that
a failure would say nothing about generality. Failure would only
indicate that more process analysis is required. Successful
promotion of transfer, on the other hand, can be interpreted as
evidence for generality, and the belief that only a general instructional effect justifies the inference that intelligence has been

trained has led many investigators to gamble on instructional tests of transfer.

C.   To Determine the Generality of Superordinate Processes....

When testing generality of subordinate processes, one will progress faster and will be able to interpret his data more completely if he has previously performed relatively complete analyses of his criterion tasks.   Still, when the focus is on subordinate processes, an investigator can choose to gamble by proceeding in the absence of well advanced process analyses.   An investigator can seek evidence of generality as soon as he has identified any subordinate process that accounts for significant criterion variance. This sort of flyer is not possible when one is seeking evidence of the generality of superordinate processes, because it is only from the changing organization of subordinates that superordinates can be inferred.   Analysis must have proceeded at least to the point of having validated two subordinate processes for each of two tasks.   Ideally, the two tasks will share no subordinate processes. This is so that any observed correlation in changes in combinations of subordinates across tasks cannot be attributed to the nature of the subordinates, but to superordinate processes.   This ideal calls for process analyses to be reasonably complete before trying to determine the generality of superordinate processes.

All of the generality tests outlined in Steps 11 and 12 presumed that superordinate processes are generalized problem solving procedures, and that generalized problem solving is central to what we mean by intelligence.   Thus, Steps 11-A and 11-B are variants of Underwood's individual difference test, which reflects the psychometric notion that intelligence is general.   Steps 12-A and 12-B include tests with novel problems, and they employ a response-to-instruction criterion of superordinate processing. The idea is that people who possess effective superordinate processes will require less subordinate process instruction to perform well on novel tasks than people with less proficient superordinate processes.   The more proficient will fill in larger gaps in instruction than will the less proficient, which is to say they will learn more from minimal instruction.

1.   Correlate quality of subordinate process selection across problems.

Step 11-A presumes two tasks that depend upon different processes for their solution, that are novel to a group of people, and from which indices can be derived of the quality of the subordinate process combinations selected by people after they have had some experience with the tasks.   The tasks must have been analyzed well enough for an investigator to specify and measure both effective and ineffective combinations of subordinate mechanisms for working the problems.   The test of generality is whether people who select effective strategies (e.g., subordinate process combinations) for one task also select effective strategies for the other. The power of the test is substantially greater when effective solutions for the problems share no subordinate processes.

2. Correlate rates of effective subordinate process selection across problems.

Step 11-B is a refinement of Step 11-A. It tests the hypothesis that rate of effective strategy selection will vary across tasks. The idea is that people who select effective combinations of subordinate mechanisms will do so at different rates, and these differences in rate will correlate across tasks that share no subordinate mechanisms. This possibility can be tested only after it is shown that effectiveness of strategy selection covaries across tasks.

3. Correlate response to subordinate process instruction with quality of subordinate process selection.

Step 12 presumes an instructional sequence graded with respect to how completely it conveys subordinate processes required for the solution of some criterion task(s). Developing such an instructional sequence will normally require considerable process analysis. Step 12-A calls for the use of graded instructions to determine the completeness of instruction required to produce excellent criterion performance by people who perform poorly prior to instruction. The assumption is that there will be individual differences in how complete the instruction must be. More effective superordinate processors should require less complete instruction. The test of generality incorporated in Step 12-A is to determine whether completeness of needed instruction on one novel task correlates with quality of selected processes on another novel task.

4. Instruct superordinate processes.

Step 12-B presumes a model of superordinate processes and ways of teaching them. It presumes too a stable of criterion tasks for each of which there has been developed a graded sequence of subordinate process instruction. The test begins by identifying a group of people who perform poorly on all of the problems for which there is a graded instructional sequence. It proceeds with superordinate training for some but not others of the people who perform poorly. The test of generality is whether fewer subordinate process instructional steps are required to promote excellent performance on all of the criterion tasks by people trained in superordinate processing than by those not trained.

References

Anderson, J. R. Language, Memory, and Thought. Hillsdale, NJ: Erlbaum, 1976. Belmont, J. M. and Butterfield, E. C. The instructional approach to developmental cognitive research. In R. Kail and J. Hagen (Eds.), Perspectives on the Development of Memory and Cognition. Hillsdale, NJ: Erlbaum, 1977, pp. 437-481.

Borkowski, J. G., Cavanaugh, J. C., and Reichart, G. J. Maintenance of children's rehearsal strategies: Effects of amount of training and strategy form. Journal of Experimental Child Psychology, in press.

Borkowski, J. G. and Wanschura, P. Mediational processes in the retarded. In N. R. Ellis (Ed.), International Review of Research in Mental Retardation (Vol. 7). New York: Academic Press, 197

Brown, A. L. The development of memory: Knowing, knowing about knowing, and knowing how to know. In H. W. Reese (Ed.), Advances in Child Development and Behavior, (Vol. 10). New York: Academic Press, 1975.

Brown, A. L. Knowing when, where, and how to remember: A problem in metacognition. In R. Glaser (Ed.), Advances in Instructional Psychology. Hillsdale, NJ: Erlbaum, 1978.

Brown, A. L. and Campione, J. C. Memory strategies in learning: Training children to study strategically. In H. Pick, H. Leibowitz, J. Singer, A. Steinschneider, and H. Stevenson (Eds.), Application of Basic Research in Psychology. Plenum Press, 1978.

Brown, A. L., Campione, J. C., Bray, N. W., and Wilcox, B. L. Keeping track of changing variables: Effects of rehearsal training and rehearsal prevention in normal and retarded adolescents. Journal of Experimental Psychology, 1973, 101, 123-131.

Butterfield, E. C. On studying cognitive development. In J. P. Sackett (Ed.), Observing Behavior: Theory and Application in Mental Retardation (Vol. 11). New York: Academic Press, 1977.

Butterfield, E. C., Wambold, C., and Belmont, J. M. On the theory and practice of improving short-term memory. American Journal of Mental Deficiency, 1973, 77, 654-669.

Chase, W. Visual Information Processing. New York: Academic Press, 1973.

Chi, M. T. H. The representation of knowledge (Review of Exploration In Cognition by D. A. Norman, D. E. Rummelhart, and the LNR Research Group). Contemporary Psychology, 1976, 21, 784-785.

Denny, D. R. Modification of children's information processing behaviors through learning a review of the literature. Child Study Journal Monograph, 3 (Whole No. 4), 1973.

Flavell, J. H. Developmental studies of mediated memory. In H. Reese and L. Lipsitt (Eds.), Advances in Child Development and Behavior (Vol. 5). New York: Academic Press, 1970.

Flavell, J. H. and Wellman, H. M. Metamemory. In R. Kail and J.
Hagen (Eds.), Perspectives on the Development of Memory and
Cognition. Hillsdale, NJ: Erlbaum, 1977.

Greeno, J. G. and Bjork, R. A. Mathematical learning theory and
the new "mental forestry." Annual Review of Psychology, 1973,
24, 81-115.

Kuhn, D. Inducing development experimentally: Comments on a research paradigm. Developmental Psychology, 1974, 10, 590-600.

Lachman, R., Lachman, J., and Butterfield, E. C. Cognitive Psychology and Information Processing--An Introduction. Hillsdale,
NJ: Erlbaum, 1979.

Miller, G., Galanter, E., and Pribram, K. Plans and the Structure
of Behavior. New York: H. Holt and Co., 1960.

Moely, B. E., Olsen, F. A., Halwes, T. G., and Flavell, J. H.
Production deficiency in young children's clustered recall.
Developmental Psychology, 1969, 1, 26-34.

Neisser, W. Cognitive Psychology. New York: Appleton-Century-
Crofts, 1967.

Newell, A. You can't play 20 questions with nature and win: Pro-
jective comments on the papers of this symposium. In W. G.
Chase (Ed.), Visual Information Processing. New York:
Academic Press, 1973.

Reitman, W. What does it take to remember? In D. A. Norman
(Ed.), Models of Human Memory. New York: Academic Press,
1970.

Schank, R. C. The role of memory in language processing. In C. N.
Cofer (Ed.), The Structure of Human Memory. San Francisco:
Freeman, 1976.

Townsend, J. T. Some results on the identifiability of parallel and
serial processes. British Journal of Mathematical and Statistical Psychology, 1972, 25, 168-199.

Townsend, J. T. Issues and models concerning the processing of a
finite number of inputs In B. H. Kantowitz (Ed.), Human Information Processing: Tutorials in Performance and Cognition.
Hillsdale, NJ: Erlbaum, 1974.

Weizenbaum, J. Computer Power and Human Reason: From Judgment to
Calculation. San Francisco: Freeman, 1976.

Footnote

The preparation of this paper and the research on which its
ideas are based were supported by USPHS grants HD-00870, HD-
08911, and HD-13029. A longer version of the paper, with descriptions of experiments to illustrate each step in the described strategy,
is available on request.

CODING AND PLANNING PROCESSES

J. P. Das

University of Alberta, Edmonton, Canada

Ronald F. Jarman

University of British Columbia, Vancouver, Canada

Part I: Relationships Between Simultaneous and Successive Syntheses and Some Existing Dichotomies

After Professor Butterfield's talk, there is no need to elaborate further on the usefulness of process theories. Such theories seem to reflect the spirit of the time. What is slightly alarming though, is the rate of proliferation of concepts of processes. Thus, there is a need to delineate clearly the distinctiveness of processes within various models. Models of cognitive processes should be similarly described so that the consumer can determine what is new in the product.

However, any new product is not entirely novel nor absolutely unique. Each new model of cognitive processes then, will share many common properties with existing notions, while its essential properties should be different from existing models if it is to maintain its identity. In addition, its explanatory powers should be demonstrated empirically.

A model of cognitive processes has been developed by Das, Kirby and Jarman (1979) and has been used to explain a variety of cognitive performances such as reading (cf. Das, 1973). In our discussion here, we propose to describe its common and essential properties, as the old logicians used to say, and then provide a close examination of one of the aspects of the model, which is modality-specificity in coding information. But before we describe the usefulness and parsimony of the model of cognitive processes, let us reflect on the processes themselves as different from abilities. It is not enough merely to reject the psychometric notions of verbal, spatial, or reasoning abilities, nor is it adequate to substitute abilities by the term processes. The abilities would still continue to connote such mental activities as memory, perception and language. We could as easily think in terms of an ability to memorize, to perceive or to use language. What is needed is a departure from the attitude of treating these abilities as fixed and immutable properties of the mind. The point of departure is provided by approaching memory or language as functional systems which have evolved and are constantly evolving in order to fulfill the needs of the individual and the society. A mental activity such as memory does not represent the function of a specific faculty in the mind or of a narrow centre in the brain. Rather, the memory system has evolved developmentally, influenced by the experiences of the individual at work and play, through the dynamics of interactions with other individuals within the social milieu.

Such a view of "process" is offered by Vygotsky, and is essentially one that Luria adopts. This view of processes is consistent also with the Piagetian concepts of operations such as concrete and formal. Processes as systems of functions may adequately describe the evolution of the relation between speech and thought, of the dynamics of the growth of inner speech from egocentric speech, of the shift from syntagmatic to paradigmatic associations, or from an enactive to a symbolic mode.

Three such functional systems have been proposed by Luria: arousal, coding and planning (Luria, 1970). While arousal is primarily in the subcortical area, the other two are located in known cortical areas of the brain. Basic cognitive processes, then, involve coding (simultaneous and successive) and making plans and decisions. Relative competence in the use of these processes can be measured by tests. We have developed tests or adopted existing tests to measure coding and planning behavior (Das, Kirby and Jarman, 1979; Ashman, 1978).

We do not intend to elaborate here on the two coding concepts of simultaneous and successive processes. However, a brief statement on each of the two coding processes should be made. Simultaneous processing involves the formation of a code which is quasi-spatial in nature, such that all parts of it are immediately surveyable. Successive processing, on the other hand, is more temporal in nature, being accessible only in a linear fashion. Simultaneous coding is linked to the broad functions of the occipito-parietal areas of the cortex, whereas successive coding is based on the frontotemporal areas. As an aside, it should be noted that Luria's work was mostly concerned with the left hemisphere.

Intelligent behavior involves coding of information, but more so, the utilization of information that has been coded for fulfilling a goal. Such behavior is purposive and organized. If appropriately coded information is not available, the individual seeks out information and codes it for his purpose. Coding as such may not describe cognitive competence adequately; it is coding for a purpose in mind, which is to utilize as best one can the information coded for goal attainment.

Such a notion of intelligence is closely associated with executive processes, which are metafunctions. Intelligent behavior is not only expressed in making good decisions and in solving problems, but also in generating problems and in creating the occasions for making good decisions. Probably, as Estes has observed, an intelligent man invents problems as well as solves them. In accordance with Luria's functional organization of cognitive processes, such metafunctions could be subsumed under planning and decision making, which are the major functions of the frontal lobes.

Studies on Coding and Planning

How well has the model worked? Does the model provide merely a new vocabulary, and is not essentially different from existing dichotomies such as verbal and nonverbal intelligence, or memory and reasoning? Are the two coding processes confounded with visual and auditory coding, and therefore how fruitful is it to relate individual differences in the metaprocesses of simultaneous and successive to specific competence in modality-matching tasks?

The first question to consider is if simultaneous and successive processing are co-existent respectively with visual and auditory modalities. The evidence suggests that this is not the case. Bickersteth studied Grade 3 children in Freetown, Sierra Leone and in Edmonton in a Ph.D. thesis, in which he gave modality matching tasks, classification tasks and measures of syllogistic reasoning. For our present purpose the results of the modality matching tasks are of particular interest. The tasks require the subject to match visually presented patterns of lights in the visual-visual condition, and auditorily presented patterns of sounds in the auditory-auditory condition. There were also visual-tactile presentations and tactile-visual presentations. The question is whether or not those children who have been identified as more proficient in simultaneous processing in comparison with those who have been identified as less proficient, would be predisposed to do better in visual-modality matching to the exclusion of auditory-modality matching. Similarly, on the basis of successive tasks, would the high successive group do better in auditory than the low successive group? In other words, generally are the two sensory modalities related to simultaneous and successive processing? As the data in Table 1 demonstrates, Bickersteth's results, both in Sierra Leone as well as in Edmonton, Canada, indicated that, in general, this is not so. Those children who were high in simultaneous processing did better in visual and auditory tasks as well as in cross-modal matching tasks. Similarly, those who were high in successive processing did better in visual and auditory tasks as well as in cross-modal matching tasks. Similarly, those who were high in successive processing did better in the same tasks. In Part II, we return to this issue in more depth, to examine whether coding varies by modality according to levels of intelligence and reading ability.

The second question which we would like to answer in order to delineate the nature of simultaneous and successive processing is as follows: is simultaneous processing another name for reasoning and successive for memory? We shall cite two studies which seem to show that this is not the case. The first one by Kirby and Das (197) examined the relationship between primary mental abilities and simultaneous and successive processing. Since this has been already published we will briefly summarize the main findings. The tests of primary mental abilities which were chosen yielded three promax factors which were: inductive reasoning, spatial and associative memory. The factor scores thus derived were correlated with simultaneous and successive factor scores. Simultaneous processing was found to be related mostly to spatial ability. It was also signiificantly related to inductive reasoning but no more so than it was to associative memory. These data confirm the spatial nature of simultaneous processing, but do not support any identification of it with inductive reasoning ability. Successive processing and associative memory were significantly related, but no more so than were simultaneous processing and associative memory.

The next is a study on levels of processing by Snart (1978), involving three groups of children, who were 6, 11, and 17 years old, and were in Grades 1, 6, and 9. The children were presented with a levels of analysis memory task according to the paradigm of Craik and his colleagues. In such a paradigm, one records the recall of words which are assumed to have been processed at different depths, so that words which have been processed at a shallow level are expected to be recalled less often than those which have been processed at a deeper level. Level of processing is manipulated by presenting the word following orienting questions which, in our case, require the subjects to attend to the physical features of the word, or to the semantic aspects of the word. All subjects were also given the target tests of simultaneous and successive processing. Recall scores were separated for the physically and the semantically tagged words, and factor-analyzed along with the scores of the target tests for simultaneous and successive processing. The question we were asking was whether

TABLE 1

Comparison Between High and Low Successive and Simultaneous Groups on Modalities Tests with Successive and Simultaneous Presentations Combined

Ranked in order of Relative Proficiency		Means			
		Successive		Simultaneous	
		High N=104	Low N=103	High N=104	Low N=104
1. Visual	M	5.46	4.58	5.43	4.61
2. Tactile	M	5.81	4.32	5.06	4.42
3. Auditory	M	4.90	3.53	4.54	3.88
4. Visual-Tactile (Cross-Modal)	M	4.25	3.45	4.16	3.56

the memory tasks would be associated only with successive rather than with simultaneous processing. In fact, Snart had hypothesized that semantic memory for the older age group would depend on simultaneous processing. Factor analysis of the data for all three age groups, as shown in Table 2, yielded three factors which were rotated to a position orthogonal to each other, and labelled as simultaneous, successive and a memory factor. The last factor had loadings mainly from the recall scores. In all three age groups, physical recall loaded only on the memory factor, and did not have loadings on either the simultaneous or the successive factors. However, semantic recall was most interesting to study. In the youngest age group, semantic recall had its major loading on the memory factor, but a secondary loading was also obtained on the successive processing factor. At age 11, semantic recall had its major loading on successive processing and minor loadings on the simultaneous and memory factors. The results of the analysis on 17 year olds however, were quite striking in that semantic recall loaded on simultaneous processing as well as, of course, on the memory factor. Thus, at age 17, simultaneous processing seems to be involved in semantic memory, whereas successive processing does not seem to contribute very much to proficiency in semantic recall.

Lastly, the question may be asked, is simultaneous processing inherently non-verbal and successive processing basically verbal? In the past, when we have factor-analyzed WISC Performance and Verbal Scale items with simultaneous and successive tests, we have noticed that WISC Verbal does not load on either simultaneous or successive factors. Obviously, since the successive tests have some resemblance to tests of short-term memory, one might be tempted to say that successive processing is verbal. However, the picture is more complex than that. For instance, in Snart's study on levels of processing, semantic memory, or recall of words which require semantic processing, had a significant loading on the simultaneous factor but not on the successive. If we remember our initial assumption that simultaneous and successive are processes to be used at the individual's option, depending on how he or she perceives the task, and the initial preference of the individual for using one or the other process, then the findings such as the loading of semantic memory on a simultaneous factor would not be surprising. In certain tasks, both processes are used. In some others one of them is predominantly used.

Let us briefly consider the use of these processes in syllogistic reasoning tasks. The study by Bickersteth in Sierra Leone and in Edmonton showed that children in both places were utilizing simultaneous and successive processes for solving three-term syllogisms (see Table 3). Those who were high in simultaneous processing did better in syllogistic reasoning than those who were low, and similarly, those who were high in successive processing

TABLE 2

Levels of Processing and Simultaneous-Successive Tasks

	Age 6.7 (N=50) Factor			Age 11.4 (N=50) Factor			Age 16.7 (N=50) Factor		
	Sim	Succ	Mem	Sim	Succ	Mem	Sim	Succ	Mem
Physical Recall			868			937			891
Semantic Recall		406	680	374	610	468	565		672
Memory for Designs	871			-643		-344	-388	-693	
Figure Copying	-825			878			875		
Visual Short-Term Memory		848			696			411	
Auditory Serial Recall		704			779			898	

TABLE 3

Comparison Between High and Low Successive and Simultaneous
Groups on Syllogistic Reasoning

	Means			
	Successive Group		Simultaneous Group	
	High N=104	Low N=103	High N=104	Low N=103
Total Syllogism	9.67	7.22	9.28	7.62
Syllogisms relevant to Sierra Leone (Sugar cane grows in hot countries. Sierra Leone is a hot country. Does sugar cane grow there?)	2.65	2.10	2.60	2.16
Syllogisms relevant to Canada (Important people in society look alike. Pierre Trudeau and Bobby Hull do not look alike. Are they important or not?)	2.32	1.93	2.29	1.96
Culturally irrelevant (Realistic) (Islands are surrounded by water. Iceland is surrounded by water. Is Iceland an Island?)	2.43	1.77	2.23	1.97
Culturally irrelevant (Artificial or unrealistic) (All dogs can fly and all elephants are dogs. Can all elephants fly or not?)	2.29	1.44	2.23	1.50

also did better than those who were low. However, Cummins has done a study on high school students in Edmonton in which he showed that three-term syllogistic reasoning loaded on a simultaneous factor (Das, Kirby and Jarman, 1975).

In a recent study relating ambiguities to the simultaneous-successive distinction, it was clearly shown that simultaneous processing was not nonverbal. Kirby and Biggs (personal communication, 1979) gave Grade 9 children tests of three kinds of ambiguities involving lexical, surface and deep structure, in order to explore their relationship to simultaneous and successive processing. The three types showed significant correlations with the simultaneous factor scores and negligible relationships with successive factor scores.

Perhaps we have given enough arguments to establish that simultaneous and successive processes are useful categories of cognitive processes, and to show that these are not redundant labels. We have also presented the case for regarding simultaneous-successive as optional processes to be used by individuals or by groups, reflecting strategies rather than abilities for utilizing information in order to solve a task at hand.

Those of us who have considered strategic behavior and whether or not strategic behavior can be taught, are optimistic. Planful behavior is possible in the case of those children who may not show that they are capable of planning, and as two participants of this conference, Butterfield and Brown, have suggested in their various writings, children can be taught to decide what plans to use and when to use these. Thus, there seems to be a consensus in recent investigations that planning and strategic behavior are probably the most important ingredients in determining cognitive competence. Planning or the adoption of strategies depends on coded information. A certain amount of coding is necessary for planning to operate. All codings on the one hand involve a certain amount of planning, but at the same time, planning can be separated as a distinct cognitive activity from coding. We think that by manipulating instructions and experimental conditions, it is possible to examine the coding and planning component in any task and subsequently to relate poor performance in the task to these components. One should be able to achieve this also by varying the samples such as comparing the deaf, autistic and the retarded as O'Connor and Hermelin have done. Such an approach is quite different from an abilities approach.

Part II: Sensory Modalities and Coding Processes

The role of sensory systems in cognition has been a topic of considerable interest for many years, but more so in recent times (e.g., Connor and Hoyer, 1976; Erdelyi, Finkelstein, Herrell, Miller and Thomas, 1976; Routh, 1976; Spitzer, 1976). A particularly pervasive issue in this research is the extent to which cognitive coding and processing is modality-specific. The clinical and remedial literature has traditionally assumed modality-specificity in processing, and ironically, because this was only an assumption, this literature appears to foreshadow some of the emerging conclusions of experimental research. Increasingly, experimental studies appear to support the modality-specificity view, with a discernible move away from single verbal-based storage systems, such as that proposed by Sperling (1963), and subsequently elaborated in various forms by other researchers (e.g., Atkinson and Shiffrin, 1968; Waugh and Norman, 1965).

The result of these trends is a full range of theoretical positions evident today concerning the functions of sensory systems in cognition, where some belief in sensory system specialization is evident, in addition to the remnants of the nonmodal theoretical position originating in early studies of memory. If one turns to the model of simultaneous and successive syntheses, however, as described by Das, Kirby and Jarman (1975, 1979), this issue is not seen as a choice between two theoretical positions, but rather a question of defining the conditions under which modality specialization occurs. The simultaneous-successive model posits different levels of modality specialization, based upon Luria's (1970, 1973a) clinical research, in which three types of cortical zones were identified, each with a different degree of modality-specificity. The primary zones are modality-specific, and are responsible for elementary registration and analysis. The secondary zones are less specific, and functionally relate information between modalities to a limited extent. Finally, the tertiary zones are responsible for higher-order analysis of information among all modalities. Thus, in structure at least, the zones posited by Luria encompass the full range of discrete theoretical positions evident today in the study of sensory modalities.

One implication of Luria's hierarchical view of sensory systems is that it is quite probable that a general answer cannot be given to the question of whether or not cognitive processes are specific to sensory modalities. It is likely that this is a substantive question (Sears, 1975), which must take into account factors such as the amount of information load (Freides, 1974), and task content, as well as the population under study. With regard to populations particularly, it is possible that variation in modality specialization may be an important parameter in the definition of intellectual deficity (Jarman, in press-a), in addition to any unique performance decrements in the modalities themselves (O'Connor and Hermelin, 1978).

Sensory Modalities, Intelligence, and Reading Ability

We will now describe two studies which have attempted to assess the degree of modality specialization in different populations of children defined by general intelligence and reading ability. The methodology in these two studies is identical, and therefore we describe this first, followed by the results.

The tasks used in the studies were cross-modal and intra-modal matching of auditory and visual input. In cross-modal matching, a stimulus pattern is given in one modality, followed by a comparison pattern in a second modality, and the subject is required to judge the equivalence of the two patterns. In intra-modal matching, both of the patterns are given successively in the same modality, followed by a judgment of equivalence. It has been suggested traditionally that cross-modal matching is a measure of sensory system integration, and intramodal matching measures the capabilities of each modality (Bryant, 1968; Rubinstein and Gruenberg, 1971).

A substantial problem with past research using these tasks, though, is that the auditory and visual modalities have been confounded with temporal and spatial input, thereby weakening their conclusions on modality integration and specialization. To circumvent this problem, the present studies used three experimental conditions to partially separate the dimensions of auditory-visual and temporal-spatial. These three conditions combine to form nine tasks as shown in Figure 1. The auditory-temporal condition consisted of 1000 Hz tones, presented successively with .15 sec and 1.35 sec pauses to create patterns (Jarman, 1977). The visual-temporal condition consisted of flashes of a 12 volt light, on identical timing to the auditory-temporal condition (Jarman, Marshall and Moore, 1979). Finally, the visual-spatial condition was comprised of a set of black dots placed in a linear pattern, with short and long spaces as in the temporal conditions. As seen in Figure 1, use of these three conditions in either or both of the stimulus and comparison positions, creates nine tasks with varying integration demands. All tasks have identical stimulus patterns, however, and are administered in balanced order within each sample to reduce the effects of learning these items, with an interval of several days between each of the tasks.

There are many ways in which data generated by these tasks may be used to assess the functions of sensory modalities. The balanced design allows within-subject questions, such as whether number of integrations is a determinant of task difficulty, whether cross-modal integrations are more difficult than intramodal integrations, and whether processing temporal information in the visual modality is more difficult than processing the same information in the auditory modality. For reasons of brevity we will confine

		COMPARISON		
		Auditory-Temporal	Visual-Temporal	Visual-Spatial
	Auditory-Temporal (AT)	Within-modal and temporal **O**	Cross-modal Within-temporal **1**	Cross-modal and spatial/temporal **2**
STIMULUS	Visual-Temporal (VT)	Cross-modal Within-temporal **1**	Within-modal and temporal **O**	Within-modal Cross-spatial/temporal **1**
	Visual-Spatial (VS)	Cross-modal and spatial/temporal **2**	Within-modal Cross-spatial/temporal **1**	Within-modal and spatial **O**

Figure 1. Type and number of information integrations in matching tasks.

ourselves here to patterns of individual differences among the tasks in the form of factor analyses in order to explore the relationships between auditory-visual and temporal-spatial processing among the sample groups.

In the first study (Jarman, in press-a), two groups of children at different intelligence levels were examined, in order to test the assumption sometimes stated by developmentalists, that growth of intelligence is characterized by increasing intersensory integration of discrete sensory systems.

There were no sex differences in the tasks and so the data were factor analyzed for boys and girls together in each IQ group. The results for the below average IQ group are given in Table 4. A summary description of these results is that the stimulus condition is a dominant influence, with the comparison condition of much less significance. A possible reason for this, which we will return to later, is that the form of presentation of the first pattern determines, to an extent, the strategy of matching information between the two patterns. With regard to interpretation of the factors, the first factor appears to be mainly visual-spatial, the second is auditory-temporal, and the third is visual-temporal.

The results for the above average IQ group are presented in Table 5, and show some differences from those for the below average group. In this case, the stimulus conditions are still a major determinant of factors, but comparison conditions also

TABLE 4

Principal Components Analysis with Varimax
Rotation: Below Average IQ Group

Task	I	II	III
AT–AT	-.144	.696	.259
AT–VT	.188	.711	.042
AT–VS	.207	.750	-.118
VT–AT	.407	.118	.560
VT–VT	-.023	-.023	.841
VT–VS	.222	.002	.692
VS–AT	.836	.164	.121
VS–VT	.826	-.097	.131
VS–VS	.682	.276	.150
Component Variance	2.160	1.680	1.638
% Total Variance	24.00	18.67	18.20

TABLE 5

Principal Components Analysis with Varimax
Rotation: Above Average IQ Group

Task	I	II	III
AT-AT	.695	.094	.490
AT-VT	.863	.140	-.070
AT-VS	.590	.570	-.116
VT-AT	.832	.081	.120
VT-VT	.712	.013	.207
VT-VS	.552	.344	-.208
VS-AT	.571	.413	-.289
VS-VT	-.007	.934	.059
VS-VS	-.113	-.038	.882
Component Variance	3.418	1.523	1.223
% Total Variance	37.98	16.92	13.59

appear to be influential.   Further, the factors themselves are
different in composition.   With the exception of the VS-AT task,
the first factor is mainly temporal, with both auditory and visual
stimulus conditions as sources of variance.   The second factor is
mainly visual-spatial, as is the third factor, with these latter
factors defined by combinations of stimulus and comparison conditions.

        The results of this study suggest then, that some modality
specialization, or lack of integration, may characterize children of
below average levels of intelligence, in contrast to the nonmodal
and predominantly temporal and spatial processing in children of
higher intelligence.

        In the second study, conducted by Marshall (1979), subjects
were matched on intelligence, and varied in reading achievement,
in order to explore modality specialization as related to reading
ability. The specialization of sensory systems has been a particularly common but curiously untested, assumption in the study of
reading disability, as evidenced in the constructs ascribed to
subtests in standard clinical assessment techniques, as well as in
the rationale given for many remedial programs.   Part of the
reason for the lack of tests of this assumption can be traced to
difficulties in the design of tasks as noted earlier, but the majority of the causes appear to be based in the uncritical acceptance
of the modal-specific position.

        The second study, which will be mentioned only briefly
here, involved 72 children of below average reading ability, and
72 children of above average reading ability.   The groups were
equally comprised of boys and girls, and were drawn from a large
population of Grade 3 children.   The reading groups were selected
by use of the Gates MacGinitie reading test, and then the final
samples were identified by matching for IQ on the Lorge-Thorndike
nonverbal battery.

        No significant effects for sex were found in analyses of
variance, and so the data were pooled for factor analyses.   The
analysis results for both groups contained two factors, using a
criterion of eigen-values greater than one.   In the results for the
below average readers, the first factor was comprised mainly of
tasks involving the VS condition, in either the stimulus or comparison position.   The second factor was comprised of tasks containing
temporal conditions, in either the auditory or visual modality.
These factors then, appeared to indicate no evidence for processing
which is specific to a sensory modality, for the major dimensions
represented in the factors were spatial and temporal respectively.

        The factor analysis results for the above average reading
ability group were similar, but clearer in composition, in that the

stimulus condition was more consistently a source of variance, and the factors divided more clearly on the temporal-spatial dimension. The first factor was a temporal factor, and the second defined spatial processing. Thus, as in the case for the below average readers, no evidence was apparent that indicated processes that are specific to a sensory modality.

The evidence presented in these studies suggests that cognitive processes may be increasingly specific to sensory modalities as related to decreasing levels of intelligence. With respect to reading, however, little evidence of modality-specificity was found, such that spatial and temporal task demands were the major sources of individual differences. These results generally support the suggestion made earlier then, that modality-specificity may be a parameter upon which different populations can be distinguished, rather than a condition of processing which generalizes to all populations. Thus, assumptions on whether cognitive processing is modality-specific or nonmodal may not represent the true state of affairs, for neither reflects an accounting of both task demands and subject characteristics. A trend which is consistent over the tasks and populations, is the distinction between spatial and temporal processing. To return to our earlier discussion of simultaneous and successive syntheses and similar dichotomies in Part I, one may ask also if these syntheses correspond to spatial and temporal processing respectively. Evidence that we have reviewed over the course of the last several years (Das, Kirby and Jarman, 1979) suggests that spatial and temporal processing may be considered as a special form of simultaneous and successive syntheses, but do not represent these syntheses completely. In particular, the dual processes in language of paradigmatic and syntagmatic associations (Jarman, in press-b) are not accommodated by the more specific spatial-temporal dichotomy These associations have been shown by Luria (e.g., 1973a, 1973b, 1976) to have a corresponding cortical basis to that of simultaneous and successive syntheses, thus elaborating the breadth of these syntheses beyond purely spatial and temporal cognitive processes.

The approach adopted the studies in Part II though, has been to minimize the effects of language through the use of non-verbal content, as well as partially separate type of stimulus presentation from modality of presentation in order to examine sensory systems specifically. This approach allows a number of theoretically interesting questions to be asked, which can be seen as extensions of the studies that we have reported here.

One of these questions is the extent to which information load affects the modality of processing. It may be, as suggested by Friedes (1974) and others, that more adept modalities are chosen or implicated under high load, but simple tasks are processed on a modality-specific basis. This would be an important question to consider obviously, in the study of learning of complex material.

A second and related question, is the extent to which different patterns of modality-specificity among different populations are influenced by, or a result of, concommitant patterns of strategic behavior. That is, to what extent are different patterns of modality-specificity in different groups based in group-specific strategies, as opposed to unique structural limitations in the groups? To return to the simultaneous-successive model, the former alternative would refer to the planning and decision-making component of the model, as discussed in Part I and proposed under other different headings by many researchers in recent years (e.g., Campione and Brown, 1978; Hunt and MacLeod, 1978; Sternberg, 1978). The latter alternative, that of structural limitations, would refer to the degree of cortical organization in the three-zone system proposed by Luria, and would imply that perhaps some groups may be less advanced in secondary and tertiary zone development.

There are various means by which the tasks used in this research may be modified, and populations selected, such that these questions may be addressed. This work is presently underway as an extension of the initial studies that we have described here. Our comments here are only an early progress report therefore, with some interesting problems apparently yet to come.

References

Ashman, A. The relationship between planning and simultaneous and successive synthesis. Unpublished Ph.D. thesis, University of Alberta, Edmonton, Canada, 1978.

Atkinson, R. C., and Shiffrin, R. M. Human memory: A proposed system and its control processes. In K. W. Spence and J. T. Spence (Eds.), The psychology of learning and motivation: Advances in research and theory. Vol. 2. New York: Academic Press, 1968.

Bryant, P. E. Comments on the design of developmental studies of cross-modal matching and cross-modal transfer. Cortex, 1968, 4, 127-137.

Campione, J. C., and Brown, A. L. Toward a theory of intelligence: Contributions from research with retarded children. Intelligence, 1978, 2, 279-304.

Connor, J. M., and Hoyer, R. G. Auditory and visual similarity effects in recognition and recall. Memory and Cognition, 1976, 4, 261-264.

Das, J. P. Structure of cognitive abilities: Evidence for simultaneous and successive processing. Journal of Educational Psychology, 1973, 65, 103-108.

Das, J. P., Kirby, J., and Jarman, R. F. Simultaneous and suc-

cessive syntheses: An alternative model for cognitive abilities.
    Psychological Bulletin,, 1975, 82, 87-103.
Das, J. P., Kirby, J., and Jarman, R. F.   Simultaneous and suc-
    cessive cognitive processes.   New York: Academic Press,
    1979.
Erdelyi, M. H., Finkelstein, S., Herrell, N., Miller, B., and
    Thomas, J.   Coding modality vs. input modality in hyper-
    mnesia: Is a rose a rose?   Cognition, 1976, 4, 311-319.
Freides, D.   Human information processing and sensory modality:
    Cross-modal functions, information complexity, memory and
    deficit.   Psychological Bulletin,   1974,   81,   284-310.
Hunt, E., and Macleod, C. M.   The sentence-verification paradigm:
    A case study of two conflicting approaches to individual differ-
    ences.   Intelligence, 1978, 2, 129-144.
Jarman, R. F.   A method of construction of auditory stimulus pat-
    terns for use in cross-modal and intramodal matching tests.
    Behavior Research Methods and Instrumentation, 1977, 9, 22-25.
Jarman, R. F., Marshall, M. F., and Moore, T. S.   Construction
    of a tape-controlled system of presenting visual-spatial and
    visual-temporal stimulus patterns for use in cross-modal and
    intramodal matching tests.   Behavior Research Methods and
    Instrumentation, 1979, 11, 363-365.
Jarman, R. F.   Modality-specific information processing and intellec-
    tual ability.   Intelligence, in press.   (a)
Jarman, R. F.   Cognitive processes and syntactical structure:
    Analyses of paradigmatic and syntagmatic associations.
    Psychological Research, in press.   (b)
Luria, A. R.   The functional organization of the brain.   Scientific
    American, 1970, 222, 66-78.
Luria, A. R.   The working brain.   London: Penguin, 1973. (a)
Luria, A. R.   Basic Problems of neurolinguistics.   The Hague:
    Mouton, 1976.
Marshall, M. F.   Auditory-visual and spatial-temporal integration
    abilities of above average and below average readers.   Unpub-
    lished Ed.D. thesis, University of British Columbia, Vancouver,
    Canada, 1979.
O'Connor, N., and Hermelin, B.   Seeing and hearing and space and
    time.   New York: Academic Press, 1978.
Routh, D. A.   An "across-the-board" modality effect in immediate
    serial recall.   Quarterly Journal of Experimental Psychology,
    1976, 28, 285-304.
Rubinstein, L., and Gruenberg, E. M. Intramodal and cross-modal
    sensory transfer of visual and auditory temporal patterns.
    Perception and Psychophysics, 1971, 9, 385-390.
Sears, R. R.   Your ancients revisited: A history of child develop-
    ment.   In E. M. Hetherington (Ed.), Review of child developme:
    research.   Vol. 5.   Chicago: University of Chicago Press, 1975
Sperling, G.   A model for visual memory tasks.   Human Factors, 196
    5, 19-31.
Spitzer, T. M.   The development of visual and auditory recall as a

function of presentation and probe modalities, serial position, and series size. Child Development, 1976, 47, 767-778.

Snart, F. D. Levels of processing and memory: A developmental approach. Unpublished Ph.D. thesis, University of Alberta, Edmonton, Canada, 1978.

Sternberg, R. J. Intelligence research at the interface between differential and cognitive psychology: Prospects and proposals. Intelligence, 1978, 2, 195-222.

Waugh, N., and Norman, D. Primary memory. Psychological Review, 1965, 72, 89,104.

PROCESS THEORIES: FORM OR SUBSTANCE? A DISCUSSION OF THE PAPERS BY BUTTERFIELD, DAS AND JARMAN

John B. Biggs

Newcastle University

Shortland, Australia

The most general common assumption underlying the papers by Butterfield, Das and Jarman--and many others presented at this Conference--implies a sharp distinction between process and performance. More specifically, competent performance is seen as the result of an interaction between task demands and various cognitive options the individual may or may not have at his disposal. Das presented one view of what those options might be; Jarman demonstrated the effect that stimulus demands have upon the range of options, differentially for high and low ability groups; and Butterfield outlined a research strategy that promises to integrate task demands and cognitive availability, not only with respect to the concerns of the other two speakers, but over a very broad front indeed.

There are so many fruitful issues here to concentrate upon. I shall nominate what seem to me to be the more important ones by asking a few questions.

1. From whence do processes derive?

The three papers present a neat line-up of replies on this point. Butterfield takes a strict operationist viewpoint: having selected one's task (on whatever grounds) one then finds out what factors correlate with task performance; and then each is varied in turn to verify that what one has is a process that fits his definition (the question of that definition is itself one that I wish to return to later).

Jarman used factor analysis as a way of determining generality of process across tasks and showed that in the low IQ group

the stimulus condition was the process source, and an inefficient
one at that, and in the high IQ group the mode of presentation,
temporal or spatial, of the comparison stimulus accounted for all
three process factors. While Das was careful not to equate simultaneous and successive synthesis with any particular modality,
the patterning observed by Jarman could be attributed to these
two forms of coding: visual-spatial as simultaneous, and auditory-
temporal and visual-temporal as successive. Hence, the process
source can be attributed for the low IQ group to the stimulus,
and for the high IQ group to a central source.

Das is quite explicit about the nature of such a central
source: it would be a physiological one, and he derives his particular model from Luria's work on brain lesions. This is very convenient. In trying to operationalise the planning and coding processes,
he can concentrate on those tasks that Luria showed were particularly impaired when there were lesions in the relevent locus in
the brain. It is very significant, then, that in the quite different
context of Jarman's experiment there is such a good line-up
between his factors and the two forms of coding.

Other process sources depend upon the kind of model used:
Sternberg (see Chapter 31 for example) turns to a computer
analogy. Whatever the particular kind of model chosen, the notion
of some source is important if one is to resolve the inevitable
hiatus in operationism. According to operationism, and paraphras-
ing Butterfield, if something works (if it correlates with performance) it's in--it's a process. If not, it's a structure. However,
as I'll be arguing below, it may not work for a multitude of
nonstructural reasons and inevitably one will end up with some
quite misleading conclusions about what is or is not a process.

This problem is obviated to a large extent by replacing this
large pragmatism with a process model in which a source of processes is hypothesized: this would have the additional benefit of
restricting the range of what would otherwise be an enormously
large universe of tasks and processes. In short, I see the
probability of a fruitful union between the comprehensive methodology proposed by Butterfield, and the more specific, substantive
model proposed by Das and Jarman. Thus, with the intervention
of theory, one can move from the subordinate task-related proces-
ses--such as those found by Jarman--to superordinate processes,
such as simultaneous and successive coding. I realise that in
proposing this I am (I think) changing the meaning of superordinate and subordinate in Butterfield's original sense, but I think
that would be a small price to pay. There is a complementarity
of form and substance in the Butterfield and Das models that can
be usefully exploited.

2.   How teachable are processes?

Oddly enough, this is not an empirical question. Butterfield defines a process factor, as distinct from a structural one, as "one whose manner of change is specified in the theory and which is manipulable." If one the evidence a factor appears nonmanipulable (Step 4C), then it is classified as a structural factor. But even later, fully to fit the model, a process must be manipulable in the particular sense of teachable (Step b). Coming from a Faculty that relies on teaching people to teach for its bread and butter, I am tempted to suggest that we shall end up with a very large number of structural factors. Butterfield does acknowledge the problem, but it is rather scary that the validation of the processes in cognitive development depends upon the assumption that another process, the teaching process, is 100% effective.

Butterfield's strategy, in fact, is an ingenious and complex version of the mastery learning paradigm from instructional psychology (e.g. Block, 1971), and it shares both its virtues and its vices. Two criticisms, or limitations at least, of mastery learning are also applicable to Butterfield's model: (i) There may be a pay-off between teachability and triviality; and (ii) The model favors content that is convergent, where the appropriate outcome can be predetermined and specified in advance.

To take the first point, it is unfortunately true that simple, less important, things are usually easier to teach than more complex things. Let us go back to the original Zeaman and House (1963) experiments. The trouble with retardates is that they don't attend to the relevant dimension. Right: signal the relevant dimension to them and they should perform as well as normals. They do. Does that mean that we have, at least for that task, "cured" their retardation? No: adequate performance is zeroing in on that relevant dimension by oneself. It's like filling in yesterday's crossword puzzle from the solution in today's paper. If mimicking good performance is all that is required, then teaching crossword puzzle solving would be easy--and trivial.

Butterfield is well aware of this problem. It applies when subordinate processes of low generality are treated. The state of play is not very promising, as his review indicates, and he himself suggests that it might be "too soon for superordinates."

I think that might be too pessimistic a view. Das and his coworkers have reported an analysis of reading skills which suggests that successive rather than simultaneous coding is most important in the beginning stages of reading, but that simultaneous coding is more important in mature, proficient readers (Das, Cummins, Kirby and Jarman, 1979; Kirby and Das, 1977). It was further found that native Indian children were very low in successive processing: they were also poor readers. The question was: Would training in successive coding result in higher reading performance? Krywaniuk and Das (1976) found the answer to be affirmative. The suggestion is that, for various reasons that may be found in Indian culture, simultaneous strategies are called out with a high degree of frequency, but not successive ones. The study suggests not that Indians were deficient in successive processing ability (which raises another issue to be dealt with below), but that their successive strategies were simply not primed. When they were, in the intervention program, the now salient successive strategy could be deployed in quite a different field, reading skills. The argument here is very similar to that used by Bryant (see Chapter 18) in accounting for conservation training.

I wonder of this kind of result would have eventuated by following through Butterfield's strategy of working from subordinate to the superordinate process? If the subordinates are trivial, as is likely, the result of meta-analysis may be a higher, more generalized level of triviality. This argument thus gets back to my earlier point that a source model is necessary to avoid the consequences of an initial bad choice of tasks or processes.

The first criticism of mastery-type strategies, then, is one that needs watching, but it is not necessarily damning, given some flexibility about sources.

The second criticism, that mastery learning is suitable only for closed content, is I think an important one. In order for mastery learning to work, one needs to specify instructional, and in this case behavioral, objectives. Now it is quite possible to specify the behaviors demanded in successive coding; and, as noted, the result, improvement in reading skill, certainly isn't trivial.

However if the process to be taught does not involve coding, but a metafunction such as planning, then prescribing the requirement in advance is to negate the whole point of the exercise. It seems to me that the Zeaman and House experiments were of that nature: choosing the relevant dimension is a planning function; as is doing your own crossword puzzles. The very point is for the individual to derive his own plan: the task is open. To close it is to turn it into a different, lower level, task. This brings me directly to the next question.

3.   What is the place for self-taught processes?

The emphasis so far has been upon processes that are taught by a Powerful Other. When we look at the broad context of cognitive development, however, many process components appear

to be spontaneously generated and deployed by the learner.
Such self-generated processing is arguably more significant in
general development than any taught process components: certainly Piaget would argue that way.

Donaldson (1978) draws attention to what appears to be a
very basic superordinate process that is increasingly significant
in cognitive development: "disembedding" task fron context.
Disembedding is not directly taught: rather, it is displayed by
the child if the task is presented sufficiently noise-free. Even
quite simple changes in wording can call out the process in otherwise conventional Piagetian tasks.   To teach the disembedded
response brings us back to crossword puzzles.

Then there are those superordinate processes that relate to
school learning and studying. Study processes are generated by
individual learners in their interaction with various academic
subject matters and teaching environments, and are observable
during high school and subsequently.   They appear to be determined by several interacting factors including personality characteristics, content and the perceived motivational context for the
learning.   These processes are not, however, formally taught, at
least not usually in high school, and when they are taught, as
they sometimes are at college level in counselling programs, the
results are very mixed. Biggs (1978) has distinguished three
broad dimensions of study process: reproducing, internalising and
structuring the content to be learned.

One hypothesis Kirby and I are currently investigating (Note
1) is that individual differences in simultaneous and successive
coding might predetermine the successful deployment of these
various study processes in 15 year old high school students.
While this does not appear to be the case in the work completed
so far, what is emerging is that process-task correlations (math
and English being the tasks) are much stronger and more frequent
in students deficient in both simultaneous and successive options.
It is as if they need to generate and deploy these more specific
processes if they are to cope, when more general coding options
are unavailable to them.

In short, then, there is evidence that in both broad and
narrow fronts of cognitive development, individuals generate their
own processes.   I think this point has much significance for
Butterfield's model.   At the best, it might mean that the course
of cognitive development could be greatly facilitated if the critical
processes are simply irrelevant to the main issues in cognitive
growth, and could even distort the course of normal development.
We are, of course, touching upon the dreaded "American Question,"
and to put the matter into a broader perspective, we should
examine the issue of developmental stage.

4.   How do process theories relate to the concept of developmental stage?
       Butterfield, in addressing a general theory of cognitive
development, should say something about the issue of developmental stage. Stages are defined by age-related boundaries, are
sequentially invariant, and place limits upon what performances
can be carried out.

There are roughly two major views on the stage question
(Siegel and Brainerd, 1978):
      (i)   That stages can be explained, indeed explained away, in
             process terms;
      (ii)   That stages can only be explained in terms of structural
             limitations that vary according to age and/or experience.

Butterfield would argue that stage-like phenomena can be
explained in terms of inadequate subordination. If all prerequisite
processes are taught, then a performance comprising those processes will be evidenced. Das says little on the stage question;
Jarman's evidence is that low grade performances are produced by
qualitatively different processes than a high grade performance.
In the former case, the processes are stimulus-dependent, and in
the latter are centrally determined. Although he does not say
that this transition from stimulus-to-central determination is a
developmental one, such a view seems plausible.

In the present context, then, Butterfield's paper has most to
say on the stage issue. His radical process theory might in fact
seem implausible in view of evidence that young learners employ
different processes to achieve a similar result than mature learners.
Case (1979), for example, shows that immature learners simply do
not attend to all the task relevant information because (essentially)
of insufficient working memory (WM) capacity. While this sounds
like a purely structural limitation, and a classic example of a
stage, Case argues ingeniously that WM capacity is invariant over
age, but because of increasing automaticity of responding over
age, functional WM does increase with age. Although the variation
in functional WM is in fact process-related, it would place strong
structural-like limitations on a person's options.

Butterfield's model is, as I see it, a methodology rather than
a theory; and it does not violate the methodology to suggest that
different processes might emerge--indeed are very likely to emerge--
according to functional WM availability. Some kind of reconciliation
between the two positions on the stage question would become
possible if some structural or quasi-structural factor, such as
WM, were used as an independent variable to be accommodated in
the model.

5.   Are the units in process analysis abilities or competencies?

In all three papers, and in others at this Conference, the
question lurks: Is intelligence or proficient performance account-
able for in terms of a few broad nomothetic abilities that individuals possess in greater or lesser degree; or in terms of strategies
of decision-making, problem solving, styles of handling information
etc., that are differentially task-effective?   In immediately topical
terms, is simultaneous coding a style or way of coding information,
or is it an ability, such as reasoning?   Despite Das's handling of
the question, it does not help to find that a power test, such as
Raven's, is used as a marker by Das for propensity for using
simultaneous coding, and by Jensen (1973) to mark an unequivocal
ability.

When then does a style become a process become an ability?
Is, for a different example, field independence as operationalised
by EFT scores a cognitive style or a visualising ability?   I think
Tyler's (1978) distinction between abilities and competencies is
helpful here: "a competency is a particular skill, something an
individual knows how to do" (op cit., p. 99).   Competencies are
criterion-referenced: they involve a task analysis in much the
same way as Butterfield's subordinate processes; they are also
more deeply involved with affective factors, such as intrinsic
interest, than broad abilities.

In an important sense, Tyler and Butterfield are saying
much the same thing in that they advocate task analysis and
behavioral interaction to define a competency for one; a subordinate
process for the other.   At this point they diverge.   Tyler is
interested in showing how competencies can be valuable in the
idiographic study of individuality;   Butterfield in piecing the
processes together to form a nomothetic account of cognitive
development.   Nevertheless, there is an important common point:
whether in building theories or in understanding individuals,
abilities in the traditional sense are sidestepped by a new research
strategy.

I would like to see in Butterfield's model more room for
different options in handling a task.   Some strategies are universal:
without them, the task will remain incomplete however one goes
about it.   Other strategies are optional: the task can be successfully completed in several different ways.   Bruner's early work
on concept attainment made this clear (Bruner, Goodnow and
Austin, 1956); as more recently, does Pask's work (Pask and
Scott, 1972) on holist and partist strategies.   What determines
which option is the best can be a whole host of things: previous
experience, WM capacity, extent of relevant background information—in fact, all those things that may enter into the A in ATI.
This question of process options is particularly relevant in cognitive

development: in Piagetian tasks successful completion often isn't the issue but rather the nature of the option chosen to get there.

So far I have not mentioned ability in this. At any point in the analysis one may ask how well or how rapidly an individual is doing the task in comparison with another individual or reference group, but this seems to me to involve quite a different sort of question.   And it is only in this context that the concept of ability becomes relevant.

The context of process analysis is quite different, and it is answering a different and to my mind frequently more important question: What does an individual need to do that he is not doing already if he is to handle the task appropriately?   Analysis into competencies, strategies and processes--however one terms it--helps directly to prescribe teaching strategy in a way that analysis in terms of abilities does not.

My answer to the question, are we dealing with abilities or competencies, is thus similar to Das's: it depends upon what one wants to do.   If one wishes to compare individuals with each other, the unit becomes abilities--and in that case I imagine one would not be very interested in process analysis.   If, on the other hand, one wishes to discover what things need doing in order that certain tasks may be successfully completed--whether for theoretical or applied reasons--then the units of analysis are most usefully expressed in terms of processes, or competencies, depending upon how nomothetic or idiographic one feels like being.

6.   What is the appropriate context for process analysis?

I ask this, my final question, following a certain relativism that seemed to be emerging from the last question.   Butterfield himself sees his task solely in terms of theory construction.   More specifically, he says that process hypothesis should only be investigated in the traditional experimental-manipulative paradigm, structural hypotheses through naturalistic observation and correlation.   This appears to be a gigantic task: some selection of tasks and processes is necessary, otherwise what one attends to becomes a matter of subjective judgment.   Although Butterfield dismisses Brown and Deloach's point that "instructional relevance be the guiding force in the initial choice of training tasks" (quoted in Butterfield's Chapter) it seems to me to be a good idea for that point to be taken.

For as soon as one does, process analysis seems to change gear. Instead of being an instrument for theory building within a conventional methodological tradition, it becomes a tool that may fit many methodological and applied contexts.   For example I can see immediate applications within the normal classroom, particularly for curriculum development. If applied to a suitable teaching subject, one could derive a hierarchical order of processes, competencies or components that are necessary in the progressive mastery of that discipline. Not only would one obtain an age-graded ordering, but the process of analysis itself looks after the teachability aspect. This would apply, however, only to some aspects of some subjects—closed tasks in fact—and a different strategy would be appropriate for open subjects. Other applications immediately come to mind: wherever, in fact, it is necessary or desirable to reduce inter-group differences on learnable-material. And that's precisely what a conference on intelligence and learning should be concerned about; and more particularly, about which the three speakers have given us so much to consider.

References

Biggs, J. B. Individual and group differences in study processes. British Journal of Educational Psychology, 1978, 48, 266-279.

Block, J.H. (Ed.) Mastery Learning: Theory and Practice. New York: Holt, Rinehart and Winston, 1971.

Bruner, J. S., Goodnow, J. and Austin, G. A. A Study of Thinking. New York: Wiley, 1956

Case, R. The underlying mechanism of intellectual development. In J. Biggs and J. R. Kirby (Eds.) Instructional Processes and Individual Differences in Learning. New York: Academic Press, 1979.

Das, J. P., Cummins, J., Kirby, J. R., and Jarman, R. F. Simultaneous and successive processes, language and mental abilities. Canadian Psychological Review, 1979, 20, 1-11.

Donaldson, M. Children's Minds. London: Macmillan, 1978.

Jensen, A. R. Genetics and Education. London: Methuen, 1972.

Kirby, J. R. and Das, J. P. Reading achievement, IQ and simultaneous-successive processing. Journal of Educational Psychology, 1977, 69, 564-570.

Krywaniuk, L. W. and Das, J. P. Cognitive strategies in native children: analysis and intervention. Alberta Journal of Educational Research, 1976, 22, 271-280.

Pascual-Leone, J. A mathematical model for the transition rule in Piaget's developmental stages. Acta Psychologica, 1970, 63, 301-345.

Pask, G. and Scott, B.C.E. Learning strategies and individual competence. International Journal of Man-Machine Studies, 1972, 4, 217-253.

Siegel, L. and Brainerd, C. (Eds.) Alternatives to Piaget. New York: Academic Press, 1978

Tyler, L. E. Individuality. San Francisco: Jossey Bass, 1978.

Zeaman, D. and House, B. J.   The role of attention in retardate
    discrimination.   In N. R. Ellis (Ed.)   Handbook of Mental
    Deficiency. New York: McGraw Hill, 1963.

Reference Notes

1.   Biggs, J., and Kirby, J. R.   Processing styles, study strate-
      gies and school performance.   Project funded by the Australian
      Research Grants Committee, 1979.

TOWARD A UNIFIED COMPONENTIAL THEORY OF
HUMAN INTELLIGENCE: I.   FLUID ABILITY

Robert J. Sternberg

Yale University

New Haven, Connecticut, U.S.A.

Abstract

A progress report on the development of a unified theory of human intelligence is presented. The report deals with that portion of the theory that concerns fluid ability, which is viewed as roughly synonymous with reasoning. The unified theory applied to reasoning comprises a number of hierarchically nested subtheories, each of which accounts for successively more specific aspects of human reasoning behavior. The basic unit of the theory is the component: It is claimed that a relatively small set of components can account for behavior in a wide range of reasoning tasks, and that individual components are general across the vertical range of the hierarchy. The components and the sub-theories in which they play a part are briefly described, and where available, data testing the subtheories are summarized. These data provide at least tentative support for the proposed theoretical structure.

During the past several years, I have been devoting a major portion of my research effort toward the development and testing of a "unified componential" theory of human intelligence. The theory deals only with intelligence narrowly defined, covering in its scope the kinds of behaviors associated with performance on conventional intelligence tests, and the kinds of behaviors these tests predict. Although there is more to intelligence broadly defined than is covered by the scope of the theory (see Zigler, this volume), the behaviors with which the theory deals seem at least to be an important subset of the broad range of behaviors associated with general intelligence.

Human intelligence is sometimes viewed as comprising at least two major kinds of abilities, fluid ability and crystallized ability (see Cattell, 1971). Snow (1978) has further distinguished a third major kind of ability, visualization ability. Fluid ability is best measured by reasoning tests such as figural analogies, abstract syllogisms, and letter series. Crystallized ability is best measured by verbal tests such as vocabulary, reading comprehension, and general information. Visualization ability is best measured by spatial tests such as mental rotation of three-dimensional objects, mental paper folding, and counting of hidden cubes. I will describe in this article only that portion of my theory that deals with fluid ability, which I view as practically synonymous with reasoning ability. The portion of the theory dealing with crystallized ability is in a less advanced state, and there currently is no portion of the theory that deals with visualization ability. In order to understand the proposed theory, it is necessary first to understand why the theory is "unified" and why it is "componential."

The proposed theory is "unified" because it attempts to explain within a single theoretical framework human information processing in a wide variety of complex tasks. The unified theory comprises hierarchically nested subtheories accounting for performance on successively more narrow classes of tasks. The hierarchical structure of the theory dealing with fluid ability, or reasoning, is depicted by the tree diagram in Figure 1. Corresponding to each node in the hierarchy is a theory or subtheory of human reasoning, and a class or subclass of tasks to which the theory applies. Theories at each level of the hierarchy include as special cases all subtheories nested beneath them.

At the top of the hierarchy is the unified theory. Under the unified theory are two subtheories, one of deductive reasoning and one of inductive reasoning. In general, the theory of deduction applies to tasks in which there is a deductively certain (logically valid) solution, whereas the theory of induction applies to tasks in which there is no deductively certain solution, but in which there is an inductively probable one.

Each of these subtheories can again be split into two subtheories. In the case of the subtheory of deduction, the two further subtheories are ones of syllogistic reasoning and of transitive inference. The theory of syllogistic reasoning deals with class inclusion (categorical) and conditional relations. The theory of transitive inference deals with transitive (linear-ordering) relations. In the case of the subtheory of induction, the two subtheories are one of analogical, classificational, serial, topological, and metaphorical reasoning, and one of causal inference.

At the lowest level of the hierarchy are specific information-processing models that describe in detail the sequencing

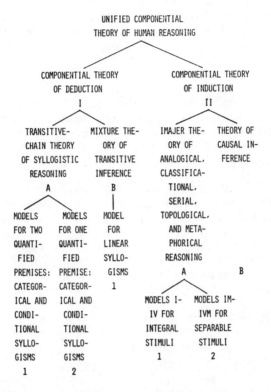

1. Hierarchical structure of unified componential theory of human reasoning.

of components used in the solution of specific types of problems. Each model is expressed in terms of a flow chart that characterizes the course of information processing from the time the problem is first perceived until the time the individual makes a response.

Although the various subtheories differ in their level of generality and in their particular contents, the structure of each subtheory (and of the unified theory) is the same. Each describes reasoning behavior at a molar level of stages, and at a molecular level of components.

At the molar level, performance on each task can be partitioned into four stages of information processing: (a) encoding, in which individuals represent the task problem in working memory, and retrieve from long-term memory information that may be relevant to problem solution; (b) combination, in which individuals interrelate various aspects of their encodings in order to generate a problem solution; (c) comparison, in which individuals test the solution from the combination stage against the available answer options; (d) response, in which individuals communicate their chosen answer. Under certain circumstances, particular stages may be bypassed. For example, comparison is not needed in problems having free-response format, and response is not needed in problems where the individual is under no compulsion to communicate his or her solution.

At the molecular level, performance on each task can be partitioned into a set of components; performance in each stage can be partitioned further in the same way. Kinds of components can be classified in two different ways: by level of generality and by function (Sternberg, Note 1).

Components can be classified in terms of three levels of generality: General components are required for performance of all tasks within a given task universe; class components are required for performance of a proper subset of tasks (including at least two tasks within a task universe; and specific components are required for the performance of single tasks within the task universe. All components considered in this article will be of the more interesting general and class varieties.

Components can also be classified in terms of five different functions they perform. Each of these five functions can be crossed with the three levels of generality, yielding 15 different types of components overall. Performance components are used in the execution of various strategies for task performance; acquisition components are skills involved in learning information from context; retention components are skills involved in retrieving information that has been previously acquired in context; transfer components are skills involved in generalizing retained information from one situational context to another; metacomponents are higher-order control processes that are used for planning how a problem should be solved, for making decisions regarding alternative courses of action during problem solving, and for monitoring solution processes. Performance components are most heavily implicated in the measurement of fluid and visualization abilities, where individuals have to carry out reasoning or spatial tasks from start to finish according to some strategy that they devise. Acquisition, retention, and transfer components are most heavily implicated in the measurement of crystallized ability, where past execution of these components has resulted in the verbal knowledge and skills that are measured by crystallized ability tests, usually several years after the knowledge and skills have been acquired. Metacomponents are implicated in the measurement of all three kinds of abilities, since accomplishment of all tasks requires planning, decision, and monitoring processes. Since this article deals specifically with fluid ability, the emphasis will be upon performance components, which are the components that have been most extensively studied in my research to date. Further discussion of the other kinds of components can be found elsewhere (Sternberg, 1979a, 1979b, Note 1, Note 2).

Each theory and subtheory in the hierarchy depicted in Figure 1 specifies at minimum six aspects of information processing at the level of performance components: (a) the <u>components</u> of response time, response accuracy, and response choice; (b) the <u>representation</u>(s) upon which these components act; (c) <u>strategies</u> (rules) for combining the components into a working algorithm for problem solution; (d) the <u>consistency</u> with which these strategies are executed; (e) <u>parameter estimates</u> corresponding to the durations, difficulties, or probabilities of component execution; and (f) theoretically-based <u>relations</u> of the components to each other and to previously <u>established</u> reference abilities (such as "reasoning" as measured by standardized tests of mental ability).

The proposed theory is "componential" because the basic unit of information processing in the theory is the component: an elementary information process that is executed in the solution of one or more problems requiring intelligence. The components of information processing pertinent to subtheories lower in the theoretical hierarchy are pertinent as well to the higher-order subtheories under which the lower-order subtheories are nested.

The components of human intelligence are explanatory as well as descriptive constructs. They are the sources not only of communalities in the performance of multiple subjects in tasks requiring intelligence, but also of individual differences in performance (see Sternberg, 1977, Chapter 4). General, group, and specific factors obtained in factor analyses of ability tests, for example, can be accounted for in terms of distributions of components across tasks: A general factor arises when one or more general components are common to all tasks in a given task universe; a group factor arises when one or more class components are common to several tasks; a specific factor arises when one or more specific components are specific to a single task. From the standpoint of the componential approach to intelligence, therefore, components are elementary units of analysis. Factors are merely constellations of these components that arise as a function of the particular mixture of components required for solution of a particular battery of items or tests subjected to a factor analysis (Sternberg, 1977).

Most, if not all, components can be split indefinitely into successively finer subcomponents. The level of division that is considered "elementary" for a given purpose is one of convenience, with convenience being determined among other things by (a) theoretical homogeneity of level of division of components within a single subtheory at a given level of the hierarchy, (b) generality of a component across tasks within a given node and at various levels of the hierarchy, and (c) univocal correlations of a component with scores on orthogonal mental ability tests. This last criterion requires that the parameter estimate corresponding to the duration, difficulty, or probability of component execution should be highly correlated with tests measuring one kind of ability, but only poorly correlated with tests measuring other kinds of abilities.

With these introductory remarks completed, it is possible to turn to the main body of the article, which will be devoted to a brief description of the contents of each node of the hierarchy.[1] In describing these contents, it will be necessary to work through the hierarchy from the bottom up in order to show how the more specific theories and tasks merge into the more general theories and tasks. Where possible, the description of each node in the hierarchy will consist of (a) a set of relevant references, (b) an example of the type of problem task that belongs at the given node, (c) a brief explication of the theory as it applies to that task, and (d) a summary of data that have been collected to test the theory. For the sake of brevity, theoretical descriptions will emphasize the level of stages, with referenced articles containing more of the details. In some cases, data have not yet been collected. Numbers and letters heading each section refer to those in Figure 1.

I.   COMPONENTIAL THEORY OF DEDUCTION

A.   Transitive-Chain Theory of Syllogistic Reasoning

1.   Models for Two Quantified Premises

References.   This section is based upon the report of Guyote and Sternberg (Note 4); see also Sternberg, Guyote, and Turner (in press), and Sternberg and Turner (Note 5).

Nature of task.   Syllogisms with two quantified premises are usually presented in categorical form, e.g., "No B are C.   Some A are B.   Which of the following conclusions can be deduced?   (a) All A are C; (b) Some A are C; (c) No A are C; (d) Some A are not C; (e) None of the above."   Our theory also applies to syllogisms presented in conditional form, e.g., "If B occurs, then C does not occur.   If A occurs, then B sometimes occurs.   Which of the following conclusions can be deduced?   (a) If A occurs, then C occurs; (b) If A occurs, then C sometimes occurs; (c) If A occurs, then C does not occur; (d) If A occurs, then C sometimes does not occur; (e) None of the above."

Theory.   During encoding, the individual forms a mental representation of each of the set relations that can be used to describe the verbal relation between the two terms in each premise. Although encoding of the possible set relations is assumed to be complete and correct, encoding of various set relations is theorized to be accomplished with differential rapidity.   During combination, the individual integrates pairs of set relations encoded for each of the premises, combining A-B and B-C relations for form A-C

relations. Certain pairs of set relations are theorized to be more
rapidly combined than other pairs. Errors in combination are asserted to be due to incomplete combination of all the possible set relations one might combine in a given problem. During comparison,
the individual compares his or her mental representation(s) of the
combined A-C set relation(s) to the verbal conclusions presented as
alternative answer options. If one of the conclusions correctly describes the mental representation, that conclusion is selected; if
not, the individual is assumed to check back over the operations of
the combination stage for one or more possible errors, and, if no
errors are found, to select the "None of the above" answer option.
Errors during this stage are due to biases on the part of the individual in favor of certain kinds of conclusions. During response, the
individual communicates his or her choice of an answer option.

Data. Data have been collected for syllogisms with abstract,
factual, counterfactual, and nonsensical content. Only categorical
problems will be considered here.

1. Means. The mean latency for syllogisms with abstract
content (the only kind for which latency data were collected) was
39.47 seconds. Mean accuracy in response choice was .59 across
content types. Factual syllogisms were easier than the other types
of syllogisms, which did not differ among themselves in difficulty.

2. Model fits. The value of R^2 for the proposed model in
fitting latency data with abstract content was .88. The value of R^2
in fitting response choice data was .92, averaged across content
types. Both values of R^2 were significantly greater than zero but
significantly less than the reliability of the data. Comparisons to
alternative models of syllogistic reasoning favored the proposed
model, regardless of content.

3. Parameter estimates. Encoding of the easier set relations
(those in which the relations between A and B and between B and
A are symmetrical, such as A overlapping with B, which is equivalent
to B overlapping with A) took 4.35 seconds; encoding of the more
difficult set relations (those in which the relations between A and B
and between B and A are asymmetrical, such as A subset of B,
which is not equivalent to B subset of A) took 5.32 seconds; combination of the easier set relations took 5.11 seconds, and combination of
the more difficult ones, 6.74 seconds. Comparison averaged 7.86
seconds, and response, 4.39 seconds. Individuals were found, on
the average, to combine just one pair of set relations .54 of the
time. They were also found to have two response biases--an atmosphere bias that encouraged the selection of a negative conclusion
when at least one premise was negative, and encouraged the selection
of a particular conclusion when at least one premise was particular
("some" rather than "all"); and a strength bias that encouraged
the selection of the most restrictive conclusion possible (e.g., "All

A are C" rather than "Some A are C" when either conclusion would
be logically permissible). When the two biases were pitted against
each other, atmosphere predominated .81 of the time; when both
encouraged selection of the same conclusion, that conclusion was
selected .92 of the time. Individuals also showed a bias toward
"None of the above" in certain cases where it was not justified.
The most salient effects of content were for factual content (a) to
increase the number of combinations of set relations made, presumably
because familiar content frees additional working memory space in
which combination takes place, and (b) to decrease susceptibility to
response biases such as atmosphere and strength, which depend
purely upon the formal characteristics of the syllogism. Presumably,
factual content results in the replacement of formal response biases
by substantive ones.

4. Relations to reference abilities. The median correlation
between proportion of syllogisms correctly answered (for factual,
counterfactual, and nonsensical syllogisms) and verbal ability was
.12 (p ⟩ .05); the median correlation between proportion correctly
answered and spatial-abstract ability (estimated as a composite) was
.42 (p ⟩ .01) for the same syllogisms. The source of the correlation
with spatial ability was localized in the combination stage: Individuals
who are higher in spatial-abstract ability combine more set relations,
on the average, than do individuals who are lower in spatial-abstract
ability. This result is consistent with the implication of the proposed
theory that because the information in syllogistic premises is represented abstractly, varying ability in manipulations performed upon abstract representations will be a source of individual differences in
syllogistic reasoning.

2.   Models for One Quantified Premise

References. This section is based upon the report of Guyote
and Sternberg (Note 3).

Nature of task. Syllogisms with one quantified premise, like
those with two quantified premises, can be expressed in either
categorical form or conditional form, e.g., "All A are B. x is an
A. Conclusion: x is a B. (a) True, (b) False" and "If A then B.
A. Conclusion: B. (a) True, (b) False."

Theory. Encoding in these problems proceeds much as in the
problems considered above, with minor adjustments made for the
second (minor) premise. Two basic strategies are involved in
combination: direct proof, whereby one seeks to determine on the
basis of the given premise information whether the proposed
conclusion is true or false; and indirect proof, used when direct
proof fails, whereby one negates the proposed conclusion, and
seeks to combine the negated assertion of the conclusion with the
conditional information in the first premise. If the result

contradicts information in the second premise, the syllogism is valid. There is no comparison stage in these problems, because there is only one conclusion, which must be either confirmed or disconfirmed. In response, the individual communicates whether the conclusion is true or false.

Data. Data have been collected only for syllogisms with abstract content.

1. Means. Mean solution latency was 13.38 seconds for categorical syllogisms, and 13.51 seconds for conditional syllogisms. Mean proportion correct was .82 for categoricals and .83 for conditionals.

2. Model fits. The values of R^2 for response times were .88 and .84 for categorical and conditional syllogisms respectively. The corresponding values for response choice were .97 and .95 for categoricals and conditionals respectively. All values of R^2 were significantly greater than zero, but significantly less than their corresponding reliabilities.

3. Parameter estimates. For categorical syllogisms, encoding time was 8.20 seconds, combination time was 5.03 seconds, and response time was 11.52 seconds. For conditional syllogisms, the respective times were 6.43 seconds, 4.61 seconds, and 11.54 seconds. The considerable estimated length of the response stage suggests confoundings of non-response parameters with the response parameter. As would be expected for these simpler problems, individuals tended to combine more set relations than they did for syllogisms with two quantified premises: The probability of combining just one set relation was only .36 for categorical and .43 for conditionals. Three other response-choice parameters involve probabilities of using indirect proof for differing numbers of negations in the syllogism's first premise. For categoricals, these probabilities were .52, .48, and .15 for zero, one, and two negatives respectively. For conditionals, the probabilities were .60, .61, and .16 for zero, one, and two negatives respectively. Thus, the presence of two negatives in the first premise seems considerably to impair an individual's ability to use indirect proof, perhaps because the additional processing capacity consumed by processing of the negatives does not leave sufficient additional capacity for a second round of indirect proof following a first round of direct proof.

4. Relations to reference abilities. The respective correlations between probabilities of correct response and verbal ability were .15 and .14 for categorical and conditional syllogisms ($p > .05$); the respective correlations with spatial-abstract ability were .60 and .54 for categorical and conditional syllogisms ($p < .01$). As in the problems with two quantified premises, the significant correlations were localized in the combination stage: The probability of

combining just one set relation and the probabilities of using indirect proof as a function of numbers of negations were all significantly related to spatial-abstract ability, but not to verbal ability.

Union of Models for One and Two Quantified Premises (IA)

A task is required that represents a union of at least most of the components that are required for the two types of syllogisms (one and two quantified premises) considered above. The following examples of categorical and conditional syllogisms, not yet investigated experimentally, seem to represent this union: "All B are C. All A are B. x is an A. Conclusion: x is a C. (a) True, (b) False." "If B then C. If A then B. A. Conclusion: C. (a) True, (b) False." These problems are like the syllogisms with two quantified premises in that they contain two quantified premises, and like the syllogisms with one quantified premise in that they contain an as sertion that is unquantified.

B.   Mixture Theory of Transitive Inference

1.   Mixed Model of Linear Syllogistic Reasoning

References. This section is based upon the report of Sternberg (in press-c); see also Sternberg (in press-a, in press-b), and Sternberg and Weil (in press).

Nature of task. A linear syllogism contains two premises and a question. Each premise describes a relation between two items, with one of the items overlapping between the two premises. The individual's task is to use this overlap to determine the relation between the two items not occurring in the same premise, and to answer the question on the basis of this determination. An example of a linear syllogism is "A is taller than B. B is taller than C. Who is tallest? (a) A, (b) B, (c) C."

Theory. During encoding, individuals read the premises, linguistically decode the comparative relation in each premise, decode the negation (if any) in each premise, and spatially recode the comparative relation so that the two terms of each premise are represented in a linear array, usually in top-down fashion. The individual must also read the question. During combination, the individual must first find the pivot (middle) term of the three terms in the series. Having found this term, the individual can combine the individual arrays from each of the two premises into a single, merged array. Next, the individual must search for the correct response in the array, and, under certain circumstances, establish linguistic congruence between the response and the adjective in the question. For example, if the solution was encoded in terms of the adjective tall, the question must be phrased (or mentally rephrased) in terms of this adjective. Finally, the individual must respond.

Comparison is not required (except in the most trivial sense), since the options merely restate the terms already in the problem (in the above example, A, B, and C).

Data. The data summarized here are from Experiment 3 of Sternberg (in press-c), except the reference ability correlations, which are from Experiments 1 to 4 combined. Three adjectives pairs, tall-short, good-bad, and fast-slow, were presented in counter-balanced order to the same individuals over three sessions.

1. Means. Mean response latency was 7.00 seconds, and mean error rate was .01. Latencies did not differ significantly across adjective pairs, but did decrease significantly over sessions.

2. Model fits. The value of R^2 for the proposed mixed model was .84 when the model was applied to the latency data. This value was significantly greater than zero, and significantly lower than the reliability of the data. The value was also higher than the values of R^2 for the alternative linguistic and spatial models to which the mixed model was compared. These patterns also held both across adjectives and across sessions. There were insufficient errors in this experiment to allow fitting of the model to the errors (but see Sternberg, in press-b).

Parameter estimates. Encoding time was estimated to be .64+ seconds; combination time was estimated to be 3.80- seconds; and response time was estimated to be 2.52- seconds. Unfortunately, confoundings in parameters resulted in small bits of encoding being confounded into combination and response. Hence, the encoding time is an underestimate, and the other two times are overestimates. At the componential level, by far the most time of any single operation is spent in combining the two single arrays into a larger, merged array (2.99 seconds).

Relations to reference abilities. Predictions were rather complex, since some of the components were linguistic and others spatial. Processing of marked adjectives is hypothesized to involve both linguistic and spatial operations, and indeed, significant correlations were obtained between marking time and both verbal and spatial abilities. Negation was originally hypothesized to be linguistic, but was found to correlate significantly with spatial but not with verbal ability. Search for the pivot term was hypothesized to be spatial, and in fact correlated significantly with spatial ability but not with verbal ability. Search for the pivot term was hypothesized to be spatial, and in fact correlated significantly with spatial ability but not with verbal ability. Formation of the combined array was hypothesized to be primarily spatial, and the pattern of correlations bore out this prediction. Response search was expected to involve only spatial ability, but was correlated significantly with both spatial and verbal ability, with the spatial correlation (nonsignificantly) higher than the verbal one. The congruence operation was hypothesized to be linguistic, and in fact its latency was correlated significantly with verbal but not with spatial ability. Response had confounded within it some linguistic processes, and hence was expected to correlate with verbal but not spatial tests; this expectation was borne out. In general, then, the pattern of correlations was consistent with a mixed linguistic-spatial model.

Mixture Theory of Transitive Inference (IB)

The mixed model of linear syllogistic reasoning is believed to be a special case of a more general mixture theory that can be applied to linear syllogisms with N terms and to other kinds of transitive inference problems. Jerry Ketron and I are currently investigating N-term series problems, e.g., "A is taller than B. C is taller than D. C is shorter than B. Who is taller, B or C?" in an attempt to extend the mixed model to more complex transitive-inference problems.

Union of Transitive-Chain and Mixture Theories (I)

We are presently analyzing data collected from performance on two tasks we believe require a union of many of the components involved in categorical and linear syllogisms. The tasks--quantified linear syllogisms--take either of the following two forms: "All C are not as tall as some B. Some A are not as short as all B. Which of the following conclusions can be deduced? (a) All A are taller than all C; (b) All A are taller than some C; (c) Some A are taller than all C; (d) Some A are taller than some C; (e) None of the above;" "All C are not as tall as some B. Some A are not as short as all B. Which are shortest? (a) All A; (b) Some A; (c) All B; (d) Some B; (e) All C; (f) Some C; (g) None of the above (indeterminate)."

II.   COMPONENTIAL THEORY OF INDUCTION

A.   IMAJER Theory

1.   Models for Integral Stimuli

References. This section is based upon Experiment 2 of Sternberg and Gardner (Note 6) and Experiment 1 of Sternberg and Nigro (Note 7). See also Sternberg (1977, 1979a) and Sternberg and Rifkin (1979).

Nature of tasks. The IMAJER theory (an acronym to be explained below) applies to analogy, classification, series completion, topological relations, and metaphor. Problems of each kind except the nonverbal topologies will be considered here (but see Sternberg, Note 3). Analogies usually take a form such as that exemplified by LAWYER : CLIENT :: DOCTOR : (a) PATIENT, (b) MEDICINE; in a classification problem, an individual may be asked which of two answer options fits better with three terms in a problem stem, as in LEAF, BRANCH, TRUNK, (a) ROOT, (b) TREE; in a series completion, an individual may be asked which of two terms best completes a series, as in TRUMAN, EISENHOWER, KENNEDY, (a) HUMPHREY, (b) JOHNSON; in a metaphorical completion task, an individual may be asked which of two answer options better completes a metaphorical statement, as in ROMANS IN THE COLISEUM WERE BEES IN A (a) SKY, (b) HIVE.

Theory. According to the IMAJER theory, performance on many inductive reasoning problems can be understood in terms of six performance components: inference, mapping, application, justification, encoding, and response (IMAJER). An individual will first encode one or more terms of the problem, perceiving the stimulus, and retrieving from long-term memory and placing in working memory attributes of each stimulus that may be relevant for problem solution. Several combination-stage operations are involved: The individual will infer the relations (nature of the similarities and differences) between two or more terms--between LAWYER and CLIENT in the analogy; among LEAF, BRANCH, and TRUNK in the classification; between TRUMAN and EISENHOWER and then between EISENHOWER and KENNEDY in the series completion; and between ROMANS and COLISEUM in the metaphorical completion. Then, the individual may map a higher-order relation between the domain and the range of the problem, if, indeed, there is a distinct domain and range, as in the analogy and metaphorical completion--in the analogy, between the legal situation of the domain and the medical situation of the range; in the metaphor, between the location of ancient Romans in the domain and the location of bees in the range. The other two problem types have a single, homogeneous domain--parts of trees in the classification, and presidents in the series completion--so that mapping is unnecessary. In the comparison stage, two operations may be involved: The individual will need to apply the previously inferred relation to each of the answer options, deciding which answer option satisfies the required relations. In the series completion, for example, the individual must select the person who was the president immediately succeeding Kennedy. In most problems, the preferred option will not be perceived as ideal, so that the individual must justify it as preferable to the other options, although nonideal. For example, if the relation in the series completion were perceived as successive elected presidents, JOHNSON would not quite fit, since he was not elected; ROOT might not be perceived as an ideal option in the classification problem, either, since a root is below ground, whereas a leaf, branch, and trunk are above ground. But in each case, the keyed option is preferable to the unkeyed one. Finally, the individual responds.

Data. In one experiment (Sternberg & Gardner, Note 6), individuals were timed while they solved analogies, classifications, and series completions formed from animal names. In the other experiment (Sternberg & Nigro, Note 7), individuals were timed while they solved analogies or matched metaphorical completions. The analogies differed from the metaphors in that the connecting terms were missing. For example, the "Romans in the coliseum" item would be presented as ROMANS : COLISEUM :: BEES : (a) SKY, (b) HIVE.

Means. In the first experiment, the mean solution latencies were 7.29 seconds for analogies, 5.47 seconds for classifications, and 6.08 seconds for series completions. In the second experiment, the means were 4.43 seconds for analogies and 4.38 seconds for metaphors (full items excluding precueing manipulation).

Model fits. In the first experiment, the values of R^2 were .77 for analogies, .61 for classifications, and .67 for series completions. In the second experiment, the values of R^2 were .72 for analogies and .86 for metaphorical completions. Reliabilities of data were in the high .80s to low .90s. All of these fits were statistically greater than zero but less than the respective reliabilities of the data sets.

Parameter estimates. Real-time models of information processing were not possible in these experiments, because the independent variables were based upon rated or multidimensionally scaled distances. Parameter estimates thus have the same meanings across parameters and tasks, but cannot be interpreted in real-time terms. In the first (Sternberg-Gardner) experiment, four parameters could be reliably estimated for all three tasks: encoding, discrimination between options (an aspect of application), justification, and response Estimates for encoding were 12.25, 7.87, and 10.01 for analogies, classifications, and series completions, respectively. The respective estimates in these tasks were -12.59, -14.13, and -13.79 for discrimination (where larger values of the independent variable are associated with faster response times, since greater distance between the correct and incorrect options facilitates rapid information processing). For justification, the estimates in the three respective tasks were 3.64, 2.42, and 1.76. And for response, they were 13.59, 29.35, and 33.58. In the second (Sternberg-Nigro) experiment, four parameters could be reliably estimated in both the analogies and metaphorical completion tasks: encoding, application, discrimination, and justification. The estimates for analogies and metaphorical completions were 5.28 and 5.39 for encoding, 6.20 and 3.43 for application, 2.53 and 3.34 for discrimination, 2.11 and 1.29 for justification.

Relations to reference abilities. In the Sternberg-Gardner experiment, small but generally significant correlations were obtained between latency scores and a reasoning factor score. Reference ability tests were not administered in the Sternberg-Nigro experiment.

2.   Models for Separable Stimuli

The models for perceptually separable stimuli are similar to those for perceptually integral stimuli, except that mapping appears not to be used, and encoding is performed attribute-by-attribute rather than holistically (see Sternberg & Rifkin, 1979).

B.   Theory of Causal Inference

Reference.   This description is based upon Schustack and Sternberg (Note 8), Experiment 3.

Nature of task.   In our causal inference task, individuals receive problems like the following, couched in abstract content (single letters as events), medical-epidemic content (see below), or stock-market content (events leading to major stock fluctuations).   An example is 1.   In City 1, it was observed that (a) a sewage line had broken, (b) the incidence of stray dogs had increased, (c) mosquito control had been abandoned for lack of funds.   An epidemic of Wilson-Barry Syndrome was reported.   2.   In City 2, it was observed that (a) the incidence of stray dogs had increased, (b) all sewage lines were intact, (c) mosquito control had been abandoned for lack of funds.   An epidemic of Wilson-Barry Syndrome was reported.   3. In City 3, it was observed that (a) a radiation leak had occurred in a nuclear reactor, (b) the incidence of stray dogs was normal (no increase), (c) a sewage line had broken.   No epidemic of Wilson-Barry Syndrome was reported.   4.   In City 4, it was observed that (a) mosquito control had been abandoned for lack of funds, (b) all sewage lines were intact, (c) incidence of measles was higher than normal.   No epidemic of Wilson-Barry Syndrome was reported. HOW LIKELY IS IT THAT A BROKEN SEWAGE LINE, IN ISOLATION, LEADS TO AN EPIDEMIC OF WILSON-BARRY SYNDROME?

Theory.   We have developed a model of response choice for how individuals assign probabilities or likelihoods, but not a model of information processing in real time.   The components of response choice occur during the combination stage of processing.   Six components enter into the model of response choice.   These estimate the weights assigned to (a) positive affirming instances (e.g., a broken sewage line has been observed and an epidemic has been reported); (b) negative affirming instances (e.g., all sewage lines were intact and no epidemic was reported); (c) positive infirming instances (e.g., a broken sewage line has been observed but no epidemic was reported); (d) negative infirming instances (e.g., all sewage lines were intact but an epidemic was reported); (e) positive infirming evidence for the two strongest distractors where these distractors are designated to be those for which there is the most affirming evidence and the least infirming evidence (e.g., an increase in the incidence of stray dogs and the abandonment of mosquito control in the problem above); (f) base likelihood, regardless of the information contained in any particular problem.

Data. Individuals solved problems with each of three types of content.

1. Means. Mean probabilities were .35, .35, and .37 for abstract, epidemic, and stock content respectively.

2. Model fits. The values of R^2 were .90 for abstract content, .91 for the medical epidemics, and .90 for the stock market content.

3. Parameter estimates. For the abstract, epidemic, and stock content respectively, parameter estimates on a 0-100 likelihood scale were (a) for positive affirming instances, 9.8, 11.9, 9.8; (b) for negative affirming instances, 3.2, 4.2, 2.9; (c) for positive infirming instances, -7.3, -7.4, -9.0; (d) for negative infirming instances, -7.4, -5.8, -5.1; (e) for the strongest alternative hypotheses (distractors), -3.4, -3.6, -3.7; (f) for the base likelihoods, 33.9, 30.3, 36.6.

4. Relations to reference abilities. An earlier experiment revealed no interesting relations between parameters and reference abilities, and so ability tests were not used in this particular experiment.

Union of IMAJER Theory and Theory of Causal Inference (II)

Brian Ross and I have collected but not yet analyzed data for a task that is the same as the causal inference task up to the question (see example above). Instead of the question, the following appears: 5. In City 5, it was observed that (a) mosquito control was operating normally, (b) a sewage line had broken, (c) the incidence of stray dogs had increased. HOW LIKELY IS IT THAT AN EPIDEMIC OF WILSON-BARRY SYNDROME WAS REPORTED IN CITY 5? This type of problem, which we call a causal classification, requires individuals to encode the terms of the problem, infer what is common to those cities in which the syndrome appears and to those in which it does not, map the differences between the two kinds of cities, apply what is learned to the fifth city and decide how likely it is that the syndrome appears in that city, and respond.

UNIFIED COMPONENTIAL THEORY OF HUMAN REASONING

A task at the very top node of the hierarchy would require a union of components from both the deductive and the inductive sides of the hierarchy. Inductive syllogisms are an example of such a task. In such syllogisms, the premises (e.g., "All A are B" or "Some B are C") must be induced from information about particular instances. Once the relationship between the two terms of each premise has been induced, the individual can apply deductive reasoning to draw a conclusion from the induced premises. Scientific reasoning, in many respects, proceeds on this basis.

To summarize, human reasoning or fluid ability can be characterized in terms of a unified theory that comprises hierarchically nested
subtheories accounting for performance on successively more narrow
tasks. At the heart of the global theory and each of the subtheories
is a relatively small set of components of various kinds that characterize the elementary information processes of fluid intelligence. The
components enter into information processing at multiple levels of
the hierarchical task structure. The data collected to date are
generally consistent with the hierarchical structure proposed here.
Although none of the accounts of reasoning are "true" in the sense
of accounting for all of the reliable variance in the data, these
accounts compare favorably with alternative ones, and in combination,
explain a fairly wide range of data in a coherent way.

Reference Notes

Sternberg, R. J. Components of human intelligence. (NR 150-412
    ONR Technical Report No. 19.) New Haven: Department of
    Psychology, Yale University, 1979.
Sternberg, R. J. The construct validity of aptitude tests: An
    information-processing assessment. (NR 150-412 ONR Technical
    Report No. 20.) New Haven: Department of Psychology, Yale
    University, 1979.
Sternberg, R. J. Toward a unified componential theory of human
    reasoning. (NR 150-412 ONR Technical Report No. 4.) New
    Haven: Department of Psychology, Yale University, 1978.
Guyote, M.J., and Sternberg, R. J. A transitive-chain theory of
    syllogistic reasoning. (NR 150-412 ONR Technical Report No. 5.)
    New Haven: Department of Psychology, Yale University, 1978.
Sternberg, R. J., and Turner, M. E. Components of syllogistic
    reasoning. (NR 150-412 ONR Technical Report No. 6.) New
    Haven: Department of Psychology, Yale University, 1978.
Sternberg, R. J., and Gardner, M. K. Unities in inductive reasoning.
    (NR 150-412 ONR Technical Report No. 18.) New Haven: De-
    partment of Psychology, Yale University, 1979.
Sternberg, R. J., and Nigro, G. Components of metaphoric compre-
    hension and appreciation. (NR 150-412 ONR Technical Report
    No. 22). New Haven: Department of Psychology, Yale Uni-
    versity, 1980.
Schustack, M. W., and Sternberg, R. J. Inferring causality: How
    hypothesized causes are evaluated. (NR 150-412 ONR Technical
    Report No. 21). New Haven: Department of Psychology, Yale
    University, 1979.

References

Cattell, R. B. Abilities: Their structure, growth, and action.
    Boston: Houghton-Mifflin, 1971.
Snow, R. E. Theory and method for research on aptitude processes.
    Intelligence, 1978, 2, 225-278.

Sternberg, R. J. Intelligence, information processing, and analogical reasoning: The componential analysis of human abilities. Hillsdale, N.J.: Erlbaum, 1977.

Sternberg, R. J. The nature of mental abilities. American Psychologist, 1979, 34, 214-230. (a)

Sternberg, R. J. A review of "Six authors in search of a character:" A play about intelligence tests in the year 2000. In R. J. Sternberg and D. K. Detterman (Eds.), Human intelligence: Perspectives on its theory and measurement. Norwood, N. J.: Ablex, 1979. (b)

Sternberg, R. J. The development of linear syllogistic reasoning. Journal of Experimental Child Psychology, in press. (a)

Sternberg, R. J. A proposed resolution of curious conflicts in the literature on linear syllogisms. In R. Nickerson (Ed.), Attention and performance VIII. Hillsdale, N.J.: Erlbaum, in press. (b)

Sternberg, R. J. Representation and process in linear syllogistic reasoning. Journal of Experimental Psychology: General, in press. (c)

Sternberg, R. J., Guyote, M. J., and Turner, M. E. Deductive reasoning. In R. Snow, P. A. Federico, and W. Montague (Eds.), Aptitude, learning, and instruction: Cognitive process analysis. Hillsdale, N.J.: Erlbaum, in press.

Sternberg, R. J., and Rifkin, B. The development of analogical reasoning processes. Journal of Experimental Child Psychology, 1979, 27, 195-232.

Sternberg, R. J., and Weil, E. M. An aptitude-strategy interaction in linear syllogistic reasoning. Journal of Educational Psychology, in press.

Footnotes

1. A more detailed but earlier version of this report is presented in Sternberg (Note 3).

TOWARD A THEORY OF APTITUDE FOR LEARNING

I. FLUID AND CRYSTALLIZED ABILITIES AND THEIR CORRELATES

Richard E. Snow

Stanford University

Stanford, California, U.S.A.

This paper focuses on the psychology of aptitude for learning in formal educational settings, and particularly on the nature of measured cognitive abilities as aptitudes. This is only a part of what is needed for a theory of aptitude, but it is perhaps the best place to start: the concept of aptitude has been connected with formal schooling almost since their mutual beginnings, and more scientific evidence is now available about the role of aptitude here, as measured by mental tests, than about aptitude, however measured, in any other natural or social situation. Whatever else it does, a theory of aptitude will need to account for the accumulated evidence about mental test performance in relation to learning from instruction.

Definition and Direction

"To keep the problem as open as possible..." aptitude has been defined ..."as any characteristic of a person that forecasts his (or her) probability of success under a given treatment." (Cronbach and Snow, 1977, p. 6). All manner of physical and psychological characteristics, then, can be thought of as sources of aptitude if they predict success in a particular situation. In this sense, an aptitude theory for educational learning must be more general than a theory of intelligence; intelligence is only one cluster among many kinds of individual differences (including, e.g., achievement motivation, relevant personal-social styles, etc.) that need to be coordinated in such a theory. In another sense, an aptitude theory needs to be more specific than an intelligence theory, because aptitude cannot be identified without specifying the performance criteria predicted and the situation in which prediction occurs. In educational learning, the defining characteristic of aptitude, then, is relation to specified learning outcomes under specified instructional conditions. It is rather like the psychometric view of predictive validity. Validity is not an inherent quality of a psychological measure; one must ask "validity for what?" Similarly, calling some construct an "aptitude" is an empty claim until one specifies "aptitude for what?" Intelligence might thus constitute aptitude for performance in one situation and not in another.

We also need to recognize that a theory of aptitude cannot be merely a theory of "traits." Individual differences in aptitude for learning have to be understood as variations in psychological processes. And, there seem to be at least two levels at which aptitude processes will need to be understood. There are cognitive processes discernable in the second-to-second and minute-to-minute changes that occur during learning or information processing activities. But there are also processes discernable in the week-to-week and month-to-month adaptation of processing activities to instructional learning, as "accretion," "restructuring," and "fine tuning" of organized knowledge and skill (Rumelhart and Norman, 1976) that occur over accumulative instruction. Aptitude process differences relevant to both levels exist before, operate through, and also are produced by, instruction to account for individual differences in learning outcome. To trace through this complex network, one needs analysis and measurement of aptitude processes, learning activities, and instructional task components operating all along the way to criterion performance requirements.

Evidence and Extensions

Figure 1 sums up schematically what is already known about cognitive aptitude tests (in the solid ellipse) and their relation to learning outcome under certain instructional conditions (the solid arrows). Further, it suggests the course that some current research is taking to elaborate this knowledge (the dashed boxes and arrows). Thus, it depicts three facts about aptitude tests, and provides an outline for the rest of this paper.

A first fact about mental tests is that they usually intercorrelate, and correlation matrices involving large numbers of such tests typically show a characteristic form. Factor analysis of such matrices usually suggests a hierarchical, general to specific, organization of ability constructs. Multidimensional scaling of the same matrices yields a similar, central to peripheral, organization of abilities (Guttman, 1965; Snow, in press). Figure 1 identifies the three major group or central factors usually obtained: fluid-analytic ability (G_f), crystallized-verbal ability (G_c), and visualization ability (G_v). The Cattell (1971) and Horn (1976) terminology is used here, without necessarily adopting all the details of their theory. Also shown are several well established

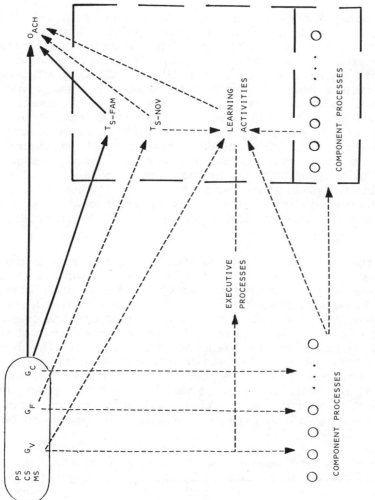

Figure 1. Schematic representation of relations between cognitive tests and achievement outcomes in alternative instruction-al treatments, with the course of future research indicated.

but more specific abilities: memory span (MS), perceptual speed
(PS), and closure speed (CS). However, while these factors can
be identified rather consistently, and while distinctions between
general and special abilities, or central and peripheral factors,
can be regularly made, there is as yet no process-based theory
that explains the nature of these factors or the relations and
distinctions among them.

The second fact is that general mental tests, and particularly
G_c tests, consistently provide strong prediction of learning outcomes across a large sample of conventional instructional environments at virtually all levels of education, from primary school to
college. With combinations of G_c predictors, validity coefficients
computed over a year or more of instruction can average well
above r=.60. The interpretation of such relations, however,
usually rests on such bland statements as "earlier learning ability
is relevant to later learning ability," or "differences in amount of
prior knowledge and skill provide a head start in continued
learning," or "measures of college achievement simply reflect
amount of prior knowledge up to the start of college learning,
plus an added unpredictable amount in college," These "explanations" really explain little, if anything at all, because they say
nothing about the psychological processes by which earlier learning
influences later learning.

Now the third fact. Even though G_c measures give good
predictive validity on average, there is usually a wide range of
coefficients across different environments. That is, instructional
treatment variables influence the aptitude-outcome relations across
different learning situations. A large number of aptitude-treatment
interactions (ATI) have been reported, involving all sorts of instructional treatment variables and aptitude measures (Cronbach
and Snow, 1977; Snow, 1977), so it is a fact that ATI exist.
Something important must be happening inside the instructional
black box if experimental treatment variables influence input-output
relationships. Unfortunately, ATI is still another fact about
aptitude, along with the general relations, that remains to be
explained.

At least one and perhaps two instructional treatment dimensions are of interest in relation to G_f and G_c. The general contrast
might be described as maximum treatment vs. minimum treatment, or
teacher structuring vs. student structuring, or student conformity
vs. student independence. As instructional treatments provide
minimal support or direction, as they allow students to organize
their own cognitive strategies, and require independent functioning, the relation of general intelligence to learning goes up. As
instructional treatments structure the task for the learner, providing maximum information-processing support, reducing student
independence in favor of conformity to teacher-set strategies, the

relation of ability to learning goes down. This seems to be the case for G_c measures; it is less clear for G_f measures unless a second treatment dimension is brought into the picture, to be thought of as familiar vs. novel learning situations. It seems to be the case that G_c is more relevant to familiar learning situations, while C_f is more relevant to novel, variable learning situations. Hence, G_c would interact more strongly with the structuring dimension in familiar environments, while G_f would interact more strongly with the structuring dimension in novel environments (T_{s-fam} and T_{s-nov} in Figure 1). This is a proposed aptitude x aptitude x treatment x treatment interaction design and potentially a four-way interaction. But it is merely a formative hypothesis for further research at present.

Thus we have several related facts and hypotheses in search of a theory. How can a process theory of aptitude for learning be constructed that will push us to a new level of understanding? The line of research represented by Sternberg's presentation, and my own present project, as well as that of several others, is the one I think will best fill in the network of dashed arrows indicated in Figure 1.

Sternberg (this volume) is building process theories for reasoning tasks that, in sum, become a unified theory of G_f. One can envision the expansion of this approach such that each task has a process model attached to it that accounts for individual differences on that test, but also suggests why various tests intercorrelate, to reproduce the hierarchical factor model or the multidimensional scaling clusters already in hand from traditional correlational research. Along this route, however, we can predict that the aptitude intercorrelations will be explained not by the presence/absence of particular processing steps or combinations of steps in particular test performance programs, but rather at the more molar level of executive processes--what I have elsewhere called "assembly and control" processes (Snow, in press). Sternberg may be making the same prediction when he talks about "meta-components." My central hypothesis is that these executive processes underly the relation of aptitude to learning, because the same or similar processes are involved in learning from instruction. But the question remains: How shall we understand individual differences in "executive," or "assembly," or "control" processes? And how shall we understand the relation of these differences to learning outcome and their interaction with instructional treatment dimensions reflecting structuring variables in familiar and novel tasks?

A Theoretical Framework

Any cognitive learning or performance test or task to which human beings must respond will require of them one or more "performance programs." A performance program is envisioned as an organized assembly of information processing activities designed to meet, as efficiently and effectively as possible, the performance demands of the test or task at hand. The analogy to computer programs is obvious. Each individual must assemble such programs, or retrieve from memory programs assembled previously, and control their operation thrughout a sustained test or task performance. These assembly and control processes must function adaptively, to adjust performance programs to variations occurring across items or subtasks within the test or task. For simplicity I refer below to items and tests only, though it should be clear that a parallel discussion is possible for instructional tasks and learning outcome measures.

Figure 2 shows some alternative assemblies schematically, for a hypothetical test item involving five processing steps. One assembly uses the sequence ABCDE (dotted path). An alternative sequence of the same steps is possible, through ADCBE (solid path). Another possibility is a route involving substitute processing steps AB'C'DE (dashed path). One can imagine an individual building and using more than one of these alternative programs, and shifting among them within or between items. Thus, in the analysis of any cognitive test performance there will be, at least potentially, three kinds of variables to consider in addition to the characteristics of the particular processing steps incorporated into a program: sequence and route variables, as suggested in Figure 2, and what have been called (Snow, 1978) summation variables, reflected in variations across multiple runs of programs like that in Figure 2, within or between items. These latter three might be expected to reflect assembly and control functions, primarily.

Individuals will likely differ with respect to these assembly and control functions, just as they will differ in aspects of the component processing activities that are incorporated into the particular performance programs used on a particular item. Individual differences in the effectiveness of all these processing functions will be reflected in item scores on the test and be accumulated into total test scores.

It is important, however, to examine more deeply just how these different processing functions might be reflected in item and total test scores. Individuals assemble performance programs based on an initial understanding of the test instructions and a few example items. If the test is of a type familiar to individuals, one or more relevant performance programs may already have been assembled and stored in memory for retrieval and application in this situation. If the test is novel, then a new assembly must be created, though it might be composed in part by some reorgani-

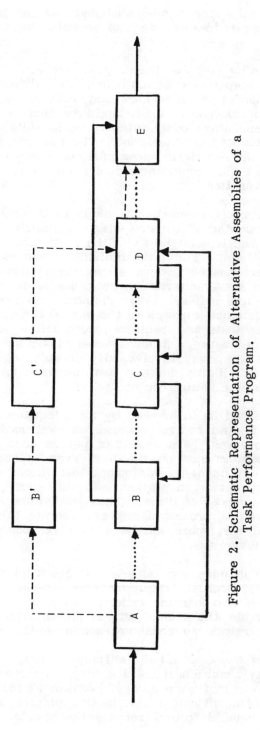

Figure 2. Schematic Representation of Alternative Assemblies of a Task Performance Program.

zation of previously stored subassemblies. As the test begins, a performance program in some stage of assembly is applied to early items.

The processing steps of the program are run off in sequence. Each component step may contribute individual difference variance to item performance; reaction time may accumulate additively across steps, for example. Or it might be that one component (say, stimulus encoding) contributes variance while another (say, stimulus feature matching) does not. The components may differ also in how crucial each is to successful performance. In any event, the adequacy of these steps, individually and in sum, will determine success on a given item.

But the program application is likely to be evaluated by the individual and adapted at an early stage, especially in a novel or complex test. And it must be fit to the particular characteristics of each item in turn. Thus, the individual must exercise active control over performance programs—monitoring, adapting, and perhaps shifting among alternative programs as characteristics of items and continuing self-evaluation dictate. Further, the performance must be sustained through to the end of the test. Relatively unspeeded power tests may require a more active flexible control and conscious endurance. A novel power test, such as Raven Progressive Matrices, may involve an especially high degree of adaptive assembly. Highly speeded tests may call for persistence, but a more repetitive, automatic control.

Item scores will be influenced by the adequacy and flexibility of these assembly and control processes as well as by particular performance processes. The result of performance on one item will influence performance on the next even without external feedback. Strategy changes, confusion, insufficient evaluation, or inflexibility in program adaptations will influence performance across whole sequences of items. Thus, individual differences in assembly and control processes may reverberate throughout the test. The total test score will be a complex summation of all these sources of variation.

Cognitive aptitudes are represented quantitatively in terms of such total test scores. Intercorrelations among such scores, and between them and learning criteria under various instructional conditions, provide the three facts about aptitude discussed earlier. Let us return to consider each of these facts again.

Measures of G_f, G_c, and G_v will appear in the central contour of a Guttman-style multidimensional scaling. Measures of PS, MS, and CS will appear in the periphery. Jensen (1970) has hypothesized that the centrality of a test in this picture, and thus its loading on the general factor, reflects the complexity of a test

problem and its associated cognitive processing. To push further,
increasing test complexity may to a significant extent be a function
of the degree to which individual differences in assembly and
control processes contribute to test score variance. This is not
to rule out the role of "primary" process variance associated with
particular performance steps; one test will differ from another in
the number or kind of performance processes it includes or ex-
cludes. It is rather to view these processes and their organization
within a more molar perspective that may be better suited to an
account of complex cognitive processing in real-world learning and
performance. Measures of PS, MS, and CS will involve fewer
processing steps, more automatically assembled and executed.
Measures of G_f, G_c, and G_v will involve more processing steps
that are also more loosely organized in complex, adaptable assemblies.
And the former, simpler measures will be more easily clustered
into independent orthogonal factors than will the latter, more
complex measures. The more central factors are often correlated
in nature and at times two or more of them may be indistinguishable.

Sometimes measures of G_c and G_f fall into the same factor.
More often, some tests of G_v (presumably) will combine with G_f
tests, leaving no separate spatial factor. British factorists
separate verbal-educational ability (G_c) from spatial-mechanical
ability (G_v) leaving no place for G_f. Recent research suggests
that G_v is least well established as a coherent central construct
(Lohman, 1979ab). It appears, in the reanalyses of old work as
well as new work, that complex power tests of spatial ability are
separable from simpler speeded tests of spatial ability; the
complex tests are mainly measures of G_f while the simpler tests
define at least one separate, more peripheral space factor, which
should probably be called "visualization speed." Speed and power
seem to be distinct psychologically, and the two sorts of factors
do not hook together neatly in a hierarchical model. This
distinction makes sense: complex spatial problems can often be
solved by logical analytic processes rather than visual image
processes. And complex fluid-analytic problems admit occasional
use of spatial visualization strategies. The more that complex G_f
and G_v tests allow a mixture of visualization and nonvisualization
processes, the more one should expect substantial correlation
between the two. A simpler "purer" spatial ability factor based
on speed of visual image processing would be somewhat less
complex, more automatic, and hence more peripheral in the
Guttman multidimensional scaling sense.

In Table 1, some hypothesized assembly and control
processes that might be involved in test performance are
identified, along with some examples of performance processes.
The latter are chosen to be at a level comparable to the information processing abilities represented in measurement batteries such
as that constructed by Rose (1978). They also serve a further

Table 1

Some hypothetical Assembly, Control, and Performance Processes Associated with Central Peripheral and Specific Variance in Mental Test Scores.

Processing Functions	Central	Peripheral	Specific Performance
Assembly			
Understand instructions and examples	XXX[a]		
Initial problem sensing and analysis	XXX		
Create new assembly as unit	XX	X	
Adapt assembly to fit test at hand	XX	X	
Retrieve existing assembly as unit	X	X	X
Control			
Autocriticism	XX	X	
Shift to another retrieved assembly	XX	X	
Refer back for new assembly	XX	X	
Persistence, endurance, tempo	X	XX	
Autopilot monitor		XX	X
Performance process			
P_1 Encode stimuli as symbols		XX	X
P_2 Look up pairing		XX	X
P_3 Substitute one symbol for another		XX	X
P_4 Memorize stimulus		XX	X
P_5 Match stimuli, same/different		XX	X
P_6 Parse stimulus field into components	X	X	X
P_7 Construct new image from stimulus analysis	X	X	X
$\vdots$			
P_{10} Code temporal order of symbol string		XX	X
P_{11} Recall stored symbol string		XX	X
$\vdots$			
P_{20} Respond		XX	X

Column heading note: *Variance Components*

[a] X's are used here as weights to suggest hypotheses about the relative contribution of different processing functions to central, peripheral, and specific components of total test score variance.

heuristic purpose here, since they are close to the kinds of processing differences thought to be reflected in tests of PS and MS, perhaps the most stable and reliable of the peripheral special factors. The table shows a hypothetical contribution of each process function to components of test score variance.

Tests can be represented as flowcharts composed of combinations of performance processing steps. It is then seen that more complex central tests tend to incorporate steps found in simpler tests, but also to add steps. In Figure 3, one such hypothetical organization is shown to suggest a hierarchical progression, from a program to perform the WAIS digit-symbol test through additions that might perform the Identical Pictures Test, the Hidden Figures Test, and the Paper Folding Test. What one does in this kind of theory construction, in effect, is to move in from the periphery along one ray of Guttman's Radex model test by test to see what must be built into each succeeding program to account for performance on it. One can imagine that the primary performance steps for WAIS digit-sympbol are P_1, P_3, and P_{20} initially. As the task proceeds, the individual learns the digit-symbol associates and replaces the "look up" step with P_4, the analogous memory step. Here might be an example of a simple assembly adaptation within a test. Identical Pictures requires P_1, P_4, and an additional step, P_5, in which a picture in memory is matched to a list of pictured response alternatives to reach a same/different response. To perform the Hidden Figures Test, the individual must add a step that produces a parsing of the complex stimulus field (P_6) so that step P_5 can be applied to separated parts in finding a match. The Paper Folding Test incorporates the still more complex step of constructing from the separate stimulus components a new image of what the correct response alternative should look like (P_7) before the alternatives are searched for a match by applying P_5.

Such an analysis is obviously too simple. It does not yet incorporate the details of other information processing analyses of these and similar tasks available in the literature (e.g., Royer, 1971; Hunt and Lansman, 1975; Chiang and Atkinson, 1976). And many of the steps shown here as "primary" might themselves be broken down to form subassembly programs of some complexity. Regions of the flowchart almost certainly can be elaborated. Some experiments in my project have sought to do this, particularly for P_7, the constructive matching loop. Using eye movement records collected during solution of Paper Folding items as well as subject introspective reports a rather complex flowchart was constructed to show the place of simpler steps as well as some major strategic process differences among subjects. (See Snow, 1978, and in press, for further discussion).

The strategic differences seem particularly important in relation to ability. One systematic strategy, called "constructive

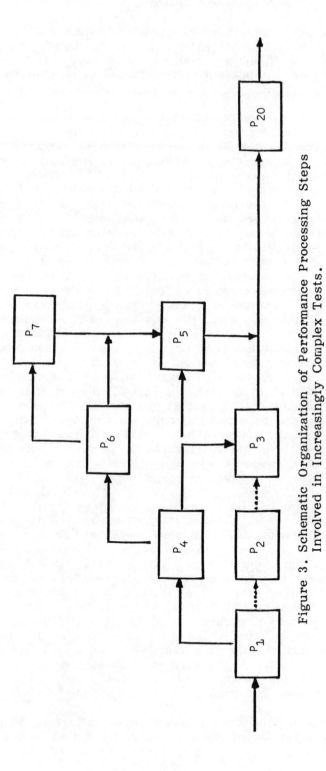

Figure 3. Schematic Organization of Performance Processing Steps
Involved in Increasingly Complex Tests.

matching" because it relies heavily on P_7 prior to a self-terminating scan of the response alternatives to find a match, is characteristic of many high general ability subjects and seems more associated with G_c than with G_f. Another strategy called "response elimination" is more characteristic of low ability subjects and seems to be a fallback strategy for the more able subjects. This involves rapid shifting between stimuli and response alternatives in search of cues that might eliminate some alternatives as incorrect. But there are many intra-individual variations in strategy from item to item and at times within an item. And there appear to be many substrategies.

Thus it seems possible to distinguish general ability levels in terms of strategic differences captured in such flow chart models, and there is a good chance also of distinguishing G_c and G_f ability patterns. High and low ability subjects appar to differ in their efficiency in assembling a systematic strategy for attacking mental tasks, their control of its application, and their flexibility in changing strategies as item characteristics demand this. A theory of individual differences will need to include these assembly and control functions along with performance process hypotheses.

Our work in this direction is progressing, but slowly. Flow-chart models of particular tests are elaborated to include performance programs for other related tests. The multidimensional scaling approach and the scheme outlined here is used to guide theory construction for families of related tests. We expect that task complexity, the degree to which a test shows variance components attributable to G_f or G_c, can be interpreted in terms of the number and kinds of processing steps assembled into the performance program, and the degree to which these steps require flexible control and reassembly as the test or task proceeds.

Instructional Task Demand

The ultimate aim is to connect process models of G_c and G_f to instructional treatment variables at a more molar level. Individual differences come into play upon situational demand. Previous ATI research suggests that the relation of G_c and G_f to learning outcome increases with instructional task demand, and with the degree of novelty vs. familiarity of this demand, respectively. There are now some new instructional studies to support this notion but space does not allow a description of these results here. (See Snow, 1977, in press; Snow, Wescourt, and Freitas, 1979.)

Instructional task demands should be understandable in the same terms as aptitude test complexity. Process models such as those derivable following the above approach should provide an outline for instructional task analysis that allows aptitude and learning differences, and the correlations between them, to be explained in common terms. And instructional variables should be found to control ATI as they alter the complexity of the perform-ance programs required for learning in the situations they define.

Crystallization and Fluidation Processes

At this point, we can at least hypothesize about process dif-ferences related to G_c and G_f at a molar educational level, aided by some old theorizing by Cattell (1963) and Ferguson (1954, 1956) and some new theorizing by J. Anderson, Kline, and Beasley (in press). Ferguson argued that abilities develop through experience as transfer functions. The more practice one receives in exercising an ability the more it develops; this exercise benefits related abilities by transfer processes so that the more similar two abilities are (i.e., the closer they are in a multidimensional scaling), the stronger the transfer relations between them. Thus for example, when the abilities involved in performance on the Terman analogies test are exercised, abilities required by the Raven matrices are benefitted more than are the abilities involved in digit span. When the performance program for the Paper Folding test is assembled and run, the program assembly for the Surface Development test is exercised more than is the program assembly for the Identical Pictures test. There is some evidence to support this notion, though it comes from research on psychomotor abilities (Heinonen, 1962). Over long learning experience, Ferguson expected that constellations of ability would appear as a result of these transfer functions and we can think of G_f and G_c (and perhaps even the strong central relations between them represented by G, or Spearman's g) as resulting from such transfer functions.

Now take each of the major cognitive aptitude factors in turn. G_c, crystallized ability, would be interpreted by Cattell (and Horn, 1976) as representing a coalescence or organization of prior knowledge and educational experience into functional cognitive systems applicable to aid further learning in future educational situations. Since this kind of ability is thought to be accumulated and structured across years of experience in conventional schooling, it is likely to be a stable individual difference, relatively unmodifiable by short-term training interventions, and applicable as aptitude in future educational settings similar in instructional demand to those experienced in the past. The transfer need not be primarily of specific knowledge but rather of organized processing strategies we think of as academic learning skills, that are in some sense crystallized as units for use in future learning whenever new learning conditions appear similar to those in which these crystallized units have been useful in the past.

Thus, G_c measures are often better predictors of learning outcome in conventional educational settings than are G_f measures, because the crystallized assemblies represented by G_c are products of past educational settings similar in processing demands to future educational settings. Again, it is not just content knowledge that accumulates and transfers, it is a transfer of skills for gaining meaning from the educational medium. In other words, both the medium and the message of conventional schooling transfer. David Olson (1974) has argued that "intelligence is skill in a medium." My variation on that theme would be that G_c aptitude is assembly and control skill in the conventional school medium.

Thus, in studies that distinguish G_c and G_f (such as those reported by Sharps, 1973, and Crist-Whitzel and Hawley-Winne, 1976; see Snow, in press, for further discussion) the relation of G_c to learning outcome is strongest in the conventional instructional treatments. This is consistent with many other ATI studies that used what amounts to G or G_c, without distinguishing G_c and G_f clearly (Cronbach and Snow, 1977). When such instructional treatments are modified to reduce the need for conventional assembly and control processes, then the relation of G_c to learning outcome often goes down, and ATI appears. The apparent effect is to help those learners whose prior educational experience has not resulted in strong development of conventional educational learning skills, while at the same time creating a situation in which those who have developed strong conventional learning abilities are less able to apply them. Such treatments do not change the medium of instruction qualitatively, but they often structure and segment instructional presentations to avoid some of the medium-related skills. They reduce the information processing burdens of conventional instruction.

What about G_f? Cattell (1963; 1971) and Horn (1976) see it as facility in reasoning, particularly where adaptation to new situatuations is required and where, therefore, G_c skills are of no particular advantage. If so, we should expect G_f to relate to learning outcome under instructional conditions that are in some sense new, unlike those that the individual learner has faced in the past. Ability to apply previously crystallized learning programs (G_c) would not be relevant here, but ability to analyze and adapt performance programs (G_f) to new kinds of learning situations would be.

J. Anderson, Kline, and Beasley (in press) have provided a detailed description, using a production system model, of how generalized ability might develop through exercise. The computer simulation assumes that a single set of learning processes underlies such development; it provides three ways by which new productions are formed, called designation, generalization, and discrimination, and shows how such new productions become integrated into the system. For the present, I prefer to think of two such processes, crystallization and fluidation, corresponding to G_c and G_f in relation to exercise in handling the demands of familiar and novel instructional situations, respectively.

To state this hypothesis in summary form then, G_c may represent prior assemblies of performance processes retrieved as a system and applied anew in instructional or other performance situations not unlike those experienced in the past, while G_f may represent new assemblies of performance processes needed for more extreme adaptations to novel situations. The distinction is between long-term assembly for transfer to familiar new situations vs. short-term assembly for transfer to unfamiliar new situations. Both functions develop through exercise, and perhaps both can be understood as variations on a central production system development.

What constitutes a "new" or variable learning situation is not really clear. But one can predict that as an instructional situation involves combinations of new technology (e.g., interactive CAI, or television), new symbol systems (e.g., computer graphics or artistic expressions), new content (e.g., topological mathematics or astrophysics), and/or new contexts (e.g., independent learning, collaborative teamwork in simulation games), G_f should become more important and G_c less important.

Closing Note

There are many other extrapolations and implications, including a cultural-developmental view of G_f and G_c, but there is no time to discuss them. And, there is no succinct summary. One has only to look back at the dashed arrows in Figure 1 to see that most of the research program is still before us. Only time and data will tell us what kind of theory of aptitude we are really entitled to. But I believe that the analysis of assembly and control as executive information processes will come closest to explaining the nature of cognitive aptitude in relation to learning.

References

Anderson, J. R., Kline, P. J. and Beasley, C. M., Jr. Complex learning processes. In Snow, R. E., Federico, P-A. and Montague, W. E. (Eds.) Aptitude, learning, and instruction: Volume 1, Cognitive process analyses of aptitude.

Cattell, R. B. Theory of fluid and crystallized intelligence: A critical experiment. Journal of Educational Psychology, 1963, 54, 1-22.

Cattell, R. B. Abilities: Their structure, growth and action. Boston: Houghton Mifflin, 1971.

Chiang, A., and Atkinson, R. C. Individual differences and interrelationships among a select set of cognitive skills. Memory and Cognition, 1976, 4, 661-672.

Crist-Whitzel, J. L. and Hawley-Winne, B. J. Individual differences and mathematics achievement: An investigation of aptitude-treatment interactions in an evaluation of three instructional approaches. Paper presented at the meeting of the American Educational Research Association, San Francisco, April 1976.

Cronbach, L. J. and Snow, R. E. Aptitudes and instructional methods: A handbook for research on interactions. New York: Irvington, 1977.

Ferguson, G. A. On learning and human ability. Canadian Journal of Psychology, 1954, 8, 95-112.

Ferguson, G. A. On transfer and the abilities of man. Canadian Journal of Psychology, 1956, 10, 121-131.

Guttman, L. The structure of relations among intelligence tests. Proceedings, 1964 Invitational Conference on Testing Problems. Princeton, N.J.: Educational Testing Service, 1965.

Heinonen, V. A. A factor analytic study of transfer of training. Scandinavian Journal of Psychology, 1962, 3, 177-188.

Horn, J. L. Human abilities: A review of research and theory in the early 1970's. Annual Review of Psychology, 1976, 27, 437-485.

Hunt, E. B. and Lansman, M. Cognitive theory applied to individual differences. In W. K. Estes (Ed.), Handbook of learning and cognitive processes; Volume I. Hillsdale, N.J.: Erlbaum, 1975.

Jensen, A. R. Hierarchical theories of mental ability. In W. B. Dockrell (Ed.), On intelligence. London: Methuen, 1970.

Lohman, D. F. Spatial ability: A review and reanalysis of the correlational literature. (Tech. Rep. No. 8.) Stanford, CA: Stanford University, Aptitude Research Project, School of Education, October, 1979.

Lohman, D. F. Spatial ability: Individual differences in speed and level. (Tech. Rep. No. 9.) Stanford, CA: Stanford University, Aptitude Research Project, School of Education, October, 1979.

Olson, D. R. Media and symbols: The forms of expression, communication, and education. Chicago: University of Chicago Press, 1974.

Rose, A. M. Information processing abilities. In Snow, R. E., Federico, P-A., and Montague, W. E. (Eds.). Aptitude, learning, and instruction: Volume I, cognitive process analyses of aptitudes. Hilldale, N.J.: Erlbaum, in press.

Royer, F. L. Information processing of visual figures in the digit symbol substitution test. Journal of Experimental Psychology, 1971, 87, 335-342.

Rumelhart, D. E., and Norman, D. A. Accretion, tuning, and restructuring: Three modes of learning. Report No. 7602, Center for Human Information Processing, University of California, San Diego, August 1976.

Sharps, R. A study of interactions between fluid and crystallized
    abilities and two methods of teaching reading and arithmetic.
    Unpublished doctoral dissertation, Pennsylvania State University, 1973.
Snow, R. E. Aptitude processes. In Snow, R. E., Federico, P-A.,
    and Montague, W. E. (Eds.). Aptitude learning and instruction: Volume I, Cognitive process analyses of aptitude, Hilldale, N.J.: Erlbaum, in press.
Snow, R. E. Research on aptitudes: A progress report. In L. S.
    Shulman (Ed.). Review of Research in education, Vol. 4.
    Itasca, IL: Peacock, 1977.
Snow, R. E. Theory and method for research on aptitude processes.
    Intelligence, 1978, 2, 225-278.
Snow, R. E., Wescourt, K. and Freitas, J. Individual differences in
    aptitude and learning from interactive computer-based instruction. (Tech. Rep. No. 10.) Stanford, CA: Stanford University, Aptitude Research Project, School of Education, 1979.

Footnote

This paper was completed while the author served as Fellow,
Center for Advanced Study in the Behavioral Sciences. The
research reported herein was supported by Office of Naval Research
Contract No. N00014-75-C-O882. The views and conclusions contained in this document are those of the author and should not be
interpreted as representing the official policies, either expressed or
implied, of the Office of Naval Research, The Advanced Research
Projects Agency, or the United States Government.

COMPARISON OF READING AND SPELLING STRATEGIES

IN NORMAL AND READING DISABLED CHILDREN

George Marsh, Morton Friedman,

Peter Desberg and Kathy Saterdahl

University of California

Los Angeles, California, U.S.A.

This paper presents evidence concerning the strategies used in reading and spelling by normal and reading disabled children. In two previous studies, the authors have proposed a developmental theory of the changes in strategies in reading and spelling (Marsh, Friedman, Welch and Desberg, in press a and b). In these studies, the development of strategies of children who were reading and spelling at grade level were compared with strategies of children who were reading disabled. However, the comparisons were done on different subjects (second grade, fifth grade and college) in two separate studies using different materials. Evidence was obtained concerning the developmental sequence of reading and spelling strategies shown in Table 1.

This study differs from the previous studies in several ways. First, the comparison of reading and spelling strategies is done on the same children using same materials. Also, the previous studies employed nonsense words based on English grapheme-phoneme correspondence rules or on irregular real words. The present study also uses these nonsense words but also uses the parallel real words to assess strategies. Finally, the spelling strategies of reading disabled children were assessed in the present study.

Method

Subjects: The subjects were 20 second and 21 fourth grade

Table 1

Reading Strategies	Response measure
Substitution	Intrusion error (incorrect real word)
Phonemic decoding	Phonemic pronunciation of irregular real and non-words
Analogy	Pronunciation of non-word by analogy to irregular real word

Spelling Strategies	
Substitution	Intrusion error (incorrect real word)
Phonemic encoding	Phonemic spelling of irregular real and non-words
Analogy	Spelling of non-word by analogy to real irregular word.

children reading at grade level and 24 reading disabled children from classes for "Educationally Handicapped" (EH) in the fourth grade who were reading two years below grade level.

Procedure: Subjects were asked to read or spell two twenty word lists. The first list contained twenty high frequency real words, one half of which were regularly spelled and the other half of which were irregularly spelled. The second list contained a transformation of each of the words in the first list into one with nonsense words. This was accomplished by changing one of the letters or sounds in the words. Subjects received the two lists as either a reading or spelling task in one of four counterbalanced orders.

In a previous study (Marsh, Desberg and Cooper, 1977) a production deficiency was found in fifth grade subjects' use of the analogy strategy in reading. In order to minimize the gap between competence and performance in the use of the analogy strategy, all subjects were told that the non-words were real words with one letter or sound changed.

Results

An overall 2x2x2x3 analysis of variance was done. The within subject factors were reading vs. spelling, real vs. nonwords, regular vs. irregular spelling patterns. Grade level (second, fourth and fourth EH) was the between subject factor. In addition, a separate analysis of order of task and list was done. The order effects were not significant (F < 1).

Performance on spelling tasks was significantly better than performance on reading tasks (F = 136, $\underline{df}$ = 1/61, p <.001). There was no significant effect of the real vs. non-word factor (F <1). Performance on regular vs. spelling patterns was significantly superior to performance on irregular patterns (F = 436, df = 1/61, p <.001). The effect of grade level was significant. (F = 15.4, df = 2/61, p <.001). Post hoc analysis showed that the difference was due to the superior performance of the normal fourth grade children. The fourth grade EH children reading at second grade level did not differ significantly from normal second grade children.

In addition to the main effects, there were several significant interactions. There was a grade by task interaction (F = 14.75, df = 2/61, p <.001) in which reading and spelling performance differences were significant with second grade and EH children and not significant with normal fourth graders. There was an interaction between grade level and real vs. non-real words (F = 3.17, df = 2/61, p < .05). This was due to the tendency for disabled readers to do better on real words than non-words while there was no significant difference in performance on these tasks in normal second or fourth grade readers. There was an interaction between reading and spelling task and the regular-irregular spelling patterns. (F = 20.72, df = 1/61, p<.001). The regularity of the spelling pattern influenced performance more on the spelling task than it did on the reading task.

In addition to the overall analysis of variance, an analysis of different response types indicating use of different strategies was done. The precentage of various response types is shown in Table 2. The second grade normal and fourth grade EH subjects

Table 2
Mean Percentage of Response Types

Grade:	2nd		4th EH		Normal	
	Reading	Spelling	Reading	Spelling	Reading	Spelling
Intrusions	33%	2%	31%	4%	10%	4%
Phonemic Equivalents	15%	56%	9%	58%	4%	47%
Analogy Responses	78%	26%	85%	20%	92%	49%

did not differ significantly on the percentage of intrusion errors indexing the substitution strategy in reading, but both groups were significantly higher than fourth grade normals. All three groups showed negligible use of the substitution strategy in spelling. The use of the phonemic encoding and decoding strategy in reading is minimal in all three groups but accounts for approximately one-half the responses in spelling in the three groups. The use of the analogy strategy in reading was not significantly different in the three groups. In spelling, the fourth grade normals were significantly superior to the other two groups in their use of the analogy strategy.

Discussion

The results indicate that the fact that a child can read a word does imply that he can spell it. The possible reasons for this decalage are numerous and include asymmetries in spelling to sound vs. sound to spelling correspondences, greater initial instructional emphases on reading, etc. By using a restricted vocabulary it would be theoretically possible to teach reading and spelling in parallel. However, the authors' previous studies and the present study suggest a more fundamental reason for this decalage. The child initially uses different and to some extent opposed strategies in reading and in spelling. In the present study the substitution strategy accounts for nearly one-third of responses in reading in normal second and Fourth grade EH subjects, but practically none of the responses in spelling. In contrast, phonemic encoding accounts for one half or more of the responses in spelling and for a negligible percentage of responses in reading. The reasons for these opposed strategies are discussed fully in the author's previous papers (Marsh et. al., in press). The lack of significant task-order effects also shows the decalage between reading and spelling. Having heard the word in spelling task did not significantly facilitate subjects reading it and having seen a word on the reading task did not facilitate subjects spelling of the word.

The fact that overall performance on real and non-words was not significantly different supports the author's previous use of non-words to assess strategies. The reading disabled children did show a slightly superior performance on real words than non-words. This suggests that these children depend more on visual familiarity than on phonemic regularity in their reading and spelling. Barron (1978) has obtained similar results with reading disabled children on a lexical decision task.

The present study demonstrates that phonemic regularity is a very important factor in spelling and, to a somewhat lesser extent, in reading for all groups. The use of the analogy strategy in reading was higher here in all three groups than in the authors' previous two studies. Baron (1979) has also shown that children in second grade can be successfully instructed in the use of the analogy strategy. The improvement in the use of the analogy strategies would therefore seem to be a function of telling the children the analogical bases for constructing the non-words. In the previous studies younger children's failure to use the analogy strategy in reading may be due to a performance deficiency rather than a competence deficiency. However, this knowledge did not help the second grade and EH children nearly as much in spelling and there may be a genuine competence deficiency in use of the analogy strategy in spelling. Finally, the pattern of results comparing the retarded readers with their peers both in chronological and mental age and "reading age" supports the authors' previous interpretation of their performance in terms of a developmental lag hypothesis.

References

Baron, J. Orthographic and word specific mechanisms in children's reading of words. Child Development, 1979, 50, 60-72.

Barron, R. W. Reading skill and phonological coding in lexical access. In M. M. Gruneberg, R. W. Sykes and P. E. Morris (Eds.), Proceedings of the International Conference on Practical Aspects of Memory. London: Academic Press, 1979.

Marsh, G., Desberg, P., and Cooper, V. Developmental changes in reading strategies. Journal of Reading Behavior, 1977, 69, 391-394.

Marsh, G., Friedman, M. P., Welch, V. and Desberg, P. A cognitive-developmental approach to reading acquisition. In T. G. Waller and G. E. MacKinnon (Eds.), Reading Research: Advances in Theory and Practice, Vol. 2. Academic Press (in press).

Marsh, G., Friedman, M. P., Welch, V. and Desberg, P. Development of Strategies in Spelling. In U. Frith (Ed.), Cognitive Processes in Spelling. London: Academic Press, 1980.

ACTIVE PERCEIVING AND THE REFLECTION-IMPULSIVITY DIMENSION

Nancy Rader and Shall-way Cheng

University of California, Los Angeles

Los Angeles, California, U.S.A.

Abstract

Zelniker and Jeffrey (1976, 1979) have proposed that perform-ance differences on the MFFT, used to index reflection-impul-sivity, stem from preferences for either detail or global informa-tion processing. Our studies, however, indicate that differences in performance reflect the tendency to make use of active percep-tual search.

Kagan and his associates (Kagan, Rosman, Day, Albert, and Phillips, 1964; Kagan, 1965) introduced the cognitive style dimen-sion known as reflection-impulsivity. A reflective person is defined as one who habitually considers all alternatives present, while an impulsive person fails to do so. To identify "impul-sives" and "reflectives" among grade-school children, Kagan et al. (1964) constructed the Matching Familiar Figures Test (MFFT). Each item of the MFFT consists of a picture of a common object (the standard) and six other pictures, one identical to the standard and each of the other five differing from it in a minute, not easily identifiable detail. The child's task is to find the picture that matches the standard exactly. Children who respond more quickly and who make more errors than the median score of a sample are considered to be impulsives; those children responding slower but more accurately than the sample median score are considered to be reflectives.

Recently, Zelniker and Jeffrey (1976, 1979) have proposed that differences in performance on the MFFT stem from differences in attention deployment. "Impulsives" are seen as having a preferred strategy of attending to global or gestalt aspects of a stimulus; "reflectives" are seen as having a preferred strategy of attending to detail aspects of a stimulus. A "preferred strategy" analysis leads to the hypothesis that individuals within each cognitive style group should perform better when the requirements of a task are compatible with their preferred strategy. To test this hypothesis, Zelniker and Jeffrey (1976) devised a modification of the MFFT that utilized two types of test items. One type of item used variants that differed from the standard in some detail inside the figure; the other type used variants that differed from the standard in contour. As predicted, impulsives were significantly more accurate on global than on detail problems whereas reflectives were significantly more accurate on detail than on global problems. However, the performance of the two groups lacked the desired symmetry. While reflectives performed significantly better than impulsives on detail problems, impulsives did not perform significantly better than reflectives on global problems.

Zelniker and Jeffrey (1979) argue for an attention deployment difference rather than a perceptual or cognitive deficit on the basis of a lack of relationship between performance on the MFFT and I.Q. Yet, while it is the case that latencies on the MFFT do not correlate with I.Q., error scores are positively correlated with non-verbal I.Q. (Messer, 1976). Also, Barrett (1977) has found that impulsives succeed less well in school than do reflectives, even though reflective and impulsive children are rated by teachers as equally motivated to learn (Ault, Crawford, and Jeffrey, 1972). Zelniker and Jeffrey (1979) argue that the poorer school performance of impulsive children results from the emphasis on tasks requiring an analytic, detailed approach. But if the reflection-impulsivity dimension reflects a "preference" for a certain type of processing, why don't impulsive children make the switch to detail processing when it is necessary? Why should a "preference" or "strategy" be so enduring when it repeatedly brings failure? An alternative explanation (as noted by Zelniker and Jeffrey, 1979) is that "impulsive" children are deficient in the skills used in processing detail information.

Both Julesz (1975) and Broadbent (1977) have argued that there are two levels of perceptual processing, the first characterized as passive, very fast, effortless, and capable of gestalt-like discrimination and the second characterized as active, more time-consuming, effortful, and necessary for perceptual search or scrutiny. Wright and Vliestra (1977) have proposed a similar distinction in the cognitive processing of "impulsive" and "reflective" children. According to their account, impulsives tend to engage in passive exploration which comprises rapid, automatic responses guided by stimulus salience, while reflectives tend to engage in active search behavior which is deliberate, goal directed, and guided by relevance rather than salience of stimulus information.

Our research set out to examine the attentional processing of children classified as reflective and impulsive. One hundred and eleven third and fourth grade children (median age = 112.3 mo.) were given the MFFT. Median performance was 13 errors with a latency of 19 sec. On the basis of a double median split of latencies and errors, 39 children were classified as reflective and 39 as impulsive. However, only those reflectives with less than 9 errors and a latency greater than 21.6 sec. and those impulsives with more than 14 errors and a latency shorter than 18.4 sec. were adopted as subjects. (This restriction followed norms reported by Messer, 1976.) Following this selection, there were 27 reflectives and 27 impulsives.

Study 1. Items from the modified MFFT (Zelniker and Jeffrey, 1976) were administered in three orders to three groups of subjects: 1) detail and global items randomly mixed, 2) detail items first, and 3) global items first. The mixed-order condition followed Zelniker and Jeffrey's procedure. In only the global first condition did reflectives make more errors on global items than detail, and this difference was not significant. Analysis of the latencies revealed a significant style x item interaction (p < .01). Both reflectives and impulsives had significantly longer response latencies on the detail than on the global items, but the difference was greater for reflectives. These findings fail to support the contention that reflectives are hindered in their processing of global items. An analysis of the types of errors showed that those variants that were difficult to distinguish from the standard for the reflectives were also difficult for the impulsives. The only difference between reflectives and impulsives was that impulsives not only made more errors on these variants but also made errors on other variants that were easier for the reflectives. Common among these variants are those having a missing or extra component.

Study 2. Pairs made up of a standard and one variant from the modified MFFT were presented in a same-different reaction-time task under two exposure conditions. The purpose was to compare the performance of reflective and impulsive children on global and detail items when they are presented with a fixed exposure duration determined by the standard duration of impulsives. According to the strategy-preference model, impulsives should perform better than reflectives on the global items under this condition because the task demands a fast processing strategy which is assumed to be compatible with the preferred attention deployment of impulsives, but incompatible with that of reflectives. A long exposure condition, based on a reflective standard, was also used. Subjects were 20 reflectives and 20 impulsives randomly drawn from the subject population described above. Subjects were told to respond as soon as they knew whether the two stimuli were the same or different; thus, subjects could respond before the end of the presentation period. Results are shown in Table 1.

TABLE 1

Mean Number of Errors and Response Latencies in Seconds on the Modified MFFT Standard-Variant Pairs Under the Short Exposure (3 Seconds) and Long Exposure (7 Seconds) Condition

		Short		Long	
		Reflective	Impulsive	Reflective	Impulsive
Modified-MFFT	Errors	2.25	2.55	1.45	2.10
Global Pairs	Latencies	3.60	3.10	4.38	3.97
Modified-MFFT	Errors	2.25	3.95	2.05	2.45
Detail Pairs	Latencies	3.80	.3.45	5.08	4.40

Analysis of errors showed main effects of exposure duration, cognitive style, and item (all $p < .05$) as well as a three-way interaction ($p < .05$). A style x item analysis for each exposure condition was performed. In the long exposure condition, while more errors were made on the detail than on the global items, and impulsives produced a larger number of errors than reflectives, none of these differences were significant. In the short exposure condition, there were not only main effects for both cognitive style and item but also a significant interaction between the two variables. Further comparisons across the two style groups indicated that impulsives made significantly more errors than reflectives on the detail items ($p < .01$), but the difference did not reach significance on the global items. Comparisons across the two exposure conditions showed that in the long exposure condition reflectives made significantly fewer errors on the global items ($p < .05$) and the impulsives made significantly fewer errors on the detail items ($p < .01$). The analysis of the latency scores yielded a significant main effect of exposure condition ($p < .01$), cognitive style ($p < .05$), and item ($p < .01$). Overall, the long exposure condition produced longer latencies than the short exposure condition, reflectives had longer latencies than impulsives, and subjects spent more time on the detail than the global items.

There were no significant interactions.

The prediction derived from a strategy perference hypothesis that, with a short exposure duration, impulsives will perform better than reflectives on the global items was not supported. Reflectives did somewhat better on the global items than impulsives, while impulsives did significantly worse on the detail items. With a longer exposure, impulsives improved their performance on detail items, suggesting the inadequacy of a motivational explanation. Reflectives improved on the global items with a longer exposure, improving on those "global" items that both impulsives and reflectives found difficult. The lack of a significant cognitive style effect in the long exposure condition suggests that a single pair is easier for impulsives to deal with than the multiple pair comparisions of the MFFT, an interpretation supported by the fact that impulsives fail to make all pair-wise comparisons on the MFFT (Drake, 1970; Ault et al., 1972). Perhaps the "impulsive" child fails to consider all alternatives, not because of a lack of reflective attitude, but because s/he is overwhelmed by the perceptual task presented, one requiring not only search within a pair but search across a number of pairs. That impulsives do perform significantly worse than reflectives with a short exposure demonstrates that an important underlying problem is the efficiency with which impulsive chlidren can detect perceptual differences involving "detail" aspects. In the standard MFFT, the problem is multiplied.

Study 3. Visual patterns, thought by Julesz (1975) to require passive or active perception, were used as stimuli. It was hypothesized that impulsives' performance would differ from reflectives' on those patterns requiring active perception. Two types of tasks were used. In Task A, subjects were shown patterns in which there could be an area of one texture embedded in an area of similar but different texture. Subjects had to decide whether the patterns were homogeneous in texture or whether a deviant embedded texture was present. Given a response of non-homogeneity, they were also asked to explain or show how the elements differed, and to cross out all the elements of one type. The patterns that were used are shown in Fig. 1. In Task B, subjects were presented with spiral-like forms and asked whether they were made up of one or two lines. The forms used are shown in Fig. 2. Twenty reflective and twenty impulsive children drawn from the population described above served as subjects.

In Task A, pattern 1, which consisted of only one type of element, was shown twice; patterns 2-4, which consisted of two types of elements, were each shown once. If a subject incorrectly said that all the elements were the same, s/he was corrected and told to search for the difference. In Task B, four spiral-like forms were used, with two identical but twice as large as the

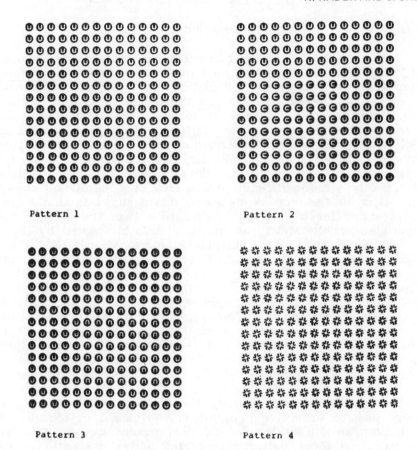

Figure 1. Patterns used in Task A, Study 3.

other two. For both Tasks A and B, latency and accuracy scores were recorded.

Since every subject responded correctly on patterns 1-3 in Task A, these patterns were excluded from analysis. On pattern 4, each subject was given an accuracy score in the following way: 1) Those subjects who perceived the difference spontaneously, correctly crossing out the elements of one type, were given 3 points. 2) Subjects who did not perceive the difference spontaneously but who could correctly cross out elements of one kind when asked to do so were given 2 points. 3) Subjects who did not perceive the difference spontaneously and were only partially correct in crossing out elements of one kind were given 1 point. 4) Subjects who could not detect the difference even after being told of its existence, being totally unable to cross out elements of one type, were given a score of 0. On Task B, 1 point was given for each correct answer. A score of 2 indicated chance

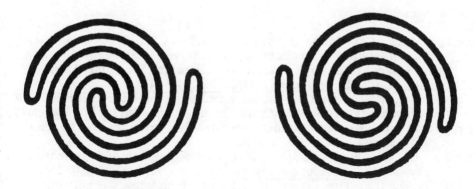

Figure 2. Patterns Used in Task B, Study 3.

level performance.

Errors and latencies were analyzed with t-tests. Impulsives performed significantly worse than reflectives in both tasks ($p <$ .005) and only in Task A did reflectives have significantly longer latencies ($p <$ .005). In Task A only two impulsives detected the embedded pattern spontaneously and only four could find the discrepant elements after being told a discrepancy existed. In contrast, among reflectives, six detected the embedded pattern spontaneously and all could find it after being told a discrepancy existed. In Task B, only 8 impulsives performed better than chance, while 18 reflectives did. Pearson product-moment correlations between accuracy scores and error scores on the standard MFFT were -0.51 for Task A and -0.54 for Task B. The correlations between accuracy scores and latency scores on the standard MFFT were 0.14 for Task A and 0.14 for Task B. These results strongly point to a difference between "impulsive" and "reflective" children in terms of their perceptual skills.

The results of the three studies reported fail to support the contention that differences in performance on the MFFT stem from preferences to attend to either global or detail aspects of a stimulus. Rather, they support the idea that impulsives are less efficient in their controlled search of detail aspects of a stimulus. This deficiency in active, perceptual skills may reflect a generally poorer ability to mobilize cognitive effort and to exercise control. This interpretation fits with findings that impulsives are less able than reflectives to inhibit their action (Messer, 1976), are less able to sustain attention (Zelniker, Jeffrey, Ault, and Parson, 1972), and perform worse than reflectives on tasks that require selective attention (e.g., Hartley, 1976; Weiner and Bergonsky, 1975).

References

Ault, R. L., Crawford, D. E., and Jeffrey, W. E. Visual scanning strategies of reflective, impulsive, fast-accurate, and slow-inaccurate children on the Matching Familiar Figures test. Child Development, 1972, 43, 1412-1417.

Barrett, D. W. Reflection-impulsivity as a predictor of children's academic achievement. Child Development, 1977, 48, 1443-1447.

Broadbent, D. E. The hidden preattentive processes. American Psychologist, 1977, 32, 109-119.

Drake, D. M. Perceptual correlates of impulsive and reflective behavior. Developmental Psychology, 1970, 2, 202-214.

Hartley, D. G. The effect of perceptual salience on reflective-impulsive performance differences. Developmental Psychology, 1976, 12, 218-225.

Julesz, B. Experiments in the visual perception of texture. Scientific American, 1975, 232, 4, 34-43.

Kagan, J. Impulsive and reflective children: Significance of conceptual tempo. In J. D. Krumboltz (Ed.), Learning and the educational process. Chicago: Rand McNally, 1965.

Kagan, J., Rosman, B. L., Day, D., Albert, J., and Phillips, W. Information processing in the child: Significance of analytic and reflective attitudes. Psychological Monographs, 1964, 78 (1, whole no. 578).

Messer, S. B. Reflection-impulsivity: A review. Psychological Bulletin, 1976, 83, 1026-1053.

Weiner, A. S. and Bergonsky, M. D. Development of selective attention in reflective and impulsive children. Paper presented at the American Educational Research Association Annual Meeting, Washington, D. C., 1975.

Wright, J. C., and Vliestra, A. G. Reflection-impulsivity and information processing from three to nine years of age. In M. Fine (Ed.), Intervention with hyperactivity. Springfield, Ill.: Charles C Thomas, 1977.

Zelniker, T. and Jeffrey, W. E. Reflective and impulsive children: Strategies of information processing underlying differences in problem solving. Monographs of the Society for Research in Child Development, 1976, 41 (5, Serial No. 168).

Zelniker, T. and Jeffrey, W. E. Attention and cognitive style in children. In G. Hale and M. Lewis (Eds.), Attention and the development of attentional skills. New York: Plenum Press, 1979.

Zelniker, T., Jeffrey, W. E., Ault, R., and Parson, J. Analysis and modification of search strategies of impulsive and reflective children on the Matching Familiar Figures Test. Child Development, 1972, 43, 321-335.

COGNITIVE STRATEGIES IN RELATION TO READING DISABILITY

C. K. Leong

University of Saskatchewan

Saskatchewan, Canada

Abstract

This paper outlines studies of cognitive patterns based on simultaneous-successive synthesis, laterality studies and discusses the role of fast, accurate decoding in reading proficiency. Differentiation of subgroups of disabled readers and relating their cognitive processing to functions of the cerebral hemispheres may provide a clue to understanding reading disability.

In this report, I will summarize some of my ongoing work on severely disabled readers ("retarded" readers or children with developmental dyslexia) with some reference also to children classified as inadequate or less skilled readers. This summary deals with: cognitive pattern studies, laterality studies and the role of fast, accurate decoding in reading.

Cognitive Pattern Studies

In a recent work on dyslexia, Mattis (1978, p. 52) implicates "a well-defined defect in any one of several specific higher cortical processes." In studies of severely disabled readers I have attempted to provide an indirect answer to the higher cognitive and also cortical processes differentiating "retarded" readers from their controls (Leong, 1976, 1976-1977, in press). The theoretical postulate is derived from Luria's (1966a, 1966b) two basic forms of integrative activity: simultaneous (primarily spatial, groups) and successive (primarily temporally organized series)

syntheses at the perceptual, memory and intellectual levels. In
Luria's terms, simultaneous-successive synthesis can be identified
with the functions of specific parts of the cortex, although conscious activity as a complex functional system is the result of
concerted working of different brain units. The Luria model has
been operationalized by Das (see review by Das, Kirby and
Jarman, 1979).

Using a battery of tasks similar to those in the Luria-Das
paradigm, I have found through different converging factor
analyses (principal component, alpha factor and promax analyses)
and factor matching that severely disabled readers performed
significantly poorer than their age--and IQ--matched non-disabled
readers in both the simultaneous and successive dimensions or
factors. There is also some evidence in qualitative analyses that
disabled readers are inefficient in using rules to solve antecedent
reading tasks such as Raven's Progressive Matrices and cross-modal

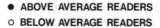

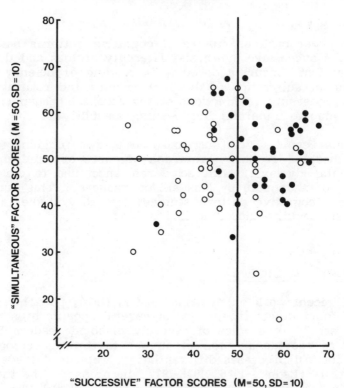

Figure 1

(auditory to visual) coding.

Das et al. (1979) have stressed the functional independence of simultaneous and successive syntheses and the importance of processes. I would further interpret simultaneous and successive modes as flexible ordering of related processes or strategies. Luria himself explains that by simultaneous synthesis is meant the "synthesis of <u>successive</u> (arriving one after the other) elements into simultaneous <u>spatial</u> schemes" and by successive syntheses is meant "the synthesis of <u>separate</u> elements into successive series" (Luria, 1966b, p. 74, italics added). Luria's terms of "verbal-logical" and "concrete-active" are reminiscent of the traditional v:ed and k:m factors found in the hierarchical structure of abilities. The flexible simultaneous and successive strategies are in keeping with generative cognitive processes of the brain.

Laterality Studies

If brain processes are implicated in reading retardation, it is possible that severely disabled readers lag behind their peers in functional cerebral development, especially the development of the left hemisphere. From different experiments with dichotic digit and letter tasks there was some evidence that "retarded" readers are less well lateralized than their age- and ability-matched controls (Leong, 1976, 1976-1977). But it is also possible, as Bakker (1973) has suggested, that the nature of the relation between ear asymmetry and reading ability is dependent on the phase of the learning-to-read process and that amongst dyslexics their reading performance may relate to left or right cerebral laterality. This postulate is supported by developmental experiments by Bakker, Teunissen, and Bosch (1976).

To further test the "two reading strategies and two dyslexias" hypothesis a recent experiment carried out in Amsterdam jointly with Dirk Bakker confirmed the null hypothesis for reading performance between "left-brained" (as shown by better right-ear dichotic digit scores) and "right-brained" (as shown by better left-ear dichotic digit results) dyslexics. From a group of 90 L.O.M. (special) school reading-disabled children, 17 predominantly left-ear children were matched on age, ability, sex with 17 right-ear children. On three reading tests, one emphasizing speed, the other accuracy and the third paragraph reading, there was no group difference in the time taken to read the words or the paragraphs and there was no speed-accuracy trade-off as shown in the total reading errors. Thus, within the group of dyslexics better left- or right-ear scores on dichotic listening do not neces-sarily reflect quantitative difference in reading performance. The difference, however, lies in the qualitative aspect. Analysis of errors shows that right-ear (presumed left-brained) dyslexics made more errors of omission and substitution probably from a

misapplication of syntactic-semantic rules and that left-ear (presumed right-brained) dyslexics made more time-consuming errors.

This study and the Bakker work lend credence to the postulate that the relation between ear asymmetry and reading ability
is dependent on the phase of the learning-to-read process. While
caution should be exercised in interpreting laterality studies (see
review by Kinsbourne and Hiscock, 1978), the question of "laterality for what" has evolved to focus attention on tasks
and on strategies adopted by the individuals. Laterality-reading
relationship is seen as the right hemisphere specializing for
wholistic and featural analysis and the left hemisphere for analytic
and naming tasks. Successful performance in early reading
involves a reciprocal contribution from both the right and left
hemispheres, in varying degrees for different individuals at
different stages of reading. A taxonomy of reading errors within
a neuropsychological context such as the notable work of Marshall
and Newcombe (1973) can with advantage be combined with pattern
studies to clearly delineate dyslexic children for more effective
remediation.

Fast, Accurate Decoding for Less Skilled Readers

In an earlier conference on the changing concept of intelligence, Perfetti (1976) emphasized rapid, automatic coding and
recoding operations as important sources of cognitive differences
in readers. Differences in reading comprehension skill are due in
the main to differences in the understanding and use of verbal
codes and the extent to which the codes are activated automatically
(Perfetti, 1977).

In our Institute L. Haines (1978) has shown in a recently
completed doctoral dissertation with grades 4, 6 and 8 skilled and
less skilled readers that both vocalization latency for predictable
and unpredictable words and pseudo words and reaction times on
lexical decision tasks differed significantly between the two groups.
The findings suggest that less skilled readers need to develop
phonological word processing skills to the automatic level and
establish flexible coding strategies. Overall, less skilled readers
show slower and less accurate word access; they lack automatized
basic decoding subskills. Their difficulty is with general language
comprehension, with organizing and integrating language units
into meaningful relationships (Perfetti, 1977). In addition, linguistic
awareness (Leong and Haines, 1979) is seen as an important
source of individual differences in reading.

Concluding Statement

Given the variability of reading retardation (see Vernon,
1979), there is a need to differentiate subgroups of "retarded"

readers, to relate their cognitive processing to neurological functions of the cerebral hemispheres and to study reading processes such as the role of fast, accurate decoding in reading comprehension. The framework for research proposed by Cummins and Das (1977) outlines the potential of simultaneous-successive processing for understanding and remediating reading difficulties. Laterality studies and information processing approaches can further explain cognitive differences in relation to reading proficiency.

References

Bakker, D. J. Hemispheric specialization and stages in the learning-to-read processes. Bulletin of the Orton Society, 1973, 23, 15-27.

Bakker, D. J., Teunissen, J., and Bosch, J. Development of laterality-reading patterns. In R. M. Knights and D. J. Bakker (Eds.), The neuropsychology of learning disorders: Theoretical approaches. Baltimore: University Park Press, 1976.

Cummins, J., and Das, J. P. Cognitive processing and reading difficulties: A framework for research. Alberta Journal of Educational Research, 1977, 23, 245-255.

Das, J. P., Kirby, J., and Jarman, R. F. Simultaneous and successive cognitive processes. New York: Academic Press, 1979.

Haines, L. P. Visual and phonological coding in word processing by grade 4, 6, and 8 readers. Unpublished doctoral dissertation, University of Saskatchewan, 1978.

Kinsbourne, M., and Hiscock, M. Cerebral lateralization and cognitive development. In J. S. Chall and A. F. Mirsky (Eds.), Education and the Brain. Chicago: University of Chicago Press, 1978.

Leong, C. K. Lateralization in severely disabled readers in relation to functional cerebral development and syntheses of information. In R. M. Knights and D. J. Bakker (Eds.), The neuropsychology of learning disorders: theoretical approaches. Baltimore: University Park Press, 1976.

Leong, C. K. Spatial-temporal information processing in children with specific reading disability. Reading Research Quarterly, 1976-1977, 12, 204-215.

Leong, C. K. Cognitive patterns of "retarded" and below average readers. Contemporary Educational Psychology, in press.

Leong, C. K., and Haines, C. F. Beginning readers' analysis of words and sentences. Journal of Reading Behavior, 1979, 10, 393-407.

Luria, A. R. Higher cortical functions in man. New York: Basic Books, 1966(a).

Luria, A. R. Human brain and psychological processes. New York: Harper and Row, 1966(b).

Marshall, J. C., and Newcombe, F. Patterns of paralexia: A

psycholinguistic   approach.   Journal of Psycholinguistic
    Research, 1973, 1, 175–199.
Mattis, S. Dyslexia syndromes: A working hypothesis that works.
    In A. L. Benton and D. Pearl (Eds.), Dyslexia: An appraisal
    of current knowledge. New York: Oxford University Press.
Perfetti, C. A. Language comprehension and the deverbalization of
    intelligence. In L. B. Resnick (Ed.), The nature of intelli-
    gence. New York: Wiley, 1976.
Perfetti, C. A. Language comprehension and fast decoding: Some
    psycholinguistic prerequisites for skilled reading. In J. T.
    Guthrie (Ed.), Cognition, curriculum, and comprehension.
    Newark, Delaware: International Reading Association, 1977.
Vernon, M. D. Variability in reading retardation. British Journal
    of Psychology, 1979, 70, 7–16.

COMPARATIVE EFFICACY OF GROUP THERAPY AND

REMEDIAL READING WITH READING DISABLED CHILDREN

V. Nel   and   H. I. J. Van der Spuy
Groote Shuur Hospital   McMaster University
Cape Town, South Africa   Hamilton, Ontario, Canada

Remedial reading teaching has been shown to have small and short-lived effects (Carroll, 1972). An alternative to this approach in the treatment of reading difficulties has therefore been to use procedures aimed at attempting to improve the child's behavioural and emotional difficulties, thereby effecting a change in his approach to learning situations at school in a way that will be more adequate and effective, with a resultant amelioration of reading skills (Bills, 1950; Lawrence, 1971 and 1972; Lawrence and Blagg, 1974). Evidence is still unclear as to which are the most effective remedial reading procedures available (Carroll, 1972). Many previous studies fail to meet the requirements of adequate experimental design, for example, inadequate sample size, unmatched control groups, inadequate controls for important variables, such as therapist effects and placebo effects. The therapist or the teacher effect may indeed be a crucial factor in the outcome (Pumfrey and Elliot, 1970). There are nevertheless indications that a "therapy" type of approach may be more effective than a traditional remedial reading approach in improving reading attainment (Lawrence, 1971). In his otherwise excellent study, in which he concluded that therapy alone was more effective in the improvement of reading skills than remedial teaching alone or remedial teaching and therapy combined, Lawrence has however not controlled for therapist effect. A study evaluating the relative efficacy of a group therapy and a remedial reading approach with retarded readers, with the evaluation of therapist effect, therefore seemed indicated.

Aim

The aim of the present study was to evaluate the relative efficacy on reading attainment, personality and adjustment of a group therapy approach, compared with the results obtained by a traditional remedial reading approach with retarded readers.

Method

The subjects were 59 white English-speaking children, between the ages of 7 and 12, with full-scale I.Q.'s of at least 85,
who were referred for reading failure. Their reading achievement
quotient (R.A./M.A. X 100) had to be 90 or below. Three
groups, matched for sex, C.A., I.Q. and R.A. were randomly
assigned to three experimental conditions: remedial reading,
group therapy and a no treatment control group. Subjects were
organized in groups of five for both group therapy and remedial
reading. Subjects in the experimental groups were exposed to
these two methods of treatment, twice a week, over four months,
for one hour sessions. Therapist effect could be assessed as the
two therapists both conducted therapy sessions as well as remedial
reading sessions on a randomly assigned basis. All subjects
remained with the same teacher or therapist for the four month
period.

Therapy procedures were directed towards a number of areas
in the child's life. These included: social relationships, attitudes
towards self, reading and school, worries and anxieties. There
was an emphasis on group problem solving. The aim was to
improve the child's attitude towards reading and to boost their
self-esteem. The techniques used were mainly reassurance, suggestion, persuasion, encouragement of group activities and discussion.

The parents and teachers of the therapy group were involved
in an attempt to give them a sympathetic understanding of the
children and modify their attitudes towards them. Parents and
teachers were seen separately on a monthly basis.

The NEALE, JEPI, CPQ AND ROGERS were administered at
the beginning and end of the experimental period. Pre/post test
difference scores were used as measures of change in reading
attainment, personality and adjustment.

Although no significant difference was shown between the
remedial reading and therapy groups, both showed significant
gains in RA over the control group at the .01 level. No significant change was shown in personality and adjustment measures.
Therapist effect was shown to vary significantly at the .01 level
in favour of the female therapist.

Discussion

The results conflict with the findings of Lawrence (1971),
who found that counselling with or without remedial reading
brought about significantly better reading improvement than
remedial reading alone. The finding that psychotherapy with

retarded readers effected significantly greater improvement in
reading attainment when compared with children in the no treatment control group is in agreement with the findings of Elliott
and Pumfrey (1972) and Lawrence and Blagg (1974).

Placebo effects were controlled in that both the remedial
reading and the therapy group children saw the therapists for
the same amount of time and with the same frequency, although
the methods of treatment were varied. Both groups of children
improved and there was no significant difference in the degree of
improvement. It can be argued that a placebo-effect or a "Hawthorne" effect might have been an important factor in bringing
about this change, especially when these experimental groups are
compared with the control group, who were not exposed to this
effect.

The female therapist's group showed significantly greater
gains in reading achievement than the male therapist's group.
This finding is in line with many authors (e.g. Meltzoff and
Kornreich, 1970; Pumfrey and Elliott, 1970; Truax and Carkhuff,
1972) who have stressed that the therapist variable should be
controlled and investigated when attempting outcome studies of
different methods of treatment. The studies of Aspy (1965) and
Aspy & Hadlock (1966), and Lawrence (1972) may be seen as
indicating similar findings in that gains in childrens' reading
achievement levels appeared to be related to teachers with particular personality characteristics.

It is impossible to conclude exactly what the active ingredient
was which caused the overall improvement. It could have been
mainly a placebo or "Hawthorne" effect. The results underline
the absolute necessity of controlling for both placebo and therapist
affects in research of this nature and to attempt to discover what
the effective ingredients are which facilitate improvement. The
fact that a method or treatment seems to work in that it has
positive results is not necessarily a validation of the theory or
techniques of that treatment, as improvement might be due to
unsuspected non-specific factors.

References

Aspy, D. N. A study of three facilitative conditions and their
    relationships to the achievement of 3rd grade students. Un-
    published doctoral dissertation, University of Kentucky, 1965.
    In Truax, C. B. and Carkhuff, R.G., Towards effective
    counselling and psychotherapy. New York: Atherton, 1972.
Aspy, D. N. and Hadlock, N. The effect of empathy, warmth and
    genuineness on elementary students' reading achievement. Un-
    published thesis, University of Florida, 1966. In Truax,
    C.B. and Carkhuff, R. B. Towards effective counselling and

psychotherapy. New York: Atherton: 1972.

Bills, R.E. Non-directive play therapy with retarded readers. Journal of Consultative Psychology, 1950, 14, 140-149.

Carroll, H.M.C. The remedial teaching of reading: an evaluation. Remedial Education, 1967, 2,1, 4-12.

Elliott, C.D. and Pumfrey, P.D. The effects of non-directive play therapy on some maladjusted boys. Educational Research, 1972, 14 (2), 158-161.

Lawrence, D. Counselling of retarded readers by non-professionals. Educational Research, 1972, 15, 1, 48-51.

Lawrence, D. and Blagg, N. Improved reading through self-initiated learning and counselling. Remedial Education, 1974, 1, 2, 61-64.

Meltzoff, J. and Kornreich, M. Research in Psychotherapy. New York: Atherton Press Inc., 1970.

Pumfrey, P.D. and Elliot, C.D. Play therapy, social adjustment and reading attainment. Educational Research, 1970, 12, 183-193.

Truax, C. B. and Carkhuff, R.R. Toward Effective Counselling and Psychotherapy: Training and Practice. New York: Aldine Atherton Press, 1972.

CODING   STRATEGIES   AND   READING   COMPREHENSION

E. Neville   Brown

Lichfield   Dyslexia   Treatment   and   Research   Unit

Lichfield,   England

Abstract

An enquiry was conducted into the constraints upon the comprehension of printed verbal material by underachieving readers. The results suggested that such readers tend to prefer single-modality mediational encoding, resulting in impaired comprehension. The phenomenon appeared to be independent of spatial and verbal ability in the children studied. A further investigation was made of the comparative effectiveness of alternative induced coding strategies for the comprehension and recognition of predictably "impossible" words by underachieving readers. The results suggested the significant superiority of a visual-to-semantic over a visual-to-acoustic-to-semantic coding path for the underachieving readers. The implications of the findings for the remediation of reading and written language difficulties in children of otherwise adequate intelligence, and also for language pedagogy in general, are discussed.

Early or "apprentice" reading is widely assumed to necessitate phonological encoding as mediational between the printed stimulus and meaning, whilst "mature" reading appears to be able to omit the acoustic-phonological encoding stage, proceeding directly from print to meaning for at least the major proportion of processing time (Goodman, 1967). LaBerge and Samuels (1974) provide a useful model of automatic information processing in reading which accommodates the three processing stages for encoding and distinguishes between automatic and attentional processing at each of the three stages. The model appears capable of extension and modification to accommodate various styles of reading and various accounts of the reading process such as "barking at print" (Brown, 1978b). To be termed "reading", of course, the order of the processing stages is invariant -- visuo-graphic to acoustic-phonological to semantic -- which suggests that caution should be exercised in generalising from the results of other kinds of experiment, such as those involving the processing of digits, to reading.

A previous study (Brown, 1975) suggested that there was a single-modality encoding preference amongst underachieving readers in the recall of disparate two-digit stimuli, simultaneously presented via the visual and auditory modalities, which tended not to be so in "normal" readers. Whilst Snodgrass et al. (1974) and Snodgrass & McClure (1975) offer strong evidence for dual coding of words and pictures in recognition memory, it was felt that the picture/word distinction needed modification in order that the dual-coding (or lack of it) hypothesis might be usefully reported here investigated the possibility that underachieving readers, that is to say those whose mechanical reading ability was markedly inferior to their language-comprehension ability, were characterized by a tendency to prefer a single encoding strategy, either acoustic-phonological or visuo-graphic, in a (reading) task that is required, albeit by conventional pedagogy, to necessitate two sensory-modality-bound codes in an invariant order as mediating to meaning or semantic. The reading behaviour of underachieving readers who tended to prefer a single or uni-modal coding path would thus be expected to be sensitive to the different characteristics of acoustic-phonological and visuo-graphic encoding.

Experiment 1. Modal Preference and Reading.

In an analogue of the dichotic listening paradigm, 149 Birmingham primary and secondary schoolchidlren were subjected, in separate conditions, to acoustic-phonological and visuo-graphic interference of possible meaningfulness whilst being required to read aloud prose passages of comparable difficulty from a standardised prose reading test. The rate of reading was recorded, and also the incidence of response to orally-administered comprehension that could be interpreted as intrusions from the interference material. A test of spatial ability was also administered.

The salience of preference for single, unimodal encoding was assumed to be indicated by (a) the magnitude of the difference between the intra-individual reading rates, in words per minute, for the two interference conditions, and (b) the incidence of intrusions from the (possible) interference materials into the responses to comprehension questions. The RATE effect was interpreted as the attempted semantic encoding of the (possible) interference material, whilst the COMPREHENSION intrusion effect was interpreted as being the successful encoding of the interfer-

ence material. The results were presented in a correlation matrix:
Although spatial ability was not correlated with underachievement
in reading, it did appear to be more highly correlated with reading
ability in the earlier stages of reading. In contrast, both the
modal-preference variables (Rate and Comprehension) were significantly correlated with underachievement in reading (p <.001).
Modal preference also appeared to be independent of age.

It was concluded that children who were achieving in reading
tended to prefer to use a single modality-bound encoding strategy,
either the acoustic-phonological or the visuo-graphic, in attempting
to impose meaning on the printed word and also when reading
aloud (mechanical reading), whereas non-underachieving children
tended not to be so characterised.

Experiment 2. Induced Coding Strategies

A body of words of a predictably very high order of difficulty
(Brown, 1978b) was taught to underachieving readers on an
individual basis by either a Path 1 or a Path 2 approach: Path 1:
Visuo-graphic————Acoustic-phonological——Semantic.   Path 2:
Visuo-graphic——————————————————————Semantic. One group
was taught the words by a Path 1 approach which is usually
termed "Phonics 2" in Britain, involving syllabification and provision
of context and examples of usage in the oral and written language.
A second group was taught the same body of words by a Path 2
approach whereby generalised actions were "attached" to "icons"
or graphic representations which were, in turn, associated with
meaningful letter groups or morphographemes (see Fig. 1). In
many long, "difficult" words, syllable boundaries do not coincide
with morpheme boundaries (Brown, 1978a, 1979a). Such boundary
incongruence has been found to correlate significantly with reading

Stage:					
(i)	"incidents"_____ (I)		"throwing action" generalized (AG) ___		semantic control
(ii)	(I) ___ (AG) ___		"icon" _____		semantic control
(iii)	(I) ___ (AG) ___		"icon" ___ morpheme ___		semantic control

Figure 1.   Coding paths for 3 progressive stages of a PATH 2
            teaching approach (Brown 1978b, p.201).

difficulty in respect of prose passages at the 9-13 age norm levels. (Brown, 1978b, Part 2). The Path 2 teaching was conducted in silence, and each of the teaching sessions was preceded by a sorting exercise designed to inihbit or suppress acoustic-phonological encoding of polymorphographemic words and pseudo-words. The experimental design also included Placebo-treatment and Non-treatment groups. Relevant interaction effects were controlled.

A comprehension test, favouring neither regimen, with "untaught" items to test generalisation of the teaching, was administered (see Fig. 2).

Question 3.	Question 7.
Oral, with visual example using empty shell cases: "A device for throwing these out of a rifle after they have been fired.	Visual example. Demonstration of building a Lego house, knocking it down and building it again.
Response choices:-	Response choices:-
expression	contactible
adjective	reconstructible (corr. resp.)
rejectible	indifferent
ejector (correct response)	irreversible
extraction	recompression
constructible	destruction
injector	contact
suppressor	conversion
dejected	destructible
extractor	distraction
expressor	substructure
injection	constructive

Fig. 2. Examples of Comprehension Test questions.

The results suggested the superiority of the Path 2 over the Path 1 teaching approach, both for the full comprehension test scores and for the scores on the untaught words element in the test:

Table 1. Analysis of Variance of Comprehension Test Scores, Path 1 and Path 2 teaching groups.

Group	Full Comprehension Test					Untaught words only				
	Mean	Mean diff.	d.f.	F	p	Mean	Mean diff.	d.f.	F	p
Path 1	10.33					1.50				
		6.11	17	24	.001		2.55	17	22	.001
Path 2	16.44					4.11				

Within the Path 2 teaching group, it was also possible to compare the performance of those who, according to the criterion from Experiment 1, were held to be visuo-graphic coding preferent with that of the acoustic-phonological coding preferent subjects:

Table 2.   Analysis of Variance of Comprehension Test Scores for Visuo-graphic and Acoustic-phonological processing preferents within Path 2 group.

Group	Full Comprehension Test					Untaught words only				
	Mean	Mean diff.	d.f.	F	p	Mean	Mean diff.	d.f.	F	p
Vis-G	18.80					4.20				
		5.3	8	9	.01		0.20	8	0.1	ns
Ac-phon	13.50				4.00	4.00				

It was felt at this stage that the relationship between ability or intelligence and modal preference was worthy of further investigation. Using the Path 2 teaching approach, subjects were matched according to intelligence as measured by the WISC (Wechsler Intelligence Scale for Children), Visuo-graphic and Acoustic-phono-logical processing preferents being paired. As matching children with similar WISC quotients but with disparate ages was inappropriate (the WISC is a deviation quotient), it was also necessary to match according to age. The results of this investigation also suggested the significant superiority of the Visuo-graphic processing preferents over the Acoustic-phonological with regard to the Path 2 teaching (p< .025, one-tailed test):

Table 3.   Analysis of variance of Comprehension Test scores, matched pairs, intelligence and age.

Group	Full Comprehension Test					Untaught words only				
	Mean	Mean diff.	d.f.	F	p	Mean	Mean diff.	d.f.	F	p
Vis-G	18.3					3.33				
		3.8	11	5.7	.025		2.0	11	7.2	.025
Ac-phon	14.5					1.33				

Conclusions and Discussion:

The results of the investigations suggest confirmation of the widely held view that mechanical reading ability does not correlate highly with intellectual ability as measured by intelligence tests such as the WISC. The WISC does not require any processing of the printed language, those sub-tests that necessitate verbal thought being conducted orally. The mode of presentation of verbal material would, further, be held to differentially affect the comprehension of that material for a considerable proportion of children. Whilst this has been recognised at times for the congenitally deaf, it has not been thought so for "normal" children. The comprehension element in the Neale Analysis of Reading Ability is also interesting in that the questions are administered orally after the reading aloud of prose passages in which the errors in sounding out the words are orally corrected by the tester. In experiment 1, it was suggested that the relationship between the Neale Comprehension quotient and the Neale Accuracy quotient is a useful diagnostic aid, particularly where a WISC may not be administered.

The results of Experiment 2 appear to demonstrate that the comprehension of long, difficult words by underachieving readers may be significantly improved by changing the processing path. Conversely, it may be argued that, for these children, an approach to reading that labours "phonics", mechanical reading aloud or "barking at print" actually inhibits the comprehension or semantic processing of the verbal material. It is accordingly suggested that distinction be made between "verbal" and "speech" when considering the processing of language. "Verbal" focuses upon the semantic level, whereas "speech" focuses on the role of one of a possible range of mediations, the most effective choice of which for a specific task may not always be under the control of the individual for "interval" and "external" (which might include pedagogic) reasons. This is particularly noticeable with regard to the reading performance of the congenitally deaf. When such children succeed in comprehending high-level texts, they appear to do so by strategies quite unknown to their teachers. The author is at present engaged on experimental teaching of reading and writing to the deaf, according to the principles described in this paper. Work is also in hand on the formulation of a theory of meaning with practical application to communication difficulties.

Finally, as the children in this study appeared to be able to use the experimental (path 2) learning strategy on words that were not in the teaching programme, with significant results, it may be suggested that the principles and procedures outline here may be worthy of wider exploitation in a pedagogic context.

References:

Brown, E.N. (1975): A study of auditory-visual modal preference
    and its relationship to underachievement in reading.
    Unpublished M.Sc. thesis, University of Aston.
Brown, E.N. (1978a): Attentional style, linguistic complexity
    and the treatment of reading difficulty. In Knights and
    Bakker (Eds.), Rehabilitation, Treatment and Management of

Learning Disorders, proceedings of the 1978 NATO International
conference at Ottawa.
Brown, E.N. (1978b): Attentional style, linguistic complexity
and the treatment of reading difficulty. Ph.D. thesis,
University of Aston. In preparation for publication by Swets,
Amsterdam.
Brown, E.N. (1979a): A procedure for studying the effects of
induced coding strategies for complex verbal material using
CNV, RT and behavioral data. Paper presented at International Neuropsychological Association conference, Amsterdam 1979.
Laberge, D. and Samuels J. (1974): Towards a theory of automatic
information processing in reading. Cognitive Psychology,
6, pp. 293-323.
Snodgrass, J.G. and McClure, P. (1975): Storage and retrieval
properties of dual codes for pictures and words in recognition
memory. Journal of Experimental Psychology, 1975. Vol 1.
No. 5, pp. 521-529.
Snodgrass, J.G., Wasser, B., Finkelstein, M. and Goldberg, L.B.
(1974): On the fate of visual and verbal memory codes for
pictures and words: Evidence for a dual coding mechanism in
recognition memory. Journal of Verbal Learning and Verbal
Behaviour, 1974. 13, pp. 21-31.

CULTURAL SYSTEMS AND COGNITIVE STYLES

J. W. Berry

Queen's University

Kingston, Canada

Abstract

The position of cultural relativism when applied to the study of cognition across cultures, leads to the view that cognitive development is likely to be relative to the cognitive problems faced by individuals in a particular cultural system. The cross-cultural study of cognitive development, then, must attend to three issues. One is the nature of the ecological and cultural context in which cognitive development takes place. A second is the kind of cognitive abilities which are developed in that context. And a third is the nature of the relationships which may exist between the cultural context and the cognitive development. To accomplish this threefold research programme, there must be a local analysis of both the context and the developed abilities; and there must also be a comparative synthesis of the patterns which may exist within each of the two domains. One implication of this strategy is that preconceived and prepackaged instruments are not adequate to the task of local analysis; that is, such gross apriori concepts as "culture" and "intelligence" cannot help in the research, and may indeed be a hindrance. A second implication is that remaining at the level of local analysis cannot yield the desired generalizations about the structure of culture, the pattern of cognitive abilities, and the systematic (perhaps causal) relation-ships between them. It is essential to search for these structures, for without them nothing may be said about panhuman features of culture or cognition. Examples of three approaches to the cross-cultural study of cognition are presented: the use of standard intelligence tests in various groups, the analysis of specific skills in local cultural contexts, and the synthesis of abilities into patterns (cognitive styles) in relation to cultural systems. It is contended that the first approach is ethnocentric (from a position outside the culture) and general, while the second is ethnocentric (but from a position inside the culture) but lacks generality; only the third can meet both goals of cross-cultural research--being cognizant of local cultural variation, while also seeking universal generalizations.

Considerable interest and debate have been devoted to the conceptualization, measurement and interpretation of cognitive differences among human populations. That there are such differences is readily demonstrated, but their meaning, of course, is far from settled. This paper is intended to be a contribution to the enquiry into the nature of these differences, and their proper interpretation.

The perspective taken is from the field of cross-cultural psychology. Given that this field has now developed to the point where it has a distinct identity and body of findings, (for example, it has its own journal, its own international association, and most recently its own Handbook, see Triandis, Lambert, Berry, Lonner, Heron, Brislin and Draguns, 1979), one might assume that its methods and perspectives would have informed the general enquiry into population differences in cognition. However, I judge that this has not been the case; rather, the enquiry and debate have pretty-well settled into a squabble about what is essentially a domestic problem in two or three countries, and only occasional (and usually misinterpreted) references are made to the wide cross-cultural literature on the topic. We begin, then, with a brief outline of the basic position of most cross-cultural psychologists, that of cultural relativism.

The Standpoint of Cultural Relativism

Cultural relativism is a scientific position which attempts to avoid descriptions and interpretations of the behaviour or culture of individuals or groups which are based upon the scientists' own culture and standards. This position has been widely accepted in anthropology as the way to avoid ethnocentric evaluations of other peoples. According to Segall, Campbell and Herskovits (1966, pp. 15-18), the position was first developed by Boas (1911) and was firmly established by the work of Herskovits (1958).

Such a position, while emphasizing local conceptions and evaluations of events (the emic approach), and avoiding the imposition of external standards (the etic approach) nevertheless assumes the existence of universals, and of the "psychic unity of mankind" (Berry, 1969; Lonner, 1979; Segall, Campbell and Herskovits, 1966, p. 17 and Wallace, 1961). This assumption permits comparisons across populations, while eschewing evaluations relative to some assumed universal standard.

A position of "radical cultural realitivism" (see Berry, 1972) goes further, and argues that for some characteristics of populations it is more appropriate not to assume psychological universals across groups. This more drastic position has been advocated for research into psychological characteristics which have been conceptualized in, and are firmly rooted in, a single (usually Western) psychological science, and where there exists wide-spread controversy surrounding the comparative use of the concept. In essence, it argues that we should "conceptually wipe the slate clean" (Berry, 1972, p. 78) and approach the question with few or no assumptions about its nature in other populations.

With respect to a concept such as intelligence or general ability the relativist position is:

1.   that human populations adapt to differing ecological and cultural contexts
2.   that the individual's cognitive development will be an integral part of that adaptation
3.   that any characterizations of that development should be relative to the particular adaptive requirements, rather than to some assumed universal dimension (such as general intelligence).

The balance of this paper attempts to demonstrate the validity of the position. Evidence is presented for the existence of varying ecological and cultural adaptive settings, and for differential development in response to these settings. Then three approaches to interpreting these variations are presented, followed by a discussion of some of their implications.

Ecological and Cultural Contexts

Little space is required to demonstrate the fact that individuals are born into and develop in widely varying environmental settings. From a cross-cultural perspective, many psychologists have been drawing upon the literature of human ecology and anthropology in order to conceptualize and measure these varying contexts; and within cultures; the works of sociologists and economists have been equally useful in our understanding of environmental variation.

For example, a recent attempt by the author (Berry, 1976) to specify varieties of ecological engagement and cultural adaptation for eleven different human populations led to the construction of scales and indices which could describe relevant features of environmental variation. Although such variation is amenable to observation and description, there has been a tendency, especially within societies with a single dominant "main stream" culture, for psychologists to ignore or underestimate even obvious differences in environment and experience (Berry, 1979). Despite this tendency, I will assume that no one will dispute the statement that human populations differ in their environmental contexts and that such variation is important and should be assessed by psychologists who are engaged in comparative work.

Perceptual and Cognitive Abilities

Little space is required, as well, to demonstrate the fact that people develop differing skills and abilities in different populations. From a cross-cultural perspective, the ability profile of peasants differs from that of hunters, and both differ from fishermen or herders; and those who have traditionally been literate differ from those who have developed without reading and writing. Within cultures, those higher in status differ from those lower in status, and individuals occupying differing economic roles exhibit ability in different performance areas.

Evidence for this assertion may be found scattered throughout the comparative literature on perception and cognition (see e.g., Berry and Dasen, 1974; Cole and Scribner, 1974; Cronbach and Drenth 1972; Dasen, 1977; Lloyd, 1972). That differences do exist is virtually without question in the literature; moreover, even greater variation would be in evidence, if only psychologists had made an effort to sample the behaviours actually manifested in the lives of these various populations, rather than limit themselves to their handy kit of Western tests. I will assume that no one will dispute the statement that human populations differ in cognitive abilities, and that such variation is important and should be assessed.

The Search for Systematic Interactions

The main question facing us is: what is the most suitable way to conceive of these variations in ecological-cultural settings, and in cognitive performances? To help us deal with this question, three issues should be kept in mind:

1. The first is whether or not we accept a cultural relativist (as opposed to a universalist) position.
2. The second is whether or not we search for systematic inter-relationships among elements of cultural contexts and of cognitive performances
3. And the third is whether or not we make comparisons across groups in order to discover more general statements about human cognition.

In the conventional approach (that of general intelligence)

there has been an a priori assumption that abilities will be inter-related in specific ways, independent of population (see e.g., Goodnow, 1973); there has been another assumption (usually implicit) that only one set of cultural experiences (namely Western, industrial) are of any importance to the research. A second approach (that of specific skills) is characterized by no assumptions about, and indeed by no interest in, systematic relationships; single features of the environment are related to single abilities, without even raising the question of relationships among cultural variables of among ability variables. A third approach (that of cognitive style, assumes no universal set of systematic relationships within either set of variables, but it is interested in searching for them.

The Interpretation of Systematic Interactions

These steps do not satisfy the normal scientific obligation to make interpretations (perhaps cause and effect interpretations) of these systematic relationships. Two broad classes of interpretation, of course, have been competing for scientific status since the beginning of intergroup contact and awareness: the learning (cultural transmission) and the genetic (biological transmission) interpretations.

Characteristics of organisms may be due to nature and/or nurture, for both biological and cultural characteristics are known to be in adaptation to the habitat of the population. The focus of this paper is not on the relative strength of these two positions, and so the debate will not be treated further except to make one observation. This is that in cross-cultural psychology (despite the term "cultural" but not "biological" in the title), there is little evidence of exclusive support for cultural transmission (e.g., Biesheuvel, 1972; Dawson 1975); most of us, I believe, accept in principle some role for the biological transmission of psychological characteristics, even though cross-group evidence may now be lacking.

Because the outcome of this particular debate is at present indeterminate, it is clear that the causal mechanisms between environment and cognitive performance must remain unspecified. Thus, this paper will limit itself to a consideration of context and performance and of the relationships within and between them. We turn now to a consideration of each of the three approaches.

General Intelligence

The classical approach to the study of cognitive differences across populations has been to take existing tests and to administer them to different populations. Of course, there has been a recognition that the test may not get any response at all until translation has been made. Typically the only modifications or additions undertaken were those necessary to get data; modifications to match the test to the cognitive life of the people have not normally been done. That is, two assumptions have usually been made: one is that the cultural life of the test developer and the cultural life of the test taker differ in only one important respect, that of language; the other is that the cognitive abilities characteristic of the cultural life of the test developer and those of the test taken differ in only one respect; that of <u>level</u> of development.

These two assumptions are illustrated in the upper portion of Figure 1. First, elements in the cultural context are treated more or less as a unit (solid boundary around elements), and, second, the cognitive abilities are assumed to be a single universally interrelated package. Both are then usually interpreted in terms of populations having bigger or smaller packages. With respect to culture, those with small packages are thought to be "deprived," while those with big ones are "enriched." With respect to cognitive

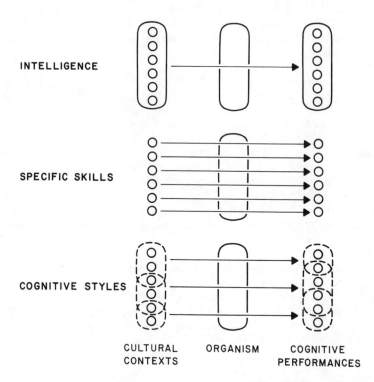

Figure 1. Schematic Diagram of Three Approaches to Relationships Between Cultural Contexts and Cognitive Performances.

performances, Vernon (1979, p. 7) has noted that it is commonly assumed that intelligence is a "homogeneous entity or mental power that, like height or weight can vary in amount or in rate of growth or decline..." His own empirical work (Vernon, 1969) illustrates these assumptions.

With respect to the first assumption, it is clear to me that cultural differences have not been taken seriously in the debate on population differences in intelligence. And with respect to the second assumption, little attempt has been made to find out what cognitive abilities are actually in place, and how they are structured. Given these two errors of omission, the great logical error of commission is then performed: if the cultures are not really different, if the abilities are not really different, then the differences in test performance must be due to different levels of development. However, from the point of view of cultural relativism, if cultural differences are real and large, and if abilities develop differentially in adaptation to these differing contexts, then differences in test performance cannot logically be claimed to be differences in levels or amount of development.

Specific Skills

An alternative to this approach is that taken by workers in "cognitive anthropology" (e.g., Cole et al., 1971). From their perspective, single features of the context (such as a specific role or a particular experience) is linked to a single performance (such as performance on a categorization task, or accuracy on a test of quantity estimation); this approach is illustrated in the mid portion of Figure 1. They contrast their "notion of culture-specific skills" with general ability theory (Cole et al., 1971, p. xiii), which often asserts that in some cultures, cognitive development is pushed further than in some other cultures. Assuming that cognitive processes are universal (Cole et al, p. 214; Cole and Scribner, 1974, p. 193), they argue that "cultural differences in cognition reside more in the situations to which particular cognitive processes are applied than in the existence of a process in one cultural group and its absence in another (Cole et. al, 1971, p. 233). This emphasis on the particular, and culturally relative, nature of cognitive skills has meant that Cole and his co-workers do not search for patterns in their data. Generally, they appear unconcerned whether performance 1 correlates with performance 2, or whether cultural element 1 tends to be experienced along with cultural element 2 by individuals in their sample. Unlike intelligence testers, they do not assume any universal pattern or structure in their skill data; indeed they seem uninterested in such a question. Similarly, they also seem uninterested in how the numerous cultural elements may be organized in a cultural system in which the individual develops. And finally, they avoid explicit cross-cultural comparisons as being inconsistent with their local (emic) emphasis.

Cognitive Styles

The two approaches to understanding the relationships between culture and cognition thus far considered have differed in their acceptance of cultural relativism, in their concern for systematic relationships and in their use of comparisons. The approach taken by <u>intelligence</u> testers ignored cultural relativism, assumed a universal structure in relationships and readily made cross-cultural comparisons; the approach taken by those interested in <u>specific skills</u> assumed the position of cultural relativism, but ignored systematic relationships and cross-cultural comparisons. The approach taken by researchers into cognitive styles also assumes the position of cultural relativism, but in addition, searches for systematic relationships among abilities, among elements of the cultural context, and between patterns of contexts and abilities across groups (see lower part of Figure 1).

One basis for this approach is in the work of Ferguson (1954, 1956) who argued that "cultural factors prescribe what shall be learned and at what age; consequently different cultural environments lead to the development of different patterns of ability" (1956, p. 121). Further, he argued that through over-learning and transfer, cognitive abilities become stabilized for individuals in a particular culture. Both cultural relativism and systematic relationships are thus implicated in this approach, and these have been adopted in much of the work on cognitive style.

A recent review of the research on various cognitive styles (Goldstein and Backman, 1978) makes it clear that while sharing a general approach, there are many important differences among the numerous research traditions. This need not be a problem here, for only one has received any substantial treatment in the cross-cultural field, that of field dependence--field independence (Witkin et al. 1962; Witkin and Goodenough, in press).

This cross-cultural work (see e.g., Berry, 1976; Witkin and Berry, 1975) is characterized by an analysis of the local cultural context (termed "ecological demands" and "cultural supports"), by attempts to assess the cognitive (and perceptual) performances of individuals in a number of groups, and by a search for systematic relationships among performances (the "style"), and between performances and contexts. No interpretation is made about <u>levels</u> of development, given that no assumptions are made about the <u>absolute</u> value of field dependency or independency; indeed, such work assumes that differing positions on the style dimension will best meet the requirements of living in differing cultural contexts (Berry, 1976). Finally, while a search is made for systematic relationships among performances to discover whether they will remain constant or vary with cultural context, there is no assumption or requirement that they should. Similarly, while a search is made for systematic relationships among cultural contexts, there is no predetermined pattern which is related to Western culture.

Implications for the Future Comparative Study of Cognition

It is assumed here that science demands value free generalizations. However, the search for generalization is often accompanied by absolute value assumptions (as in the approach taken by intelligence testing); and the effort to avoid ethnocentric value judgements is often accomapnied by a disinterest in comparison and generalization (as in the specific skill appraoch). It is claimed here that the middle road, that of cognitive style research, can best meet the need for culturally relative and systematic statements about cultural systems and pan-human cognition.

It is not claimed, though, that the work on cognitive style accomplished to date has met these needs well. Critics from the other two approaches have tried to identify cognitive style work with levels of general ability (e.g., Cole and Scribner, 1977), or with a combination of a specific skill and general ability (e.g., Vernon, 1972). And workers employing the cognitive style approach themselves have seen the need to develop culturally more sensitive test materials (Van de Koppel, 1977), to examine more closely the specific cultural context (Okonji, in press), and to consider some of the developmental implications of the approach (Dasen, Berry, and Witkin, 1979).

The repetition of the finding, study after study, that the intelligence of individuals in a particular group is lower than in another group, or that yet another specific skill can be linked to some specific cultural context, cannot help in our search for value free generalizations. What we require is:

1. a truly comparative study (i.e. many groups, not just two), in which
2. the position of cultural relativism is employed in the
3. specification of characteristics of each groups' ecological and cultural contexts
4. and of each groups' cognitive life;
5. in which cognitive tasks are developed and employed to sample their cognitive abilities, and there is a search for systematic relationships:
6. among cognitive abilities within each group (i.e., cognitive styles),
7. among cultural contexts within each group (i.e., cultural systems), and
8. between these cognitive styles and the cultural systems in which they are found.

Until such a programme is carried out, the comparative

study of cognition will remain bogged in ethnocentric value judgements and in repetitious statements of limited generality. Moreover, we will not be able to say anything valid about either the
nature of group differences (or similarities), or about their
origins. Until all eight features are included in our studies, any
attempt to allocate causality to biological transmission, or to
environmental transmission, or to allocate relative contributions to
one environmental feature or another, will fall short of our commonly accepted standards of research.

References

Berry, J. W.   On cross-cultural comparability. International
    Journal of Psychology, 1969, 4, 119-128.
Berry, J. W.   Radical cultural relativism and the concept of intelli-
    gence. In L. J. Cronbach and P. Drenth (Eds.) Mental Tests
    and Cultural Adaptation.   The   Hague: Mouton, 1972.
Berry, J. W.   Human Ecology and Cognitive Style: Comparative
    Studies in Cultural and Psychological Adaptation.   New York:
    Sage/Halsted, 1976.
Berry, J. W.   Implications of cross-cultural methods for research
    in multicultural societies.   Journal of Cross-Cultural Psy-
    chology, 1979, 10, 415-434.
Berry, J. W. and Dasen, P. (Eds.)   Culture and Cognition,
    London: Methuen, 1974.
Biesheuvel, S.   An examination of Jensen's theory concerning edu-
    cability, heritability and population differences. Psychologia
    Africana, 1972, 14, 87-94.
Boas, F. The Mind of Primitive Man. New York: Macmillan, 1911.
Cole, M., Gay, J., Glick, J. and Sharp, D.   The Cultural Con-
    text of Learning and Thinking. New York: Basic Books,
    1971.
Cole, M. and Scribner, S.   Culture and Thought.   New York:
    Wiley, 1974.
Cole, M. and Scribner, S.   Developmental theories applied to cross-
    cultural cognitive research. Annals of the New York Academy
    of Sciences, 1977, 285, 366-373.
Cronbach, L. J. and Drenth, P. (Eds.) Mental Tests and Cultural
    Adaptation. The Hague: Mouton, 1972.
Dasen, P. R. (Ed.). Piagetian Psychology: Cross-cultural Contrib-
    utions. New York: Gardner, 1979.
Dasen, P. R., Berry, J. W. and Witkin, H. A.   The use of develop-
    mental theories cross-culturally.   In L. Eckensberger, Y.
    Poortinga and W. Lonner (Eds.).   Cross-Cultural Contributions
    to Psychology.   Amsterdam: Swets and Zeitlinger, 1979.
Dawson, J. L. M. B.   Psychological Effects of Bio-social Change in
    West Africa.   New   Haven: Hraflex   Press,   1975.
Ferguson, G. A.   On learning and human ability, Canadian Journal
    of Psychology, 1954, 8, 95-112.
Ferguson, G. A.   On transfer and the abilities of man.   Canadian

Journal of Psychology, 1956, 10, 121–131.

Goldstein, K. and Backman, S. Cognitive Style: Five Approaches and Relevant Research. New York: Wiley, 1978.

Goodnow, J. J. The nature of intelligent behaviour: questions raised by cross-cultural studies. In Resnick, (Ed.). Nature of Intelligence. 1973.

Herskovits, M. J. Some further comments on cultural relativism. American Anthropologist, 1958, 60, 266–273.

Lloyd, B. Perception and Cognition: A Cross-Cultural Perspective. Harmondsworth: Penguin, 1972.

Lonner, W. J. The search for psychological universals. In H. C. Triandis and W. W. Lambert (Eds.). Handbook of Cross-Cultural Psychology Vol. 1, Perspectives. Boston: Allyn and Bacon, 1979.

Okonji, J. Cognitive styles across-cultures. In N. Warren (Ed.) Studies in Cross-Cultural Psychology, Vol. 2. London: Academic Press, in press.

Segall, M., Campbell, D. T., and Herskovits, M. T. The influence of Culture on Visual Perception. Indianapolis: Bobbs-Merrill, 1966.

Traindis, H. C., Lambert, W. W., Berry, J. W., Lonner, W. J., Heron, A., Brislin, R., and Draguns, J. (Eds.). Handbook of Cross-Cultural Psychology, Vols. 1-6. Boston: Allyn and Bacon, 1979.

Van de Koppel, J. A preliminary report on the Central African differentiation project. In Y. Poortinga (Ed.). Basic Problems in Cross-Cultural Psychology. Amsterdam: Swets and Zeitlinger, 1977.

Vernon, P. E. Intelligence and Cultural Environment. London: Methuen, 1969.

Vernon, P. E. The distinctiveness of field-independence. Journal of Personality, 1972, 40, 266–391.

Vernon, P. E. Intelligence: Heredity and Environment. San Francisco: Freeman, 1979.

Wallace, A. F. C. The psychic unity of human groups. In B. Kaplan (Ed.). Studying Personality Cross-Culturally. Evanston: Row, Peterson, 1961.

Witkin, H. A. and Berry, J. W. Psychological differentiation in cross-cultural perspective. Journal of Cross-Cultural Psychology, 1975, 6, 4–87.

Witkin, H. A., Dyk, R. B., Goodenough, D. R., and Karp, S. A. Psychological Differentiation. New York: Wiley, 1962.

Witkin, H. A. and Goodenough, D. R. Essence and origins of cognitive styles: Field-dependence and field-independence. Psychological Issues, in press.

CULTURE, COGNITIVE TESTS AND COGNITIVE MODELS:

PURSUING COGNITIVE UNIVERSALS BY TESTING

ACROSS CULTURES

S. H. Irvine

Plymouth Polytechnic

Devon, England

Abstract

Trends in the use of tests across cultures include stabilizing true-score variance in performance and paper-and-pencil tests. Comparisons across cultures have also been made, but not without controversy, reflecting the lack of consistent theoretical frameworks for defining test scores. Developments in paper and pencil testing during the past twenty years are evaluated. Progress in combining theory and methods so that cognitive function assessment may proceed systematically has been slow. Current work assumes that test scores are complex dependent variables. Preliminary results indicate that independent process variables in established information processing models affect test-score performance. Such independent variables may prove consistent in their effects on test scores from any culture.

Cross-Cultural Psychology and Construct Validity

Among the many activities of cross-cultural psychologists, two are most frequently mentioned. The first is to use variations in human behaviour that are associated with culture differences to verify the status of an assumed psychological law. The second is to determine what cultural variables in fact are; and how, for example, these relate to ethnic or geographical variation, since culture, ethnicity and geography are not one and the same. If and when cultural variables are identified, the use of these to extend, or restrict, the range of cultural variation so that its effect on assumed psychological universals may be observed, is an appropriate scientific procedure. These two main goals of cross-cultural psychology become interdependent in construct definition and validation for the discipline of psychology at large.

Nevertheless, this verification procedure is most effective
when the "universal law" produces a constant effect on behaviour
in whatever matrix of cultural variables it operates.   When the
effect of the assumed universal varies with one or more cultural
variables, the scientist often has a decision to make for which
proof may be absent. He has to decide, from the results, whether
the variability may be attributed to cultural differences, or whether
some experimenter bias, artifact, or intervening variable has
engineered the result.   When differences occur, attribution to
culture is a complex matter, (Thoday, 1969;   Irvine and Sanders,
1972;   Poortinga, 1971) that may never, when doubt arises, be
resolved.

Inconsistency in the face of cultural variation is a much more
difficult finding to explain than consistency;   and variations
attributed to cultural contexts are perhaps no more dramatically
observed, and debated, than in the field of mental testing.   In
fact, the variability of performance by different ethnic groups on
mental tests has been so pronounced that the scientist is faced
with some unenviable choices.   It could be that there are, indeed,
no universal relationships governing cognitive operations;   and
that their pursuit is irrational.   That the laws of cognition seem
to operate when test scores are correlated with work and school
performance in studies that span all continents is, however, a
reasonable inference from the consistent and predictable correlations of tests with both these criteria.   Ord's (1972) monograph
gives rise both to the certainty and consistency of these findings.
Such correlations are indeed statistical universals: but, as they
lack a theory to explain them the correlations represent the limits
of our knowledge.   Given the necessary logical assumption that
universal cognitive functions exist, one is faced with the realization that tests as they are presently used and analyzed across
cultures offer only limited evidence for the definition or consistent
operation of such laws.   Like the users of bows and arrows, we
know that the technology works, but we do not yet seem to have
the scientific knowledge to explain why.   This paper addresses
the task of using a systematic scientific framework for the construction of tests across cultures in the search for universals.   It
follows from a critique of how cross-cultural scientists have used
tests and how they have compared the scores derived from them.

Cultural Groupings and Test Scores: Independent or Dependent Variables?

Whereas the goals of cross-cultural psychology are quite
explicit, their statement operationally is much more tentative.
Definition of the independent variable in the construct validation
of test scores is, for example, an extremely difficult task.   Many

have attempted to use "cultural variables" as if they were independent variables, in order to examine variability in test scores as
dependent variables. <u>Within</u> any one geographical region, variables
such as spoken language, socio-economic status, ethnicity, family
size, family position, season of birth, and attributes such as sex,
retinal pigmentation, or onset of myopia, have been regarded as
capable of influencing or moderating test scores. <u>Across</u> geographical regions, observed differences in food accumulation habits,
child rearing practices, and pressure to conformity have steadily
progressed to independent variable status. All of those are,
strictly speaking, complex dependent variables themselves. They
are proxies for the contribution of many influences on how information shall be encoded, remembered, processed and produced in
the brain. Correlation, rather than cause seems to be as much
as one safely can venture with cultural variables. To insist on
cause for a complex dependent variable, such as ethnicity or
socio-economic status, is to invite well-deserved criticism, since
any one of a number of unidentifiable causal agents may lurk
within it.

The second major problem of attribution or definition is in the
meaning attached to a group average, or any single test score, or
any derived variable from a cluster of test scores (a factor, or
factor score). Such measures have, in psychometric theory, been
made to do duty as, or perhaps stand in for, inferred dispositional
qualities (e.g., intelligence, creativity, verbal ability, etc.).
Theories of ability based on correlations have always invoked dispositional constructs to account for those very correlations.
Thoseconstructs, in turn, have been given determinant status
more often than not. Test scores, however, are the product of
far too complicated a series of mental events to be regarded as
independent variables. Both the complexity of the processes
behind the test score and the status of the score as a measure
relative to a group mean put special constraints on its scientific
interpretation. Mental operations, particularly strategies, are as
uncertain in their potential for cause as cultural variables are.

It seems, therefore, that scientists who use group membership
within and across cultures to examine the nature of individual differences in test scores have severe epistemological problems to
solve. Both culture variables and test scores are dependent
variables of ranging degrees of complexity. Nevertheless, empirical
pursuits of test score meaning have taken place, many of them
the by-product of the use of ethnic and other cultural groupings
<u>per</u> <u>se</u> for comparing means and variances in test scores. There
are severe limits to the comparison of attributes within or across
genetic or environmental conditions. Thoday (1969) wrote the
definitive comment on these limits and no social scientist has
issued a rebuttal, or, indeed, is able to. In spite of Thoday's
incontestable logic, various techniques have been tried in the

attempt to compare test scores derived from different groupings of subjects. These attempts are not to be dismissed lightly, since they have helped to question the very theories implicit in the use of cultural variables and test scores. Nevertheless, they have not, until now, resolved any scientific issues because the very nature of the tests and culture variables used have forbidden generalization. What has been achieved can be judged by a review of procedures involving these variables.

Using Test Scores to Compare Performances of Static Groups

Assertions have been made about the comparability of test scores, and the dispositional qualities in persons that they operationally define, by subjecting test scores to various forms of statistical treatment. Perhaps the simplest, and most trusting, procedure is to classify different groups by some "cultural variable" and then compare mean scores on a number of tests. The comparison of mean profiles (Figures la and lb) is a step in the argument associated with Lesser (Lesser, Fifer, and Clark, 1965; Stodolsky and Lesser, 1967) that "patterns" of abilities are invariant across socio-economic status within ethnic groups, but highly distinctive across ethnic groups. The statistical finding may be exact, but there are two perennial problems associated with explaining the finding. First, the precarious step from observation of mean profile differences to an assumption of differential underlying "patterns" of abilities: and second, the question of whether the proxy variable ethnicity alone produces such findings. The second question has not received much attention, although the first has often been debated. Mean scores from a number of measures (Verbal, Reasoning, Numerical and Social Studies Tests), Irvine, (1966, 1969) applied to male and female African adolescent students Rural Boarding (RB) Rural Day (RD) and Urban Day (UD) schools in Mashonaland, Zimbabwe, in 1962, are presented in Figures 2a and 2b for comparison with Lesser's work. Two observations can be made. First, schools are associated with mean profile differences; and so too are sex categories. Here, though, males are superior in all aspects of performance; and female dominance in North American elementary schools is not to be construed as a universal. Significantly, school differences may yield profiles that are similar but are separated by level (Rural Boarding vs. Rural Day). Sometimes the profiles associated with school type seem to be completely different (Urban Day). Sex differences are pronounced: but if girls are superior in performance to boys in North American elementary schools and inferior to boys in African schools, neither genetic nor environmental explanations can be said to fit the results produced by using sex as an attribute by which to examine test variance in other cultures. In short, subtle learning variables implicit in school quality categories can act both like ethnicity and socio-economic status.

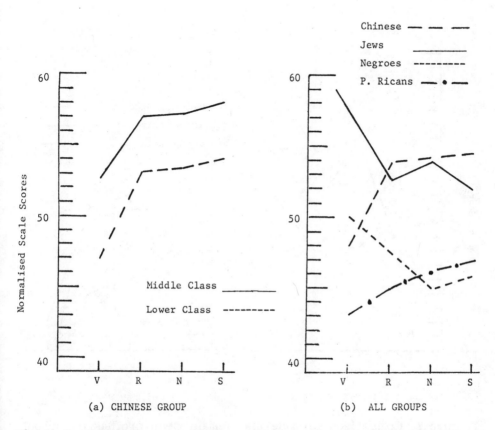

<u>Figure 1</u>. Comparison of socio-economic status and ethnic group mean profiles (Lesser, Fifer, and Clark, 1965).

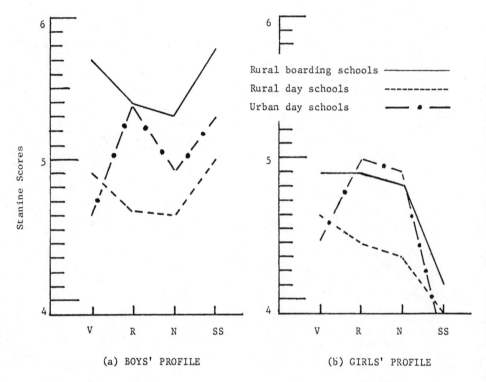

(a) BOYS' PROFILE   (b) GIRLS' PROFILE

Figure 2. Comparison of male and female mean profiles by school
type (Irvine, 1964 and 1973).

If school and sex differences are associated with the kinds of mean profile differences that have been attributed to ethnicity, then ethnicity has to be proven to be a variable not associated with different school practices and different sex roles. Of course, it is not possible to do this. Lesser offers no evidence that the proxy variable "ethnicity" in his work is identifiable apart from school and sex role differences. Cultural variables are no less complex than test scores. What variance is present in test scores is, in any inference from the test score to its underlying dispositional quality, quite crucial. The next examples show how it is almost impossible to offer proof for assumptions of test equivalence across cultures.

While the last example compared means over four variables for three types of schools, this next example compares only two variables in two groups. The cross-cultural ideal is an infinite number of variables over an infinite number of samples each with an infinite number of subjects. The method is adopted by Jensen (1974) who had used the comparison of regression line slopes for two variables that claim to sample, respectively, intelligence and associative memory. Ethnicity is once more the static group criterion. He has calculated these regression slopes from the scores of subjects from two ethnic groups, North American Whites and Blacks. Figure 3a and 3b show Jensen's results. A cross-over effect in Figure 3a shows that the same Lorge-Thorndike non-verbal test score produces, at the lower end of the Lorge-Thorndike scale, higher average memory scores for Blacks than Whites. At the upper end of the scale the reverse is true. However, there is considerable memory-score overlap throughout the intelligence scale. Figure 3b shows, to the contrary, that from the same memory span test score, one could predict very much greater average performance on Lorge-Thorndike for Whites than Blacks. Jensen concludes "There is no point on the scale of memory scores at which equated groups of whites and blacks obtain the same intelligence scores...When equated for intelligence, on the other hand, Whites and Blacks are considerably more alike in memory...In other words it appears that if the subjects have the intelligence, then they have the memory, while if they have the memory they do not necessarily have the intelligence." (op. cit. p. 106). The reader may doubt that regression analysis on two variables permits this remarkable degree of generalization. The next example may help define some limits.

Further analysis of the same Mashonaland data illustrates what happens when Raven's Progressive Matrices test scores are correlated with a test called Geography and Nature Study--a test that tested whether Mashonaland children diligently learned by rote the contents of the syllabus--an associative memory course for different scholars--in English, a foreign language. Perhaps this may be regarded as a severe test of associative memory.

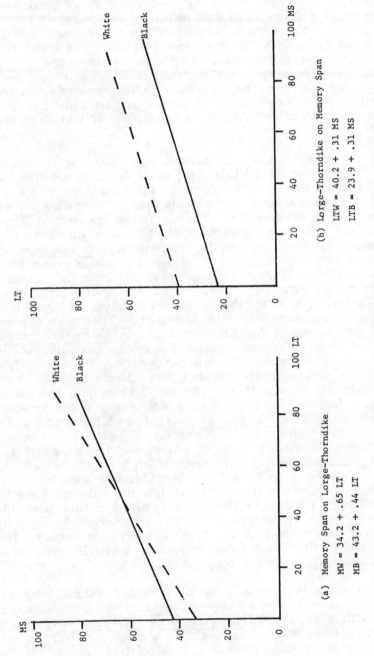

(a) Memory Span on Lorge-Thorndike

MW = 34.2 + .65 LT

MB = 43.2 + .44 LT

(b) Lorge-Thorndike on Memory Span

LTW = 40.2 + .31 MS

LTB = 23.9 + .31 MS

Figure 3. Regression lines for White and Black Groups on Lorge-Thorndike and Memory Span Table (after Jensen, 1976).

Regression analysis revealed a rather surprising result if one were to expect Jensen's finding to be replicated; or even if his generalizations about "intelligence" and "memory" were to hold as a universal. Figures 4a and 4b present the results.

The Jensen results showed regression slope crossover when Black and White groups were ordered along the scale provided by the "intelligence" score, and a large gap with almost parallel lines when they were placed in groups along the "memory" score scale. The Mashonaland results simply reverse this finding. The "same" intelligence score at any point on the scale would reveal, first, great differences among children, grouped and compared according to schools, in their ability to remember their second-language lessons in Geography and Nature Study; and, within each school type marked differences between boys and girls are evident. Figure 4b, on the other hand, shows boys and girls, and schools, to be pretty much alike in the "Intelligence" test scores if first they are grouped according to their scores on the Geography and Nature Study--or rote memory--test.

The conclusion from this example could be stated thus: "There is no point in the scale of intelligence test scores at which equated groups of males and females in the same school type obtain equal memory scores. When matched for memory, on the other hand, boys and girls are considerably more alike in intelligence. In other words, it appears that if the subjects have the memory they have the intelligence, while if they have the intelligence they do not necessarily have the memory." Not an usual finding in schools where achievement is a function of teacher quality and learning opportunity, which the correlations in Table 1 demonstrate. The parody is possible because of gross over-generalization and of committing the cardinal psychometric sin of assuming that a single test adequately samples "intelligence" any better than a single test adequately samples "memory." My own choice of a rote-learned, second-language, achievement test as a "memory" test is no better than my choice of Raven's Progressive Matrices test as an "intelligence" test. Both choices have something to recommend them, but they do not permit me to make general inferences about the intelligence level of all African boys and girls within all schools, any more than they permit me to generalize about the quality of teaching or learning of Geography and Nature Study in the schools now or then. The step from regression slope differences to trait level differences is simply an unidentifiable assumption.

The results underline that cross-cultural replication itself, within the techniques of psychometrics, is often enough to limit the generalizability of conclusions. The influence of schooling is the generalizable result in all of Figures 2 and 4. School quality, and sex differences within school types, are strongly apparent.

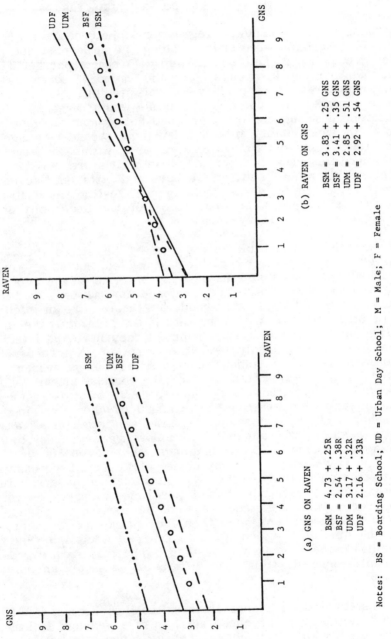

(b) RAVEN ON GNS

BSM = 3.83 + .25 GNS
BSF = 3.41 + .35 GNS
UDM = 2.85 + .51 GNS
UDF = 2.92 + .54 GNS

(a) GNS ON RAVEN

BSM = 4.73 + .25R
BSF = 2.54 + .38R
UDM = 3.17 + .32R
UDF = 2.16 + .33R

Notes: BS = Boarding School; UD = Urban Day School; M = Male; F = Female

Figure 4. Regression Slopes of Raven and Geography/Nat. Study
          Tests by sex and school type (Masbanaland, 1962).

The third comparison method that has received some attention within psychometric technology is the method of examining the stimuli of tests rather than the total scores. Items represent arrays of stimuli ordered by difficulty. Irvine (1964) first produced evidence based on item difficulty correlations to support the contention that test score meanings in Africa might be similar, if not identical to, those assumed for the same test in Britain. Later (1969b) he rejected the notion that comparative identity could be assumed for visually identical stimuli, basing this on factor analysis of Raven's Matrices item difficulty values calculated from separate groups in different cultures, and on factor analysis of subjects who had all scored the same score, exactly half, the items in Raven's Matrices. He concluded that residual cultural variance could be attributed to item difficulties, and that not all subjects used the same strategies to produce "equal" scores. Poortinga's (1971) definitive work on item difficulties in learning tasks and also Mellenbergh's (1972) paper, using the Rasch Model, made the assumption of construct congruence from evidence of similar item-index correlations seem tenuous. Again, there are just too many unverifiable and untestable loose ends in the argument, in particular, the idiosyncrasy of subject strategies, and the effect of second-language instruction on these.

Lastly, the traditional tool of psychometricians has been factor analysis of test score correlations. Without going into detailed technical arguments that have been presented elsewhere (Irvine and Sanders, 1972; Irvine and Carroll, 1980; Irvine, 1979) it has become clear over the years that comparison of data matrices from correlation coefficients derived from two, or even three or four samples can not, by itself, offer proof for the assumption of construct comparability. The simplest argument against this procedureis the one that most psychometricians advance for the scientific use of tests in the first place. Tests simply sample intellectual activities and are subject to sampling error. Even "objective" analytical procedures are subject to experimenter bias (in the choice of procedure) as are the fixing of factor analytic axes in space. The key argument is, of course, an infinite progression. We do not know what may happen to the structure if more and more measures of different functions are taken.

If a dramatic example is needed, the correlations of tests with non-test variables in Mashonaland show that before test scores could logically mean the same, the non-test variables would have to stand in the same relation to tests in Africa as they did outside it. Table 1 shows the failure of measures of socio-economic status, family size, birth order, number of languages spoken to correlate consistently or significantly with achievement and aptitude measures. When the net of measures is extended, the relationship of test scores to environmental (or cultural) variables is inconsistent.

TABLE 1

Correlation of Environmental Variables with Some Aptitude Tests (N=1615)

MASHONALAND, ZIMBABWE, 1962.

Aptitude Tests	Environmental Variables*										
	1	2	3	4	5	6	7	8	9	10	11
Raven's P.M.	02	17	17	-19	09	-06	04	09	13	-09	04
Vocabulary	16	24	26	-10	18	-03	-00	02	00	-15	-04
Arithmetic (operations)	08	20	28	-15	09	-04	-02	03	00	-11	-05
Arithmetic (problems)	15	22	29	-19	21	-03	-04	-05	-05	-14	-06
Geography & Nature Study	25	35	41	-14	32	-04	-00	-04	-07	-14	-05
Headmaster's Estimate	21	37	46	-16	24	-06	-04	-03	-05	-21	-06

*Key:

1. Rural/Urbal Select.
2. Inspect. Est.
3. Teaching Efficiency Est.
4. Age of Pupil.
5. Sex (male high).
6. No. of Siblings.
7. Birth Order.
8. Fath. Occup.
9. Yrs. of Urban Res.
10. Yrs. at School.
11. No. of Lang. Spoken.

Vernon's (1969) exploratory studies in many cultures revealed the same inconsistency of environmental variable-test variable correlations. The assumption of meaning remaining constant--when meaning itself is a function of the similarity of correlation among different classes of dependent variables--cannot be upheld whenever there are other unknowns that have been omitted.

Lastly, and I do not wish to dwell too much or too long on this point, Mashonaland shows what happens when one splits up large groups into smaller, but still viable samples and then performs factor analysis on them? This analysis was conducted on 13 ability and achievement measures administered to the total sample (Irvine, 1964, 1966). One concludes, from the proportions of variance extracted for each factor, that (a) progressively less variance is extracted in the more selective school systems and that the factor analysis of the female scores produced more variation than among the male groups. First factor variance is smallest in the most selective schools. Students of factor analysis would not find this unusual. Once again, the school one attends may make not just a difference to achievement as measured by a test total, but individual differences themselves may be fashioned differently. Knowing how to pass examinations (strategies) may be what is learned in the better schools. Such algorithms may be tools of considerable power and many uses. And the schools' approach to learning may be determined by many influences. In Mashonaland, religious rural/urban, and boarding/day school differences, as well as sex and selection ratios, all played their part (Irvine, 1966).

To sum up, there have been a number of methods used to compare the performance of groups on tests. These include the analysis of mean profiles, regression slopes, item indices and test-score correlations. None of these methods of dealing with dependent variables can resolve the sharp debate that has chara cterized the use of tests with non-white subjects. The limits of argument from attempts to test-score variance by groupings of subjects have been reached. What alternative, and theoretically justifiable avenues are open?

Finding Independent Variables

If debate about test scores is to move from the podium to the laboratory, independent variables by which to control experimentally the performance of groups of subjects on tests will have to emerge from what we know to be consistent findings arising from sustained use of tests in all cultures. A recent review (Irvine, 1979) of 91 factor analyses of tests used around the world revealed some consistend findings. These consistencies emerged in spite of cultural contexts, age, educational and language variation among subjects and also in spite of experimenter bias in testing method, test selection and analysis. These similarities, revealed by secondary source analysis, included typically correlated factors, and verbal (long-term memory), visual (short-term memory) perceptual speed, numerical and "other memory" groupings of tests. Basic intra-hominem structural and control processes in cognition were held to be responsible for the emergence of consistencies in spite of maximal opportunity for diversity. A rationale for using distributive memory theory after Miller (1956), Atkinson and Shiffrin (1968), Hunt, Frost and Lunneborg (1973), and Carroll (1974) to define structural and control processes implicit in perceptual speed factors resulted in an experiment by Gorham (1978) that revealed conclusively how verbal ability, size of task in short-term memory (STM) and strategy transfer all influenced performance on the Wechsler Symbol Substitution Test (WSST). Subjects had three trials on the task, in which code size varied (sets of five, seven or nine digits were randomly assigned to two groups of high and low verbal subjects). One half of the subjects had received previous practice on an analogue of the SST, using the same code set size that they would encounter on the actual test. This practice version was different in both symbol and its substitute (or code) to permit the inference that transfer was confined to strategies for dealing with the task. The four-way analysis of variance of Gorham's data yielded significant main effects for size of task in short-term memory, strategy transfer, and repeated trials. All three main effects interacted significantly with learning, but no other interactions revealed significance. Detailed analysis revealed that many subjects independently stopped using the key always printed at the top of each page and substituted search in memory (the isomorphic search of Shepard and Metzler, 1971). The mean scores of those who recalled accurately all the substitutes given the symbols (at the end of three two-minute trials), were a linear function of the size of the set of symbols to be coded. This coincided neatly with Sternberg's (1966, 1969) findings that the latency to retrieve an item from memory set in STM is a linear function of memory set size. When the SST code is learned, each new digit becomes a probe in a memory set. Subjects using such isomorphic search strategies achieved higher scores than subjects using physical search. Referring to the key at the top of the page took longer and, of course, this search did not involve extended STM capacity. These two self-determined subject groupings had scores that were not equivalent, as far as strategies were concerned. The dependent variable still contained variance unaccounted for by the design of the experiment.

From this experiment, it was clear that constructs such as limits imposed by brain architecture, strategy-transfer and learning during the test were valuable for the scrutiny and interpretation of test performance. The study was therefore replicated entirely in Samoa (Stanko, 1979) with children of the same educational and age range as those in Fort Erie, Ontario. No exact matching of subjects was sought since we considered the independent variables applied in the Fort Erie test to be theoretically robust. The same hypotheses were entertained and the four-way analysis of variance showed that all four main effects revealed significance, but that interactions with trials were confined to the code-size and practice variables. High and Low Verbal students were more clearly identified by Samoan than by Canadian teachers and no interaction between trials and the Hi-Lo Verbal condition was observed. The closeness of fit of the effect of the independent variables on the SST scores give some confidence in the assumption of universal control processes in cognition. However, both experiments under-line how changeable test scores are likely to be in their demands; and how greatly learning involves a re-structuring of these demands. In particular, the role of individual differences in language skills in learning and re-structuring information would seem to warrant more investigation, judging from the Samoan and Canadian results.

Some insight into the relation of language to control proces-ses can, of course, be gained from studies of bi-linguals. A suitable experiment for replication with bi-linguals is the Clark and Chase (1972) task of matching pictures with sentences. In that experiment Clark and Chase demonstrated that the time to encode and compare a verbal message and a pictorial representation of it to determine whether the words and pictures mean the same thing (true condition) or are different (false condition) was predictable from the sum of the various times in separate stages. They also showed that positive sentences took less time to process than negative sentences, that the preposition above in a sentence took less time to encode than below, and that the true condition took less time than the false condition. It seems reasonable to expect that all these findings could be replicated in bilinguals, irrespective of the language used to state the proposition in the sentence. It is also entirely reasonable to expect that the original mother tongue should be a faster vehicle for processing than the second language. The hypothesis receives more than intuitive backing with Carroll and White's (1973) studies of the latencies to recognize pictorial nouns. The earlier the word is learned, the faster the time to recognize it. Subjects whose second language has been acquired after complete literacy in the vernacular can be expected to demonstrate slower times to identify pictorial nouns in that second language. They are probably slower "translators" of information also, which any rusty second-language speaker will verify. Six high-caste female Gujerati-speaking subjects who were completely literate in Gujerati but who had learned to speak, but not write, English after arrival in Africa were the subjects. They were tested by a psychology student of the same caste and sex in their own homes, using portable apparatus (see Note 1). Language and order of presentation of sentences were randomized. Particular care was taken to establish rapport and confidence in the subjects.

In the main, all predicted effects occurred, except that the above-below differences were not in the expected direction for the True-Gujerati sentences. Effects of presentation in first (Gujerati) and second (English) language were most clearly observed in the False-Verification condition. Times were not clearly differentiated in the True-Verification condition, particularly in the "simplest" comparison cases where a sentence with the preposition above was found to be truly represented by the picture that followed it. Hence, a second language presentation in the "simplest" condition made little observable difference. Language of presentation interacts with the meaning of the sentence. This, though, is a statistical finding, and there is little or no theory as yet to explain why. Cognitive operations in a second language, involving positive and negative prepositions and control processes dependent on prepositions from this experiment, predictably carry extra latencies for coding and comparison processes in the second language. Few speeded tests, particularly those involving "reasoning" in short-term memory, whether figural or verbal, avoid this problem if they were originally standardized and administered in a second language. Thus, comparison of means or correlations derived from tests administered to bilingual groups in a second language requires a particularly demanding rationale.

The problems of investigating cognitive control processes on a large scale, of measuring individual differences in these, and of relating these differences to performance on traditional tests, are generated by these experiments. A final example suggests how one might begin in the traditional "perceptual speed" domain, well documented in factor analytic studies around the world (Irvine, 1979). One experiment has been ventured and others are in progress. The Clark and Chase experiment leads us to expect that short-term memory operations involving comparisons of meaningful cognitive material will take less time in simple (unmarked, or same) conditions and longer in complex (different, or marked) conditions. For example, if we compared two arrays of symbols looking for similarities, the operation ought to take less time than looking at the same arrays seeking differences. These are typical task demands of perceptual speed tests. The work of Posner (1969) also leads one to expect that identifying symbols that look alike (AA) should take less time than identifying symbols that mean the same (Aa) but look different. An extra encoding step is needed before any comparisons can be made of symbols that look different but mean the same. By combining the theory of Posner with that of Clark and Chase, one can produce from the theory a series of cognitive tasks varying in coding demands and in comparison demands. If the theory is exact, the latency to complete the most complex task should be predictable, in additive fashion, knowing times for baseline and intermediate conditions, on the assumption that coding precedes comparison. One hundred and four adolescent subjects were randomly assigned to one of four conditions in an experiment that required visual search of two parallel arrays each of four letters in order to compare them. Half of the subjects were instructed to search for and record the number of similar vertical pairs in each array; the other half were instructed to search for and record the number of different vertical pairs. Randomly distributed throughout were booklets with letters all in the same (upper) case, and booklets with letters in which upper or lower case was randomly set on the top or bottom line for each item. Four two-minute trials were given. Analyses of variance showed significant main effect due to case and type of comparison (same-different). No trials effect and no interactions were found.

The results showed clearly that the simplest task is finding same pairs within the same case condition. Both the different case condition and the extra semantic coding step increase time per item. Same-Different comparisions are constant within each case condition. Time for the most complex condition (different pairs, different case), is an additive function of base time (same pairs, same case) and difference between base time and times for comparison within cases and encoding cases within comparisons. The fit between predicted and observed times was close, the model accounting for 99.3 percent of the variation among group means. Individual differences in completing the task were pronounced. At present, studies are in progress to relate individual differences in performance under each of these four conditions to scores on traditional ability tests. One ought to find that individual differences in the most complex condition correlate most closely with ability test scores involving STM operations. The next step after that involves replicating the Canadian findings in different language groups, involving bilinguals (see Note 2). Extrapolation of the principles of semantic encoding, or recoding, and complexity of task demand should allow, perhaps, the construction of test items from theoretical first principles. When that happens, psychometric testing may make theoretical progress beyond the limits imposed by correlating gross dependent variables.

Conclusions

This paper began with the argument that cultural and test variables were both complex and dependent; and that their use to ascribe cause was probably unscientific and possibly illogical. Four traditional ways of comparing test performance were outlined and illustrated. These illustrations underlined the need for new approaches to cross-cultural research using test scores. Some synthetic approaches have been suggested, based on distributive memory theory and the study of test score behaviour under conditions in which verifiable control process operates. Experiments involving strategies, variable loads on short-term memory and learning confirm the consistency of effect on these conditions on symbol substitution tests in Canada and Samoa. The effect of first or second language in encoding verbal and visual stimuli is observable although a theoretical frame for such observations remains to be produced. Finally, the advent of group measures of encoding and comparison processes that verify experimental findings may bring closer a scientific study of individual differences in such processes. To predict behaviour of test items constructed from theoretical principles, and also to predict what cognitive behaviour may be observed in groups of persons differentiated by cultural variables seems closer now than it was ten years ago. The use of psychometric tests in such enterprises, however, can be justified only when their cognitive characteristics have been discovered, verified experimentally, and used as independent variables. And, of course, this paper has not addressed itself to the equally compelling task of finding theoretically derived procedures by which cultural variables can undergo the same relentless verification.

References

Atkinson, R. C. and Shiffrin, R. M. Human memory: a proposed system and its component processes. In K. W. Spence and J. T. Spence (Eds.), Advances in the psychology of learning and motivation: research and theory Vol. 2. New York: Academic Press, 1968, pp. 89-195.

Carroll, J. B. Psychometric tests as cognitive tasks: a new structure of intellect. Research Bulletin 74-16. Princeton, NJ: Educational Testing Service, 1974.

Carroll, J. B. and White, M. N. Word frequency and age of acquisition as determiners of picture-naming latency. Quarterly Journal of Experimental Psychology, 1973, 25, 85-95.

Clark, H. H. and Chase, W. G. On the process of comparing sentences against pictures. Cognitive Psychology, 1972, 3, 472-517.

Gorham, R. Verbal ability, previous practice and load on short-term memory as determiners of differences in a complex learning task: an experimental study. Unpublished M.Ed. Thesis, Brock University, St. Catharine's, Ontario, Canada, 1978.

Hunt, E., Frost, N., and Lunneborg, C. Individual differences in cognition: a new approach to intelligence. In G. Bower (Ed.), The Psychology of Motivation and Learning, Vol. 7.

Irvine, S. H. A psychological study of selection problems at the end of primary school in Southern Rhodesia. Unpublished Ph.D. Thesis, University of London, England, 1964.

Irvine, S. H. Towards a rationale for testing attainments and abilities in Africa. British Journal of Educational Psychology, 1966, 36, 24-32.

Irvine, S. H. Culture and mental ability. New Scientist, 1969,

42, 230-231.

Irvine, S. H.   Figural tests of reasoning in Africa.   Studies in the use of Raven's Matrices across cultures.   International Journal of Psychology, 1969, 4, 217-228(b).

Irvine, S. H. and Sanders, J. T.   Logic, language and method in construct identification across cultures.   In L. J. Cronbach and P. J. D. Drenth (Eds.), Mental Tests and Cultural Adaptation. The Hague: Mouton Press, 1972.

Irvine, S. H. and Carroll, W. K.   Testing and assessment across cultures: issues in methodology and theory.   In H. C. Triandis, W. W. Lambert, and J. W. Berry (Eds.), The Handbook of Cross-Cultural Psychology.   Boston: Allyn and Bacon, 1979.

Irvine, S. H.   The place of factor analysis in cross-cultural methodology and its contribution to cognitive theory.   In L. Eckensberger (Ed.), Cross-Cultural Contributions to Psychology. Lisse, Netherlands: Swets and Zeitlinger, 1979 (in press).

Jensen, A. R.   Interaction of Level 1 and Level 2 abilities with race and socio-economic status.   Journal of Educational Psychology, 1974, 66, 99-111.

Lesser, G. S., Fifer, G., and Clark, D. H.   Mental abilities of children from different social-class and cultural groups.   Child Development Monographs, 1965, 30, No. 4 (whole number).

Mellenbergh, G. J.   Applicability of the Rasch model in two cultures. In L. J. Cronbach, and P. J. D. Drenth (Eds.), Mental Tests and Cultural Adaptation.   The Hague: Mouton, 1972, 453-458. Miller, G. A.   The magical number seven, plus or minus two: some limits on our capacity for processing information.   Psychological Review, 1956, 63, 81-97.

Miller, G. A.   The magical number seven, plus or minus two: some limits on our capacity for processing information. Psychological Review, 1956, 63, 81-97.

Ord, I. G.   Testing for Educational and Occupational Selection in Developing Countries.   Occupational Psychology, 1972, Vol. 46, Number 3, Monograph Issue.

Poortinga, Y. H.   Cross-cultural comparision of maximum performance tests: some methodological aspects and some experiments with auditory and visual stimuli.   Psychologica Africana, Monograph Supplement No. 6, 1971.

Posner, M. I., Boies, S. J., Eichelman, W. H. and Taylor, R. J. Retention of visual and name codes of signal letters.   Journal of Experimental Psychology Monograph, 1969, 79, No. 1, Part 2.

Shepard, R. N. and Metzler, J.   Mental rotation of three-dimensional objects.   Science, 1971, 171, 701-703.

Stanko, M.   Unpublished M.Ed. Project Report, Brock University, St. Catherine's, Ontario, Canada.

Sternberg, S.   High speed scanning in human memory.   Science, 1966, 153, 652-654.

Sternberg, S.   Memory scanning: mental processes revealed by reaction-time experiments.   American Scientist, 1969, 4, 421-457.

Stodolsky, S. and Lesser, G.   Learning patterns in the disad-
     vantaged.   Harvard Educational Review, 1967, 37, 546-593.
Thoday, J. M.   Limits to genetic comparision of populations.
     Journal of Biosocial Science, (Supplement) 1969, 1, 3-14.
Vernon, P. E.   Intelligence and cultural environment.   London:
     Methuen, 1969.

Reference Notes

1.   Nisha Patel carried out this experiment as a project for a
      psychometrics course with some advice from me, during a
      sabbatical leave I spent in the University Department of
      Psychology. Adam Latif carried out the same experiment with
      Khachi (Moslem) subjects, with much the same results. His
      study was unique because the Khachi language has no known
      orthography.

2.   At the time of writing, the following groups have been tested:
      Afrikaans speakers; African, Asian and coloured groups. The
      research was carried out in cooperation with the Human Scien-
      ces Research Council, South Africa. Preliminary results are
      promising.

HUMAN AGEING AND DISTURBANCES OF MEMORY CONTROL PROCESSES UNDERLYING "INTELLIGENT" PERFORMANCE OF SOME COGNITIVE TASKS

Patrick Rabbitt

University of Oxford

Oxford, England

This essay is an attempt to suggest lines along which we may begin a new kind of analysis of some of the cognitive operations which humans find necessary in order to carry out simple cognitive tasks. It is based on the general premise that most cognitive tasks which people perform in everyday life—that is most tasks which require the sort of behaviour which we call "intelligent"—depend on the efficiency with which people can briefly remember, and rapidly retrieve, information about events which they have just experienced, actions which they have just performed, or information about the results of computations which they have just completed. A second premise is that individual differences in cognitive efficiency may be partly related to individual differences in the efficiency with which information about recent events, actions, and computations can be stored, indexed and retrieved in immediate memory systems, and may also be related to differences in the efficiency with which information held in immediate memory can be used to control, to access, or to "index," other information which is held in long term memory. A last premise is that neither the processes underlying most cognitive skills, nor differences in the efficiency with which individuals carry out these processes can be properly understood if we continue to consider immediate memory simply as an elementary, passive buffer for the storage and retrieval of information. We must rather learn to consider immediate memory as an active system in which new items of information are continually received, transformed, up-dated, re-indexed and re-combined. When these

427

premises are thus presented as abstract generalisations they are, no doubt, irritatingly vague and convey to the reader no more than a sense of particular, ill-formed prejudices. They are perhaps better introduced in terms of four distinct classes of experiments in terms of which the author and his associates have tried to give them concrete definition.

1. Performance on paced serial addition tasks

From 1975 to 1977 Caroline Thomas and the author made a series of investigations of the after effects of mild concussion on cognitive performance (see Thomas, 1977). Previous investigations, by Gronwall (1977) and Gronwall and Sampson (1974) among others, had suggested that performance on a task involving paced serial addition (PASAT) was very sensitive to the after effects of mild closed-head injury. In this simple task strings of single digits are played, one at a time, at rates ranging from 1/1.5 secs to subjects who have to mentally add each digit to its immediate predecessor in the string and then to call out each resulting sum in turn. Note that each digit, except the first, is thus added, in turn, to the one before it and to the one after it. A series of successive digits, with correct answers, can be represented as follows:

Presented series of digits: 4, 9, 3, 7, 1, 2, 6 . . . etc.

Correct answers: -13, 12, 10, 8, 3, 8 . . . etc.

Gronwall (1977) and Gronwall and Sampson (1974) found that patients suffering from the after-effects of closed head injuries performed worse than normal controls, particularly when strings of digits were presented at fast rates. Thomas (1977) was unable to replicate these findings, possibly because the patients she tested had suffered much milder concussions than those investigated by Gronwall and associates. However, Thomas carried out a scrupulous and insightful analysis to distinguish between the various kinds of errors which her patients and their controls made while attempting to perform this simple task. Most errors were simple omissions. That is, subjects appeared to be unable to keep up serial additions at fast presentation and would gradually lag behind until they were forced to omit one or more successive additions so that they could catch up again. Most other errors were simple failures of arithmetic. However two particular classes of errors, though rare, were of particular interest because they showed that the demands made on immediate memory by continuous serial addition were not met by any simple, passive, short-term buffer system. Subjects occasionally made mistakes because they added a current digit to the last <u>total</u> which they had announced (that is, to their own last response rather than to the last digit which they had heard (type A errors). Thus:

Presented series of digits: 4,   9,   3,   7,   1,   2,   6

Answers: - 13, 16, 10,   8, 10,   8 etc.

An even more interesting type of error occurred when subjects failed to add a current digit to the last digit presented and instead added a current digit to the digit which had preceded it two places earlier in the sequence (type B errors), thus:

Presented series of digits   4,   9,   3,   7,   1,   2,   6

Answers: - 13,   7, 10,   8,   9,   8, etc.

In Thomas' (1977) careful study, since head injured and hospitalised control patients had much the same overall error rates, it was not surprising that they also did not differ in terms of the proportions of different kinds of errors which they committed. However the fact that these kinds of errors should occur at all is of considerable theoretical interest. In order for a type A error to occur a subject must have correctly identified the last and the preceding digit (because he had already correctly added it to the last-but-one digit to give a correct answer). He also must have correctly identified the current digit, because he evidently adds that digit, and no other, to his previous answer in order to achieve a particular, identifiable, erroneous report. His mistake must therefore lie in selecting the wrong number out of several simultaneously held in short term memory in order to use it as a component in his new addition. This could be because his last answer had occurred more recently in time than the last digit with which he had been presented. A simple, passive theory of immediate memory might thus suggest that his mistake occurs because he selects a (more recent) item from short term memory, with a (consequently) "stronger trace" rather than a more remote item, the memory trace for which has become unavailable because its "strength" has declined due to lapse of time and interference from intervening cognitive operations (e.g., the act of addition and the act of reporting a new total): see Posner and Rossman (1965).

This rather clumsy line of explanation cannot serve in the case of type B errors. Here subjects appear to add a current digit to a more remote item, which they have just heard and correctly registered. In this case, as with type A errors, we know that the last item must have been correctly perceived because the subject has used it to arrive at his correct answer on the previous trial. Thus we cannot suppose that such errors represent simple substitutions of "stronger" for "weaker" traces in retrieval from a passive, temporary buffer storage system.

They must rather represent intermittent failures in the operation of a complex control process which operates somewhat along the lines of that sketched in Figure 1. That is, they represent failures of a process which continuously indexes and updates items in an active working memory (e.g., see Baddeley, 1976) in which items are held, re-labelled, and discarded according to moment-to-moment changes in the demands of a simple, repetitive cyclical sequence of necessary operations.

The present author and others of his associates undertook further series of experiments in order to see which particular factors might "drive" subjects to commit larger proportions of these kinds of errors in comparison to the proportions of simple failures of arithmetic. Changes in presentation rate do not substantially increase the number of errors due to failures in arithmetic although omissions increase dramatically as digit sequences are presented at faster and faster rates. Two factors were found to do this--distraction from secondary tasks and greater chronological age of subjects tested. When subjects performed a PASAT task and a distracting, serial, self-paced choice reaction task simultaneously, all categories of errors increased, but type A and type B errors increased more than others. Similarly, fit, community-resident elderly (70 to 82 year old) people made higher percentages of type A and type B errors (7.0% to 8.0%). than did comparison groups of young subjects (18 to 22 years old; 3% to 4%). Subjects were pair-matched for socio-economic class and adjusted IQ scores on Raven's matrices and Mill Hill Vocabulary tests. This comparison suggests that a

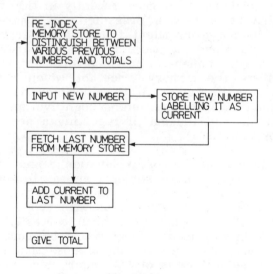

Figure 1

distracting secondary task cannot simply reduce the information channel capacity available for rapid mental arithmetic (see lucid discussions of limitations to dual task performance in these terms set out by Poulton, 1970). Nor can the secondary task simply accelerate the rate of decay of item traces in immediate memory consequent on reduction of channel capacity necessary for rehearsal (see Posner and Rossman, 1965, Posner and Konick, 1966). The secondary task seems rather to interfere specifically with complex control processes which are necessary to index, update and select among events registered in working memory. The same points apply when we consider the nature of the decline in performance which accompanies advancing age. In other words, as age advances, the efficiency of control of working memory declines, and this decline in control is at least partly responsible for decline in efficiency at a particular cognitive task. Of course, this is not to say that the effects of advancing age and the effects of distraction are identical. However, both age and distraction reveal limitations in the same, complex, control process, so that we see that the effects of neither age nor distraction can be fully understood except in terms of models which include descriptions of this control process, and of the various ways in which it can fail.

2. Multiple indexing of information in working memory

The results which we have just discussed suggest that some limitations to performance in cognitive tasks occur because of failures in the efficient control of working memory. They also suggest that such failures may be associated with failures to update, and to index successive events during a continuous, complex sequence.

In her useful examination of cognitive changes following mild concussion Thomas (1977) found evidence for another kind of failure in the efficiency of active working memory. This can best be described as a failure to index a series of successive events in more than one way so that they can be recalled, at option, in at least two different orders.

Thomas (1977) employed a "keeping track" task in which subjects sorted packs of 56 cards. Each of these cards carried a representation of an item from one of 4 categories. Cards (and so categories) were in random order. Categories were "numbers" (digits 1 to 10), "letters" (from A to K excluding I), "Colours" (colour names written in the corresponding coloured inks) and "shapes" (heart, diamond, club, spade, rectangle, square, cross, star and triangle drawn in black ink). Cards were sorted into four piles corresponding to these categories. Each card was inspected for sorting and then placed face-down so that the symbol on it was no longer visible. At random intervals, as the pack was sorted, 16 "question" cards appeared. Each question card interrogated a particular category pile. On encountering such a question card the subject stopped sorting and immediately tried to recall as many items as possible from the required category (i.e., pile), in the order in which they had been sorted onto that pile. Thomas found that this task was more sensitive to the after-effects of mild head injury than any other in the battery which she employed.

Rabbitt and Vyas (1978, unpublished) adapted the task for comparisons between young (19 to 30 years) and elderly (65 to 74 years) people. In this case subjects were not matched on IQ tests, though groups were of closely comparable socio-economic status. This choice was made because it was considered that the most appropriate way in which old and young subjects could possibly be matched was in terms of task performance. Subjects were shown sequences of coloured shapes presented on a Sony domestic colour TV set coupled to an Apple II microprocessor. Random sequences of coloured shapes appeared, one every 3 seconds in random order, at any of 4 locations of a rectangular 2 x 2 matrix as illustrated in Figure 2. During a pre-test session subjects were simply asked to recall the shapes, and their colours, in the temporal order of their appearance, irrespective of changes in the locations at which they appeared. This allowed particular pairs of old and young people to be matched in terms of their "temporal order immediate memory span" for successively presented coloured shapes. Subjects were then run on two further conditions. In the first they were told, before each trial, whether they were to be asked to recall which coloured shapes had occurred, in temporal order, on specific display locations, or whether they were rather to be simply asked to recall all shapes in the temporal order in which they had occurred, irrespective of the locations in which they had appeared. In the second condition they were not told, before

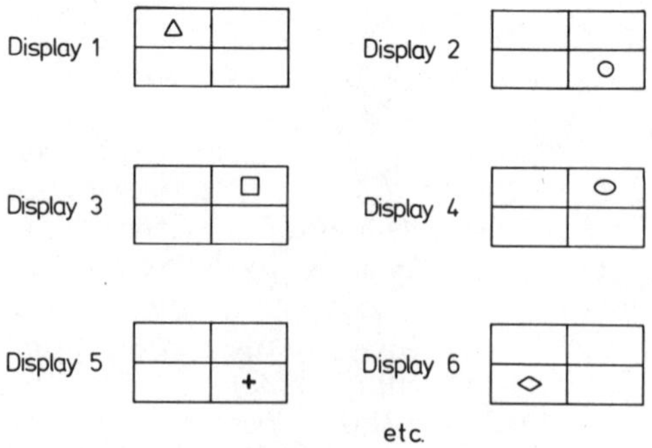

Figure 2

each trial, whether they would be asked to recall by location or by temporal order.

The results were complex, but are broadly consistent with the general statement that when young and old subjects are equated on their straightforward, simple memory span for successively presented coloured shapes the old nevertheless do worse than the young if they are required to recall by display location. This difference between temporal recall and location recall is accentuated when subjects do not know, before each trial, which they are to use.

Once again it would be clumsy to describe this change in efficiency in cognitive tasks as simply the result of an accelerated decay of memory traces over time, or as the result of increased vulnerability of memory traces in a "passive" intermediate term storage register to interference from concurrent activity. The change in performance which takes place as people grow older seems, at least partly, to reflect a decline in the efficiency with which they can "cross index" a series of events in working memory to allow themselves the option of retrieving them in terms of more than one possible order (e.g., in these experiments in both spatio-temporal order and in temporal order alone).

It is interesting to consider that performance on a wide range of cognitive tasks must depend on this ability to index information about the recent past so that it can be retrieved in more than one way. The implications of a particular event may not be immediately apparent at the moment at which it occurs and is perceived. These implications may only gradually become apparent as subsequent events supply a context in which it can be interpreted. Any reduction in the efficiency of multiple indexing of events must therefore present a corresponding decline in the ability to seek and discover multiple implications in a series of recent events.

3. Variations in the efficiency of control of learned sequences of responses by reference to information stored in long-term memory.

Consider how a man carries out some very familiar sequence of responses, such as repeating the alphabet aloud. He has learnt the alphabet many years ago, and can repeat it whenever he wishes, so we must assume that he has stored the information necessary to do this in long-term memory. However he would hardly manage to get through the alphabet unless he could also remember, at least momentarily, what he has just said (in other words unless he has, at least, momentary short term memory for his last response). It is not too fanciful to compare the representation of the entire alphabet in his long term memory to a linear, non-branching programme of instructions to repeat this particular

series of letters. But such a programme must be "indexed" at
every response he makes so as to obtain the information necessary
in order to decide what to say next.

The alphabet, considered as a linear programme, is
particularly easy to index because each letter is different and
always occurs in the same unique location. This means that in
order to discover which letter to say next a man need only
remember which letter he has just said. The memory of this
unique letter is sufficient information to guide him to the correct
unique location in the programme at which he can discover the
next response. Not all sequences of operations which a man must
learn and use to guide himself through the world are of this
simple type. Consider a sequence of the type

A, B, C, D, B, E, F, C, B, G, H, I   etc. etc.   (No. 1)

In this case in order to know what to do after he has said
either B or C a man must at least remember one previous
response. It is obvious that sequences containing repeats can
become indefinitely demanding of immediate memory, so that a man
may have to correctly remember from 1 to N previous responses
in order to be sure which point he has reached during a long
string of responses. As there are many ways in which this
memory load can be increased so there are also many coding
devices a man may use in order to reduce the load or economise
on it. For example in the finite sequence:

A, B, C, A, B, D, A, B E, A B,F   (No. 2)

a man can encode his sequence as the repetitive sequence A, B
plus one member at a time of the sequence C, D, E, F, -
provided that he can count and update four successive cycles of
responses. A more laborious alternative to some such trick of
reduction coding would be to remember a backwards sequence of
as many as 10 individual items in order, so as to be able to
recognise the point at which the last letter, F, followed the final
appearance of the letter B.

When items in a sequence are repeated it will always be
necessary to carry some short-term memory load in order to be
sure that one can always find one's place. Thus the relative
difficulty of indexing sequences may vary in terms of the length
of the backward span (or in terms of the complexity of the
necessary coding tricks) which they require. There is another
source of difficulty; the degree of "embeddedness" of repetitive
strings within a sequence. Consider the two sequences that
follow:

A, B̲, C, D, B̲, E, F̲, G, H, I, F̲, J, K, L ... etc. (No. 3)
A, B̲, C, D, F̲, G, H̄, I, F̲, J, B̲, K, L ...   etc. (No. 4)

In both sequences strings of letters intervening between repeated
letters can be regarded as "loops" which a man must traverse in
his pathway through a sequence (see also No. 3) and in terms of
which he must remember his current position. In sequence No. 3
the two loops --the loop between the two occurrences of the letter
B and the loop between the two occurrences of the letter F-- are
independent, and are separated by the string of other letters, G,
H and I. In sequence No. 4 the string of letters between the
two Fs falls within the string of letters between the Bs. We can
consider the F loop as being embedded within the B loop. Note
that embedded loops do not, per se, mean that it will be necessary
to remmeber more or fewer letters in order to index all sequence
points (in both sequences the same span of one backward letter is
all that is necessary to locate either occurrence of B or F). But
if subjects try to use more complex strategies of "loop counting,"
rather than the straightforward strategy of simply registering the
minimum necessary backward span of letters in order to locate
their position, any increase in the degree to which loops in a
sequence are embedded within each other, or overlap with each
other, may make their task more complex.

Rabbitt and Heptinstall (1976, unpublished) carried out
series of experiments to compare performance with sequences of
responses in which all elements were unique and non-recurrent
with sequences in which the loop-structure was more or less
embedded. Before the experiment sequences of 15 to 40 responses
were learnt to a criterion of 300 successive, flawless repetitions.
Elderly (60 to 80 year-old) and young (20 to 30 year-old) subjects
were matched in terms of family (they were grandparents and
grandchildren) and in terms of verbal IQ (Mill Hill scores). They
then ran through sequences under conditions of distraction from a
secondary task (counting backwards in sevens). Subjects of all
ages made more errors on sequences containing repetitions than
on sequences composed of uniquely occurring items. Performance
was also worse when loops within a repetitive sequence were
embedded (as in No. 4) than when they were separated (as in
No. 3 above). Elderly subjects were more affected than the
young by distraction, and they were relatively more affected than
the young by the occurrence of any repetitions, by increases in
the memory load necessary to deal with repetitions, and by increases
in the degree of embeddedness.

Again it seems that the effects of distraction on performance
of a simple cognitive task must be evaluated in terms of a model
for an active control process by means of which people briefly
hold information in immediate memory and use it to index information which is permanently available in long term memory. Differ-

ences in the relative difficulties of various sequences (tasks) can
be interpreted in terms of differences in the load which they
place upon these memory control processes, and in terms of the
relative ease or difficulty of the reduction-coding mechanisms
which they allow subjects to use in order to reduce this load. It
seems that some changes in cognitive efficiency which accompany
advancing age can also be interpreted in terms of reduction of
efficiency of the processes by which information stored in an
active, working memory, including information obtained by active
cognitive encoding devices such as counting cycles, can be used
to index more enduring programmes, plans, or rules for guidance
through familiar tasks.

In simple terms it seems that old people sometimes make
mistakes during complex repetitive tasks because, although they
can accurately hold in memory the rules which they have to follow
(in more picturesque terms, although they retain the benefits of
previous experience) they cannot access, or index, these available
rules (experience) when under stress from a simultaneous task.
It remains a matter for conjecture whether, even when they are
not stressed by distracting tasks, old people gradually become
incapable of indexing as complex sequences, or of employing as
subtle or complex coding rules to index these sequences, as are
the young. In other words, they retain their valuable life experience but are gradually denied useful access to it!

4.   Memory for previous attempts at solution of complex problems

When problems are complex people are characteristically
unable to find a solution at the first attempt. Indeed they sometimes only gradually appreciate the nature of a problem by successively trying out, and discarding, series of attempts at a solution.
In this case the ability to remember, or to profit from, attempted
solutions which have been identified as inadequate will be an
obvious advantage. If a man, or a computer programme, does not
keep a record of all attempts at a solution which have been made
and found inadequate, there is an obvious danger that the same
unsatisfactory attempts might be endlessly repeated so that no
advance is made. It also follows that the better the memory for
previous attempts the more efficient performance will be, since
partially successful solutions may be merged to provide effective
answers. An opportunity occurred to study how old and young
subjects employ their memory for previous attempted solutions to
guide their play at chess against a small computer.

An Apple II computer was programmed with a chess-playing
programme written by Peter Jennings, now commercially available
as "Microchess 2". An excellent graphical representation of a
chessboard appears on a TV screen and the human player makes
his moves by typing out algebraic notation coordinates on a

keyboard. The designated "piece" on the display board then moves, and, after a pause, the machine replies. The machine takes more or less time to reply depending on which of 8 levels of "look ahead" the programme is instructed to employ. The programme can generally be beaten by quite mediocre chessplayers, but offers a challenge for players whose chess has become rusty or who have never learned to play well. Differences in the levels of competence of the chess playing programme were used to match pairs of young (17 to 22 year-old) and elderly (70 to 85 year-old) amateur chessplayers. A player's "level" was taken as that programme level <u>below</u> the one at which he lost against the machine on about 80% of games. It conveniently fell out that at this level players would win from 40% to 60% of games against the machine. This allowed us to closely match young and old players for comparability of competence in play against the computer programme.

All players greatly enjoyed playing against the machine, and were easily persuaded to discuss their strategy, move by move, describing to the experimenter the problems which they saw as each position developed, and their plans to deal with these problems as they came up. For present purposes we shall consider only one feature of the data. When planning each move players were encouraged to consider, and describe aloud, a number of different possible moves, and analyse in turn the consequences of each, as far as they could work them out. They would, of course, eventually have to decide on one of these possible moves and make it. A first point was that our matched old and young pairs did not differ in terms of the number of moves ahead for which they could project any particular line of analysis. The number of moves ahead to which they habitually analysed their possible moves correlated well with their level of competence in play against the machine (as of course, given the nature of the programme, it should have done). Nevertheless two differences between younger and older players were striking. First, young players tended to be more adventurous, or "creative," and to consider the possible outcome of a much larger (2.4x) number of possible moves in each position. More interestingly, players of all ages made an interesting category of avoidable errors in their selection of moves. These occurred when a player would consider and analyse a superficially tempting move and rapidly discover, and correctly announce, that it came to nothing or was actually dangerous. He would then go on to consider a number of other possible moves. If none of these proved promising he might then return to reconsider the move he had earlier analysed and rejected, fail to analyse its consequences as thoroughly as he had on his previous inspection, and make the move. Often to instantly, and vocally, repent of his oversight! Errors of this type were four times as common among older players as among their young, pair-matched controls. Our suggestion is that the older players were not noticeably deficient in power of depth of analysis of a sequence of moves,

but that they tended, more often than the young, to forget
outcomes of their earlier analyses or solutions. This handicapped
them in two ways: As we have seen it made them more likely to
impulsively accept moves which they had already considered and
rightly rejected. Equally unfortunately, it limited them in their
consideration of moves, so that they could not relate lines of play
which had emerged from analysis of one sequence to lines of play
which emerged from consideration of another. Thus an attempted
attack or defence was often bungled because they failed to perceive
that two or more moves necessary to complete it were closely
related in the logical structure of the position in as much as they
were both necessary to reach a goal, but would work if made in
one order, and not if made in any other. For our present purposes this served as a simple demonstration that the power of
problem-solving, and the efficiency of strategic approaches towards
a solution of a problem, may rest very heavily upon the efficiency
of immediate memory, and on the flexibility with which two or
more different lines of analysis may be held in immediate memory
and become available for re-combination as an effective solution.

Conclusions

A rude aphorism current among cognitive psychologists is
that psychometricians know all about intelligence and nothing
whatever about the way in which human beings go about solving
problems. A suitable rejoinder might be that cognitive psychologists do not even know anything about intelligence. The exchange
is undignified, but there is some point in considering what lies
behind it. With recent noteworthy exceptions (see Hunt and
Sternberg in this volume) psychometricians have contributed little
to our understanding of the functional processes which underlie
intelligent behaviour. In spite of their lip service to the desirability of such models, cognitive psychologists have also not
contributed a great deal to our grasp of these processes. Cognitive psychologists may have been particularly handicapped
because they have, until recently, refused to adapt their models
to consider individual differences of any kind. The very pertinent
insights of psychometricians into the nature of such differences
have passed unused because of this failure in communication.
Cognitive psychologists, and psychometricians, may have been
equally handicapped because they have set their minimal requirements for a theory of problem solving at an unrealistically high
level. It will certainly be a great day when we can point to a
"unified field theory of intelligent behaviour." Most of us are not
optimistic that such a day will arrive within our working lifetimes.
Until then we may have to be content with more modest models for
the particular functional operations necessary to carry out particula
simple tasks. This essay has attempted to see how far we can
take a simple premise that efficiency of immediate memory is
necessary for successful performance of many tasks--even of some

tasks in which such success is usually accepted as a sign of "Intelligence." Our first development of this premise has showed us that individual differences in performance of such "intelligent" tasks may be partly related to differences in the efficiency of immediate "working" memory. In studying the role of immediate memory in these tasks we have been obliged to go beyond models based on the idea of immediate memory as a "passive" temporary storage register for necessary items of information, and to develop the idea that immediate memory also involves complex, active control processes which allow us to meaningfully organise our responses to the environment from moment to moment in relation to events which have just occurred. Perhaps this modest gain will be sufficient to encourage more able investigators to take the matter further.

References

Baddeley, A. D. (1976). The Psychology of Memory. Harper and Row, New York, San Francisco, London.

Gronwall, D. M.A. (1977). Paced auditory serial addition – A measure of recovery from concussion. Perceptual and Motor Skills, 44, 367-373.

Gronwall, D. M. A. and Sampson, H. (1974). The Psychological effectsof concussion. Auckland University Press/Oxford University Press, Oxford, England.

Posner, M. I. and Konick, A. F. (1966). On the role of inter-ference in short-term retention. Journal of Experimental Psychology, 72, 221-231.

Posner, M. I. and Rossman, E. (1965). Effect of size and location of information transforms upon short-term retention. Journal of Experimental Psychology, 70, 496-505. Poulton, E. C. (1970). Environment and Human Efficiency. Charles Thomas, Springfield, Illinois.

Thomas, Caroline M. (1977). Deficits of Memory and Atten-tion following closed head injury. Unpublished MSc thesis, University of Oxford.

Poulton, E. C. (1970). Environment and Human Efficiency. Charles C Thomas, Springfield, Illinois.

Thomas, Caroline M. (1977). Deficits of Memory and Attention fol-lowing closed head injury. Unpublished MSc thesis, University of Oxford.

Footnote

The author wishes to express thanks to the Social Sciences Research Council who supported this work under grant No. HR 6258.

ABILITY FACTORS AND THE SPEED OF INFORMATION PROCESSING

Marcy Lansman

University of Washington

Seattle, Washington, U.S.A.

Abstract

Measures based on two models from cognitive psychology, the Clark and Chase model of sentence verification and the Shepard and Metzler model of mental rotation, were related to ability factors of the Horn-Cattell theory of fluid and crystallized intelligence. Analysis of individual differences was also used to test the internal consistency of the two cognitive models. The individual differences analysis cast serious doubt on the validity of the sentence verification model, and measures derived from that model were only very weakly related to the ability factors. On the other hand, the model of mental rotation was supported by the individual differences analysis, and a strong relationship was found between parameters of the model and the visualization factor of the Horn-Cattell theory. Possible reasons for the failure to find process explanations for verbal ability factors are considered.

The theoretical constructs of experimental and differential psychology are derived from different kinds of analyses. Evidence for the information processing models of cognitive psychologists is based on the effects of experimental manipulations on group means. The dimensions of human ability hypothesized by differential psychologists are based on analysis of performance differences between subjects. One point where there seems to be some correspondence between the dimensions of cognition as defined by cognitive and differential psychologists is in the distinction between verbal and spatial processes. In this study, we attempted to relate process and ability measures in each of these two areas. We looked at the relationship between ability factors associated with the Horn-Cattell model of fluid and crystallized intelligence and process parameters associated with two well-known experimental paradigms: the sentence verification paradigm used by Clark and Chase (1972), and the mental rotations paradigm introduced by Shepard and Metzler (1971). The main purpose of the study was to find out whether the reaction time parameters derived from cognitive models of the two experimental tasks were correlated with factor scores based on a series of paper and pencil measures of the ability factors. A second purpose was to use individual differences analyses to test the internal consistency of the models for the two experimental tasks.

Our battery of psychometric tests was chosen to define four factors from the Horn-Cattell theory of fluid and crystallized ligence: fluid intelligence (Gf), crystallized intelligence (Gc), visualization (Gv), and clerical and perceptual speed (CPS). The crystallized intelligence factor reflects the influence of acculturation and education, especially in the area of verbal skills. Measures of verbal knowledge, such as vocabulary tests and tests of general information, typically load on this factor. Fluid intelligence involves the ability to solve new problems for which the subject has no learned strategy. Tests of fluid intelligence generally involve what we would call complex reasoning. An example is the letter   series test, in which the subject is asked to study a series of letters which was formed by some rule and to decide which letter comes next in the series. The visualization factor is closely related to what we more commonly call "spatial ability," and involves the ability to manipulate a visual image. We included a number of tests expected to load on the Gv factor in our battery, since we wanted to find out whether measures derived from the Shepard and Metzler mental rotations tasks were specifically related to tests of visualization. We also felt it was important to include in our battery tests of the clerical and perceptual speed factor, since most of our process measures involved reaction time.

Our subjects were 84 college students, 42 male and 42 female. Since we were interested in sex differences in spatial ability, male and female subjects were each selected to represent a stratified sample of the college population based on the distribution of scores for students of that sex on a college entrance examination of spatial ability. Each subject participated for 10 days, one hour each day. The first four days were devoted to paper and pencil testing. The following six days were devoted to computerized measures of sentence verification, mental rotation, and several other tasks. The mental rotation task was by far the most time-consuming, involving four one-hour sessions of computerized testing.

Discussion of specific methods and results will be orga..ized as follows: I will first discuss analysis of the psychometric battery, then present method, group results, internal consistency analysis, and correlations with psychometric factors for each of the two experimental tasks separately.

The 16 psychometric tests were subjected to an initial principal components factor analysis. As expected, a four factor solution best described the data. Squared multiple correlations were then inserted in the diagonals, and the four principal factors were subjected to a Varimax rotation. The resulting factors were quite clearly interpretable. Table 1 shows the factor loading

Table 1

Factor Loading Matrix for Psychometric Tests

	Factor 1 (Visualization)	Factor 2 (Crystallized Intelligence)	Factor 3 (Clerical & Perceptual Speed)	Factor 4 (Fluid Intelligence)
Letter Series	-.02	.05	.05	.56
Matrices	.11	.03	.09	.34
Common Analogies	.08	.22	-.18	.39
Remote Associations	.01	.36	.26	.07
Esoteric Analogies	.06	.75	.00	.20
Vocabulary	.02	.70	-.09	.12
General Information	.20	.66	.02	-.06
Form Board	.63	.10	.08	.05
Surface Development	.76	.00	-.09	.19
Paper Folding	.50	.06	-.26	.20
Cards	.76	.05	.20	.02
Figures	.71	.05	.37	-.03
Cubes	.71	.15	.24	-.02
Identical Pictures	.17	.11	.55	.00
Cancelling Numbers	.10	-.12	.49	-.08
Finding A's	.00	.05	.58	.32

matrix.   Factor I can easily be identified with Gv, visualization, since all six tests of spatial ability loaded highly on this factor. The principal loadings on Factor II were Vocabulary, General Information, and Esoteric Analogies (a test involving analogies drawing upon sophisticated knowledge of a variety of content areas), with a test called "Remote Associations" also loading on this factor.   Thus Factor II is clearly identified as crystallized intelligence, Gc.   Factor III was identified with clerical and perceptual speed, CPS, since all three tests of clerical speed loaded on this factor.   Factor IV had moderate loadings on Letter Series, Matrices, and Common Analogies (a test involving easy words but subtle relationships), so it was identified as fluid intelligence.   Fluid intelligence was the least well-defined of our four factors.

To avoid capitalizing on sampling error, we computed scores on each factor as the unweighted sums of the standardized scores on the tests which loaded most highly on that factor.   The tests that were summed to form each of the factor scores are underlined in Table 1.

The Sentence Verification Task

The computerized sentence verification task was the same as that used by Clark and Chase (1972).   The subject first saw a fixation point, which served as a warning signal, for 500 milliseconds.   Following the warning interval, a sentence and a picture appeared simultaneously.   For example, the subject might see the sentence, "Star below plus," and a picture of a star below a plus. Th subject had to decide if the sentence was a true description of the picture and respond by pressing one of two keys with either right or left index finger.   Feedback consisted of the subject's reacton time if the response was correct and the word "no" if the response was in error.   The sentence could be stated affirmatively or negatively, and the correct response could be either true or false.   Equal numbers of sentences contained the unmarked preposition "above" and the marked preposition "below." Each subject completed four blocks of 80 trials each.   Data from only 72 subjects was available from the sentence verification task.

We first analyzed the data to find out how well the group means fit two well-known models of the sentence verification task, the Clark and Chase (1972) model and Carpenter and Just (1975) model.   The Clark and Chase model is illustrated in Table 2.   According to this model, subjects first encode sentence and picture in propositional form. They then perform two comparisons: They compare the embedded strings of the two propositions, then the embedding strings.

Table 2

The Clark and Chase Model of sentence verification.
(Adapted from Clark and Chase, 1972.)

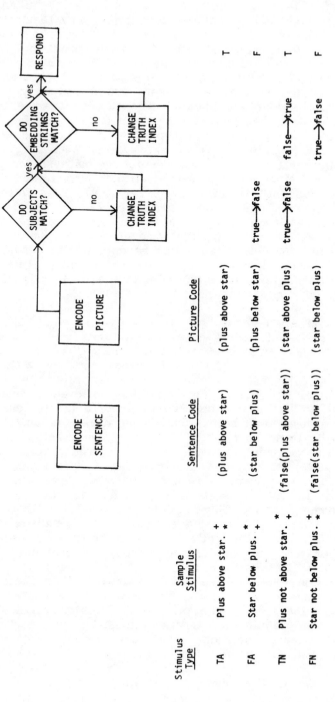

Stimulus Type	Sample Stimulus		Sentence Code	Picture Code			
TA	Plus above star.	+ *	(plus above star)	(plus above star)			T
FA	Star below plus.	* +	(star below plus)	(plus below star)	true→false		F
TN	Plus not above star.	* +	(false(plus above star))	(star above plus)	true→false	false→true	T
FN	Star not below plus.	+ *	(false(star below plus))	(star below plus)		true→false	F

When a pair of strings does not match, they change the "truth index" for the sentence-picture pair. The Clark and Chase model explains reaction times in terms of four parameters: a) time needed to encode an unmarked versus a marked preposition; b) time required to change the truth index when the embedded strings don't match, called "falsification time"; c) the extra time needed to process a negatively stated sentence, which includes time to encode the negation and time to change the truth index when the embedding strings don't match, and is called "negation time"; and d) a base time parameter, which includes all processes not included in the other parameters. According to this model, true negative (TN) sentence-picture pairs should take longest to process since neither embedded nor embedding strings match. Next longest are false negatives (FN), then false affirmatives (FA), and finally true affirmatives (TA). In the Carpenter and Just constituent comparison model, negation time and falsification time are explained by the same process parameter. Processing of the four sentence types differs only in the number of times this process is repeated. The order of reaction times predicted by the Carpenter and Just model is the same as that predicted by the Clark and Chase model.

Figure 1 shows the fit of the Chase and Clark and the Carpenter and Just models to our data. Both models accounted for 97% of the variation between conditions. Obviously the group data provided no basis for choosing one model over the other. However, John Palmer, now a graduate student at University of Michigan, proposed that individual differences analysis might be used to distinguish between the models and to test certain implications common to both. This general type of analysis was suggested by Underwood (1975), who proposed that patterns of individual differences might provide a means of rejecting inaccurate cognitive models. Underwood asserted that if, according to a certain model, two measures reflect the same underlying process, then these two measures should be highly correlated across individuals. According to the Clark and Chase model, the difference in reaction time between FA and TA sentences and the difference between TN and FN sentences both reflect the same model parameter, falsification time, or the time to process a mismatch between embedded strings. Similarly, FN-TA and TN-FA are both measures of negation time. If the model is correct, then these pairs of measures should be highly correlated. The Carpenter and Just model makes the same prediction plus the further prediction that negation time and falsification time should be highly correlated, since they both reflect repetitions of the same process. To test these hypotheses, we did the following analysis.

For each subject, we obtained two measures of falsification time:

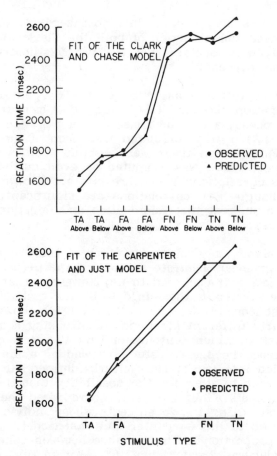

<u>Figure 1.</u> Fit of the a) Clark and Chase (1972) and b) Carpenter and Just (1975) models to the mean reaction times over all subjects in the sentence verification task.

$$\text{Falsification I} = \text{FA} - \text{TA}$$

$$\text{Falsification II} = \text{TN} - \text{FN}$$

and two measures of negation time:

$$\text{Negation I} = \text{FN} - \text{TA}$$

$$\text{Negation II} = \text{TN} - \text{FA}$$

The reliabilities of these two measures were all .7 or above. The correlation between the two negation measures was .51 (p < .01), suggesting that the two measures may indeed reflect duration of the same process. But the correlation between the two falsification measures was only -.05. This near-zero correlation casts doubt on an essential assertion of both the Clark and Chase and Carpenter and Just models: that the two falsification measures reflect the same process.

In order to test the assertion of the Carpenter and Just model that negation time and falsification time measure repetitions of the same process, a combined measure of negation time ((FN + TN) - (TA + FA)) was correlated with each of the falsfication measures. (Falsification times were computed on odd numbered trials and negation time was computed on even numbered trials to avoid spurious correlations.) The results are shown at the top of Table 3. Although both correlations are significant, neither is high enough to suggest that falsification and negation time measure the same process.

These analyses suggest that neither of the two sentence verification models gives an accurate account of the processes used by this group of subjects. It occurred to us, however, that subjects may ha differed in the strategies they used to do the task. A recent study in our lab (MacLeod, Hunt, and Mathews, 1978) showed that subjects adopt two clearly different strategies in attacking a similar sentence verification task in which sentence and picture were presented sequer tially. Data from the present study showed no evidence that subjects could be divided into two or more clearly defined strategy groups. However, in order to allow for the possibility that the models account for some but not all of our subjects, we repeated the individual differences analysis using the 25 subjects whose individual data most closely conformed to the Carpenter and Just model. For each of thes subjects, the proportion of the variance between conditions accounted for by the model was greater than .95. Within this group the correla tion between the two negation measures was .84 (p< .01) and the correlation between the falsification measures was .21 (n.s.). The correlations between the two falsification measures and the combined negation measure, shown at the bottom of Table 3, are not significant Thus even within this group, whose pattern of mean reaction times within subjects conformed most closely to the model, the correlational analysis suggests the model is inaccurate.

In spite of these problems, we proceeded with the original purpose of the study, which was to relate parameters of the model to ability factors. The parameters used were: a) TA reaction time, which should reflect all processes not measured by the other parameters, b) the two falsification parameters, and c) the combined negation parameter. Correlations between these four parameters and the factor scores are shown in Table 4. The only significant correlations are those that involve TA reaction time, and these are very weak. We must conclude that this part of the study failed to isolate parameters of the sentence verification task that are related to ability factors. One obvious explanation is that the models themselves are inade-

Table 3

Correlations Between Falsification and Negation Times

All Subjects (N = 72)

	Negation Time (TN+FN)-(FA+TA)
Falsification I (FA-TA)	.29**
Falsification II (TN-FN)	.24*

Subjects Best Fit by the Carpenter and Just Model (N = 25)

	Negation Time (TN+FN)-(FA+TA)
Falsification I (FA-TA)	.33
Falsification II (TN-FN)	.30

*p < .05

**p < .01

quate, as suggested by individual differences analyses of the parameters. Another is that the "essence" of verbal ability lies not in the type of manipulations represented by the falsification and negation parameters of this task, but in more elementary "encoding" processes. A number of other individual differences studies suggest that there is at least a weak relationship between encoding and verbal ability (Hunt, 1978; Jackson and McClelland,

Table 4

Correlations Between Experimental Task Parameters and Factor Scores

	Gv	Gc	CPS	Gf
Sentence Verification				
True Affirmative Reaction Time	.00	-.19	-.23*	-.22*
Falsification I (FA-TA)	.01	-.13	-.04	-.10
Falsification II (TN-FN)	.03	-.17	-.07	-.05
Negation Time (TN+FN)-(TA+FA)	.09	-.21	.12	-.11
Mental Rotations				
Slope	-.50**	.03	-.02	-.08
Intercept	-.35**	.12	-.14	-.24*
Mean RT	-.57**	.09	-.05	-.28**
% Errors	-.30**	-.01	.00	-.01

$*$ p .05

$**$ p .01

1979; Sternberg, 1977). The fact that TA reaction time, which in this study was the only parameter that reflected encoding, was correlated with the ability factors supports the hypothesis.

The Mental Rotations Task

The situation is clearer when we turn to measures of spatial

ability. In the Shepard and Metzler mental rotations task, subjects
were asked to determine whether two pictures show the same
object in different orientations, or whether they show two different
objects. The task can best be explained with reference to Figure
2. The first pair of figures can be brought into congruence by
rotation in the picture plane. The second pair requires rotation
in depth around a vertical axis. The third pair of pictures represents two different objects. Rotation will not bring them into
congruence. Shepard and Metzler found that reaction time to determine that the two members of a pair were the same was a strikingly
linear function of the number of degrees one figure had to be
rotated in order to match the other. Furthermore, the function
relating reaction time to angular disparity was almost identical for
depth and picture plane rotations. They concluded that subjects
solved the problems by mentally rotating one of the objects to see
if it could be brought into congruence with the other. According to
their reasoning, the slope of the reaction time function is a measure
of the speed with which subjects can mentally rotate the objects.

Shepard and Metzler used as subjects students who had high
scores on a test of spatial ability. One question to be answered
by this study was whether a broader student population could do
this very difficult task with a reasonably low error rate. Another
was whether the original finding of identical linear relationships
between reaction time and angular disparity on depth and picture
plane rotations would hold up in this population. The main
purpose was to find out how the slope parameter was related to
the psychometric factor scores.

In our replication, we used both depth and picture plane
rotations randomly intermixed. The angular disparity between
"same" pairs was 20, 60, 100, 140, or 180 degrees. The combination of the same and different, depth and picture plane, and five
angular disparities provided 20 trial types. Five different slides
were made for each of these 20 trial types. Subjects saw each of
these slides twice, for a total of 200 trials, on each of four days.
The first day was considered practice.

The mean results across all subjects are shown in Figure 3,
along with the lines of best fit for depth and picture plane rotations.
Although the slopes of the lines are similar, there is a definite
departure from linearity for depth rotations of 140 and 180 degrees.
This same pattern of results was found for all subgroups of
subjects (including very high spatial ability subjects similar to
those of Shepard and Metzler) on all days. It may be a result of
a selection of particularly hard slides for the 140-degree depth
rotation. Error rate as well as reaction time indicates the difficulty
of these trials.

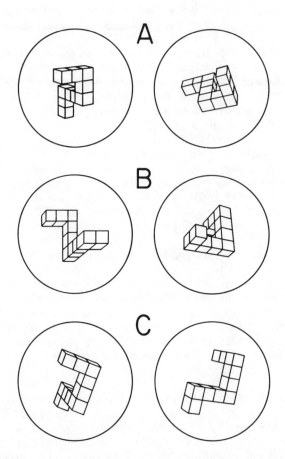

Figure 2. Examples of stimuli from the Shepard and Metzler (1971)
mental rotations task: a) "same" pair differing only in
rotation in the picture plane; b) "same" pair differing
only in rotation in depth; c) "different" pair which cannot be made identical by rotation.

For every subject we computed four measures based on
depth trials and four measures based on picture plane trials:
slope and intercept of the reaction time function, mean reaction
time over all trials, and percent errors. The reliabilities of these
measures are shown in Table 5, along with the correlations between
corresponding measures from depth and picture plane trials.

According to Shepard and Metzler's model for the mental rotations task, the same process accounts for reaction times in both depth and picture plane conditions. The very high correlations between corresponding measures from the two conditions are consistent with this aspect of the model.

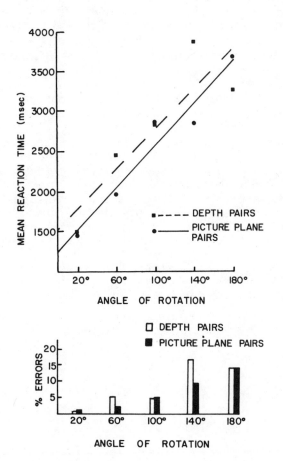

Figure 3. Mean reaction times (over all subjects) and lines of best fit for depth and picture plane trials separately for the mental rotations task.

Since the correlations between corresponding measures were so high, data from depth and picture plane conditions were combined to compute the correlations between factor scores and parameters of the mental rotations task. These correlations are shown in Table 4. The most interesting measure, the slope measure, is highly correlated with Gv, the spatial visualization factor, and uncorrelated with any of the other factors. If we accept Shepard and Metzler's interpretation of the slope parameter, we can conclude that there is a strong relationship between speed of rotation and the Gv factor. However, the relationship between performance on the mental rotations task and Gv is not limited to this slope parameter. The intercept and total reaction time are significantly correlated with Gv, as well as Gf, and accuracy is correlated with Gv.

It is obvious from Table 4 that performance on the mental rotations task is much more strongly associated with the visualization factor than with any of the other factors. Although reaction time is the measure of principal interest on this task, correlations with the clerical and perceptual speed measures are close to zero. There are no significant correlations with crystallized intelligence, and relatively weak correlations with fluid intelligence. Various psychologists have argued that the linear relationship between reaction time and angular disparity in the mental rotations task was not conclusive evidence for an analogue, non-propositional

Table 5

Reliabilities and Intercorrelations of Mental Rotations Measures

Measure	Reliability: Depth	Reliability: Picture Plane	Correlation: Picture Plane and Depth
Slope	.82**	.79 **	.71 **
Intercept	.91**	.89**	.79 **
Mean RT	.99**	.99 **	.92 **
% Errors	.86**	.92 **	.81 **

** $p < .01$

reasoning process. Palmer (1975), for instance, has proposed a propositional solution process that would produce the same results. We feel that the fact that the mental rotations parameters are strongly associated with the visualization factor, and not with the factors representing verbal processes, strengthens the argument that mental rotation involves a different sort of process than verbal reasoning tests.

In summary, it seems that the mental rotations data provides a example of a situation where experimental and individual differences analyses form a coherent picture and where the individual differences data can be used to support the model designed to account for the experimental results. The sentence verification data, on the other hand, illustrate a situation where individual differences analyses cast serious doubt on an information processing model that seems to account quite well for group data.

Most of the research relating process parameters of cognitive models to ability measures has involved verbal processes. In some cases, parameters based on paradigms reported in the information processing literature have been correlated with more conventional verbal ability measures (Hunt, 1978; Hogaboam and Pellegrino, 1978; Jackson and McClelland, 1979). In other cases, cognitive models have been developed to explain performance on the ability measures themselves (Sternberg, 1977; Sternberg and Weil, Note 1). However, with the exception of low correlations involving encoding parameters, no clear pattern of results has emerged that provides an explanation of verbal ability in terms of cognitive processes. This failure represents something of a paradox. At first glance it seems tautological to say that the ability to do well on tests of verbal ability must be related to the efficiency of the component verbal processes. One possible explanation for the failure to find process explanations of verbal ability is that our process models of verbal tasks are inaccurate. This seems to be the case with respect to the sentence verification models discussed in this paper. However, there are at least two further possibilities.

The first is that performance on complex verbal tasks is not so much a matter of the efficiency of component processes, but how subjects combine these processes. Studies in our laboratory and in Robert Sternberg's (MacLeod, Hunt, and Mathews, 1978; Sternberg and Weil, Note 1) have indicated that there are important individual differences in strategies even on quite simple verbal tasks, and that strategy mediates the relationship between cognitive processes and ability factors. It is also possible that theoretical concepts from attentional research may be necessary to relate verbal ability factors and cognitive processes (Lansman, Note 2). The most popular theories of attention at the present time assert that all cognitive processes compete for attentional capacity (Kahneman, 1973; Norman and Bobrow, 1975). According to these theories, most complex cognitive tasks require subjects to divide their attentional capacity between several simultaneous mental processes, or at least to switch attention rapidly between processes. It is possible that important differences between individuals lie not within any particular process, but rather in the characteristics of the attentional system--e.g., total capacity, ability to switch attention between processes, strategies of dividing attention between simultaneous processes. Both strategy and attention represent possible links between the cognitive processes of experimental psychology and the ability factors of differential psychology.

Another possibility, which is particularly relevant in the case of crystallized intelligence, is that verbal ability measures are best explained not in terms of current processing differences, but in terms of processes which took place long before the test. For example, the most important factors determining performance on a vocabulary test may be a) processes involved in incidental learning of word meanings during reading and conversations, and b) the person's past verbal environment. Cognitive tasks that have been designed explicitly to eliminate the effects of knowledge, such as the sentence verification task, cannot be expected to elucidate individual differences in such ability tests. Analysis of performance on such tests would have to include analysis of subjects' knowledge structures.

The study reported here is one of very few that have tried to relate process parameters of a spatial task with spatial ability measures (Snyder, Note 3; Tapley and Bryden, 1977). The results suggest that it may be easier to isolate the cognitive processes responsible for subject variation on spatial tests than has proved to be the case with tests of verbal ability.

References

Carpenter, P. A. and Just, M. A. Sentence comprehension: A psycholinguistic processing model of verification. Psychological Review, 1975, 82, 45-73.

Clark, H. and Chase, W. On the process of comparing sentences against pictures. Cognitive Psychology, 1972, 3, 472-517.

Hogaboam, T. W. and Pellegrino, J. W. Hunting for individual differences in cognitive processes: Verbal ability and semantic processing of pictures and words. Memory and Cognition, 1978, 6, 189-193.

Hunt, E. Mechanics of verbal ability. Psychological Review, 1978, 85, 109-130.

Jackson, M. D. and McClelland, J. L. Processing determinants of reading speed. Journal of Experimental Psychology: General, 1979, 2, 151-181.

Kahneman, D. Attention and effort. Englewood Cliffs, NJ:

Prentice-Hall, 1973.

MacLeod, C. M., Hunt, E. B., and Mathews, N. N. Individual differences in the verification of sentence-picture relationships. Journal of Verbal Learning and Verbal Behavior, 1978, 17, 493-507.

Norman, D. A. and Bobrow, D. G. On data-limited and resource-limited processes. Cognitive Psychology, 1975, 7, 44-74.

Palmer, S. E. The nature of perceptual representation: An examination of the analog/propositional debate. In R. C. Schank and B. L. Nash-Webber (Eds.), Theoretical issues in natural language processing. Arlington, VA: Tinlap Press, 1975.

Shepard, R. N. and Metzler, J. Mental rotation of three-dimensional objects. Science, 1971, 171, 701-703.

Sternberg, R. J. Intelligence, information processing, and analogical reasoning: The Componential Analysis of human abilities. Hillsdale, NJ: Erlbaum, 1977.

Tapley, S. M. and Bryden, M. P. An investigation of sex differences in spatial ability: Mental rotation of three dimensional objects. Canadian Journal of Psychology, 1977, 31, 122-130.

Underwood, B. J. Individual differences as a crucible in theory construction. American Psychologist, 1975, 30, 128-134.

Reference Notes

1. Sternberg, R. J. and Weil, E. M. An aptitude-strategy interaction in linear syllogistic reasoning. Yale University Technical Report, 1979.

2. Lansman, M. An attentional approach to individual differences in immediate memory. University of Washington Technical Report, 1978.

3. Snyder, C. R. R. Individual differences in imagery and thought. Unpublished doctoral dissertation. University of Oregon, 1972.

THE DESIGN OF A ROBOT MIND: A THEORETICAL APPROACH TO ISSUES IN INTELLIGENCE

Earl Hunt

The University of Washington

Seattle, Washington, U.S.A.

As I write sounds drift into my room. I have finished my morning coffee. Will my writing be successful?

What would a properly programmed computer need to know to answer this question? Some information about relatively permanent, historical facts would be of use. What were my school grades and intelligence test scores? Temporary information would be needed. Did I sleep well last night? What are those outside sounds? Was the coffee decaffeinated, or perhaps Irish?

A comprehensive psychology of cognition should deal with all such influences on thinking. Calls for such approaches have been made before (Cronbach, 1957; Underwood, 1975). This paper is an attempt to go further, by sketching a general theoretical framework for thinking about thinking, and using it to develop some hypotheses about individual performance.

The approach that will be taken is based upon the proposal of Newell and Simon (1972; Newell, Shaw, and Simon, 1958) that thinking be modeled by computer simulation. The reader is asked to envisage a population of robots whose minds are, indeed, controlled by simulation programs of the type considered by Newell and Simon. The approach taken here is different from the Newell and Simon approach in that, instead of being interested in the logic of the programs contained in the robot, interest will be focused upon some of the machinery that the robot might contain for executing the programs. Some psychological assumptions about that machinery will be made, building upon ideas proposed earlier both on psychological grounds (Hunt, 1976; Hunt and Poltrock, 1974) and with the design of artificial intelligence systems in mind (McDermott and Forgy, 1978). A claim will be made that variations in the efficiency of the resulting machinery could account for some of the important phenomena observed in individual differences research.

A General Theory of Thought

Newell and Simon (1972) propose that the <u>production</u> be the basic step in a simulation program. A production is a rule, written in the form

(1) $\underline{L} - \underline{R}$

where $\underline{L}$ is a pattern recognition rule for classifying input as being acceptable or unacceptable, and $\underline{R}$ is the action to be taken when an acceptable input is found. The input to a production is the current state of active memory, including input from the sensory system and from the arousal of long term memory records. The action of a production is a command to do something, such as rearranging the contents of working memory or issuing an order to make a physical action.

Productions are not models of problem solving, they provide a notation in which models can be written. Newell and Simon used the notation to construct specific models for very complex problem solving, such as finding the solution to a mathematical logic problems. As has been indicated, this paper will focus on the design of machinery for executing productions. Newell and Simon say little about this.

Production execution involves a pattern recognition and an action phase. Pattern recognition can be further broken down

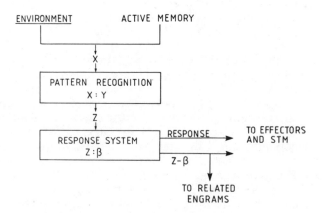

Figure 1.

each production can independently determine how closely its expectations are satisfied. Input (x) to the productions in long term memory (LTM) arises both from the environment and the short term memory (STM) system. The possible complexities of the latter system will not be discussed. Input X is compared to the internally stored record, Y, to provide a comparison index, Z. Z is then compared to the production's current threshold, B. If Z is greater than B the response system, R, is activated, otherwise it does not. Activating R does not mean that R necessarily happens, but rather that a request for action goes to an interpreter that can cause things to happen, because the interpreter has control over effectors. In this context, an "effector" is any brain mechanism that can do something, including altering the contents of STM. The term is not limited to mechanisms in the motor system.

As the productions are matched to active memory independently, two or more productions may issue simultaneous but incompatible commands to the effectors. At this point an interpreter must control the conflict resolution phase. The interpreter must be restricted in its scope of action, so that it can be understood. In particular, the actions of the interpreter must not depend upon an interpretation of the current state of active memory. Otherwise it would be necessary to develop a psychology of the homunculus inside the interpreter.

The interpreter proposed here combines some of the basic ideas of Selfridge's (1959) PANDEMONIUM system for pattern recognition with the psychological notion of spreading semantic activation (Collins and Loftus, 1975). To explain its action a more psychological terminology is useful. Individual productions will be referred to as engrams, and are assumed to be resident in the long term memory (LTM). Active short term memory (STM) is assumed to be physically distinct from the LTM system. The contents of STM can be changed either by input from the sensorium, or by input from LTM, or by the execution of one of a small set of transformations that can alter material in STM. To capture the flavor of this assumption, the transformations hypothesized by Podgorny and Shepard (1978) to account for the data from mental rotation studies would be examples of STM transformations. Any changes in STM not due to input from the external environment are the result of activation of the R stage of some engram. Thus, the ability to control STM through commands to the effectors makes it possible for a production to feed forward information to affect the sequence of firing of engrams in the immediate future.

The relations hypothesized are shown in Figure 2. The contents of active memory are broadcast throughout LTM. At any instant each engram will be engaged in a signal detection exercise, trying to match its L part to some part of the signal from active into a <u>match</u> and a <u>conflict resolution</u> stage (McDermott and Forgy, 1978). In the match stage each production compares its expected pattern (its L rule) to the current contents of active memory. How this might occur is shown in Figure 1. It is assumed that each production can independently determine how closely its expectations are satisfied. Input (x) to the productions in long term memory (LTM) arises both from the environment and the short term memory (STM) system. The possible complexities of the latter system will not be discussed. Input X is compared to the internally stored record, Y, to provide a comparison index, Z. Z is then compared to the production's current threshold, B. If Z is greater than B the response system, R, is activated, otherwise it does not. Activating R does not mean that R necessarily happens, but rather that a request for action goes to an interpreter that can cause things to happen, because the interpreter has control over <u>effectors</u>. In this context, an "effector" is any brain mechanism that can do something, including altering the contents of STM. The term is not limited to mechanisms in the motor system.

As the productions are matched to active memory independently, two or more productions may issue simultaneous but incompatible commands to the effectors. At this point an interpreter must control the conflict resolution phase. The interpreter must be restricted in its scope of action, so that it can be understood. In particular, the actions of the interpreter must not depend upon an interpretation of the current state of active memory. Otherwise it would be necessary to develop a psychology of the homunculus inside the interpreter.

The interpreter proposed here combines some of the basic ideas of Selfridge's (1959) PANDEMONIUM system for pattern recognition with the psychological notion of spreading semantic activation (Collins and Loftus, 1975). To explain its action a more psychological terminology is useful. Individual productions will be referred to as <u>engrams</u>, and are assumed to be resident in the long term memory (LTM). Active short term memory (STM) is assumed to be physically distinct from the LTM system. The contents of STM can be changed either by input from the sensorium, or by input from LTM, or by the execution of one of a small set of transformations that can alter material in STM. To capture the flavor of this assumption, the transformations hypothesized by Podgorny and Shepard (1978) to account for the data from mental rotation studies would be examples of STM transformations. Any changes in STM not due to input from the external environment are the result of activation of the R stage of some engram. Thus, the ability to control STM through commands to the effectors makes it possible for a production to feed forward information to affect the sequence of firing of engrams in the immediate future.

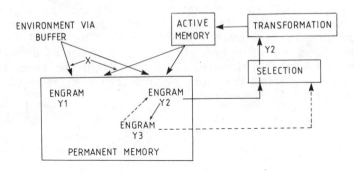

Figure 2.

The relations hypothesized are shown in Figure 2. The contents of active memory are broadcast throughout LTM. At any instant each engram will be engaged in a signal detection exercise, trying to match its L part to some part of the signal from active memory. Whether or not a particular engram will respond depends upon the correlation between the noisy signal it receives from active memory, its stored record, and its current threshold value.

When an engram does respond, the response will consist of two parts; a request for effector action and a confidence level (Z in Figure 1) indicating the extent to which the engram's threshold has been exceeded. The pairs of responses from the various activated engrams will be forwarded to a selector (analogous to the decision demon in Selfridge's PANDEMONIUM) which determines the strongest signal input to it, and permits the engram sending that signal to control the effectors. The selected engram can then plant signals for related engrams in STM, thus providing a first mechanism for the coherent execution of productions over time.

The spreading activation concept provides a second mechanism for the execution of a coherent set of productions. Assume that there exists an assymetrical relationship R(A,B) between engrams A and B, and that R(A,B) may vary from zero to one. R(A,B) will be called a sequencing relationship. Two engrams will be said to be strongly sequenced if R(A,B) approaches one and

weakly sequenced if it approaches zero. The learning mechanism
for producing R(A,B) will not be discussed here. The spreading
 activation assumption is that when engram A fires its confidence
level (Z-B) will be communicated to all related engrams B, in an
amount proportional to the sequencing relationships between A
and the members of B. This occurs regardless of whether or not
A is permitted control over the effectors. The signal from A to
B will be called an activation signal. The effect of an activation
signal is to lower the threshold values (B) in the receiving engrams, thus rendering the more sensitive to the presence of their
L patterns in active memory.

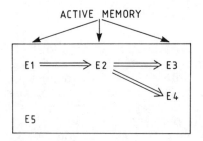

Figure 3.

Figure 3 illustrates how spreading activation can control the
power of the environment over the order in which engrams are
activated. The figure shows a set of engrams, El to E4, with
strong sequencing relations between them. When El is activated,
E2 is almost certain to be activated, since no other engram receives
an activation signal. When E2 is activated both E3 and E4 will
have their thresholds lowered, so the exact state of active memory
will determine which path is taken in further engram activation.
The "choice" of sequences, however, is almost completely restricted
to E3 and E4, as engrams outside of the sequenced set (e.g.,
engram E5) have not received any sensitizing activation signal.

How should the selector mechanism work? The simplest selection system would record its inputs at fixed time intervals and then compare them. But what would be the clocking mechanism that determined the time interval? To avoid the need for an internal digital clock, a selector system of the type shown in Figure 4 will be postulated. The basic idea is that the selector receives noisy signals continuously over time. Each engram corresponds to a line into the selector system. The system must determine that engram which has the consistently strongest signal over some time interval, $t0-t1$, which is both long enough to permit reliable selection of the correct engram and short enough so that the robot mind can keep up with the environment. There are a number of ways of designing selection systems that might work like this. All of them appear to have the following characteristics:

(a) So long as the engrams' responses display the same relative strengths, accuracy of selection will increase with longer time samples.

(b) The selector itself must have some form of memory, so that it can accumulate information about a signal over time. The size of this memory will determine the longest time period over

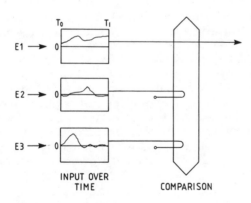

Figure 4.

which selection can take place, and thus will determine the maximum accuracy of selection.

The basic concepts of the robot mind have now been presented. Many details need to be filled in, particularly with respect to learning and development. Since the focus of this paper is on individual differences, however, these topics will not be discussed. Instead some thought will be given to how individual differences would arise in a population of mature robots who have completed their education.

The Physiology and Education of Robots

Four classes of influence on the robot mind will be discussed. Two are "biological," as they depend upon the efficiency of physical operations in the system, and two are "educational," as they depend upon the experience of the robot.

Structure: We consider first the effects of changes in the mechanical structure of LTM and STM.

LTM consists of a dispersed set of independently acting pattern recognizers, the engrams. Deficiencies in engram arousal would arise from defective pattern recognition. Such deficiencies would be quite specific, since the pattern recognition capability has been dispersed to the engrams rather than being retained in a central pattern recognition system. The concept of spreading activation also assumes a mechanism for transmitting and receiving activation signals. This mechanism has also been dispersed to individual engrams, and hence deficiencies in it would be restricted to influencing the engrams involved.

Structural limits clearly apply to STM. Given a fixed scheme for coding information in active memory, the size of STM clearly limits the amount of information that can be contained in the signal broadcast to LTM. The effect of such a limit, however, would clearly depend upon the effectiveness of the scheme used to represent information in STM. Two robots might vary considerably in their effective use of STM, even though they had identical structural capacities, if they used different codes for representing the external world.

Any malfunction of a mental effector would also limit thought, but the limit would only apply to the action of engram systems that used the effector in question.

Attention: Many of the phenomena associated with limits on attention can be understood by considering the limits on cognitive behavior that are due to engram selection. Selection, and the concomitant maintenance of STM, is a volatile process, and hence more likely to be dependent upon neural firing than upon posses-

sion of a static memory record, such as would be required for
pattern recognition. Thus engram selection should be affected by
procedures that alter the efficiency of neural firing.

Schneider and Shiffrin (1977) made a distinction between
"controlled" processes and "automatic" processes. In their terminology automated tasks are tasks that are highly overlearned, and
relatively impervious to disruption by the introduction of distractors.
Translating to the terminology of the robot mind, automated tasks
would be executed by highly sequenced sets of engrams, and
thus only minimally guided by conflict resolution. Controlled
tasks, on the other hand, are those that require close monitoring
of active memory, which means that conflict resolution will be
important and that the tasks will be subject to disruption by
irrelevancies introduced into active memory.

A similar explanation can be offered for task interference.
If two tasks are maximally sequenced and do not compete for the
same effectors, then there will be no need to resolve conflicts
between engrams and the tasks can be executed simultaneously.
To the extent that the tasks are not sequenced, so that information placed in active memory by one task may disrupt the sequence
of engrams involved in the other task, conflict resolution will
become important. Similarly, conflict resolution will be important if
incompatible orders for effector action are issued.

To summarize, structural differences affecting the pattern
recognition process and affecting STM size will determine the
capacity of a robot when operating at maximum efficiency. The
limits imposed by structural capacities should be specific to the
cognitive tasks that utilize the engrams involved. A limit imposed
by the selector mechanism, on the other hand, would be quite
general. It would be imposed on any task that was not highly
overlearned, or that required frequent examination of the current
state of active memory. A task that might appear to be automated
when executed alone could appear to be controlled--i.e., dependent
upon the conflict resolution process--if it were executed in conjunc-
tion with another task that competed for the same effectors.

"Educational" differences will be considered next. These are
the differences in robot problem solving capacity due to differences in the information available to different robots, and due to
differences in the way that information is organized.

Process variation: The term process will be used to refer to
a (possibly branching) sequence of engrams used to solve a
problem. Most problems can be solved by several processes.
This can confuse a simple correlational analysis. Suppose that we
are studying a population of robot problem solvers in which,
unknown to us, some of the robots use one problem solving

process and some use another. Suppose further that the two
processes utilize different structures for information handling.
The average correlation between a measure of structural ability
and problem solving performance, computed over the entire population, may give a quite false picture of the true state of affairs.
An illustration of this sort of effect is found in a series of experiments that my colleagues and I have conducted on sentence verification (Lansman, 1980; MacLeod, Hunt and Mathews, 1978; Mathews,
Hunt, and MacLeod, in press). The subject's task is to decide

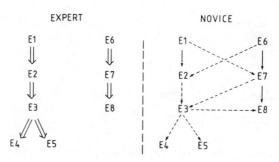

Figure 5.

whether a sentence accurately describes a picture. This is a
laboratory analog of the general problem of co-ordinating linguistic
and non-linguistic representations of the world. We have found
that some people translate the picture into a sentence and do the
bulk of their reasoning in a linguistic mode, while others form a
mental image on the basis of the sentence, and compare this image
to the picture, thus doing the bulk of their reasoning in a spatial-
imaginal mode. Not surprisingly, the relation between a subject's
performance in the miniature linguistic task and measures of
verbal or spatial abiiity depends upon strategy being used.

Knowledge: It is easy to imagine two robots (or people) who have identical mechanistic capacities for information processing and who know the same pieces of problem relevant information, but who differ in their problem solving capacities. Both expert and embryo chess players know the rules, but the expert has a firmer grasp of their implications. What does this distinction mean to the robot mind?

A robot would be well tuned for problem solving if it had the right engrams, organized into an appropriate strong sequence. Consider Figure 5, which shows "expert" and "novice" organizations of an abstract problem solving process. To give the novice every chance, the same engrams are shown in each process. The difference is that the expert's engrams are strongly sequenced and hence will need less guidance from active memory. At points at which guidance is needed, the choices are more clearly defined for the expert. So long as choice is not required, the expert's processes will be less affected by the action of the selector mechanism. At choice points, the situation reverses, for the expert will demand that a discrimination be made between two or more strongly sequenced engrams.

This analysis suggests the following contrasts between expert and novice performance in mental tasks:
(i) Expert performance will be closer to structural limits, as during most phases of problem solving the expert does not engage the selector mechanism.
(ii) When the problem solving situation does not involve a choice the expert will have available attentional resources that can be used to monitor problem irrelevant activity.
(iii) At choice points the expert will engage the selector mechanism fully in the problem solving process, and will be less able than the novice to monitor problem irrelevant stimuli.
(iv) Because of the stronger sequencing in experts, the style of problem solving within an individual expert will be more consistent than it will within an individual novice.

Individual Differences and the Robot Mind

In this section some of the issues that arise in individual differences research are examined from the viewpoint of the robot mind.

The structure-intelligence relationship: There is ample evidence for substantial contributions to intelligence that seem to be associated with biological structure, i.e., permanent biological characteristics (Willerman, 1979). Three classes of structure were identified in the robot brain; engram arousal, STM capacity, and engram selection. The latter was associated with attention, and will be discussed separately.

To investigate the relationship between pattern recognition and thought we need to identify some cognitive behavior that shows important individual differences and that involves the automatic arousal of memory for well learned stimuli. Since reading is a highly overlearned skill in adults, and since rather complex reading is well within the capability of young children, the ability to recognize the basic stimuli of written language, letters and words, can be evaluated as a test of pattern recognition efficiency. Two paradigms have been used in this line of research. In the stimulus matching paradigm an observer is shown two different graphemes and asked if they name the same thing. For example, the letter pair (A,a) is name identical (NI), while the pair (A,A) is both name identical and physically identical (PI), and the pair (A,B) is different. The dependent variable in a stimulus matching study is the time required to make the appropriate identification. It has been found that when the two letters are presented simultaneously more time is required to match an NI pair than a PI pair. There are several models that could account for this observation (Posner, 1978), but all of these models agree that the difference arises from the need to arouse more engrams in memory in order to make the NI response. Thus the difference between the time required to make an NI and a PI identification can serve as a crude index of the time required to extract highly overlearned information from LTM. The NI-PI difference in reaction times can be illustrated with either letters or words (Goldberg, Schwartz, and Stewart, 1977; Palmer et al., note 1), as can the individual differences results to be discussed below.

In a lexical identification paradigm the observer is asked to indicate whether or not a stimulus is a common word (e.g., BAT, BAD) or a non-word conforming to normal English structure (e.g., BAK, BAS). The individual differences data from lexical identification is consistent with that from the stimulus matching task, although not nearly so extensive (Palmer et al., Note 2). This is unfortunate, as the lexical identification task has somewhat more face validity as a measure of pattern recognition.

Figure 6 summarizes results from a number of stimulus matching studies using letter stimuli. In subjects of normal to above average ability there is a correlation of about -.3 between the NI-PI difference and measures of verbal ability, a small though reliable effect. In terms of absolute effects, the difference score is about 65 msecs. for bright university students, while people of the same age recruited outside the university have scores of about 110 msecs. (Hunt, 1978). The picture changes drastically if we study the lower ranges of conventional test scores. Elderly individuals (over 65) produce scores closer to 200 msecs., and educable mental retardates show differences of over 350 msecs. even though they make no more errors than do college students. This pattern of scores is not consistent with

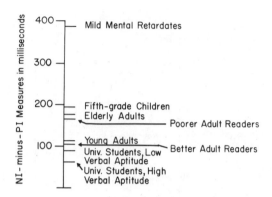

Figure 6.

the statement that the low correlations at the upper end of the scale are due solely to restrictions in variance, although that is undoubtedly a factor. It appears that there is a non-linear relation between the reaction time score and conventional psychometric measures. Taking a bold (irresponsible?) step with data, Figure 7 replots the NI-PI scores of Figure 6 against the "estimated verbal IQ" scores of the groups in question. The IQ scores were taken from a variety of studies of norms for equivalent groups, and were not obtained by measuring the actual participants in each study, so Figure 7 needs to be regarded with considerable skepticism. Nonetheless, the striking suggestion of non-linearity is worth considering. This is particularly the case because the non-linearity seems to appear when we consider another structural measure, the size and speed of STM. Schwartz (1980) has observed that, if anything, there is less evidence for a correlation between verbal aptitude and short term memory performance than there is for a correlation between verbal aptitude and measures of long term memory access, so long as one restricts consideration to average or above average individuals. Inadequate short term memory capacity does appear to limit mental performance in the elderly, young children, and the mentally retarded.

The evidence for non-linearity is entirely too tenuous. More studies are needed, using more paradigms and investigating wider ranges of ability within a single study. The issue has been raised because, seen through the robot mind, the issue is important. A non-linearity between structural measures and measures of relative standing in the general intellectual population (which is

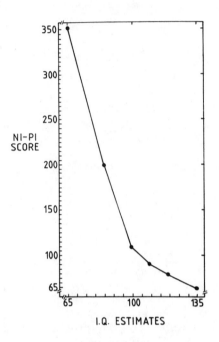

Figure 7.

what an intelligence test is) suggests that the role of structural
limitations is different in people of above or below average intelligence. The above average may have more than sufficient structural capacity to deal with most of the intellectual challenges that
they meet. Their deficiencies, then, will be more related to im-
proper selection of process, or to distraction of attention. The
below average, and especially those whom we define to be defectives, may simply not have the structural capacity to cope with
our culturally defined mental tasks, even when they are operating
at full efficiency.

The nature of general intelligence: The evidence for a general
intelligence factor is well known. In this section Cattell's (1972)
and Horn's (1978) notion of a distinction between a general ability
to apply culture specific, learned solutions to problems (crystallized intelligence, or Gc) and a general ability to deal with new
or unusual problems (fluid intelligence, or Gf) will be considered.

It will be argued that this distinction has a rough parallel in the distinction between structural and attentional effects in the robot mind.

Carroll and Maxwell (1979) observed that verbal aptitude tests typically have high Gc loadings, and speculated that the relation between lexical pattern recognition performance and verbal ability might be due to both tasks being indicants of Gc. We have obtained preliminary evidence that the NI-PI scores in a stimulus matching task do indeed have high Gc loadings, but much more work remains to be done. Theoretically, it seems more appropriate to regard LTM pattern recognition efficiency as an underpinning of Gc rather than the other way around, as the one is a process and the other is a statistical abstraction, defined from the common covariance over tasks. It would also be of interest to know whether stimulus matching tasks using non-verbal stimuli would be related to Gc. Such a study would address the question of specificity of the pattern recognition ability to particular types of stimuli, on the one hand, and would also address the question of whether or not Gc is anything other than a renaming of "verbal ability." At present almost all our experimental psychological data about stimulus matching and our psychometric data about Gc depends upon the use of verbal tasks.

Horn and Cattell's fluid intelligence (Gf) factor is defined by tasks that are either novel to the participant or that require monitoring of the stimulus situation. Such tasks would depend heavily upon the action of the robot's selector mechanism. Thus we should expect to find that paradigms designed to measure the person's ability to distribute attention and make discriminations between stimuli could be used as indicants of Gf.

We have obtained a small amount of data that bears upon this question (Hunt, in press). One piece of evidence depended upon the secondary task method. In this paradigm performance on a simple task is monitored as the subject attempts to solve a concurrent, more complex problem. Deterioration of performance on the simple task is supposed to measure the amount of attentional effort required to execute the complex task. Colene McKee and I asked people to solve Raven Matrices problems (an indicator of Gf) while simultaneously balancing a small lever between two posts. Problems were presented in ascending order of difficulty, as determined by population norms. We found that an individual's lever balancing performance began to deteriorate, "on the average," just prior to that person's making an error on the increasingly difficult matrix problems. This is consistent with the argument that tasks that make high demands on Gf also involve high demands on attention. To make this argument stronger, though, we need also to show that tasks which make equivalent demands on Gc, as determined by population norms, do not interfere with the simple psychomotor task.   There is no evidence on this point.

A "non-intellectual" measure of the ability to make discriminations between stimuli can be obtained by observing how fast a
person can discriminate the occurrence of one of n simple stimuli;
e.g., indicate which of n possible lamps has been lighted.   The
choice reaction time (CR$\overline{T}$) in this situation increases logarithmically
with the number of stimuli present, and the slope of this function
can be regarded as a measure of the time required to resolve a
decision between two stimuli.   Jensen (1979) has reported substantial
correlations between CRT and measures of Gf, notably the Raven
Matrix test. His data are somewhat puzzling, though, since the
correlation seems to be produced by a drop in the <u>intercept</u> of
the function relating CRT to the logarithm of the <u>number</u> of
alternative choices, whereas the theory behind the CRT task
makes the slope the measure of "mental speed."   In our own
laboratory we have found a correlation of -.39 between the slope
measure of the CRT task and the Raven Matrix scores of college
students.

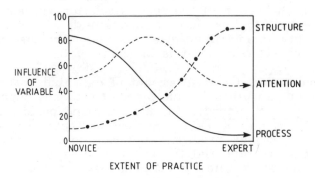

Figure 8.

The effects of practice: Good tests are traditionally defined to
be tests that do not show practice effects, or tests that are
scored only after performance has been stabilized by extensive
training.   Simon (1976) pointed out that this is a questionable
procedure if general intelligence is associated with the ability to
develop problem solving strategies.   A robot mind analysis suggests
that Simon's concern is realistic.   Behavior in a transient learning
state should be more closely related to measures of attention
allocation than should problem solving behavior either before or
after extensive practice.

        To see this, imagine that robots are working on unusual
problems, but problems not completely unrelated to those attempted
before.   Initial performance will be dominated by individual differences in the relevance of previously learned strategies.   The
situation changes as practice continues, in a way summarized by
Figure 8.   As deficiencies in inital strategies are revealed the
robots will begin to hunt for strategies appropriate to the particular situation.   During this stage maximum demands will be placed
upon the selector system, since the developing engrams will
neither be finely tuned to the environment or strongly sequenced
to each other.   Thus the relative importance of individual differences in prior knowledge decreases and the relative importance of
attention differences increases.   Now suppose that there is a
single optimal strategy, and that all robots find it.   As this
process becomes more strongly sequenced each robot will become
an "expert," and individual differences will reflect structural
variations between individuals on those structures used by the
optimal strategy.

Concluding Comments

        The robot mind provides a framework for thinking about
thinking, but does not dictate a model for any one task.   This
approach to theory building contrasts sharply with the style of
theory building in both experimental and differential psychology.
Differential psychologists define intelligence by refining our
observations of behavior that, by consensus, is thought to demonstrate mental competence.   Processes are then inferred from the
refined observations.   This is what "naming factors" is all about.
Here, by contrast, we begin with assumptions about the process
of thinking and define intellectual behavior in terms of these
assumptions.

        The robot mind analysis, although superficially a collection
of concepts derived from studies in experimental psychology, is
based upon a much different philosophy.   The robot mind analysis
draws together our reasoning about different components of
cognition, but since it does not specify the exact nature of any
component, the analysis is not directly falsifiable by data.   There

is an analogy to Newell and Simon's approach to the simulation of
thinking by computer programming, which is an idea distinct from
the rightness or wrongness of a particular simulation.   Current
wisdom in experimental psychology is to resist such broad theories,
and to concentrate upon the analysis of isolated components of
thought.   Mandler (1975, p. 15) states the case eloquently:

> ...theories of perception, learning, sensation, psycho-
> pathology, attitude formation, and so forth need not be
> deducible from a general theory of learning or perception
> or whatever.   Indeed, these subsystems and minitheories
> can exist in their own right, and it is not even encum-
> bent on the theorist to show how his minitheory of acoustic
> information processing, for example, is parallel to or
> tied in with a theory of speech production.   Such an
> outcome is highly desirable, but it is not necessary,
> and the last 30 years of the history of psychology have
> shown the utility of this approach.

I admit the practical utility of Mandler's approach, but I am
less impressed with the results of the last 30 years of psychology.
Perhaps there is room for an effort to do something that is highly
desirable.   This is particularly true in the study of individual
differences for, after all, thinking does go on in just one head at
a time.   Even if we can predict behavior by computing linear
combinations of scores, does anyone seriously believe that behavior is produced in this way?   A model of the thinking process has
been offered.   It will be a success if it is useful, or if it is
replaced by other models of similar scope.

References

Carroll, J. B. and Maxwell, S. E.   Individual differences in
     cognitive abilities.   In M. E. Rosenzweig and L. W. Porter
     (Eds.), Annual review of psychology (Vol. 30).   Palo Alto,
     Ca.:   Annual Reviews, Inc., 1979.

Cattell, R. B.   Abilities: Their structure, growth, and action.
     Boston: Houghton Mifflin, 1972.

Collins, A. M., and Loftus, E. F.   A spreading activation theory
     of semantic processing.   Psychological Review, 1975, 82,
     407-428.

Cronbach, L.   The two disciplines of scientific psychology.
     American Psychologist, 1957, 12, 671-684.

Goldberg, R. A., Schwartz, S., and Stewart, M.   Individual
     differences in cognitive processes.   Journal of Educational
     Psychology, 1977, 69, 9-14.

Horn, J. L.   The nature and development of intellectual abilities.
     In R. T. Osborne, C. E. Noble, and N. Weyl (Eds.), Human
     variation: The biopsy of age, race, and sex.   New York:
     Academic Press, 1978.

Hunt, E.   Varieties of cognitive power.   In L. Resnick (Ed.),

Nature of intelligence. Hillsdale, New Jersey: L. Erlbaum
    Associates, 1976.
Hunt, E. Mechanics of verbal ability. Psychological Review,
    1978, 85, 109-130.
Hunt, E. B. Intelligence as an information processing concept.
    The British Journal of Psychology (in press).
Hunt, E. B., and Poltrock, S. The mechanics of thought. In
    B. Kantowitz (Ed.), Human information processing: Tutorials
    in performance and cognition. Potomac, Maryland: L.
    Erlbaum Associates, 1974.
Jensen, A. R. g: Outmoded theory or unconquered frontier?
    Creative Science and Technology, 1979, 2, 16-29.
Lansman, M. Ability factors and the speed of information proces-
    sing. In M. Friedman, J. P. Das, and M. O'Connor (Eds.),
    Intelligence and Learning. New York: Plenum Press, 1980.
MacLeod, C. M., Hunt, E. B., and Mathews, N. N. Individual
    differences in the verification of sentence-picture relation-
    ships. Journal of Verbal Learning and Verbal Behavior, 1978,
    17, 493-507.
Mandler, G. Mind and emotion. New York: Wiley, 1975.
Mathews, N. N., Hunt, E., and MacLeod, C. Strategy choice and
    strategy training in sentence-picture verification. Journal of
    Verbal Learning and Verbal Behavior (in press).
McDermott, J., and Forgy, C. Production system conflict resolution
    strategies. In D. A. Waterman and F. Hayes-Roth (Eds.),
    Pattern-directed inference systems. New York: Academic
    Press, 1978.
Newell, A., and Simon, H. Human problem solving. Inglewood
    Cliffs, New Jersey: Prentice Hall, 1972.
Newell, A., Shaw, J. C., and Simon, H. Elements of a theory of
    human problem solving. Psychological Review, 1958, 65, 151-166.
Podgorny, P., and Shepard, R. N. Functional representations
    common to visual perception and imagery. Journal of Experi-
    mental Psychology: Human Perception and Performance, 1978,
    4, 21-35.
Posner, M. I. Chronometric explorations of mind. Hillsdale, New
    Jersey: L. Erlbaum Associates, 1978.
Schneider, W., and Shiffrin, R. M. Controlled and automatic
    human information processing: I. Detection, search and at-
    tention. Psychological Review, 1977, 84, 1-66.
Schwartz, S. Verbal ability and the controlled-automatic processing
    distinction. In M. Friedman, J. P. Das, and M. O'Connor (Eds.),
    Proceedings from the 1979 international NATO conference on
    intelligence and learning. New York: Plenum Press, 1980.
Selfridge, O. Pandemonium: A paradigm for learning. In D. L.
    Blake and A. M. Uttley (Eds.), Proceedings of the symposium
    of the mechanisation of thought processes. London: HM
    Stationary Office, 1959.
Simon, H. A. Identifying basic abilities underlying intelligent
    performance of complex tasks. In L. B. Resnick (Ed.),

The Nature of intelligence.   New   York: Wiley, 1976.
Underwood, B. J.   Individual differences as a crucible in theory con-
    struction.   American Psychologist, 1975,   30,   128-134.
Willerman, L.   The psychology of individual and group differences.
    San Francisco: W. H. Freeman   and   Company, 1979.

Reference Notes

1.   Palmer, J. C., MacLeod, C. M., Hunt, E., and Davidson, J.
     Some relations between visual information processing and
     reading. University of Washington Technical Report, 1979.

COGNITIVE PSYCHOLOGY AND PSYCHOMETRIC THEORY

Alan D. Baddeley

MRC Applied Psychology Unit

Cambridge, England

For the purpose of discussion I would like to consider the issue of what a cognitive psychologist can contribute to the development of psychometrics. I shall be primarily concerned with the more general issues of the relationship between cognitive psychology and psychometrics. Dr. Lansman's paper described some intriguing experiments, but it is probably true to say that her work illustrates the contribution of psychometric technique to experimental psychology rather than the reverse. I shall say relatively little about Dr. Rabbitt's paper primarily because at the time of writing this discussion it is not available.

In order to explore the relationship between psychometrics and psychonomics I shall move one step back to the motivation of the two disciplines. Psychometrics has its origins as an applied science concerned with producing a technology of measuring human performance or potential. Frequently the measurement had a very specific practical end in view such as selecting among candidates for particular types of jobs or particular types of educational institution, or assisting the educational or clinical psychologist in the problem of diagnosing and possibly advising and treating clients. Under such circumstances, it is important to come up with an answer that works. This may be contrasted with the essentially academic nature of most psychonomics. Here the aim is to produce a better theory of some selected phenomenon; the time-scale is elastic and there is no outside customer who must be convinced and satisfied.

Since the psychometrician was concerned with making predictions about performance in very complex situations in the outside world where he is likely to have minimal control over the relevant

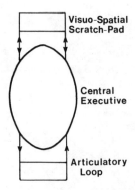

<u>Figure 1</u>. A simplified representation of working memory. A multi-
purpose Central Executive takes advantage of a number of
slave systems. These include the Visuo-spatial Scratch
pad involved in manipulating spatial imagery and the
Articulatory Loop, involved in inner speech.

I shall argue that the theoretical approach typified by our
reséarch on working memory can be of value to the psychometrician both by offering a deeper understanding of current psychometric concepts, and by helping to explain existing psychometric
data in ways that are not suggested by current psychometric
theory.

I would like to suggest that our investigation of the visuo-
spatial scratch pad (Baddeley & Lieberman, in press) provides
potentially useful insights into the way in which subjects perform
tasks aimed at measuring visuo-spatial ability. Using the selective
interference technique we have been able to present evidence for
the operation of a temporary spatial storage system. The system
is spatial since it is disrupted by a spatial but non-visual task,
tracking a moving sound source, but not visual in the peripheral
sense, since it is not disrupted by a visual but non-spatial bright-
ness judging task. Phillips and Christie (1977) present evidence
suggesting that the system is dependent on the central executive
since performance can be impaired by a non-spatial auditory
arithmetic task. The system does appear to be involved in comprehending spoken directional information (Wright, Holloway and
Aldrich, 1974), and in the operation of spatial imagery mnemonics,
but is not responsible for the greater memorability of high imagery
words, a phenomenon which is probably attributable to the long-
term semantic characteristics of the words (Baddeley and Lieberman,
in press). I would like to suggest that our investigation of this
system might be of value to the psychometrician concerned with
visuo-spatial abilities.

variables, he inevitably tended to concern himself with large and robust effects. During the last years of the 19th and early years of the 20th century, there were two types of approach to the problem. J. M. Cattell initially used a laboratory-based approach, and attempted to predict performance outside the laboratory using such measures as choice reaction time, while Binet opted for a more shotgun approach in which the subject was presented with a wide range of tasks which might plausibly be assumed to test a range of cognitive abilities. The Binet approach proved the more robust and successful, and has dominated much of psychometrics ever since.

It is clear that very large differences between people do exist, and that intelligence tests such as the Stanford Binet do allow one to classify individuals in a reasonably satisfactory way; indeed it is probably true to say that psychometrics has been responsible for the greatest practical impact that psychology has made on Western society over the last 50 years. Interestingly, the differences in the efficiency of a given individual as a function of changes in either his external or internal environment appear to be much less. Subjects are remarkably good at maintaining their performance despite loud noises, quite large amounts of alcohol or fluctuations in body temperature. This area of psychometrics has tended to abandon the Binet type intelligence test and look for more precise and analytic measures of performance. As such, it has tended to rely much more heavily on experimental psychology for both techniques and theory. It is an interesting area of overlap of interests from which I believe lessons can be learnt for the case of psychonomics and psychometrics in general, but there is insufficient time here to go into the issue further.

Psychometrics coupled its initial broad spectrum approach with a set of techniques which, by selecting objective criteria, allowed it to progressively refine its tests. The strength of such a process of natural selection is that it does not rely too heavily on the tester's theoretical assumptions as to how the subject is performing a particular task; if a task predicts performance well, then it is retained, if not, it is replaced.

While from a practical point of view however psychometrics has been extremely successful, inherent in the approach were a number of problems. (1) In contrast to the considerable technology involved in filtering out and refining test items, there is no adequate way of ensuring that the appropriate tests go into the test battery in the first place. (2) The appropriate source for such tests would presumably be from some growing and developing model of human cognition. Unfortunately, the type of model which tends to be produced by psychometrics is one that is essentially a recategorization of the data; a classification system rather than a model in the sense in which it would be understood

in psychonomics or, I suspect, in most sciences. Classification is
an important first step, but there is, I believe, a limit to the
usefulness of producing increasingly sophisticated classification
systems that are based on tasks which themselves are not understood.   (3) Finally, an important problem that faces psychometrics is the need to develop and maintain population norms.
Such norms are absolutely crucial for the practical tasks that a
psychometrician must perform, but inevitably they must lead to
conservatism; if you already have norms on 10,000 people, you
need to be very convinced before you decide to change your test
in any way. These three problems are not presented as criticisms
of psychometrics, merely as constraints place on the theoretical
development of the concept of intelligence by the very success of
psychometric technology.

Can the Cognitive Psychologist Help?

To what extent can a cognitive psychologist working in a
psychonomic tradition help the psychometrician? The obvious
answer, on which Professor Hunt and I agree, is in providing
potentially helpful theories. I would like in the present paper to
say a little about Professor Hunt's approach, before going on to
discuss a complementary approach which I myself favour.

Traditionally psychometrics has reached its theoretical conclusions inductively by attempting to produce more efficient and
satisfying conceptual descriptions of a detailed mass of data.
Professor Hunt has adopted exactly the opposite procedure,
namely that of starting with the assumption of a complex functioning system, and deducing what its relevant characteristics are
logically likely to be. The approach will be familiar to experimental psychologists as that adopted by the cognitive science approach
to theorizing. It differs from cognitive psychology in relying less
on experimentation and depending much more on computer simulation as a method of exploring and testing its concepts. Its
strength is that it is prepared to tackle important but difficult
problems, often in novel ways. Its weakness lies perhaps in its
tendency to over-ambition, and to problems of evaluating the
theories it produces.

The problem of evaluation is particularly acute in the case of
Professor Hunt's approach since it is very much in its infancy.
How, for example, should one decide whether the particular
concepts formulated are the most appropriate, indeed does it
matter if they are not? Suppose one has two cognitive science
based models, how does one decide which is preferable? These
problems may of course solve themselves if, for example, only one
satisfactory formulation proves possible. Alternatively, two
separate formulations may turn out to have very crucial common
elements. At present then all one can say is that an overall

conceptualization of human thought is a very worthwhile challenge, and to wish Professor Hunt and his theory the best of luck.

While accepting the potential usefulness of global top-down theories of a cognitive science variety, I would like to suggest that cognitive psychology can also offer some theoretical assistance to the psychometrician at a rather more mundane level. My own theoretical approach is to try to break down cognitive performance into the operation of sub-systems. Although it relies heavily on standard laboratory tasks, the aim is to investigate the underlying system, not the task, and we therefore make use of the method of converging operations whereby the same hypothetical sub-system is studied using a range of different tasks and procedures. Since the sub-systems are not linked in a simple linear way, interactions between components do occur, raising problems that are difficult, but not, I believe, intractable. Theorizing at such an intermediate conceptual level is not of course inconsistent with Hunt's more global top-down approach. Development of his approach will subsequently demand a more detailed analysis of sub-components, while a consideration of specific sub-systems makes implicit or explicit assumptions about the role of such components in a more global conception of human cognition.

My main current interest concerns the role of short-term memory in other information-processing tasks such as reading, learnning and reasoning. Graham Hitch and I have developed a conceptual system we term Working Memory to account for existing data and guide future research. We found the assumption of a simple unitary STM system increasingly implausible and at an early stage decided to split off two "slave" systems, one concerned with temporarily maintaining speech information through subvocalization (the articulatory loop), the other maintaining spatial information through a labile image (the visuo-spatial scratch pad). Current research suggests we may be able to justify splitting off further sub-systems, but for the present purposes, Fig. 1 gives a reasonable summary of the Working Memory model. The heart of the system is the Central Executive which has both storage and attentional capacities. It is responsible for selecting and switching strategies, and is probably closely associated with consciousness and maintaining what James called "the spacious present." I would expect the central executive to be heavily involved in tasks that are assumed to measure fluid intelligence. Graham Hitch and I have shown that working memory is involved in verbal reasoning, learning and comprehension (Baddeley and Hitch, 1974; Hitch and Baddeley, 1978), and the concept has subsequently proved useful in investigating both arithmetic (Hitch, 1978) and reading (Baddeley, 1979). An overview of this work is presented by Hitch and Baddeley (1978).

My second illustration is concerned with the implications of
the concept of an articulatory loop for a problem which overlaps
the areas of psychometrics, psychonomics and education. It may
be recalled from the papers given earlier in the proceedings by
Chase (Chapter 13) and by Nicolson (Chapter 16) that one of the
keystones of the concept of the articulatory loop is the word-length
effect. Memory span for words is a function of their spoken duration,
span being roughly equivalent to the number of words that can
be spoken in two seconds (Baddeley, Thomson and Buchanan,
1975).

Ellis and Hennelly (in press) noted that the digit-span for
Welsh children on the Welsh Children's Intelligence Scale were
reliable lower than the equivalent U.S. norms on the WISC.   This
difference is shown in Fig. 2.

This might of course reflect a verbal inferiority of the
Welsh, compensated no doubt by superior performance on tests
involving choral singing or the passing of rugby balls.   Ellis and
Hennelly however suggested that the result might stem from the
word length effect, since although Welsh digit names have the
same number of syllables as English, they tend to have longer

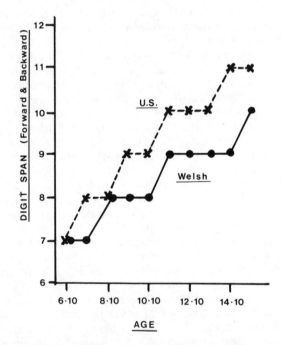

Figure 2. Digit span (combined forward and backward span) for US
          and Welsh children. The data are based on Ellis and
          Hennelly (in press).

vowel sounds. They went on to test a group of bilinguals who
had Welsh as their dominant language and observed that their
mean span for English digits (6.55 items), was significantly
greater than their span their span in Welsh (5.77). As predicted,
their digit reading rate was also significantly faster in English
(321 ms/digit) than in Welsh (385 ms/digit). Converting their
digit span into time scores gives equal spans in the two languages;
2.22 seconds' worth of digits in Welsh and 2.10 seconds' worth
in English. As a final check of the articulatory loop hypothesis,
they compared Welsh and English digit span under conditions of
articulatory suppression. Here the subject is continuously required
to articulate an irrelevant sound such as "the," thereby preventing
the use of the articulatory loop, and abolishing the word length
effect (Baddeley et al, 1975). Under these conditions, English
and Welsh span showed no reliable difference. Subsequent work
by Ellis (personal communication) has capitalized on Hitch's (1978)
demonstration of the role of working memory in arithmetic, and
has shown the predicted increase in errors when bilingual subjects
carry out the computation in Welsh, in contrast to their performance
in English.

Ingenious though they are, the results of Ellis and his
co-workers are not of course of major practical importance. They
are, I believe, significant however in providing concrete evidence
that concepts such as that of working memory and its articulatory
loop can make a genuine contribution to understanding of both
psychometric data and real, if minor, educational problems. As
such they encourage us to share the belief of the other three
speakers in this session that the experimental cognitive psychologist does have something to contribute to the development of
psychometric theory.

References

Baddeley, A. D. Working memory and reading. In: Kolers,
    P.A., Wrolstad, M. E., and Bouma, H. (Eds.), The Processing
    of Visible Language. New York: Plenum Press, 1979.
Baddeley, A. D., and Hitch, G. Working memory. In: Bower,
    G. A. (Ed.), The Psychology of Learning and Motivation,
    vol. 8. New York: Academic Press, 1974.
Baddeley, A. D., andLieberman, K. Spatial working memory. To
    appear in: Nickerson, R. (Ed.), Attention and Performance
    VIII. Hillsdale, New Jersey: Lawrence Earlbaum Associates,
    in press.
Baddeley, A. D., Thomson, N., andBuchanan, M. Word length
    and the structure of short-term memory. Journal of Verbal
    Learning and Verbal Behavior, 1985, 14, 575-589.
Chase, W. G. Individual differences in memory span. In:
    Friedman, M. (ed.), Intelligence and Learning. New York:
    Plenum Press, in press.

Ellis, N. C., and Hennelly, R. A. A bilingual word length effect:
    implications for intelligence testing and the relative ease
    of mental calculation in Welsh and English. British Journal
    of Psychology,   in   press.
Hitch, G. The role of short-term working memory in mental
    arithmetic. Cognitive Psychology, 1978, 10, 302-323.
Hitch, G., and Baddeley, A. D. Working memory. In: Cognitive
    Psychology: Memory, Pt. 1. D.303, Block 3, Units 13-15.
    Milton Keynes: Open University Press, 1978.
Nicolson, R. The relationship between memory span and
    encoding speed. In: Friedman, M. (Ed.), Intelligence and
    Learning. New York: Plenum Press, in press.
Phillips, W. A., and Christie, D. F. M. Interference with
    visualization. Quarterly Journal of Experimental Psychology,
    1977, 29, 637-650.
Wright, P., Holloway, C. M., and Aldrich, A. R. Attending to
    visual or auditory information while performing other concur-
    rent tasks. Quarterly Journal of Experimental Psychology,
    1974, 26, 454-463.

A COMPARISON OF PSYCHOMETRIC AND PIAGETIAN ASSESSMENTS OF SYMBOLIC FUNCTIONING IN DOWN'S SYNDROME CHILDREN

Lorraine McCune-Nicolich   and   Patricia Munday Hill

Rutgers University   New Jersey Dept. of Education

New Brunswick, N. J., U.S.A.   Trenton, N. J., U.S.A.

Abstract

A comparison is made between the results of psychometric and Piagetian assessment of the symbolic functioning of young Down's syndrome children. Complementary information is gained from the two assessment paradigms. Symbolic play intervention is suggested as one approach to stimulating cognitive growth for retarded subjects.

In the past several decades two parallel themes in the study of infant intelligence have been apparent. The psychometric approach has evolved from the tradition of mental testing which began with Binet, and is exemplified by such infant tests as the Bayley Scales of Infant Development (Bayley, 1969). The Piagetian approach is based on the search for antecedents in infancy to the logical processes of thought that are evident in adults. Assessment in this tradition focuses on the development of sensorimotor skills which have been shown to evolve in hierarchical fashion in such domains as object permanence, imitation, and means-end relationships. The psychometric approach has the advantage of empirical validation, that is, standardized procedures which elicit aspects of infant behavior consistently indicating developmental progress. The Piagetian approach has the advantage of a strong theoretical orientation to guide interpretation of results. Infant intelligence tests are customarily used for diagnostic purposes. Piagetian measures have been largely confined to research, although some diagnostic uses have been made. Important benefits will accrue from the interaction of these two traditions (McCall, Eichorn and Hogarty, 1977). In interpreting results of multivariate analyses of the Berkeley Growth Study data, which is illustrative of a psychometric approach, these authors specify patterns of test performance which characterize the onset of symbolic functioning. According to Piaget (1962) the symbolic ability (or semiotic function) develops at the end of the sensorimotor period allowing the simultaneous development of symbolic play, deferred imitation and language.

A recent study of symbolic abilities in Down's syndrome children (Hill, 1978) which provides the background for the present paper illustrates the interactive potential of psychometric and Piagetian assessment for improving diagnosis. Previous research had supported a general relationship between cognition and symbolic development demonstrated in play, without specifying the nature and degree of the relationship. Hill included the Bayley Mental Scale and Infant Behavior Record, a psychometric measure, and Piagetian measures of object permanence and symbolic play in her design. In addition the language performance of the subjects was compared to the other measures. Down's syndrome subjects were selected (a) to facilitate comparison of the correlation between symbolic play and mental age with the correlation between symbolic play and chronological age and (b) to determine the characteristics of play in this population.

Method

Subjects were 30 Down's syndrome children between 20 and 53 months of age with a range of mental ages from 12 to 26 months. Each child was seen at home with the mother or primary caretaker present. A 1/2 hour play session was videotaped, followed by administration of the Bayley (Mental Scale and Infant Behavior Record) and the object permanence task (Corman & Escalona, 1969).

The videotapes were transcribed and divided into episodes based on the child's object contacts. Each episode was assigned a symbolic level. Subjects were then assigned a symbolic play level based on their highest play performance independently and consistently demonstrated. The levels of play were defined as follows. Level 1 play is presymbolic and does not involve pretend. Here the child demonstrates recognition of an object's function by gesture. In Level 2 play the child engages in simple acts of self pretend. Level 3 games, like those in Level 2 are also single acts of pretend but here the symbolism is extended beyond the child's own body and daily activities. Level 4 play includes combinations where the same action scheme is repeated with several objects (4.1) and combinations of several action schemes (4.2). The highest level, Level 5 involves games that are planned prior to performance.

Results And Discussion

Analysis of play behavior supported a four level scale based on structural properties identified by Nicolich (1977). The four levels which scaled were as follows: Level 1, Presymbolic Scheme; Levels 2 and 3 pooled, single pretend acts; Level 4, Combinatorial Pretend; and Level 5, Planned Pretend. The scale analysis yielded coefficients of reproducibility (.98) and scalability (.88) well above the minimum values required for an ordinal scale. Subjects were grouped according to the highest symbolic play level observed for further analysis.

Symbolic play level was more highly correlated with mental age (.74) than with chronological age (.44). Performance on the Infant Behavior Record which "assesses the child's social and objective orientations toward his environment" (Bayley, 1969, p. 4) was also highly correlated with symbolic play level. (The canonical correlation was .97.)

Both symbolic play and object permanence were related to productive language. Subjects who spoke in single words all showed Symbolic Play Level 2 or higher. Three of these subjects exhibited Stage 5 object permanence, the other 22, Stage 6. Twenty-two of the twenty-four subjects who had attained Stage 6 object permanence used at least single words. Only four children in the study used two-word sentences. These subjects had entered Stage 6 object permanence and showed multi-scheme combinations in play (Level 4.2).

Separate discriminant analyses were performed relating the Bayley Mental Scale and the Infant Behavior Record to symbolic play level. Four sets of cognitive (Mental Scale) skills were influential in discriminating symbolic play groups: doll behaviors, means-end skills, cube behaviors and a set of behaviors reflecting language comprehension competence (93% discrimination). These items reflect both sensorimotor skills as identified by Piaget and items noted by McCall et al. (1977) as defining a major transition in normal symbolic development. Productive language items which were prominent in the McCall et al. analysis did not influence the discrimination, reflecting the specific deficit of this population.

A perfect discrimination of subjects into symbolic play groups (100% discrimination) was achieved by analysis of Infant Behavior Record results. The following behaviors were most influential in the discrimination: general object orientation, social responsiveness to persons, general emotional tone, overall status during the testing, goal directedness, and fine motor coordination. This result suggests a strong interdependence between affective development and symbolic play ability.

The results of this study show a convergence between results of psychometric and Piagetian assessment. Item analysis of infant test results can be related to sequences of cognitive development and used to pinpoint specific deficits in sensorimotor functioning which may be preventing the continued cognitive development of the child. Symbolic play assessment can be used in supplementary fashion to determine non-linguistic symbolic functioning.

Based on such assessments attempts can be made to induce developmental milestones. Play intervention may be of particular importance in stimulating symbolic functioning. Many retarded people fail to move beyond sensorimotor functioning. Following the sensorimotor period Piaget (1962) describes a "symbolic period" during which time the child internalizes the sensorimotor knowledge gained in some form of mental representation. Children who fail to exhibit higher symbolic play behaviors may be showing a general symbolic deficit which prevents the transition to the pre-operational level and the learning of language rules. Further study of symbolic abilities in normal subjects as well as impaired populations is required before the success of play intervention could be securely predicted.

References

Bayley, N. Bayley Scales of Infant Development. New York: The Psychological Corporation, 1969.

Corman, H. and Escalona, S. Stages of sensorimotor development: A replication study. Merrill-Palmer Quarterly, 1969, 15(4), 351-361.

Hill, P.M. An analysis of the relationship of cognitive development to symbolic play behavior in Down's syndrome children. Doctoral Dissertation, Rutgers University, 1978.

McCall, R.B., Eichorn, D.H., & Hogarty, P.S. Transitions in early mental development. Monographs of the Society for Research in Child Development, 1977, 42(5, Serial No. 173).

Nicolich, L. McCune. Beyond sensorimotor intelligence: Assessment of symbolic maturity through analysis of pretend play. Merrill-Palmer Quarterly, 1977, 23, 2, 89-99.

Piaget, J. Play, Dreams, And Imitation. New York: Norton, 1962.

A COMPARISON OF THE CONSERVATION ACQUISITION

OF MENTALLY RETARDED AND NONRETARDED CHILDREN

Dorothy Field

University of California, Berkeley

Berkeley, California, U.S.A.

This paper reports results of a continuing study of the cognitive development of mildly mentally retarded and nonretarded children in which Piagetian conservation training has been the primary investigative tool. I will argue that this complex cognitive training is a particularly useful way to increase our understanding of the similarities and differences between children of normal and subnormal intelligence. I will further suggest that this training may be useful as a diagnostic tool to distinguish children who suffer from retardation from those whose learning disabilities stem from other causes.

Method

One hundred eighty children were trained in six studies: 87 of normal intelligence, MA 3-1 to 7-6, CA 3-0 to 5-9, and 93 cultural-familial retarded children with no known organic defects, MA 3-10 to 11-0, CA 6-8 to 14-2. Sixty natural conservers, 30 retarded and 30 nonretarded, have been examined as well. As described elsewhere (Field, 1974, 1977, 1978), the procedure in all studies included a pretest, three training sessions, and a posttest. Children were seen individually for 15 to 25 minute sessions in which number and length concepts were trained in an oddity format. Five quantities, number, length, mass, liquid, and weight, were included in the posttest. Only nonconservers were included in the experimental groups. All groups were matched for MA, CA, sex, and school. Materials were the same throughout. Only type of training, MA, CA, and type of subject varied among studies. Control children received no training but experienced similar amounts of individual attention and reinforcement. Only the results of Verbal training will be reported here, for it was more successful than Learning Set training for all groups of children. In four of the six studies a second posttest was given after 2½ to 16 months. Both posttests included quantities that had not been trained. The studies were able, then, to test for (a) generalization of conservation understanding and (b) permanence of acquisition.

Results: Similarities of Retarded and Nonretarded Children

Training. Figure 1 shows that retarded and nonretarded children did not differ in their training scores, although MA differences were apparent.

Posttest 1. Figure 2 shows that there were few differences in the number of quantities mastered by retarded and by nonretarded children. Only among the youngest children did retarded and nonretarded differ significantly in posttest conservation. In all studies, control group members made very little progress.

Generalization. A surprising number of retarded children conserved three or more quantities on the posttest, showing that they had generalized their conservation mastery to quantities not included in training, even though they had been complete nonconservers when the study began. On the posttest, 25% of the retarded children conserved all five quantities, and an additional 19% conserved four.

Posttest 2. The delayed posttests given after 2½ to 16 months showed considerable similarities among the groups of children. In all studies, verbally trained children maintained their conservation mastery over time. Of the 92 children who took two posttests, 69

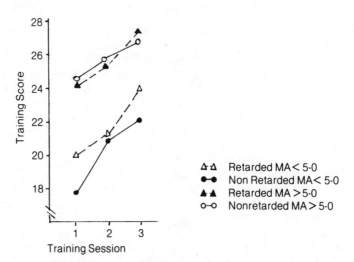

Figure 1. Training scores of retarded and nonretarded children, by mental age.

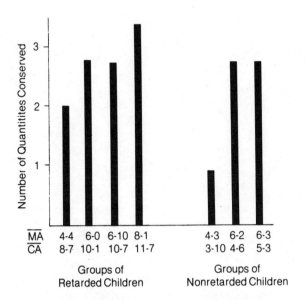

Figure 2. Group means: Posttest 1.

did not change their conservation status between posttests;   16 increased their mastery to become generalizers.   Only seven of the children who generalized on the first posttest failed to do so on the second;   only one of these was retarded, and four others were less than five years old.

Justifications.   At posttest, each time the children judged whether a quantity was still the same amount, they were asked "How can you tell?"   Resulting justifications were examined, as were the responses of the natural conservers, yielding conservation tests of 204 children.   Figure 3 shows the very similar patterns of responses. Trained or naturally conserving, mentally retarded or normal, the groups differed only in one instance: addition/subtraction justifications, the most analytical of the identity justifications, were given only by the naturally conserving normal children.   This may well indicate a true difference in the quality of conservation mastery shown by the groups.

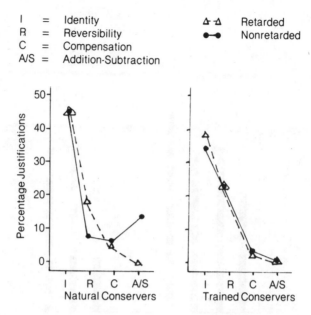

Figure 3. Percentage of Justifications of each type offered in all posttests.

Multiple justifications. A child might say, for example, "Well, of course it's still the same amount of plasticene; here, I'll roll it back into a ball and show you." This response was scored as including both identity and reversibility justifications. Of the trained children, 60% of the retarded and 39% of the normal gave such multiple justifications; of the natural conservers, 43% of the retarded and 13% of the normal produced these more complex responses. Retarded children were significantly more likely to give multiple justifications than nonretarded children, and trained children offered significantly more multiple justifications than natural conservers. Familiarity with the test situation may account for the differences between trained and natural conservers; the importance of motivation to retarded children's performance will be discussed below.

Variety of justifications. More information was gained when the investigator took the trouble to inquire at length, giving children further opportunity to display their understanding. Eighty children participated in posttests in which four transformations were given for each quantity. Different justifications were offered on the third and fourth transformations by 41% of the retarded and 33% of the nonretarded children. It seems surprising to find such a large number of children, especially the retarded children, enterprising enough to invent new and different explanations as each "game" progressed.

Discussion

How can it be that these cultural-familial retarded children were able to acquire conservation through verbal training quite as well as children of normal intelligence, when the groups were matched for MA? They conserved the trained quantities; they generalized to untrained quantities; they maintained their conservation understanding over long periods; their verbal responses were as varied and often more complex.

The games format contributed to the effectiveness of training; the broad variety of materials captured the children's interest and attention; spaced practice is known to be effective; the variety of transformations facilitated generalization; individual attention and praise increased rapport and motivation. Motivation is crucial. The retarded children enjoyed the "games"; they valued the time away from the hurley-burley of their classrooms, the individual attention and the praise they so rarely received elsewhere; and they engaged in activities that prolonged the sessions, such as participating actively in the games and giving many and varied justifications.

This behavior was not typical of these children on initial contact. At the beginning they were, for the most part, passive spectators. But as rapport was established and the pattern of the sessions became understood, they became more active in the experimental situation. They manipulated the materials and sometimes invented quite acceptable transformations in the oddity games format. This behavior seems to be in accord with Piaget's (1952) emphasis on the necessity of action on the part of the child in order to accommodate new schemata. With guidance, these children did act on their environment, and their generalizations reflect this. Why had the retarded children not evolved this strategy on their own? Were they initially passive because past failure experiences had depressed their expectations and performance? Or is this passiveness inherent in their retardation? Piaget (1974) suggested that subjects who passively received information from the adult would no longer learn anything without help.

How was it possible for some of the retarded children to acquire such a complete understanding of conservation in such a short time? Vygotsky (1962) was concerned with the relation between a given level of development and a child's potentiality for learning, which he called the "zone of proximal development." As he pointed out, the level of the child's actual development is the result of particular experiences and training, as well as of his or her maturation level. Most kinds of tests--school achievement, IQ, and so on--evaluate only what the child has already learned. Vygotsky wished to evaluate the capability of the child's zone of proximal development. With the help of imitation, primary cues, guiding questions, and so on, the child can do much more than he could do independently. That which the child can do with guidance throws light on the processes that are in the course of being established and can serve to predict the child's potential performance under optimal conditions.

The training of conservation tasks appears to be a particularly good tool to distinguish between true retardation and learning disabilities from other sources. The verbal method used in these studies is a complex conceptual activity and makes the training most promising for diagnosis. This training should be able to predict the child's potential in great breadth. The verbal require- ments of the conservation justifications seem to encourage the child to formulate a more advanced concept of invariance and to general- ize this concept to many quantities. Many of these retarded chil- dren have shown that their learning potential is far greater than their usual performance. Further experiments in this series will seek to determine what other characteristics may accompany the differences revealed so far.

References

Field, D. Long-term effects of conservation training with education- ally subnormal children. Journal of Special Education, 1974, 8, 237-245.

Field, D. The importance of the verbal content in the training of Piagetian conservation skills. Child Development, 1977, 48, 1583-1592. Field, D. How children in ESN schools in London learn conservation skills. In G. L. Lubin, M. K. Poulsen, J. F. Magary, and M. Soto-McAlister (Eds.). Piagetian Theory and its Implications for the Helping Professions, Pro- ceedings of the Seventh Inter-disciplinary Conference (Vol. 2). Los Angeles: University of Southern California Press, 1978.

Piaget, J. The Child's Conception of Number. London: Routledge and Kegan Paul, 1952. Piaget, J. Forward. In B. Inhelder, H. Sinclair, and M. Bovet. Learning and the Development of Cognition. London: Routledge and Kegan Paul, 1974.

Vygotsky, L. S. Thought and Language. Cambridge: MIT University Press, 1962.

GENERALIZATION OF A REHEARSAL STRATEGY

IN MILDLY RETARDED CHILDREN

R.W. Engle, R.J. Nagle and M. Dick

University of South Carolina

Columbia, South Carolina, U.S.A.

Abstract

Two groups of EMR children were given a series of free recall tasks. One group was trained to use a strategy designed to induce deeper level semantic encoding and a "no training" control group received standard free recall instructions. Subjects received either related or unrelated lists during training and related or unrelated lists during two posttests (immediately following and one week after training). Semantic strategy usage was retained at posttesting and also generalized to word lists unlike those used during training.

Several interpretations of the memory deficits associated with mild retardation propose that these deficits are not a result of reduced memory capacity or other structural deficits, but result from either a failure to use any rehearsal strategy during acquisition or the use of an ineffective strategy (Ellis, 1970; Belmont & Butterfield, 1969). One consequence of this viewpoint has been the evolution of what Belmont and Butterfield (1977) call the "instructional approach" which is directed at discovering and developing effective training methods for improving memory skills in the retarded.

While most of the early research on strategy training with the mentally retarded involved the use of simple strategies such as cumulative rehearsal (Brown, Campione & Murphy, 1974), our research has been directed at training strategies likely to establish more durable and easily retrieved memories. Engle and Nagle (1979) used a "Semantic" strategy in which subjects in a free recall task were instructed to think of any functions the object might have, to think of any personal experiences he or she might have had with the object and to try to remember other objects in the list that were related or similar in any way. These instructions were directed at making the material more meaningful to the child by inducing what Craik and Lockhart (1972) have called deeper levels of encoding. Engle and Nagle found that Semantic instructions greatly facilitated recall and categorical clustering compared to two control groups designed to mimic strategies commonly used by children.

The present research was designed to investigate further the benefits of the Semantic strategy and the generality of the strategy usage to stimuli different from those used during training. The study consisted of training subjects on the Semantic strategy or a Neutral strategy consisting of standard free recall instructions with no advice on how to insure optimal performance. These two training conditions were crossed with whether the subjects received related or unrelated lists during training and whether they received related or unrelated lists on the two posttests.

Methods

Subjects. The subjects were 51 children with IQ scores in the 50 to 75 range with a mean CA = 11.5 years and SD = 1.2 years.

Materials. Items were chosen from 23 common taxonomic categories and were all common responses from the Posnansky (1974) norms. The two types of 20-item lists were (1) related lists consisting of random arrangement of four words from each of five different categories and (2) unrelated lists containing one item from each of 20 different categories.

Conditions. Subjects were assigned to one of 8 conditions derived by factorily manipulating three factors in a completely balanced design. These factors were (1) the type of instructions given to the child -- Semantic strategy instructions or Neutral, i.e., standard free recall instructions (2) related or unrelated training lists during training and (3) related or unrelated lists at posttesting. The children were assigned to conditions based on a pretest consisting of two free recall trials on a list of 20 unrelated items. The mean pretest scores did not differ for the eight groups.

Training Procedure. Training was carried out in two sessions. For children assigned to the Semantic strategy condition, the training lists were preceded by instructions and practice on the use of the strategy. They were told that the best way to remember a list was to think about each item in the following terms: (1) its function, (2) personal experiences with it, and (3) other items from the list that were related to it. During the instructional period, the experimenter stressed the importance of the child's active participation in applying these three criteria to each of the items in the list. The amount of prompting decreased over training sessions.

For both the Semantic and Neutral instructional groups, the first of the 20-item lists was presented at 15 sec/item while the second list was presented at 10 sec/item. A third training list was given in the second session which was followed by one post-test list immediately afterward and a second posttest one week later. During posttesting no reference was made to either group as to any strategy or method to be used to remember the items.

Results and Discussion

The dependent variables of interest were mean number of words recalled and the amount of categorical clustering (for the related lists) or subjective organization (for both unrelated and related lists).

Recall. Analysis of the data from the training trials (displayed in Figure 1) showed that the Semantic groups that received training on related lists recalled more items per trial (M = 15.8) than the Semantic group with unrelated lists (M = 10.4), or either of the Neutral groups (10.9 and 10.1). These data suggest that Semantic strategy instructions enhance memory performance only on related lists and that superior performance on related lists is accomplished only when accompanied by Semantic strategy training. The postest data, however, suggest otherwise.

For several reasons the first posttest data are not the best index for the generalization of the Semantic strategy from one type of material to another. Looking at the data from posttest two, however, we can observe several interesting trends. For one thing, every group that received Semantic training yields a higher performance than the corresponding Neutral group. Secondly, this is true regardless of whether the lists during training were related or unrelated and whether the lists during the posttest were related or unrelated. While the SUR group, given Semantic training on unrelated lists but tested on related lists showed about 11% transfer on PT2, the SRU group, given Semantic training on related lists but tested on unrelated lists, showed 42% positive transfer. This strongly suggests that the Semantic strategy does generalize to novel stimuli, particularly if the training involved lists of clusterable items.

The data for organization at recall reflect a pattern of data similar in many ways to the recall data. The training trial performance seemed to be greater only if the groups received both Semantic training and related lists but the posttest data indicated

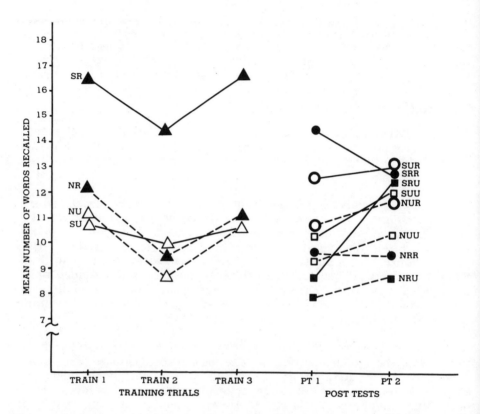

Figure 1. Mean number of words recalled for each of the eight group

facilitation for the Semantic groups regardless of type of material received during training or testing. That is, the Semantic groups organized their material at output more on the second posttest regardless of whether the material was related or unrelated. The ITR for both types of list is shown in Figure 2 and this clearly demonstrates the enhanced organization for the Semantic groups on PT2.

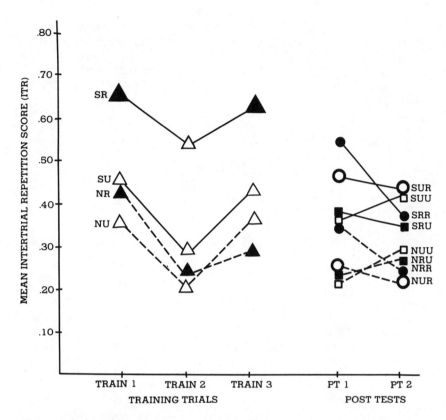

Figure 2. Mean ITR for each of the eight groups (both related and unrelated lists).

References

Belmont, J.M. and Butterfield, E.C. The relations of short-term memory to development and intelligence. In L.P. Lipsitt and H.W. Reese (Eds.), Advances in child development and behavior (Vol. 4). New York: Academic Press, 1969.

Belmont, J.M. and Butterfield, E.C. The instructional approach. In R.V. Kail and J.W. Hagen (Eds.), Perspectives on the development of memory and cognition. Hillsdale, N.J.: Erlbaum, 1977.

Brown, A.L., Campione, J.C. and Murphy, M.D. Keeping track of changing variables: Long-term retention of a trained rehearsal strategy by retardate adolescents. American Journal of Mental Deficiency, 1974, 78, 446–453.

Craik, F.I.M. and Lockhart, R.S. Levels of processing: A

framework for memory research. Journal of Verbal Learning
and Verbal Behavior, 1972, 11, 671-684.

Ellis, N.R. Memory processes in retardates and normals. In N.R.
Ellis (Ed.), International review of research in mental
retardation (Vol. 4). New York: Academic Press, 1970.

Engle, R.W. and Nagle, R.J. Strategy training and semantic
encoding in mildly retarded children. Intelligence, 1979, 3,
17-30.

Posnansky, C.J. Category norms for children. Behavior Research
Methods and Instrumentation, 1974, 6, 373.

COGNITIVE PROCESSING IN LEARNING DISABLED AND NORMALLY
ACHIEVING BOYS IN A GOAL-ORIENTED TASK

D. Kim Reid   and   Iris Knight-Arest

University of Texas at Dallas   New York University

Richardson, Texas, U.S.A.   New York, New York, U.S.A.

Abstract

Ten normally achieving and ten learning disabled boys were
studied as they performed a block balancing task. Some of the
blocks had obvious and some had hidden weights. The learning
disabled boys performed very much like younger children: they
placed the blocks randomly and then adjusted them proprioceptively
or they placed the blocks at their geometric center or in the spot
where the previous block had balanced. Many had no theories
about how things balance or described specific instances rather
than rules. Inefficient language and stress-related avoidance
behaviors were evident. The two learning disabled boys who
expressed initial theories, however, performed more like their
normally achieving peers.

Introduction

Research applying a genetic epistemological framework to the
study of learning disabled children has confirmed their normal
progress through the stages, often with developmental delays
(Reid, 1978). What appears to differ significantly in the learning
disabled is the process they use to achieve normal progress
(deAjuriaguerra, Jaeggi, Guignard, Kocher, Macquard, Paunier,
Quinodoz, and Siotis, 1963). Because traditional "Piagetian" tasks
designed to classify stage-related behaviors have not revealed these
processing differences (Reid, Knight-Arest, and Hresko, in press),
we chose to examine the interplay between children's "theories-in-
action" and their spontaneous organizing activity. The task

(Karmiloff-Smith and Inhelder, 1975) involved balancing blocks with obvious and hidden weights, so that using the center as the balancing point was not always successful. The developmental sequence observed in young, normal children was not stage-linked.

Until about 39 months of age, children placed the blocks randomly on the bar and either let go or used a finger to hold them up. Older children (up to about six) proceeded with additional attempts to balance the blocks and/or became diverted by the subgoal of discovering the properties of the blocks. Children from about six to nine employed the "theory-in-action" that things balance in the middle. They placed the blocks either at the geometric center or in the place in which the previous block had balanced. Later when asked to balance the blocks a second time, many proved incapable. They had become so certain that "things balance in the middle" that they failed to make use of proprioceptive information. Finally, children from seven to nine demonstrated implicit understanding of the relations between length and weight. Unsuccessful trials were followed by continuous, rapid adjustments in the right direction.

Since previous research indicated that learning disabled children often adopted inefficient problem-solving strategies (failing to comprehend the links between their actions and the states of objects) (Reid, in press), we anticipated immature cognitive processing. Since the learning disabled often exhibit language difficulties (Cf. Wiig and Semel, 1976), we expected their verbal explanations to be less precise. Finally, we expected that failure to recall and anticipate the effects of their activities (Inhelder and Siotis, 1963), would interfere with the ability of the learning disabled to modify ineffective "theories-in-action."

Method

Subjects and Procedure. Ten normal and ten learning disabled boys were randomly selected. All of the boys were of average or better intelligence, from middle socioeconomic class suburban schools and were between 10 and 12 years old. The learning disabled had, in addition, a psychoeducational evaluation leading to a diagnosis of specific learning disability.

Each child was first asked to explain how things balance. As each proceeded through the task, he was asked to explain what he was doing. Questions were used to determine what was being learned from activity. Blocks were presented in the order in which they appear in Figure 1. Each session lasted 15 to 20 minutes and was videotaped. Prior to the block balancing, the clay-ball tasks of conservation of substance and weight were administered, but results were unrelated to findings.

Results and Discussion

All of the normal and nine of the learning disabled boys succeeded on all tasks. One boy failed to balance the last block,
which required the use of counterweights. All of the children made
explicit reference to weight (none the compensatory effects between
weight and length), but in many ways the processing of the learning
disabled boys resembled what Karmiloff-Smith and Inhelder described
for younger children. Only five of the learning disabled children
adopted the strategies of placing the block near its geometric center
or in the place of the previous block 75% or more of the time (only
half of the normal boys used this second strategy). Although
none of these children argued that blocks that couldn't be balanced
in the middle couldn't be balanced at all, their performances were
otherwise very similar to those of the normal, Genevan six and
seven year olds. The other five placed the blocks randomly and
apparently used proprioceptive cues to make them balance--a
strategy characteristic of normals younger than six.

Our initial question about how things balance provided some
insight as to why many of these children proceeded proprioceptively.

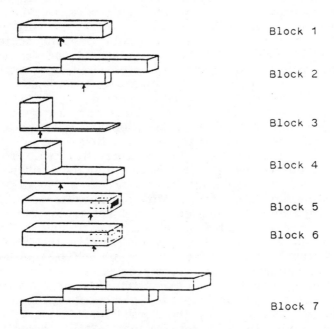

Figure 1. Blocks with obvious and hidden weights. The arrow under-
neath each block indicates the point of contact with the bar
when the block is in equilibrium. (Adapted from Karmiloff-
Smith and Inhelder, 1975).

Nine of the normal, but only two of the learning disabled boys expressed an initial theory as a generalization (e.g., "to make the weight the same" or "to make them equal"). These two learning disabled boys expressed theories, but in perceptual terms: "make a fat thing stay on top of a skinny thing" or "...say you have a scale and one clay is really big and the other..." Because they had theories to test, the normally achieving boys tended to gain more information from their activities. When a block didn't balance as predicted, all ten were able to revise their theories correctly. Of the learning disabled children the two who had initial theories were able to do so. Six placed blocks off center, balanced them, and insisted that they were balanced in the middle! When the examiner argued that the blocks didn't balance in the middle, they resorted to perceptual data for explanation: "there's a notch there" or there "is liquid inside" or "this block has been drilled and plugged." The normally achieving children said only that there was more weight in one end than the other, usually after a pause, and did not scrutinize the blocks to look for the explanations. Furthermore, the learning disabled boys expressed inaccurate theories: "these blocks balance in different places, because this one is heavy and this one is light."

Explanations were more difficult for the learning disabled. They used gestures in place of language, more demonstration, pronouns without antecedents, and poor syntax. Finally, avoidance behaviors were used by one normally achieving and all ten learning disabled boys: they built towers, addressed the camera, stared into space, made hostile gestures, or changed the task (e.g., one boy used his head to support the blocks).

Overall, the normally achieving boys were more efficient strategizers. As with the children in the Genevan study, what they did with the blocks depended on the theories they held and the feedback from their activities in turn affected their theories. The learning disabled tended to have fewer theories and even when they did have them, they did not seem to use them to guide their behavior. Their thinking appeared to be situation-specific and substance-specific. They had not adopted general rules, nor did they stop to think when confronted with a problemmatic finding. They sought explanations in the perusal or manipulation of the blocks. Our findings confirm those of Inhelder and Siotis (1963) with respect to the ability of children to subordinate delayed figurative functions to operations once they have been constructed. The two boys who expressed theories at the beginning of the study used strategies comparable to those used by the normally achieving children and were able to revise their "things balance in the middle" theory after balancing the weighted blocks. Although we have known for some time that the language of the learning disabled is often incorrect and imprecise, this is the only study known to these authors that suggests that their spontaneous organizing activity may be equally inefficient.

References

deAjuriaguerra, J., Jaeggi, F., Guinard, F., Kocher, F., Maquard, M., Paunier, A., Quinodoz, D., and Siotis, E. Organisation psychologique et troubles de development du language. (Etudes d'un groupe d'enfants dysphasiques). Problems de psycholinguistiques. Paris: Presses Universitaires de France, 1963, 109-140.

Inhelder, B., and Siotis, E. Observations sur les aspects operatifs et figuratifs chez les enfants dysphasiques. Problems de psychollinguistiques. Presses Universitaires de France, 1963.

Karmiloff-Smith, A., and Inhelder, B. If you want to get ahead, get a theory. Cognition, 1975, 3, 195-212.

Reid, D. K. Genevan theory and the education of exceptional children. In J. M. Gallagher and J. A. Easley (Eds.), Knowledge - and development Volume II: Piaget and education. New York: Plenum, 1978.

Reid, D. K. Learning and development from a Piagetian perspective: The exceptional child. In I. E. Sigel, R. M. Golinkoff, and D. Brodzinsky (Eds.), Piagetian theory and research: New directions and applications. Hillsdale, New Jersey: Erlbaum Associates, in press.

Reid, D. K., Knight-Arest, I., and Hresko, W. P. The development of cognition in learning disabled children. In J. Gottlieb and S. S. Strichart (Eds.), Current research and application in learning disabilities. Baltimore: University Park Press, in press.

Wiig, E. H., and Semel, E. M. Language disabilities in children and adolescents. Columbus, Ohio: Charles E. Merrill, 1976.

HOME ENVIRONMENT, COGNITIVE PROCESSES,

AND INTELLIGENCE: A PATH ANALYSIS

Robert H. Bradley and Bettye M. Caldwell

University of Arkansas at Little Rock

Little Rock, Arkansas, U.S.A.

Information from longitudinal studies conducted in the last 20 years indicates a strong link between cognitive development and the quality of stimulation available in the early home environment. To date, however, primary attention has been devoted to the direct relationship between environmental stimulation and various cognitive outcomes (i.e., IQ, achievement test scores, language performance). Very few data are available on the relations between environmental quality and the cognitive processes which may facilitate intellectual attainment.

Research by Lewis (1971), Yarrow, Rubenstein, Pederson, Jankowski (1973), and Wachs, Uzgiris and Hunt (1971) shows that such essential learning processes as attention, goal orientation, and foresight are significantly related to environmental quality. However, the precise mediating role that such variables play in cognitive outcomes is less well documented. It is the purpose of this study to investigate several models involving home environment, cognitive processes, and intelligence in order to establish the most likely direction of effect among the three sets of variables.

Method

Subjects

Ninety-three children (31 white, 62 black) residing in Little Rock, Arkansas, participated in the study. The children were part of a longitudinal study conducted by the Center for Child Development and Education. The parent volunteered to participate and was paid a small fee for each testing session. Families were predominantly lower to lower middle income.

Instrumentation

Assessment of environmental quality. In order to assess the quality of stimulation available to children during infancy, the home of each child was assessed with the HOME Inventory when the childrer were 6 months old. This instrument is an observation/ interview procedure composed of 45 items scored in a binary fashion. The items are clustered into six subscales: (1) emotional and verbal responsivity of mother, (2) avoidance of restriction and punishment, (3) organization of the physical and temporal environment, (4) provision of appropriate play materials, (5) maternal involvement with child, and (6) opportunity for variety in daily routine. In terms of reliability, raters were trained to achieve a 90% level of agreement. Internal consistency coefficients for subscales range from .44 to .88. Considerable validity data also exist for the instrument (Bradley and Caldwell, 1976).

Assessment of cognitive processes. In order to assess the cognitive behaviors of participants, the Bayley Scales of Infant Development were administered to each child at ages 6 and 12 months. Three differentiated clusters of items from the Bayley were identified as measuring cognitive functions by Yarrow, Rubenstein, Pederson, and Jankowski (1973): (1) goal-directedness, (2) social responsiveness, (3) language use. Split-half reliabilities for the clusters ranged from .74 to .84. Data reported by Yarrow and his colleagues indicate moderate correlations between these variables and early social stimulation.

Assessment of intellectual capability. In order to assess the intellectual capability of participants, the Stanford-Binet Intelligence Test was administered to each child at age three.

Procedure

Since the primary question addressed was the role played by early cognitive capabilities in the relation between environmental quality and cognitive attainment, three models of early experience were tested using path analysis. Specifically, path analysis was used to study the direct and indirect effect of environmental quality on IQ. HOME scores were treated as "exogeneous" variables in the path analyses, intelligence scores were treated as "endogenous" variables, and cognitive process scores were treated as "exogenous" variables in some analyses and "endogenous" variables in others (see Kerlinger & Pedhazur, 1973 for a discussion of this technique).

Results

Six-month scores on the HOME Inventory showed negligible to moderate relations with social responsiveness, goal directedness and

TABLE 1
Correlations Between Home Environment, Cognitive Process and IQ Scores

Environment Variables	Social Responsiveness		Goal Directedness		Language		IQ
	6 mo.	12 mo.	6 mo.	12 mo.	6 mo.	12 mo.	
Maternal Responsivity	.00	-.12	.23*	-.07	-.03	.00	.07
Avoidance of Restriction and punishment	.23*	.24*	.02	.22*	.06	.33**	.38**
Organization of the Environment	.23*	.23*	.13	.32**	.08	.37**	.56**
Toys and Materials	.17	.08	.20	.24*	.08	.26*	.55**
Maternal Involvement	.12	.14	.10	.18	.05	.19	.42**
Variety of Stimulation	.07	.08	.14	.11	-.03	.18	.29*

*p<.05

**p<.01

language items from the Bayley Scales (see Table 1). Only two
HOME subscales significantly correlated with social responsiveness:
avoidance of restriction and punishment and organization of the
environment. Path diagrams involving HOME subscales, social responsiveness and IQ revealed that most of the relation between 6-month
HOME scores and 3-year IQ scores is not mediated through an impact
on social responsiveness as measured at 6-months or 12-months.

With respect to goal directedness, only one significant correlation was obtained (.23 for maternal responsivity) for 6-month HOME
scores. However, 12-month goal directedness scores were significantly correlated with three HOME subscales: avoidance of restriction
and punishment, organization of the environment, and provision of
appropriate play materials.

Path diagrams involving HOME subscales, goal directedness,
and 3-year IQ indicated that goal directedness as measured at 12-
months does mediate the relations somewhat. An indirect effect of
.05 or better through goal directedness was noted for four environmental variables: avoidance of restriction and punishment, organization of the environment, provision of appropriate play materials,
and maternal involvement.

Six-month language scores showed no significant relation to
HOME scores, but 12-month language scores were correlated with
three HOME subscales: avoidance of restriction and punishment,
organization of the environment, and provision of appropriate play
materials.

Path diagrams involving HOME, language competence, and IQ
showed that language competence measured at 12-months seemed to
mediate all relations except for maternal responsivity.

Discussion

Of major interest to this study was the finding that the three
behavioral clusters derived from the Bayley Scales when assessed at
6-months do not appear to mediate the relation between home environment and IQ. The path coefficients between HOME scores and IQ
indicate that most of the correlation could be considered a "direct
effect." When 12-month Bayley scores were examined to determine
the extent to which they might mediate the relation between HOME
and IQ, a somewhat different picture emerged. First, all three
Bayley behavior clusters showed significant correlations with IQ.
Second, the path coefficients between HOME scores and IQ indicated
a measurable indirect effect for both goal directedness and language
competence. For five of the six HOME subscales, an indirect effect
greater than .05 was observed.

In sum, it appears that the relation between early environmental experiences and later IQ is mediated to a modest degree through several diverse capabilities manifested during infancy. Of course, it is important to remember that path models do not allow for strict causal interpretations. In particular, from the present study it is not possible to rule out a mutually facilitative effect between HOME scores and Bayley scores since the alternative models using HOME scores as endogenous variables were not investigated. In addition, the generally low stability of early developmental measures makes it difficult to draw strong conclusions from the results.

Of interest in the investigation are the differential relations between the six HOME subscales and later IQ. The two subscales showing the strongest relation are organization of the environment and provision of appropriate play materials (.56 and .55). The effects of these two environmental variables appear to be mediated through goal directedness, language competence, and to a lesser degree, social responsiveness. The relations with IQ for avoidance of restriction and punishment and maternal involvement show a similar pattern, although the strength of the relationship is not as great.

References

Bradley, R. and Caldwell, B. The relation of infants' home enrivonments to mental test performance at fifty-four months: a follow-up study. Child Development, 1976, 47, 1172-1174.

Kerlinger, F. & Pedhazur, E. Multiple Regression in Behavioral Research. New York: Holt, Rinehart, and Winston, 1973.

Lewis, M. Individual differences in the measurement of early cognitive growth. In Hellmuth, J. (ed.), Exceptional Infant, Vol. 2. New York: Brunner/Mazel, 1971.

Wachs, T., Uzgiris, I., & Hunt, J. Cognitive development in infants of different age levels and from different environmental backgrounds: an explanatory investigation. Merrill-Palmer Quarterly, 1971, 17, 283-317.

Yarrow, L., Rubenstein, J., Pederson, F., & Jankowski, J. Dimensions of early stimulation and their differential effects on infant development. Merrill-Palmer Quarterly, 1973, 19, 205-219.

INDUCING FLEXIBLE THINKING: THE PROBLEM OF ACCESS

Ann L. Brown   and   Joseph C. Campione

University of Illinois   University of Illinois

Champaign, Illinois, U.S.A.   Champaign, Illinois, U.S.A.

Abstract

We began by illustrating that the concept of accessibility was central to many theories of psychology from quite disparate domains. A distinction similar to Pylyshyn's of multiple and reflective access also seems to be, at least implicitly, part of many theories. Given that accessibility is a core concept in so many current disputes, we suggest that no theory of intelligence can be complete unless provision is made for the operation of second-order knowledge, i.e., knowledge about what we know (reflective access) and flexible use of the routines available to the system (multiple access).

In the second part of the paper we consider the evidence that diagnosis of retarded and learning disabled children's learning problems based on process theories are fundamentally diagnoses of restricted access. Training studies, whether successful or not at inducing transfer, provide rich support for the hypothesis that the slow learning child has peculiar difficulty with the flexible use of knowledge. In the final section we consider the implications of the position for the design of training programs to alleviate the problem of accessibility. Here we address the developing technology we have for programming transfer of training and the importance of interpersonal settings, particularly Socratic tutoring, as cognitive support systems for learning.

I. Introduction

One of the traditional games played by developmental psychologists is the training study, the aim of which is to induce flexible thinking. The purpose of this enterprise is twofold. First, because our subjects fail to display many of the skills used by more mature learners, it is interesting to see if we can induce these skills by providing instruction. For those who work with impaired thinkers, such as retarded children, the enterprise is much more than interesting; it is essential for remediation. If a child cannot, or will not, invent clever learning ploys for himself, perhaps he can be helped by others more knowledgeable than he.

The training study for the developmental theorist is more than an exercise in applied psychology, however. It serves a purpose very similar to that of the computer models of artificial intelligence or computer simulation. If one's aim is to instill intelligent behavior into a machine, it is necessary to explicitly program what one thinks this is. But to program one must understand. Similarly, for the developmental psychologist who wishes to understand flexible thinking in children, or its absence in special populations, the training study is a device for making explicit what we think intelligence is. Sutherland's (1978, p. 116) claim that at present "computer programs are the only tool we have for giving rigorous expression to psychological models, for proving their formal adequacy and consistency, and for investigating their formal limitations" may be true. We would argue, though, that training studies could be used to serve very much the same function.

We have argued elsewhere that central to any theory of learning are three core concepts: competence, induction and access (Brown, 1979). By competence, we refer to the complex issue of the special "belongingness" or compatibility of certain learning activities, a compatibility that is often species-specific with important survival value. Developmentalists tend to address this problem with a consideration of naturalness, and the special value of early learning. By induction we refer to the acquisition of new competence and the transition mechanisms accounting for growth. By access we refer to the ability to access competence, to use flexibly and appropriately the information available to the system. We argue that the training study is an invaluable tool for uncovering problems of competence, induction and access. As our space is limited, we concentrate primarily in this paper on the topic of accessibility and its importance both in theories of intelligence and in prescriptions for remediation.

II. Accessibility

The concept of accessibility of knowledge is a central one for many theories of intelligence. To illustrate the centrality of the point we will describe, briefly, a few quite disparate psychological areas where the question of access is paramount. These examples are not meant to represent an exhaustive overview or even a current position statement. The main point of this section is to highlight the notion that some general concept of accessibility is explicitly a central tenet of theories in a variety of domains that differ widely in their methods but share a concern with the nature of intelligent systems, biological or mechanical.

A. Cognitive Ethology

The area of cognitive ethology appears to be a blossoming one but for our purposes here we will concentrate mainly on an imaginative paper by Paul Rozin (1976) concerned with the evolution of intelligence. Rozin considers intelligence as a complex biological system, hierarchically organized, and consisting of a repertoire of adaptive specializations that are the components or subprograms of the system. Throughout the animal world there exist adaptive specializations related to intelligence that originate to satisfy specific problems of survival. Because they evolve as solutions to specific problems, these adaptive specializations are originally tightly wired to a narrow set of situations that called for their evolution. In lower organisms the adaptive specializations remain tightly constrained components of the system. This form of intelligence is tightly prewired; although it can sometimes be calibrated by environmental influence, it is pretty much preprogrammed (birdsong development is probably the most elegant illustration of the interplay between pre-wired components and environmental tuning; Marler, 1970). Rozin's theory is that in the course of evolution, cognitive programs become more accessible to other units of the system and can therefore be used flexibly in a variety of situations. This flexibility is the hallmark of higher intelligence, reaching its zenith at the level of conscious control, which affords wide applicability over the full range of mental functioning.

Rozin refers to the tightly wired, limited access components in the brain as the cognitive unconscious, and suggests that

> "...part of the progress in evolution toward more intelligent organisms could then be seen as gaining access to or emancipating the cognitive unconscious. Minimally, a program (adaptive specialization) could be wired into a new system or a few new systems. In the extreme, the program could be brought to the level of consciousness, which might serve the purpose of making it applicable to the full range of behaviors and problems." (Rozin, 1976, pp. 256-257.)

Just as part of the progress in evolution toward more intelligent organisms can be seen as gaining access to the cognitive unconscious, so too the progress of development within higher species such as man can be characterized as one of gaining access. Intelligent behavior is first tightly wired to the narrow context in which it was acquired and only later becomes extended into other domains. Thus cognitive development is the process of proceeding from the "specific inaccessible" nature of skill, to the "general accessible."

There are two main points to Rozin's accessibility theory that are of special interest to developmental psychologists. First is the notion of welding (Brown, 1974, 1978), that is, intelligence components can be strictly welded to constrained domains, i.e. skills available in one situation are not readily used in others, even though they are appropriate. Rozin uses this concept to explain the patchy nature of young children's early cognitive ability, which has been described as a composite of skills that are not necessarily covarient. Young children's programs are "not yet usable in all situations, available to consciousness or statable" (Rozin, 1976). Development is the process of gradually extending and connecting together the isolated skills with a possible ultimate extension into consciousness.

Closely connected is the second notion of awareness or knowledge of the system that one can use. Even if skills are widely applicable rather than tightly welded, they need not necessarily be stable, statable and conscious. Rozin would like to argue that much of formal education is the process of gaining access to the rule-based components already in the head, i.e. the process of coming to understand explicitly a system already used implicitly. As Rochel Gelman (1979) points out, linguistic (and possibly natural number) concepts are acquired very easily, early and universally, but the ability to talk and the ability to access the structure of the language are not synonymous. The ability to speak does not automatically lead to an awareness of the rules of grammar governing the language.

In his commentary in the special issue of Behavioral and Brain Sciences devoted to consciousness in nonhuman species, Pylyshyn (1978a) makes a similar point when he distinguishes between multiple access and reflective access. Multiple access to the representational components governing behavior is shown by the ability to use knowledge flexibly, i.e., a particular behavior is not delimited to a constrained set of circumstances (the welding argument). Similarly knowledge is informationally plastic in that it can be "systematically varied to fit a wide range of conditions which have nothing in common other than that they allow the valid inference that, say, a certain state of affairs holds" (Pylyshyn, 1978a, p. 593). Reflective access refers to the ability to "mention as well as use" the components of the system, a situation that would demand that the representational system be available for purposes other than those directly determining the immediately relevant behavior, such as inferring representational states in others, or comparing various desired end states.

In his commentary in the same issue, Garner (1978) also makes a distinction similar to the one of multiple and reflective access. Garner suggests that the hallmarks of intelligence are: a) generative, inventive, and experimental use of knowledge rather than preprogrammed activities (multiple access) and b) the

ability to reflect upon one's own activity (reflective access).
However, Garner makes the point that no organism ever reaches a
level of "total consciousness, full awareness, and constant intentionality" for these are "emergent capacities" useful as indices for
comparative purposes both within and between species, but never
perfectly instantiated even in the mature human. To the extent
that organisms come to exhibit more and more of the qualities of
reflective and multiple access, we tend to say that they exhibit
intelligent behavior.

B.   Cognitive Psychology

     In the limited space available, we obviously cannot begin to
review the major use of the accessibility notion in mainstream
cognitive psychology. Here we would just like to point out that
such a concept has traditionally been central to theories of memory
and learning. Tulving's classic distinction between availability and
accessibility, and his theory of encoding specificity, have been
incorporated within the levels of processing theories to explain a
great deal of the recent process oriented literature on adult
memory (Tulving, 1978). We have a great deal of evidence that:
1) people frequently store information that they are unable to
retrieve when needed; 2) the presentation of appropriate retrieval
environments leads to access to material previously "forgotten";
3) different testing situations provide different retrieval environments and therefore, assessments of the availability of knowledge
vary as a function of retrieval support in the testing context;
and 4) the compatibility between encoding and retrieval contexts
is vitally important as a determinant of the ability to access
previously stored materials. All these arguments concern the
optimal conditions for making information in memory accessible
when needed; it is not sufficient to simply store information, for
unless it can be activated when needed it is of little use.

     It would appear that the memory system can be quite inflex-
ible unless careful planning for retrieval is undertaken, a notion
that is reflected in Bransford's (1979) theory of transfer-appropriat
e processing which stresses the compatibility between the learning
activity and the goal of that activity or the purposes to which the
information must be applied. Learning activities are purposive
and goal directed, and an appropriate learning situation must be
one that is compatible with the desired end-state. One cannot,
therefore, discuss appropriate learning activities unless one
considers the question of "appropriate for what end?" Again the
guiding principle of these arguments is one of accessibility--how
to ensure, by preplanning, the flexible use of knowledge available
to the system.

     The second major concept in mainstream cognitive psychology
that is pertinent to our argument is the controversial notions of
executives, head-demons, interpreters, homunculi, central proces-

sors or "the single, conscious high-level mechanism that guides
the conceptual processing" (Bobrow & Norman, 1975). The development of these concepts was inspired by the emergent field of
artificial intelligence, and, therefore, we will address them under
that heading.

C.   Artificial Intelligence

Researchers concerned with the creation of intelligent behavior in machines are forced to make explicit exactly what they
think constitutes intelligence, hence the fascinating controversies
surrounding the problem of how intelligent machines are now (or
could be in the future). The issues raised by these controversies
are central to our conception of mind (Flores and Winograd, 1978;
Pylyshyn and following commentaries, 1978a and 1978b). We will
restrict ourselves to the problems of accessibility and knowledge
of knowledge.

Moore and Newell (1974, pp. 204-204) made a succinct statement of the welding problem when they defined the essence of
machine understanding in reference to two criteria. First, "S
understands K if S uses K whenever appropriate"; second, this
"understanding can be partial, both in extent (the class of appropriate situations in which the knowledge is used) and in immediacy
(the time it takes before understanding can be exhibited)." We
judge as intelligent the flexible, appropriate and rapid application
of the knowledge available to the system.

A more stringent criterion of understanding is that knowledge be available to consciousness and perhaps be statable (Garner,
1978; Rozin, 1976). An intelligent system must have the capability
to be aware of itself. This second-order knowledge, knowing
about what we know and what we can know, is a thorny problem
for the designers of machine intelligence (Winograd, 1975).
Ignoring the complexities, most theories of machine intelligence
assume some form of executive bookkeeping, a system that plans
and guides cognitive activities; keeps track of the activities of
subordinate processes; determines their success, failure or appropriateness; generates new subprocesses; and allocates resources.
This central system must in some sense have "awareness" of its
own processes and of the information sent to it by lower order
mechanisms. In other words the intelligent machine must have
access to and control of its own attempts to be intelligent. "Man
not only has consciousness, but he knows that he has it" (Katz,
1939). Of issue to cognitive ethologists is the question, do animals
know? Of issue to those in the field of artificial intelligence is
the question, can machines know? Of issue to those who would
build a theory of intelligence is the centrality of the concepts of
accessibility.

D. Developmental Psychology: Metacognition

One of the most influential trends in developmental cognitive psychology is the growing interest in problems subsumed under the heading metacognition (Flavell, 1978; Brown, 1978). Metacognition has always been a controversial term referring to an imprecise concept with fuzzy boundaries, and many of the controversies reflect some of the persistent problems of psychology, e.g., the nature of consciousness, intentionality, cognitive homunculi and epistemic mediation. The area shares therefore, an affinity with cognitive ethology and artificial intelligence in confronting the problems of second order knowledge.

The term has been used in the developmental area to refer to two somewhat separate phenomena and we would like to make this separation explicit here. Flavell (1978) defined metacognition as "knowledge that takes as its object or regulates any aspect of any cognitive endeavor." Two (not necessarily independent) clusters of activities are included in that statement--knowledge about cognition and regulation of cognition.

The first cluster is roughly concerned with a person's knowledge about his own cognitive resources and the compatibility between himself as a learner and the learning situation. Prototypical of this category are questionnaire studies and confrontation experiments, the main purpose of which are to find out how much a child knows about certain pertinent features of thinking, including himself as thinker. The focus is on measuring the relatively stable information that the learner has concerning subject, task, and strategy variables (Flavell, 1978) involved in any cognitive task. This information is stable in that one would expect a child who knows pertinent facts about the total learning situation--(e.g., that organized material is easier to learn than disorganized material)--to continue to know these facts if interrogated appropriately. These are stable forms of knowledge which develop with age and experience but are information sources available to the learner whenever needed. This type of information is also statable, by definition, as the measure of awareness used is almost always verbal justification and explanation (Brown, 1978).

The second cluster of activities are those concerned primarily with self-regulatory mechanisms during an ongoing attempt to learn or solve problems. These indices of metacognition such as checking, planning, monitoring, testing, revising, and evaluation (Brown, 1978), are not stable features in the sense that the degree to which they will be available to the system depends upon other aspects of the learning situation. These "executive functions" are resource demanding and are most likely to occur when the subprocesses that they control are relatively familiar or automatized. The executive competes for workspace with the subroutines it controls and the degree to which these monitoring activities will be engaged in depends very critically on the nature of the task, the expertise of the learner, and the resultant pressures on central processing capacity. Thus, these activities are not necessarily stable, because they will appear or disappear depending on the familiarity and difficulty of the problem, the child's motivation, etc. They are also not necessarily statable as a great deal of selecting, monitoring, inferring, etc. must go on at a level below conscious awareness.

The issues of metacognition have been examined at length, some might say ad nauseum, elsewhere. For our purposes here we emphasize that once again the underlying problems are those of appropriate use of, or access to, knowledge. This emphasis is illustrated in the attempts to use a metacognitive explanation of transfer of training (Brown and Campione, 1978), and the extensive research devoted to uncovering the child's awareness of the knowledge available to the system (Flavell, 1978).

Given the pervasiveness of the concept of accessibility, we are convinced that no theory of intelligence can be complete without ceding it a central place, and no serious discussion of what intelligent behavior is could occur without mention of the difficult issues elicited by the family of ideas implied by the term, i.e., awareness, intentionality, consciousness, automatic vs. deliberate processing, etc. We argue that multiple and reflective access to knowledge is the hallmark of intelligent activity. Elsewhere we have detailed a theory of intelligence in terms of executive control processes (Brown, 1974, 1978; Brown & Campione, 1978; Brown & French, 1979; Campione & Brown, 1978), as indeed have others (see Butterfield this volume), and we will not repeat the argument here.

III. Implications for a Theory of Retardation

The recent increase in both the extent and quality of theoretical and empirical work concerned with learning in retarded individuals affords greater security to those who would assert the locus (loci) and magnitude of academic deficits in the intellectually impaired; at least this holds true for the use of strategies to solve common memory and problem solving tasks. Within this domain we are confident that multiple and reflective access to available knowledge present particular difficulty. Specifically, lack of multiple access to the fruits of learning is reported so often that "welding" has been described as a characteristic feature of the learning of retarded children by both Soviet and American researchers (Brown & French, 1979), not to mention parents and teachers.

Our current knowledge about the performance of retarded children on common learning and memory problems can be summarized as follows. These children perform poorly on a variety of problems that demand the use and control of strategies for adequate solution. With intensive, well-designed training their performance improves dramatically, particularly when the training concentrates on both inculcating the desired strategies and providing detailed instructions concerning self-regulation. Retarded children experience difficulty primarily in transferring the results of any training to new situations, and this diagnostic transfer failure is particularly likely to occur if explicit instruction in self-regulatory mechanisms is not provided. When training does include instruction in both the use and control of the desired skill(s), training attempts are successful (Brown, Campione & Barclay, 1979). Another technique that is showing early promise is training in multiple contexts (Brown, 1978), a procedure that makes explicit the fact that the trained behavior is transsituationally applicable.

Recent successes at inculcating transfer has been taken as evidence to weaken the claim that generalization of the effects of instruction is a major, if not the major, drawback to academic efficiency in the mildly retarded. We disagree and suggest that, transfer successes not withstanding, the training literature provides a rich illustration of the centrality of the access problem for such children. The limited number of successful studies to date rest on extensive, explicit instruction in how to approach the problem, based on detailed task analysis that are provided by the experimenter (no invention on the part of the learner is required). In addition, explicit, detailed instruction in the multiple uses and control of the trained skill may be required. We would argue that in order to find significant transfer effects in retarded learners, one must make explicit what average children can induce.

A traditional definition of intelligence is the speed and efficiency of learning (Thorndike, 1926) and one must consider the efficiency of training attempts in this light. How readily do the subjects respond to training? And, how efficiently do they transfer the information, where efficiency is measured in terms of Moore and Newell's (1974) criteria of extent (broad generalizations) and immediacy (without additional prompting and training)? Resnick and Glaser (1976) also argue that intelligence is the ability to learn in the absence of direct and complete instruction, and Brown and French (1979) identify this as the crux of Vygotsky's theory of proximal distance or potential development.

Rejecting phylogenetic discontiguity theories, Garner (1978) uses similar criteria for comparisons between species: "Just where we ultimately draw the line between human and infrahuman capacities will depend on the ease with which, and the extent to which, other animals acquire the kind of cognitive, linguistic and symbolic behavior which human beings universally acquire." (Garner, 1978, p. 572.)

He argues further that these are suitable criteria for those who would make ontogenetic comparisons. Flexible, inventive and playful behaviors in the absence of complete programming are the essence of intelligence.

"Conversely, to the extent that behaviors (1) appear only when elicited by strong training models, (2) recur in virtually identical form over many occasions, (3) display little experimental playfulness, (4) exhibit restricted coupling to a single symbolic system, or (5) fail ever to be used to refer in "meta" fashion to one's own activities, we are inclined to minimize their significance." (Garner, 1978, p. 572.) (As indices of intelligent behavior).

To the extent that the above definition of restricted coupling, welding, etc. is a reliable description of retarded children's learning, i.e. they tend to employ strategies only if someone else invents them and programs their appropriate use, they are by definition displaying evidence of limited intellectual capacity. To date training studies, whether successful or not, support the original diagnosis of a fundamental problem of accessibility underlying the pervasive learning problems of retarded children.

IV. Implication for a Theory of Remediation

A thorough understanding of the nature of retarded children's problem solving activities should enable us to design programs that will alleviate their characteristic difficulties. If we accept that restricted access to acquired knowledge is an adequate diagnosis, how then would this influence our design of training programs? Also, what kind of cognitive support systems can we offer the immature as a prop for their learning activities? In this section we concentrate on two main technologies designed to overcome the problem of "welding" or lack of multiple access, and to provide a scaffolding for the emergence of executive control on the part of the child. First, we deal very briefly with the design of adequate training programs in terms of task analysis and programming self-regulation and generalization. Second, we deal with the interpersonal nature of the problem solving and the importance of social settings as cognitive support systems.

A. Programming Transfer

Detailed prescriptions concerning ideal training programs to overcome the problem of multiple access exist elsewhere (Brown, 1974, 1978; Brown & Campione, 1978; Butterfield, this volume;

Meichenbaum, 1977). In previous papers we identify seven features that a training procedure must include if generalization of the effects of training is the desired result: 1) careful selection of the cognitive skill to be examined; 2) sensitivity to the actual beginning competence of the learner; 3) stringent analyses of the requirements of the training and transfer tasks so that transfer failures may be interpreted properly; 4) training in multiple settings to alleviate the problem of "welding"; 5) direct feedback concerning the effectiveness of the trained skill; 6) direct instruction concerning the generalization of the trained skills; and 7) direct instruction in self-management routines (see previous papers, especially Brown 1978 and Brown and Campione 1978 for full details of these steps).

B.   Other-Regulation to Self-Regulation

The most available cognitive support system for the developing child is that provided by interaction with significant others, initially the parents and then teachers and peers. There are some who claim that the primacy of social support for intellectual activity is true also of adults (Cole, Hood & McDermott, 1978). Studies of mother-child dyads solving problems provide a rich picture of the interactive nature of learning. It is not simply the case that the mother models and the child imitates. The interactions are far more elaborately orchestrated. The mother appears to tailor her intervention to the child's "region of sensitivity to instruction" (Wood & Middleton, 1975), or "level of potential development" (Brown & French, 1979; Vygotsky, 1978), i.e., just one step beyond the child's current operational level. If, following such help, the child succeeds, the mother is less explicit on the next attempt. If the child fails she repeats the help or becomes more explicit. The choreography of the dynamic interaction reveals a great deal of interpersonal sensitivity on the part of both mother and child. The successful mother extracts from the child not only optimal performance but, more importantly, she elicits autonomy by ceding executive control to the child.

Wertsch's (1978) study of mother-child dyads suggests just such a gradual progression from other-regulation (mother) to self-regulation on the part of the child. The assumption is that through such interactions the child develops self-regulation by gradually assuming the regulatory role first adopted by the mother. Initially, the mother directs, but her instructions do not guide the child's behavior. An intermediate stage then follows where the mother successfully adopts the role of executive, guiding and regulating the problem solving activity of her child. Finally, the mother cedes control to the child and functions primarily as a sympathetic audience. These mother-child interactions are prototypical of other ideal interpersonal learning situations, such as Socratic teaching. A novice is led to mastery and autonomy by the sensitive intervention of another who is more skillful.

Parents are by no means the only social agents to perform the function of fostering self-regulation. Teachers, tutors, and master craftsmen in traditional apprenticeship situations all function ideally as promotors of self-regulation by nurturing the emergence of personal planning as they gradually relinquish their own direction. Effective teachers are those who engage in continual prompts to get children to plan and monitor their own activities. In a recent study of effective teachers, Schallert and Kleiman (1979) described four general strategies used to facilitate children's learning; tailoring the message to the child's existing level of understanding, activating relevant schemata (prior knowledge), focusing attention on relevant and important facts and monitoring comprehension by means of such Socratic ploys as invidious generalizations, counterexamples, and reality testing (Brown, 1978; Collins, 1977). In short, the expert teacher provides much of the executive control for the child, executive functions that the child must internalize (Vygotsky, 1978) as part of his own problem solving activities if he is to develop effective problem solving strategies.

Just as the tutoring situation is one form of social support system for learning, groups may also relieve some of the personal responsibility of control from the individual members. Indeed, in their classic review of group problem solving, Kelley and Thibaut (1954) put forward an internalization theory very similar to Vygotsky's, and a social psychologist's description of group functions sounds very like a description of executive control.

"Qualitatively group discussions seemed to be adequately characterized by the traditional analyses of individual thinking, e.g., stated by Dewey as: 1) motivation by some felt difficulty, 2) analysis and diagnosis, 3) suggestion of possible solution or hypothesis, and perhaps 4) an experimental trying out, before 5) accepting or rejecting the suggestion." (Dashiell, 1935, p. 1311.)

Most of the activities seem to be variants of the basic transsituational skills of predicting, checking, monitoring, and reality testing (Brown, 1978). But, in spite of the evidence that the basic elements of self-regulation become part of a child's repertoire via the process of internalizing that which was originally social (Vygotsky, 1978) most studies concerned with training self-regulation have not used social interactions as a vehicle for training, and most studies of metacognition have been concerned with self-regulation during individual problem solving. The child is typically told to check, monitor, or self-test by an experimenter who invents the program for him; he has no chance to take part in a dynamic social interaction where experts (adults or peers) display executive functions in the normal course of problem solving. The natural situation of the expert unobtrusively adopting, then gradually relinquishing control as the novice gains mastery seems to be an ideal training model to follow if the aim is to encourage autonomy.

The management of such dynamic interplay is by no means simple. A crucial problem facing the tutor is deciding at what level to intervene. In effect, the tutor must engage in continuous diagnosis of the present state of learning so that intervention can be tailored to the child's current needs. In peer problem solving, the participants must divide up the responsibility of performing subparts and accepting control. In the classroom, the problem is even more difficult as ideally the teacher should be sensitive to the level of understanding of several children at once. The basic aim of all those activities is to train the child to think dialectically, in the sense of the Socratic teaching method, where the teacher constantly questions the student's basic assumptions and premises, plays the devil's advocate, and probes weak areas. The desired end-product is that the student will come to perform the teacher's functions for himself via self-interrogation and self-regulations. We realize the difficulty of mounting training programs based on naturally occuring tutoring situations. But in view of the pervasiveness of the retarded child's problems with multiple and reflective access, intensive training in the laboratory that aims at mimicking the cognitive support systems believed to be responsible for the natural development of self-regulation seems to be a worthwhile endeavor.

References

Bobrow, D. G. and Norman, D. A. Some principles of memory schemata. In D. G. Bobrow and A. Collins (Eds.), Representation and Understanding. New York: Academic Press, 1975.

Bransford, J. D. Human cognition: Learning, understanding and remembering.

Belmont, CA: Wadsworth, 1979. Brown, A.L. The role of strategic behavior in retardate memory. In N. R. Ellis (Ed.), International review of research in mental retardation, Vol. 7. New York: Academic Press, 1974.

Brown, A. L. Knowing when, where, and how to remember: A problem of metacognition. In R. Glaser (Ed.), Advances in instructional psychology. Hillsdale, N.J.: Lawrence Erlbaum Associates, 1978.

Brown, A. L. Constraints on learning. Human Development, 1979, in press.

Brown, A. L. and Campione, J. C. Permissible inferences from the outcome of training studies in cognitive development research. Quarterly Newsletter of the Institute for Comparative Human Development, 1978, 2, 46-53.

Brown, A. L., Campione, J. C. and Barclay, C. R. Training self-checking routines for estimating test readiness: Generalization from list learning to prose recall. Child Development, 1979, in press.

Brown, A. L. and French, L. A. The zone of potential development: Implication for intelligence testing in the year 2000. Intelligence, 1979, in press.

Campione, J. C., and Brown, A. L. Toward a theory of intelligence: Contributions from research with retarded children. Intelligence, 1978, 2, 279-304.

Cole, M., Hood, L. and McDermott, H. Ecological niche picking: Ecological invalidity as an axiom of experimental cognitive psychology. Unpublished manuscript. Rockefeller University, 1978.

Collins, A. Processes in acquiring and using knowledge. In R. C. Anderson, R. J. Spiro, and W. E. Montague (eds.), Schooling and acquisition of knowledge. Hillsdale, N. J.: Lawrence Erlbaum Associates, 1977.

Dashiell, J. F. Experimental studies of the influence of social situations on the behavior of individual human adults. In C. Murchison (Ed.), Handbook of social psychology. Worchester: Clark University Press, 1935.

Flavell, J. H. Cognitive monitoring. Paper presented at Conference of Children's Communication. University of Wisconsin, October, 1978.

Flores, C. F. and Winograd, T. Understanding cognition as understanding. Unpublished manuscript, Stanford University, 1978.

Garner, H. Commentary on annual awareness papers. Behavioral and Brain Sciences, 1978, 4, 572.

Gelman, R. The preschool child's understanding of number. Paper presented at the American Education Researchers Association. San Francisco, April, 1979.

Katz, D. Animals and men: Studies in comparative psychology. London: Longman Green, 1939.

Kelley, H. H. and Thibaut, J. W. Experimental studies of group problem solving and process. In G. Lindsey (Ed.), Handbook of social psychology, Vol. 2. Reading, Mass.: Addison-Wesley, 1954.

Marler, P. A comparative approach to vocal learning; song development in white crowned sparrows. Journal of Comparative and Physiological Psychology,, Monograph 71(2), Part 2, 1-25, 1970.

Meichenbaum, D. Cognitive behavior modification: An integrative approach. New York: Plenum Press, 1977.

Moore, J. and Newell, A. How can Merlin understand? In L. W. Gre (Ed.), Knowledge and cognition. Hillsdale, N.J.: Lawrence Erlbaum Associates, 1974.

Pylyshyn, Z. W. When is attribution of beliefs justified? Behavioral and Brain Sciences, 1978a, 1, 592-593.

Pylyshyn, Z. W.   Computational models and empirical constraints. Behavioral and Brain Sciences,   1978b,   1,   93-100.

Resnick, L. B. and Glaser, R.   Problem-solving and intelligence. In L. B. Resnick (Ed.), The nature of intelligence.   Hillsdale, N.J.:   Lawrence Erlbaum Associates, 1976.

Rozin, P.   The evolution of intelligence and access to the cognitive unconscious.   Progression in Psychobiology and Physiological Psychology, 1976, 6, 245-280.

Schallert, D. L. and Kleiman, G. M.   Some reasons why the teacher is easier to understand than the textbook.   Reading Education Report Series, Center for the Study of Reading, University of Illinois, 1979.

Sutherland, N. S.   Task constraints and process models.   Behavioral and Brain Sciences, 1978, 1, 116.

Thorndike, E. L.   Measurement of intelligence.   New York: Teacher's College, Columbia University, 1926.

Tulving, E.   Relation between encoding specificity and levels of processing.   In L. S. Cermak and F. I. M. Craik (Eds.), Levels of processing and human memory.   Hillsdale, N.J.: Lawrence Erlbaum Associates, 1978.

Vygotsky, L. S.   Mind in society: The development of higher psychological processes.   In M. Cole, V. John-Steiner, S. Scribner, and E. Souberman (Eds.), Cambridge, Mass.: Harvard University Press, 1978.

Wertsch, J.   Untitled manuscript, 1978.

Wood, D. and Middleton, D.   A study of assisted problem-solving. British Journal of Psychology, 1975, 66, 181-191.

HEMISPHERIC INTELLIGENCE: THE CASE OF
THE RAVEN PROGRESSIVE MATRICES

Eran Zaidel

University of California

Los Angeles, California, U.S.A.

Abstract

Neuropsychological findings show that different brain regions and particularly the two cerebral hemispheres mediate different aspects of intelligence. Which hemisphere, then, is richer in general intelligence or "g"? Several versions of Raven's Progressive Matrices, a nonverbal test of intelligence said to be loaded on g, give similar IQ estimates to the left and right hemispheres of commissurotomy and hemispherectomy patients. The mean IQ for three left hemispheres was 87 (range 74 to 103), and for the corresponding three right hemispheres it was 83 (range 74 to 93). However the left and right hemispheres excelled in different parts of the tests. The right hemispheres were also less sensitive than the left to item difficulty as defined by test progression, and less able to benefit from trial and error. But several attempts failed to define a priori the problems that yield left as against right hemisphere superiority, such as conceptual vs. perceptual or in terms of earlier vs. later stages of cognitive development. A proposal is made to split g into g_L, and g_R and redefine some primary factors of intelligence in terms of tests that index left vs. right hemisphere abilities.

Introduction

Brain and Intelligence

For too long psychological theories of intelligence have ignored neurological evidence of critical importance. A factor analytic theory, such as Spearman's, Thurstone's or Guilford's, can increase in or acquire validity when its independent factors are shown to be localized in distinct cortical areas and to be dissociable from each other by focal lesions. In turn, an ontogenetic theory of intelligence, such as Piaget's, is compelling if it can be shown that the dissolution of a cognitive skill due to circumscribed cerebral lesions traces in reverse the order of stages that occurs in the normal acquisition of the skill.

On the basis of a brief literature review Guilford (1967, p. 368) did conclude that the right hemisphere (RH) is involved with "figural" abilities and the LH with "semantic" functions as defined in his Structure of Intellect Model. But no direct empirical test of this has ever been undertaken. Neither has any other factor-analytic theory of intelligence received a systematic neuropsychological analysis. Lansdell's work is one of the few exceptions (1970, 1971). Working primarily with temporal lobectomy patients he has shown the lateralizing significance of several mental abilities, such as verbal abilities in the LH and visual closure in the right. Poeck's group in Aachen has also applied some of Thurstone's tests of Primary Mental Abilities to hemispherically-damaged patients and concluded that Closure Flexibility (Gestalt completion) lateralizes to the right parietal region (Orgass, Poeck, Kerschensteiner, and Hartje, 1972). Reitan's studies on laterality effects in the WAIS seem inconclusive. Left brain-damaged (LBD) adults do show a larger deficit on the verbal scale of the WAIS and RBD patients are more impaired on the performance scale but only in the acute post-traumatic stage and not in cases of slowly developing or chronic static lesions (Klove, 1974). The brief foregoing survey probably exhausts most attempts to interface psychometric theories of intelligence with neuropsychological evidence. This is nothing short of a scandal.

It now seems clear that there is no simple association between the usual primary factors in a multivariate theory of human intelligence, such as verbal or spatial, and the cerebral hemispheres (E. Zaidel, 1978), although it is still natural to conceive of different primary abilities as being sustained by distinct cerebral regions. In any case the question then arises as to the left-right status of general intelligence or "g" as conceived by the British School of Intelligence (cf. Piercy, 1969). What is a good measure of g that is at the same time free of other abilities with known laterality biases? The Raven Progressive Matrices seems an ideal candidate. Spearman considered the Raven one of the best of all nonverbal tests of g (Spearman and Jones, 1951) and Vernon described it as one of the purest tests of g available (1961).

The Raven Matrices in Neuropsychology

The test has been a popular measure of intelligence in neurologic patients with focal lesions. This is because it requires neither verbalization nor skilled manipulative ability so that it is ostensibly relatively insensitive to the presence of speech and motor deficits due to aphasia and apraxia. Even verbal instruction is kept to a minimum and the progression of the test items serves as training. However, accumulating data show that performance on the Raven Matrices is sensitive to the presence of aphasia and apraxia (Table 1) whereas some even regard it a test of visuo-spatial discrimination.

The neuropsychological evidence for hemispheric specialization on the matrices is conflicting and inconclusive (E. Zaidel, D. W. Zaidel, and R. W. Sperry, in press). As Table 1 shows, there is even no agreement on whether the presence of a focal unilateral lesion depresses scores on the test. There is also disagreement about asymmetry of deficit with side of lesion and about the selective effect of aphasia. But there is some consensus that severe receptive aphasics with posterior localization and constructional apraxics with lesions to either side show selective deficit on the Progressive Matrices. The confusion here is symptomatic. First there are the usual difficulties of assessing and matching the extent, location, nature, and chronicity of hemispheric lesions. Second, the brain-damaged syndrome combines and confounds residual function in the damaged region, compensatory takeover by the undamaged hemisphere, and possibly pathological inhibition of the healthy side by the diseased side. These multiple influences can be teased apart by comparing the performance of hemisphere-damaged patients with the positive competence of each hemisphere in the split-brain syndrome.

In the present paper I will report some results of administering the usual book forms as well as a board form of the RPM to the disconnected and isolated hemispheres of selected commissurotomy and hemispherectomy patients. I will use this as a case study illustrating the application of a neuropsychological analysis to the study of human intelligence.

Method

Subjects

The subjects included two complete commissurotomy patients of Drs. P. J. Vogel and J. E. Bogen of Los Angeles, L.B. and N.G., believed to have minimal extracallosal damage relative to the whole Vogel-Bogen split brain group (Sperry, Gazzaniga, and Bogen, 1969). L.B., a male, was three and a half when symptoms started, 13 when the operation was performed, and 20-24 when tested. His presurgical WISC IQ at age 13 was 113 (119 verbal, 108 performance) and postsurgically at age 14 he obtained a post-surgical WAIS IQ of 106 (110, 100). N.G., a female, was 18 when seizures started, she was operated on when 30, and tested when

TABLE 1

Study	Test	Predom. etiology	Effects of lesion				Effects of aphasia (A)			
			Effect of brain damage	Laterality effect	Effect of severity of lesion	Measure of severity of lesion	Effect of presence of A	Effect of type of A	Effect of severity of A	Measure of aphasia
Costa & Vaughan, 1962	RCPM 20 problems	CVA, tumor, trauma	L,R	none					yes	Mill-Hill vocabulary
Archibald et al., 1967	RCPM	CVA	none	RH			none	yes[1]		LMTA
Gainotti, 1968	RCPM	CVA, neoplastic	-	-	-	-	-	-	-	-
Costa et al., 1969	RCPM	CVA, neoplastic	L,R	RH[2]			yes[1]	yes[3]		Weisenburg & McBride
Basso et al., 1973	RCPM <10' time limit	CVA, neoplastic	U	none	yes	duration of illness; RT for ipsi-lateral hand	yes	no[1]	none	TT and visual naming; standard A examination
Kertesz & McCabe, 1975	RCPM	CVA, tumor, trauma		LH?			none	yes[2]	none[1]	WAB;AQ and comprehension subscore
Costa, 1976	RCPM	CVA, neoplastic	L,R	RH						
Denes et al., 1979	RCPM modified, untimed	CVA	-	none[1]	-[2]	sensory-motor and visual field deficits	-	-	-	-
Piercy & Smith, 1962	RSPM	tumor, CVA	?		?		none			
Meyer & Jones, 1957	RSPM untimed	temporal lobe epilepsy	none	none						
Urmer et al., 1960	RSPM	CVA	U,R							
Zangwill, 1964	RSPM	CVA?					yes	yes[1]	yes	clinical evaluation
Arrigoni & De Renzi, 1964	RSPM 40' time limit	CVA, neoplastic	U?	LH		RT	none			
De Renzi & Faglioni, 1965	RSPM 40'	CVA, neoplastic	U	none			none			
Colonna & Faglioni, 1966	RSPM 40	CVA, neoplastic		none	yes	RT	yes			
Boller & Vignolo, 1966	RSPM 40'	-	U	none			none		none	TT
Newcombe, 1969	RSPM untimed	gunshot wounds	none (U,L,R)	none			?			
Van Harskamp, 1973	RSPM untimed	CVA, trauma, tumor		RH					yes	TT
Van Dongen, 1973	RSPM	CVA, traumatic, neoplastic	L				none			TT word fluency
Messerli & Tissot, 1973	RSPM	CVA, tumor		RH			none			

U = total unilateral brain damage effect; Pts. = patients; RBD = right brain damage pts.; RT = reaction time measure; TT = Token Test (Boller & Vignolo, 1966); N = controls; A = aphasia; C = constructional apraxia; A* = very severely aphasic; C = severely constructional apraxic; LMTA = Language Modalities Test for Aphasia; WAB = Western Aphasia Battery.

TABLE 1 (Continued)

Effects of constructional apraxia (C)

Effect of presence of C	Effect of severity of C	Laterality in C	Measure of C	Comments
		L,R	Kohs blocks	Trend RBD < LBD. High intercorrelation between RPM and Kohs blocks and Knox cubes in RBD but not in LBD.
				RBD < LBD; RBD-A = LBD-A; RBD+A < RBD-A. 1. Nontalking (global) aphasics.
-	-	-	-	Much higher neglect of responses in half-space contralateral to lesion in RBD than in LBD.
L^1,R			Kohs blocks	1. Significant? 2. EEG evidence for more extensive and more posterior lesions in RBD. 3. Mixed or receptive aphasics.
				1. Fluent vs. nonfluent aphasics.
L	L		drawing	1. By "exceptional case" of Broca's A but + correlation of RPM with AQ (aphasia quotient) or comprehension. 2. Comprehension aphasics.
				For all BD pts. scores on problem sets $A>A_B>B$. Posterior RBD were selectively poor on set A_B. LBD superior to RBD on sets $A_B>B>A$.
-	-	-	-	Test modification: a. missing piece placed on side ipsilateral to lesion; b. choices arranged in columns under complex figure. 1. Test-retest showed improvement by LBD on set A, and by RBD on set B. 2. Controlled to match populations.
L,R	L,R	RH	copying Kohs blocks, object assembly	RBD+C < LBD+C < RBD-C < LBD-C. Parietal pts.
				No significant decrease in RSPM scores following temporal lobectomy on either side but bigger trend in RBD.
				Aphasics excluded. UBD made qualitatively different errors than N.
U			Kohs blocks, form board, drawing, copying	Patients included 2 motor A with comprehension deficits. Improved speech → improved RSPM. 1. Amnesic < jargon < motor.
U,L		RH	copy drawings, 3-dimensional blocks, copy token model	Significantly more frequent but not more severe C in RBD. More severe BD in RBD by RT.
				Trend LBD < RBD in spite of more extensive lesions in RBD as measured by RT(?).
				Trend RBD < LBD. No difference between BD and N on sensitivity to item difficulty.
				RBD = LBD+A < LBD-A < N. Studied only expressive aphasics.
				None of the aphasics had severe receptive deficits. No lesion localization effect on RPM.
Yes			drawing from memory	$A^+ + C^+ < A^+ + C^+ = A^+ + C$. When include pts. with diffuse and brainstem lesions get LBD = RBD.
	L		no. of elements in drawings	There is a significant correlation of number of elements in drawings with RSPM score in A only, but there is no correlation between number of elements and severity of A.

39-43. Her preop Wechsler-Bellvue IQ was 76 (79, 74) at age 30 and her postop WAIS was 77 (83, 71) at age 35. The other two subjects had hemispherectomy for post infantile lesions. D. W., a patient of Dr. I. G. Gill, was left-handed prior to right hemispherectomy for encephalitis, but a presurgical sodium amytal test showed left hemisphere dominance for speech. His symptoms started at age six and a half, the surgery was performed about a year later and testing started when he was 16. He is reported to have had a Stanford-Binet IQ of 125 at age 3.5 and at age 15.5 his WISC IQ was 67 (80, 60). R. S. was a formerly right-handed dominant hemispherectomy patient of Drs. Bogen and Vogel. Symptoms started at age 8, left hemispherectomy for tumor was performed at age 10, and testing started at age 13. Her Kuhlman Anderson IQ was 100 at age 8 and her WISC IQ was 56 (63, 55) at age 13. For further clinical information see Sperry et al. (1969), Bogen and Vogel (1975), and Gott (1973).

Procedure

The two commissurotomy patients used a right-eyed contact-lens system designed to permit continuously lateralized visual presentation with free hemispheric ocular scanning of the stimuli as well as manual guidance on a board in the subject's lap (Zaidel, 1975).

The RSPM and RCPM consist of five and three problem sets, respectively, with each containing 12 progressively more difficult items. The book form of the RCPM (Raven, 1962) was administered to each commissurotomy patient first in the left visual half-field with left-hand pointing to the one out of six choices and, a week later, in the right visual half-field with right-hand pointing to the answers. This order was fixed for both subjects since a long experience of testing these patients has shown that the left hemisphere is less likely to interfere with right hemisphere performance when it is ignorant of the task. The book form of RSPM (Raven, 1958) was administered in the same sequence a month later. Free vision testing followed the lateralized versions. The book forms of the tests were presented in free vision in the standard manner to each of the two hemispherectomy patients.

The board form of the test was administered to the same patients in the same order from four to three years later (except for R. S. who had died meantime due to recurrence of tumor at the age of 17). In this form each problem appears on a board with the missing part physically removed. The answers consist of six movable pieces, each of which exactly fits the space in the board. The subject is encouraged to use a trial and error approach by fitting different pieces in the empty space until s/he is satisfied with the answer, using the hand ipsilateral to the stimulated visual half-field. As usual, each commissurotomy patient served as his/her own control and the two hemispherectomy patients were compared with each other.

Results and Discussion

Laterality

Results with the book form of the Coloured Matrices showed a nonsignificant RH advantage (number correct out of 36, N.G.: RH = 25, LH = 20, free vision = 20; L.B.: RH = 36, LH = 35, free vision = 36; R.S. = 18; D.W. = 21). This confirms previous data on a tactile-visual modification of a subset of the Coloured Matrices (D. Zaidel and Sperry, 1973).

Results with the board form showed that the disconnected LH but not the RH benefited from error correction through trial and error. Both hemispheres performed worse in the first solution attempt with the board form than in the book form, suggesting that they had adopted new strategies (N.G.: RH = 16, LH = 18; L.B.: RH = 32, LH = 30; D.W. = 18). However, in N.G. (and D.W.) who did not show ceiling effects the final solution with the board form resulted in a significant improvement relative to the book form only for the LH (N.G.: RH = 25, LH = 28; L.B.: RH = 35, LH = 36; D.W. = 23).

Results with the book form of the Standard Matrices showed a nonsignifcant LH advantage (number correct out of 60, N.G.: RH = 16, LH = 16, free vision = 19; L.B.: RH = 36, LH = 46, free vision = 50; R.S. = 17; D.W. = 19). The IQ estimates for the two hemispheres, based on the conversion table of Burke (1972), are remarkably similar: 74 for the left and right hemispheres of N.G., 103 as against 93 for the LH and RH of L.B., respectively, 75 for R.S. (RH) and 78 for D.W. (LH). Thus, the complete Matrices tests failed to show strong and consistent laterality effects (Zaidel, Zaidel and Sperry, 1979).

Item analysis

A laterality index was computed for each 12-item problem set in all verions of the RPM, $f = (L_c - R_c)/(L_c + R_c)$ if $L_c + R_c <$ 100%, and $f = (L_e - R_e)/(L_e + R_e)$ if $L_c + R_c >$ 100%, where $L_c (L_e)$ = percentage correct (erroneous) left hemisphere responses. The resulting indices varied radically across sets (Zaidel et al., in press) showing that the RPM is not homogeneous with respect to hemispheric factors. There was a progressively larger RH dominance for the subsequent sets A, A_B, and B of the RCPM but a heterogeneous spread in the RSPM. The sets common to different versions (A, A_B, B) showed consistent relative laterality indices.

This result is supported by a factor analysis of the five problem sets in RSPM (Rimoldi, 1948). Five of the six identified factors which differentiated the problem sets lend themselves to a priori interpretation in terms of hemispheric specialization. Rimoldi's identification of the factors (and my hypotheses about their presumed laterality) is as follows: α: Perhaps analytic activity leading to the formation of rules or principles (LH, Levy-Agresti and Sperry, 1968); β : Perception of relations in space necessary for the construction of a whole (RH, Nebes, 1974); γ: Difficulty in constructing a gestalt when there are disturbing forces (LH, Zaidel, 1978); ϵ: Immediate digit and sequential memory (LH, Albert, 1972); ζ: Perceptual speed, i.e., quick perception of detail (LH, Zaidel, 1978). Higher saturation on β was associated with lower saturation on and vice versa. Table 2 shows the hypothesized laterality and actual loadings of the factors on problem sets A-E of RSPM. The a priori assignment of laterality to these factors, together with the simplifying assumption of equal weights assigned to each, yield the following ranking of RSPM problem sets in order of decreasing right hemisphere involvement, A>D>C>B>E. This is in substantial agreement with the ordering (D>A>C>B>E), obtained from the unilateral scores of the commissurotomy and hemispherectomy patients using the measure mean percent $(R - L)/(R + L)$.

There is another line of indirect evidence that the two hemispheres solve the matrices in different ways. In addition to shifting hemispheric asymmetries with test (colored vs. standard), form (book vs. board), and problem set, the RH is also less sensitive than the LH to item difficulty. First, there is an expected larger LH dominance on the more difficult six items within each problem set of 12. Second, however, the difference in the competence of a hemisphere between the hard and easy items in each set, increases with progressively more difficult sets for the LH but not for the RH (Zaidel et al., in press).

A yet third hint of RH "idiosyncrasy" in problem solving is provided by an error analysis. Raven (1965) has classified the alternative choices in each matrix problem in terms of the following types of prevalent errors. (a) Repetition of pattern: alternatives presenting figures already on the matrix. (b) Incomplete correlates: alternatives which are wrongly oriented or incomplete. (c) Inadequate individuation: alternatives contaminated by irrelevancies and distortions, or alternatives which are whole or half the pattern to be completed. (d) Difference: alternatives with no or irrelevant figure. The data show that the same rank ordering of error frequency by type on RCPM which occurs for the disconnected hemispheres, is also found in patients with unilateral brain damage (Costa, Vaughan, Horwitz, & Ritter, 1969), children, and old people (Raven, 1965). In order of decreasing error rates this is: repetition of pattern > incomplete correlates > inadequate individuation difference.

TABLE 2

Factor Analyses (Loadings) of Problem Sets in RSPM (Rimoldi, 1948)
and Hypothesized Laterality of the Factors (see Text)

Factor	α	β	γ	ε	ζ	Predominant predicted laterality	Predominant observed laterality
Hypothesized laterality	LH	RH	LH	LH	RH		
RSPM Problem Set							
A	.34	.5				RH	RH
B	.54	.31	.2			LH	LH
C	.55	.23				LH	LH
D	.68	.28			.28	LH	LH
E	.45			.32	.29	LH	LH

So far there is no hemispheric difference.

It has also been said, however, that the percentage of repetition errors increases with increased competence on the test (Raven, 1965) and that there may be reason to consider them relatively "high quality errors" (Costa et al., 1969). The one exception to this rule is the RH which has a relatively low rate of repetition errors incommensurate with its relatively high score. Of particular interest is the discrepancy in error rates between the disconnected or isolated LH, and patients with right sided damage. It would seem again that the trauma affects performance qualitatively in such a way that it is invalid to assume that the unaffected RH simply takes over the function. In fact the error pattern of the disconnected LH most resembles that of 8.5-year old children. The disconnected RH, on the other hand, does not follow any developmental pattern at all.

We can ask how the observed hemispheric profiles on the Raven problem sets compare with those expected of normal sub-jects, children or adults who obtain the same total score on the test? Figure 1 shows that there is a greater tendency for the RH than for the LH to deviate from normal score profiles. On the book form of RCPM the obtained unilateral profiles show higher

scores than, but similar deviations from normal profiles to, those
of patients with corresponding unilateral lesions (Costa, 1976).
For both hemispheres and both patient groups the deviation is
larger on problem sets A and B than on A_B. On the RSPM, RH
deviation from normal profile is especially marked for sets A and
B. Again, contrary to intuition, it is not the more difficult sets
that dissociate best unilateral from normal competence.

In spite of the recurrent hint that the two hemispheres solve
the same problems in different ways and excel in different types
of problems, further attempts to analyze these differences largely
failed. First of all the mean rank ordering of problem difficulty
for the LH and RH in all sets of RCPM and RSPM, except A and
A_B, correlate positively and significantly (Table 3). Secondly,
attempts to characterize the problems that best discriminate between
the LH and RH according to some a priori criterion were consistently frustrated. Even casual inspection of the RCPM problems
reveals that they can be naturally classiied into two groups:
those that can be solved through visual pattern completion or
Gestalt closure and those that require additionally the coordination of two abstract rules for their solution. The abstract problems include at least items A_{12}, A_{B12}, B_8-B_{12}; the rest of the
problems are then perceptual. The scores were analyzed to check
the hypothesis that the RH is more likely to solve problems which
allow Gestalt closure, whereas the left is more adept at solving
problems that require simultaneous coordination (multiplicative
classification) of abstract rules. Surprisingly, the converse
result obtained (Figure 2). The RH tended to fail more perceptual items and the LH to fail more abstract ones! Moreover, both
hemispheres of N.G. failed to solve the abstract problems (chance
= 83% errors) but both of L.B.'s hemispheres solve them without
error.

Alternative classification of RCPM items on the basis of
presumed (a priori) strategies necessary for their solution or on
the basis of their factorial groupings, similarly failed to yield
hemispheric asymmetries in the form of hemisphere by item-group-
ing interaction in our data. Thus, a partition of RCPM problems
into "conceptual" items requiring abstract reasoning (A_{11}, A_{12},
$A_{B9}-A_{B12}$; B_6-B_{12}) and "perceptual" items that can be solved by
pattern completion (A_2-A_{10}; $A_{B1}-A_{B8}$; B_1-B_5) (Carlson & Goldman,
1974) though, again, suggestive of left and right hemisphere strategies, respectively, nevertheless also failed to show left-right interaction or test form effects (book vs. board). A slightly modified
classification of RCPM items into "perceptual" (A_7-A_{12}; $A_{B4}-A_{B7}$)
and "conceptual" (A_{B12}; B_6-B_{12}) (Carlson, Goldman, Bollinger,
and Wiedl, 1974) again shows no significant hemispheric interactions
in our data (Figure 2).

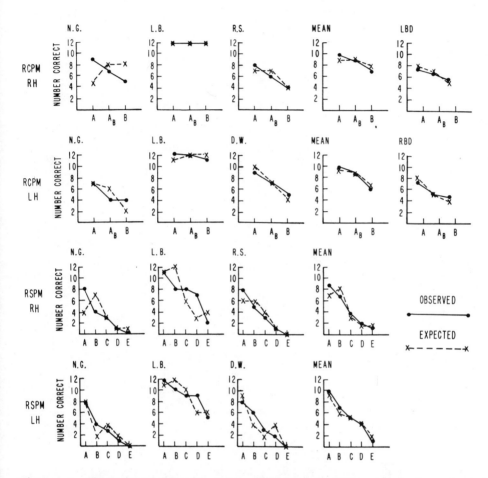

Figure 1. Mean hemispheric scores compared with expected scores as a function of problem set on Raven's Progressive Matrices. Expected problem set scores are found in the norms for normal subjects with the same total score. LBD (RBD) = left (right) brain-damaged patients of Costa (1976). LH (RH) = mean score of the left (right) hemispheres. RCPM = Coloured Matrices. RSPM = Standard Matrices.

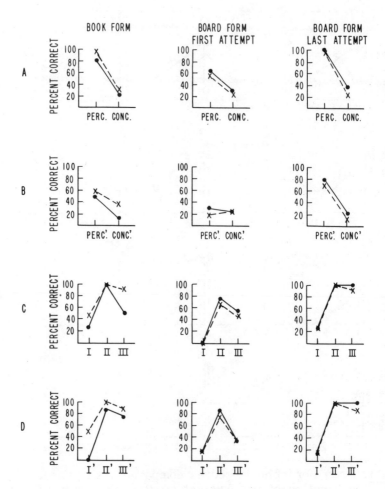

Figure 2. Illustrative hemispheric scores of commissurotomy patient
N.G. on several factorial partitions of items in the Colored
Matrices. A: Conceptual-Perceptual partition (Muller,
1970; Carlson and Goldman, 1974). B: A Modified Conceptual-Perceptual partition (') (Carlson et al., 1974). C: The
three factors of Weidl and Carlson (1976); I. Concrete and
abstract reasoning. II. Concrete and discrete pattern completion. III. Pattern completion through closure. D: A
modified three-factor partition (') (Carlson and Weidl, 1976).
·———· = Left hemisphere, x---x = Right hemisphere.

TABLE 3

Spearman's Rank Order Correlation Coefficients Between Mean
Unilateral Left and Right Hemisphere Scores on RPM Item Sets
(Items were ranked in ascending order of difficulty)

Test		Problem set					
		A	A_B	B	C	D	E
RCPM	r_s	.32	-.11	.52			
	p	N.S.	N.S.	< .05			
RSPM	r_s	.38		.70	.75	.60	.51
	p	N.S.		< .05	< .01	< .05	< .05

Lastly, Wiedl and Carlson (1976) have proposed a three-factor
grouping of RCPM problems based on a factorial study with
first, second, and third grade children. They labeled the first
factor (marked by items A_{B8}, A_{B9}, B_8-B_{12}) "concrete and abstract
reasoning," the second factor (marked by items A_2, A_3, A_5, A_6,
A_7; A_{B2}; B_1, B_2, B_3) was labled "continuous and discrete pattern
completion," and the third factor (items A_4, A_9, A_{10}; A_{B3},
A_{B5}-A_{B8}; B_3-B_5) was called "pattern completion through closure."
However, these dimensions as well as a modified assignment of
items to the three factors (I: A_{B9}; B_8-B_{12}. II: A_2, A_3, A_5-A_7;
A_{B2}; B_4, B_5. III: A_4, A_9, A_{10}; A_{B3}, A_{B5}-A_{B7}; B_1, B_2) again
yield similar patterns for left and right hemisphere scores:
factor I is most difficult, factor II is least difficult, and factor III
is of intermediate difficulty for both hemispheres alike (Figure 2).
Thus, our data suggest that from the point of view of hemispheric
specialization these various factorial classifications of the Colored
Matrices are completely confounded with item difficulty. Since
these factors have been correlated with Piagetian tests of concrete

operational stages in intellectual development, it follows that their hemispheric symmetry carries some positive conclusions about the development stages of the two hemispheres (see Discussion below).

Developmental Analysis

The Progressive Matrices have particular developmental interest insofar as they are purported to measure the ability of the subject to coordinate two abstract rules concurrently ("multiplicative classification"). Inhelder and Piaget (1969) consider this ability a most important prerequisite for the stage of concrete operations when conservation of matter and number is acquired by the child, around age 7. Our results record age estimates greater than seven for all hemispheres of all patients. This is true not only according to Raven's norms (1965, 1960) but also according to the more up-to-date norms of Carlson and Goldman (1974) and of Carlson and Wiedl (1976). Yet, as already pointed out, neither N.G.'s two hemispheres nor the patients with left and right hemispherectomy could actually solve the problems which seem insoluble by Gestalt and asymmetry closure alone and require abstract coordination of rules. Both of L.B.'s hemispheres, however, could solve these problems. Thus the fact that a hemisphere reached a certain mental age in some task is no guarantee that it reached the corresponding "stage" in normal cognitive development.

The fact that many or even most of the problems in the RCPM can be solved "perceptually" through pattern completion is not surprising and was already noted by Inhelder and Piaget (1969, p. 153). It is more perplexing, however, that some of the factorial studies with children, which found the RCPM to correlate positively and highly with Piagetian multiplicative classification and conservation tasks, found this to be just as true or stronger for the "perceptual" items in the test as for the "conceptual" ones (e.g., Carlson and Goldman, 1974; Carlson and Wiedl, 1976).

Piaget and Inhelder (1969, p. 174) suggest that in order to distinguish those solutions to matrix problems that use figural strategies based on perceptual symmetries from operational solutions using double classification, it is enough to ask the subject to justify his choice and to test his commitment to it by rejecting alternative solutions. Using similar criteria Carlson (1973), cited in Carlson and Goldman (1974), deduced what our analysis also confirmed, that the concrete operational component was specific to only certain subsections of the RCPM, particularly part of set B. The board form of RCPM may provide a related measure of operativity: does a higher incidence of trial and errors indicate a more mature ability to reject incorrect solutions and thus a stronger commitment to one final solution, or, on the contrary, is willingness to consider alternative solutions a sign of poor commitment to

a solution?   Interestingly, there is no significant difference
between the hemispheres in either the number of problems that
involved one or more corrections or in the total number of corrections during the test.   Thus, N.G.'s LH used 37 corrections in 15
different problems whereas her RH corrected itself 38 times in a
total of 15 problems.   In L.B. LH performance included 13 corrections in 6 problems and the RH executed 11 corrections in 9 different problems.

But as mentioned above it was the LH that benefited more
from these self corrections.   In that regard the LH resembles
second grade children rather than fourth graders.   Second graders
do and fourth graders do not benefit from trial and error with
the board form of the test.   Indeed, comparable improvements in
RCPM scores relative to the standard book administration was
obtained for the second graders (a) by the board form and (b)
by an administration which required verbalization of the strategy
during and after the solution (Carlson and Wiedl, 1976, Figure 3).
Thus the board form may reinforce and selectively reward a LH
solution strategy in children as well as in our patients.

On the Neuropsychological Method

Taken together, the diverse studies of patients with hemispheric lesions are consistent with the commissurotomy data: the
Raven Matrices do not yield strong hemispheric asymmetries and the
test is therefore ill-suited to detect laterality effects in focal brain
damage.   However, when a specific syndrome, such as aphasia or
constructional apraxia, occurs in isolation, it may create an
overwhelming deficit on the RPM which is due to the pathology of
a hemispherically specialized "control" center, and which masks
the potential contribution of the undamaged hemisphere.

An example of "pathological" dominance effects where a
unilateral lesion impairs the performance of the whole brain is the
particularly strong tendency among patients with right hemsiphere
lesions to prefer response alternatives positioned in the right half
of the visual field and consequently to perform especially poorly
on the Raven Matrices (Piercy and Smith, 1962; Gainotti, 1968;
Costa et al., 1969; Basso et al., 1973).   In contrast, an analysis
of side preferences in response by the disconnected hemispheres
shows that, in general, there is no consistent and significant
neglect of ipsilateral visual space in unilateral presentations, i.e.,
there is no preference for response alternatives on that side of
the page which is contralateral to the working hemisphere (Zaidel
et al., in press).   This verifies that unilateral neglect must
include pathological inhibition of competence in the healthy
hemisphere.

Conversely, with the benefit of split-brain data on the

pattern of laterality effects across different RPM problem sets we can now attribute some recently observed laterality effects on RCPM in unilaterally brain-damaged patients to genuine positive effects of hemispheric specialization. Denes, Semenza, and Stoppa (1979) found no difference in RCPM scores between acute LBD and RBD patients when the effect of unilateral neglect of space was minimized. However, a retest two months later showed selective improvement by the RBD patients on set A and by the LBD patients on set B. This is just what our data show: a RH advantage on set A and a LH advantage on set B (see Table 2).

The Neuropsychology of "g"

In a recent study Basso et al. (1973) administered the RCPM under a time constraint as a measure of "g" to patients with uni-lateral brain damage and concluded that "g" is sustained by a LH region which is partially coextensive with the classical language area. The logic used by the Italian workers has been used in other neuropsychological studies to localize higher order functions by their association with other perceptual or cognitive functions with known localization. Thus the sensitivity of RCPM scores to the presence of aphasia was taken as evidence for coextensive localization, and the independence of RCPM scores from severity or type of aphasia justified the dissociation between "g" and language.

Our data support the dissociation between RPM and language but reject the concept that g is measured by RPM and is sustained by a LH region. We have seen that the RPM are heterogeneous with respect to loading on hemispheric factors. Hence if the RPM are indeed a pure measure of g it would follow that hemispheric specialization is a more fundamental or elementary concept than g. This would argue against the concept of g as a superordinate factor and support the American Primary Abilities Model (Thurstone, 1938) in lieu of the British Hierarchical Special Abilities Model (Spearman, 1946). At the least one would need to distinguish two subspecies of g, i.e., g_L and g_R, representing the LH and RH information processing styles respectively. Whatever they are, g_L and g_R are not simply identical with the primary verbal and spatial factors, respectively (Zaidel, 1978).

Of course, it could still be argued that general intelligence underlies to various degrees all the primary mental abilities and that a g factor could be obtained as a second-order factor among the primaries. Cattell took this one step further when he proposed (e.g., 1971) that g be split into two broad or general ability factor, fluid intelligence, "g_F," and crystallized intelligence, or "g_C." g_C is said to correspond in content to many traditional IQ tests with heavy loadings on the primaries of "verbal ability," "numerical ability," and "reasoning." It operates in areas where judgment has been taught systematically or experienced before. g_F, on the other hand, operates where the sheer perception of complex relations is involved and where stored experience is irrelevant.

Is there a simple assignment of g_C to one hemisphere and g_F to the other? Three considerations are relevant. First, g_F has a higher loading than g_C on spatial ability and on closure flexibility (Gestalt completion) tests where g_C has a much higher loading on verbal ability. Thus g_F seems to correspond to the RH and g_C to the LH. Second, Cattell showed in longitudinal studies that g_F peaks at a young age (around 20) and declines steadily thereafter whereas g_C does not drop until old age. This could constitute an attractive model for predicting laterality changes with age. Third, Cattell argues that any brain lesion, incurred at any age, will produce a loss in g_F whereas damage to certain specific localizations result in loss of specific abilities making up g_C. This is reminiscent of Semmes' observation (1968) about diffuse representation of function in the RH, and focal representation in the LH. Furthermore, Cattell believes that brain damage before the maturity of abilities making up g_C produces permanent deficit in these abilities, whereas later damage leads to their considerable recovery (cf. Hebb, 1949, whose "intelligence A" and "intelligence B" clearly correspond to g_C and g_F, respectively).

Unfortunately, there are several reasons why it is not simply the case that g_C is in the LH and g_F is in the RH. For one thing, both g_C and g_F load on tests of spatial ability and the superiority of g_F on this factor declines from 10 to 13. For another, Cattell believes that g_F is more important in the functioning of all areas of the cortex in the young but that more and more behavior shifts to g_C with age. On the contrary, some recent views have it that RH abilities mature later than LH abilities (e.g., Carey and Diamond, 1977). In fact, Cattell views g_F similarly to Burt's view of g, i.e., as distributed throughout the cortex (1971, p. 189). Indeed, since for Cattell the Raven Matrices remains a marker test of g_F, it follows that neither his theory nor our data support the association of g_F with the RH. Do our data negate Cattell's view that g_F "is a function of the total, effective, associative, cortical cell mass" (ibid.)? Not necessarily. The difference between the two hemispheres, when it exists (as in L.B. on the RSPM) is not large and may be attributable to non-g_F impurities in the Raven Matrices.

Nevertheless, the two general intelligence factors may have some other illuminating interpretation in terms of hemispheric specialization. Consider the following conjecture: the specialization of the RH is in storing and applying conventional rules to behavior, be those rules perceptual, cognitive, or social. This would make the RH rather like g_C in the sense of storage of conventional, rule-bound experience, both visuo-spatial and verbal.

Conclusion

This study as well as factor analysis and other neuropsychological evidence all converge on the conclusion that the RPM taps abilities and strategies from both hemispheres so that either one can solve a substantial part of the test. Comparing the presurgical WISC IQ (113) and postsurgical WAIS IQ (106) of commissurotomy patient L.B., say, to the IQ estimates from his RSPM scores (111 in free vision, 103 for the LH, 93 for the RH) we see that there seems to be no substantial loss of intelligence following cerebral disconnection. The "intelligence" of the disconnected RH is fairly similar to that of the LH which, in turn, approximates the intelligence of the whole brain--split or intact.

It is instructive to compare this pattern in split-brain human patients with data on unilateral vs. bilateral problem solving capacity in split-brain and intact monkeys. Briefly, the monkey experiments show that a chronically split-brain macaque working with one hemisphere in monocular vision has about half the problem solving ability or intelligence (measured by trials to criterion) as the whole brain (in binocular vision). However, while showing no hemispheric specialization for intelligence, the split-brain monkey who has recovered long enough will perform normally in binocular vision (Hamilton, 1976). Thus each monkey hemisphere is as "smart" as the other but only half as "smart" as the whole brain. By contrast, one of the human hemispheres is often better than the other, as on the RSPM, and either hemisphere is roughly as "smart" and sometimes even "smarter" than a normal person. This contrast highlights the role of hemispheric specialization in creating functional redundancy and autonomy in the cerebral organization of higher cognitive functions in humans.

References

Albert, M. L. Auditory sequencing and left cerebral dominance for language. Neuropsychologia, 1972, 10, 245-248.

Archibald, Y. M., Wepman, J. M., and Jones, L. V. Nonverbal cognitive performance in aphasic and nonaphasic brain-damaged patients. Cortex, 1967, 3, 275-294.

Arrigoni, G.,and De Renzi, E. Constructional apraxia and hemispheric locus of lesion. Cortex, 1964, 1, 170-197.

Basso, A., De Renzi, E., Faglioni, P., Scotti, G., and Spinnler, H. Neuropsychological evidence for the existence of cerebral areas critical to the performance of intelligence tasks. Brain, 1973, 96, 715-728.

Bogen, J. E., and Vogel, P. J. Neurologic status in the long term following complete cerebral commissurotomy. In F. Michel and B. Schott (Eds.), Les Syndromes de Disconnexion Calleuse Chez l'Homme. Lyon: Hopital Neurologique, 1975.

Boller, F., and Vignolo, L. A. Latent sensory aphasia in hemisphere-damaged patients: An experimental study with the Token Test. Brain, 1966, 89, 815-830.

Carey, S., and Diamond, R. From piecemeal to configurational representation of faces. Science, 1977, 195, 312-314.

Carlson, J. S., and Goldman, R. D. The relationship between multiplicative classification and inductive reasoning. Journal of Genetic Psychology, 1974, 125, 265-272.

Carlson, J., Goldman, R., Bollinger, J., and Wiedl, K. H. Der Effekt von Problemverbalisation bei verscheidenen Aufgaben gruppen und Darbeitungsformen des "Raven Progressive Matrices Test". Diagnostica, 1974, 20, 133-141.

Carlson, J. S., and Wiedl, K. H. Modes of presentation of the Raven Coloured Progressive Matrices test: Toward a differential testing approach. Trier Psychologische Berichte, Band 3, Heft 7. Trier University, 1976.

Cattell, R. B. Abilities: Their structure, growth, and action. Boston: Houghton Mifflin, 1971.

Colonna, A., and Faglioni, P. The performance of hemisphere-damaged patients on spatial intelligence tests. Cortex, 1966, 2, 293-307.

Costa, L. D. Interset variability on the Raven Coloured Progressive Matrices as an indicator of specific ability deficit in brain-lesioned patients. Cortex, 1976, 12, 31-40.

Costa, L. D., and Vaughan, H. G. Performance of patients with lateralized cerebral lesions. I. Verbal and perceptual tests. Journal of Nervous and Mental Diseases, 1962, 134, 162-168.

Costa, L. D., Vaughan, H. G., Horwitz, M., and Ritter, W. Patterns of behavioral deficit associated with visual spatial neglect. Cortex, 1969, 5, 242-263.

De Renzi, E., and Faglioni, P. The comparative efficiency of intelligence and vigilance tests in detecting hemispheric cerebral damage. Cortex, 1965, 1, 410-429.

Denes, F., Semenza, C., and Stoppa, E. Note: Selective improvement by unilateral brain-damaged patients on Raven Coloured Progressive Matrices. Neuropsychologia, 1979, 16, 749-752.

Gainotti, G. Les manifestations de negligence et d'inattention pour l'hemispace. Cortex, 1968, 4, 64-91.

Gott, P.S. Cognitive abilities following right and left hemispherectomy. Cortex, 1973, 9, 266-274.

Guilford, J. P. The nature of human intelligence. New York: McGraw-Hill, 1967.

Hamilton, C. R. Investigations of perceptual and mnemonic lateralization in monkeys. In S. Harnad, R. W. Doty, L. Goldstein, J. Jaynes and G. Krauthamer (Eds.), Lateralization in the nervous system. New York: Academic Press, 1977.

Hebb, D. O. The effect of early and late brain injury upon test scores, and the nature of normal adult intelligence. Proceedings of the American Philosophical Society, 1942, 85, 275-292.

Inhelder, B., and Piaget, J. The early growth of logic in the - child: Classification and seriation. New York: Norton (originally published in French), 1969.

Kertesz, A., and McCabe, P. Intelligence and aphasia: Performance of aphasics on Raven's Coloured Progressive Matrices (RCPM). Brain and Language, 1975, 4, 387-395.

Klove, H. Validation studies in adult clinical neuropsychology. In R. M. Reitan and L. A. Davidson (Eds.), Clinical neuropsychology: Current status and applications. New York: Halsted Press, 1974.

Lansdell, H. Relation of extent of temporal removals to closure and visuomotor factors. Perceptual and Motor Skills, 1970, 31, 491-498.

Lansdell, H. Intellectual factors and asymmetry of cerebral function. Catalog of Selected Documents in Psychology, 1971, 1, 7.

Lebrun, Y., and Hoops, R. (Eds.) Intelligence and aphasia. Amsterdam: Swetz and Zeitlinger, 1974.

Levy-Agresti, J., and Sperry, R. W. Differential perceptual capacities in major and minor hemispheres. Proceedings of the United States National Academy of Sciences, 1968, 61, 1151.

Messerli, P., and Tissot, R. Operational capacity and aphasia. Unpublished paper presented at the Brussels Conference on Intelligence and Aphasia, 1973. (See Lebrun and Hoops, 1976.)

Meyer, V., and Jones, H. G. Patterns of cognitive test performance as functions of the lateral localization of cerebral abnormalities in the temporal lobe. Journal of Mental Science, 1957, 103, 758-772.

Muller, R. Eine kritische emprische Untersuchung des "Draw-a-Man-test" und der "Coloured Progressive Matrices." Diagnostica, 1970, 16, 138-147.

Nebes, R. D. Hemispheric specialization in commissurotomized man. Psychology Bulletin, 1974, 81, 1-14.

Orgass, B., Poeck, K., Kerschensteiner, M., and Hartje, W. Visuo-cognitive performances in patients with unilateral hemispheric lesions. Zeitschrift fur Neurologie, 1972, 202, 177-195.

Newcombe, F. Missile wounds of the brain; a study of psychological deficit. London: Oxford University Press, 1969.

Piercy, M. Neurological aspects of intelligence. In P. J. Vinken and R. W. Bruyn (Eds.), Handbook of clinical neurology (Vol. 3). Amsterdam: North Holland, 1969.

Piercy, M., and Smith, V. O. G. Right hemisphere dominance for certain non-verbal intellectual skills. Brain, 1962, 85, 775-790.

Raven, J. C. Standard progressive matrices: Sets A, B, C, D and E. London: H. K. Lewis, 1958. (Originally published in 1938.)

Raven, J. C. Guide of the standard progressive matrices sets

A, B, C, D and E.   London: H. K. Lewis, 1960.
Raven, J. C.   Coloured progressive matrices sets A, A$_B$, B.
    London: H. K. Lewis, 1962. (Originally published in 1947,
    revised order in 1956.)
Raven, J. C.   Guide to using the coloured progressive matrices
    sets A, A$_B$, B. London: H. K. Lewis, 1965.
Rimoldi, H. J. A.   Study of some factors related to intelligence.
    Psychometrika, 1948, 13, 27-46.
Semmes, J.   Hemisphere specialization: A possible clue to
    mechanism. Neuropsychologia, 1968, 6, 11-26.
Spearman, C.   The abilities of man. London: Macmillan, 1923.
Spearman, C.   Theory of general factor. British Journal of Psy-
    chology, 1946, 36, 117-131.
Spearman, C., and Wynne-Jones, L. L. Human ability. London:
    Macmillan, 1951.
Sperry, R. W., Gazzaniga, M. S., and Bogen, J. E. Interhemi-
    spheric relationships: The neocortical commissures; syndromes
    of hemisphere disconnection. In P. J. Vinken and G. W.
    Bruyn (Eds.), Handbook of clinical neurology (Vol. 4).
    Amsterdam: North Holland, 1969.
Thurstone, L. L.   Primary mental abilities. Chicago: The Uni-
    versity of Chicago Press, 1938.
Urmer, A. H., Morris, A. B., and Wendland, L. V. The effect
    of brain damage on Raven's Progressive Matrices. Journal of
    Clinical Psychology, 1960, 16, 182-185.
Van Dongen, H. R. Impairment of drawing and intelligence in
    aphasic patients. Unpublished paper presented at the Brussels
    Conference on Intelligence and Aphasia, 1973 (See Lebrun
    and Hoops, 1974).
Van Harskamp, F. Some considerations concerning the utility of
    intelligence tests in aphasic patients. Unpublished paper
    presented at the Brussels Conference on Intelligence and
    Aphasia, 1973 (See Lebrun and Hoops, 1974).
Vernon, P. E.   The structure of human abilities. London:
    Methuen, 1961.
Wiedl, K. H., and Carlson, J. S. The factorial structure of the
    Raven Coloured Progressive Matrices test. Educational and
    Psychological Measurement, 1976, 36, 409-413.
Zaidel, D., and Sperry, R. W. Performance on the Raven's
    Coloured Progressive Matrices test by subjects with cerebral
    commissurotomy. Cortex, 1973, 9, 34-39.
Zaidel, E. A technique for presenting lateralized visual input with
    prolonged exposure. Vision Research, 1975, 15, 283-289.
Zaidel, E. Concepts of cerebral dominance in the split brain. In
    P. Buser and A. Rougeul-Buser (Eds.). Cerebral correlates
    of conscious experience. Amsterdam; Elsevier, 1978.
Zaidel, E., Zaidel, D. W., and Sperry, R. W. Left and right
    intelligence: Hemispheric performance on the book and board
    forms of Raven's Progressive Matrices following brain bisec-
    tion and hemi-decortication.   Cortex, in press.

Zangwill, O. L.   Intelligence in aphasia.   In A. V. S. De Reuck
    and M. O'Connor (Eds.), <u>Ciba foundation symposium on dis-
    orders of language</u>.   London: Churchill, 1964.

INDIVIDUAL DIFFERENCES IN THE PATTERNING OF CURVES

OF D.Q. AND I.Q. SCORES FROM 6 MONTHS TO 17 YEARS

C. B. Hindley

University of London

London, England

Introduction

The main points that I want to make, arising from our longi-tudinal research, are:
firstly, and this is not new, though it is often overlooked, that
I.Q. scores do not remain constant during development;
secondly, and more importantly, that the changes in score which
can be detected from age to age cannot be regarded as mere
random fluctuation, but have to be seen as representing
systematic trends, characteristic of particular individuals;
thirdly, that these systematic trends are concealed by the methods
of analysis commonly employed with developmental measurement
data.

The basic data consist of a succession of scores from each S
across k ages (Table 1). The columns consist of a frequency
distribution of scores at each age, and the means of the sample, or
sub-sample, will reveal any average trend across ages. Each row
contains one S's successive scores, from which his mean score and
s.d. can be obtained, indicating his average status and variability
across all ages. Any systematic trends in an S's scores are
concealed.

Correlation, the commonest method of examining stability of
relative scores (Wohlwill, 1973), relates measures at two arbitrary
points in time, from two columns, and provides an average measure
of consistency for the sample. Alternatively, differences in scores
at the two ages are computed, and mean or median differences
obtained. With either method, individual differences in consistency
tend to be ignored, but, more seriously, neither provides any
means for examining the form of each S's array of scores.

Table 1

Scheme of Measures Available on n Subjects at k Ages

Subjects	Ages 1	2	3	4 - - - k		Subject Means, s.d.'s	
S1	X_{11}	X_{12}	X_{13}	X_{14}	X_{1k}	$X_{1\cdot}$	1
S_2	X_{21}	X22	- - - - - - - - -			$X_{2\cdot}$	2.
	- - - - - - - - - - - - - - - -					- - - - -	
S_n	X_{n1}	X_{n2}	- - - - - - -	X_{nk}		$X_{n\cdot}$	n.
Sample: Means	$X_{\cdot 1}$	$X_{\cdot 2}$	- - - - -	$X_{\cdot k}$		X_{nk}	nk
s.d.'s	.1	.2	- - - - -	.k		nk	

A number of previous workers have called attention to the fact that curves of I.Q. scores against age differ considerably in different subjects (Dearborn and Ruthney, 1941; Honzik et al., 1948; Bayley, 1956; Sontag et al., 1958), but they did not subject the curves to systematic analysis. Up to now McCall et al. (1973) have gone furthest in classifying I.Q. curves, but by clustering component scores, rather than by subjecting each individual curve to analysis. Our approach (Hindley and Owen, 1979) is one that has been used for many years in the study of physical growth (Tanner, 1951; Goldstein, 1976), namely that of fitting mathematical expressions to each individual's curve.

In our case, successive terms of the polynomial equation ($y_x = a + bx + cx^2 + dx^3 + - - -$) were fitted to each individual's array of scores (Hindley and Owen, 1979). This approach has several virtues. In the first place it provides an objective means of characterizing the shape of each curve. An absence of significant fit indicates that the subject's curve is best regarded as horizontal. Whether successive terms, linear (b), quadratic (c), cubic (d), etc., account for a significant reduction in error, therefore indicating that the null hypothesis of no significant slope must be rejected, is tested against the residual error around the fitted curve. Thus, each subject's curve may be characterized as horizontal, linearly sloping, or curvilinear, with varying degrees of complexity. In the second place, the parameters of the fitted curves can be compared across individuals, or groups.

Data and Subjects

Our data come from a longitudinal sample of 109 subjects validly tested at 7 ages from 6 months to 14 years, and 84 subjects up to 17 years. The tests used were: at 6 and 18 months, the Griffiths Scale of Infant Development (Griffiths, 1954), broadly derived from Gesell's scales; at 3, 5, 8, and 11 years the Stanford Binet; and, at 14 and 17 years, the AH4 (Heim, 1970), a test with verbal and non-verbal sub-scales. The subjects came from a wide variety of social class backgrounds (Hindley and Owen, 1978, 1979). It would be nice if it had been possible to have used the same test throughout, but that is a practical impossibility. The tests used, particularly the Stanford Binet, have the merits of having been widely used, and of measuring general intelligence, which has been the subject of most of the heredity $\underline{v}$. environment debate.

Findings

An idea of the gross amounts of change in scores is provided by distributions of change scores across pairs of ages (Hindley and Owen, 1978). These are of similar order to those of Honzik et al. (1948) and Pinneau (1961). Median amounts of change from baby-test scores at 6 or 18 months to scores at 17 years are, perhaps not surprisingly, as high as .84 s.d. units or more. Leaving aside the babytests, median changes from 3 to 17, 5 to 17, 8 to 17, and 11 to 17 years, amount to .51, .52, .49 and .58 s.d. units. Even over intervals between Stanford Binet tests alone, median changes from 3 to 11, and 5 to 11 years, amount to .55 and .43 s.d. units. As regards greater shifts, from 3 or 5 years to 17 years, with correlations of .53 and .61 respectively, a quarter of the sample change by 1.09 s.d. units or more; from 8 to 17 years, when r is .74, a quarter shift by .86 s.d. units or more; and from 11 to 17 years, when r is .68, by .95 s.d. units or more. Individual subjects, of course, can display much larger changes, with a maximum after 3 years of 3.54 s.d. units (from 3 to 11 years).

In examining the form of individual trends of scores, curve-fitting yielded several results of interest (Hindley and Owen, 1979).

1. In Table 2, within-subject regression and residual variance is summed over the sample as a whole. Fitting of linear terms to each subject's curve accounted for highly significant amounts of variance (p 0.001) over each of the periods 6 months - 17 years, 3 years - 17 years, and 3 years - 11 years. This indicates that I.Q.'s cannot be regarded as fluctuating randomly, as would be required by the thesis of I.Q. constancy, but that there are systematic upward and downward trends of individual's scores.

Table 2

Significant Fits of Polynomials to Individual's Curves, for
Whole Sample: Analyses of Variance of Regressions over
Different Age Spans, (S.d units, N = 84)*

Source	Sums of Squares	d.f	Mean Square	F	P (of improvement)
6m - 17 yr					
Linear	107.3	84	1.28	3.41	***
Linear + Quadratic	171.2	168	1.02	3.32	***
Linear + Quadratic + Cubic	211.2	252	0.84	3.12	***
Residual	84.9	336	0.25		
Total (within individuals)	296.1	588	0.50		
3 yr - 17 yr					
Linear	54.9	84	0.65	3.00	***
Linear + Quadratic	83.8	168	0.50	2.84	***
Residual	44.3	252	0.18		
Total (within individuals)	128.1	420	0.31		
3 yr - 11 yr					
Linear	37.7	84	0.45	2.77	***
Residual	27.2	168	0.16		
Total (within individuals)	64.9	252	0.26		

* From Hindley and Owen (1979)

2. Over the longer periods, curvilinear terms are required to
    adequately characterize the individual curves.   Thus, while
    only the linear fit is significant over 3 - 11 years, from 3 - 17
    years the quadratic term is also significant, and over 6
    months - 17 years the cubic term is significant in addition.

3. The curves of the subjects can be classified according to the
    polynomial terms which yield a significant fit.   Thus, from 6

months to 14 years, on the larger sample (N 109), the curves of 54% of subjects yield a significant fit at the 0.05 level: linear in 27 cases, quadratic in 17, cubic in 9, and a mixture of terms in 6 cases. Over shorter periods, the number of significant fits drops: with 28% over 3 to 14 years, and 18% over 3 to 11 years (compared with chance expectation of 5%), linear only in 13 subjects, quadratic only in 11, cubic in 3, and mixed in 3; and 18% over 3 to 11 years (compared with chance expectation of 5%), linear only in 13 subjects, and quadratic only in the remaining 7 subjects.

4. It is to be noted that over the longer period of 6 months to 14 years (Table 3) approximately half the total variance in scores is attributable to differences in the mean amplitude of subjects' curves (between-subjects: 47% of variance) and half to within-subjects variance (53%). Over shorter periods the proportion of variance attributable to within-subject variation falls to 31% over 3 to 14 years, and 26% over 3 to 11 years. Most of this within-subject variation cannot be considered random as between 58% and 75% is accounted for by the fitted curves ("regression" in Table 3).

An alternative, and very simple approach, which might have been used at the start but was only used after individual differences in shape of curves had been firmly established, is that of classifying the curves visually. Seven categories of curve were developed: Up, up with hump, hump, horizontal, u-shaped, down with u, down. Substantial numbers of curves were allocated to each

Table 3

Estimated Proportions of Total Variance: Between and Within Subjects, Regression and Residual.* (N. = 109)

	6m – 14 yr	3 yr – 14 yr	3 yr – 11 yr
Between S's	47	69	74
Within S's	53	31	26
Regression	40	23	15
(Linear	20	14	15)
(Quadratic	13	09	–)
(Cubic	07	–	–)
Residual	13	08	11

* From Hindley and Owen (1979)

category, after discussion in cases of disagreement. Thus, over 3 years - 17 years (N = 84) only 29 were judged horizontal, 16 up, 17 down, and the rest in intermediate categories. Multivariate analyses of variance, comparing the seven groups of curves, confirmed the effectiveness of this classification, in that it accounts for 61% to 75% of the systematic regression variance (Hindley and Owen, 1979).

Differences in trend have also been examined according to sex and social class. Overall sex differences are small, but there are substantial differences in the mean curves of three major social class groups. However, when the individual curves of the subjects in each group are examined, it becomes evident that the mean curves may characterize only a minority of the subjects in that group.

In evaluating our results, it has to be recognized that different abilities are being assessed at different ages, so that maintenance of a high score, for example, indicates that a subject continues to be of high status on whatever abilities are measured at the succession of ages. Obviously somewhat different curves would have been obtained had different tests been employed, but insofar as we used tests of a type that have been commonly used there would seem little reason to doubt the generality, in principle, of our findings, which are not inconsistent with those of McCall et al. (1973).

Conclusions

1.   Methods of seeking group trends in I.Q. scores, by averaging, or the use of correlations, conceals what is going on in the individual subject, and is liable to lead to an underestimation of the extent of individual variations in trend.

2.   The significance of the fitted curves indicates:
      a) untenability of the doctrine of constancy of the I.Q.;
      b) substantial individual differences in systematic trends of I.Q. curves against age, which cannot be regarded as simply due to random variation. With mental age, or "absolute scale" units, as these can be derived from the I.Q. scores, it would follow that individual curves based on these measures would also differ.

3.   It also follows that an isolated I.Q. measure on an individual can only be regarded as derived from a curve of scores of unknown shape.

4.   More generally, it is concluded that not only does the nature of intelligence change with age, as Piaget and others have convincingly demonstrated, but also the relative status of

individuals on typical intelligence tests, largely in systematic ways.

Summary

The case is made that commonly used methods of analysis conceal the extent, and more particularly the form, of individual curves of I.Q. scores. Curve-fitting to longitudinal data reveals highly significant individual differences in curves over periods from 6 months to 17 years. Implications are discussed.

References

Bayley, N. Individual patterns of development. Child Development, 27, 45–74, 1956.

Dearborn, W.P. and Ruthney, J. W. M. Predicting the child's development. Cambridge, Mass.: Dearborn SciArt Publishers, 1941.

Goldstein, H. Approaches to the statistical analysis of longitudinal data. Paper presented at the Seminar of Longitudinal Studies, organised by the SSRC Survey Unit, Cambridge, 25–26 March, 1976.

Griffiths, R. The abilities of babies. London: University of London Press, 1954.

Heim, A. W. Manual for the AH4 Group Test of General Intelligence. Slough: NFER, 1970.

Hindley, C. B., and Owen, C. F. The extent of individual changes in I.Q. for ages between six months and seventeen years, in a British longitudinal sample. Journal of Child Psychology and Psychiatry, 19, 329–350, 1978.

Hindley, C. B., and Owen, C. F. An analysis of individual patterns of D.Q. and I.Q. curves from 6 months to 17 years. British Journal of Psychology, 70, 273–293, 1979.

Honzik, M. P., Macfarlane, J. W., and Allen, L. The stability of mental test performance between two and eighteen years. Journal of Experimental Education, 17, 309–324, 1948.

McCall, R. B., Appelbaum, M. I., and Hogarty, P. S. Developmental changes in mental performance. Monogram of Social Research and Child Development, 38, No. 150.

Pinneau, S. R. Change in intelligence quotient from infancy to maturity. Boston: Houghton Mifflin, 1961.

Sontag, L. W., Baker, C. T., and Nelson, V. L. Mental growth and personality development: a longitudinal study. Monogram of Social Research and Child Development, 23, No. 2.

Tanner, J. M. Some notes on the reporting of growth data. Human Biology, 23, 93–159, 1951.

Wohlwill, J. F. The study of behavioral development. London: Academic Press, 1973.

THE SOCIAL ECOLOGY OF INTELLIGENCE IN THE

BRITISH ISLES, FRANCE AND SPAIN

Richard Lynn

University of Ulster

Londonderry, Northern Ireland

Abstract

The social ecology of intelligence is concerned with the relation between the mean IQ of populations and a variety of social and econ- omic phenomena. Data are presented for the British Isles, France and Spain. It is shown that there are regional variations in the mean population IQ in all three countries. These mean IQs are closely related to measures of intellectual achievement, income, un- employment and infant mortality. It is proposed that the intelligence didfferences are causal to the social and economic differences. Data are also presented to show that selective migration between regions have been an important factor in bringing about contemporary dif- ferences in regional mean IQs.

Introduction

The social ecology of intelligence is concerned with the relation between the mean IQ of populations and a variety of social and economic phenomena. In my inquiries in this area I have worked with a three stage causal chain model in which it is envisaged that selective migration has given rise to differences in mean IQ between regions. These mean IQ differences are in turn partly responsible for regional differences in the output of people of intellectual distinction, per capita incomes, rates of unemploy- ment and rates of infant mortality. The model is shown in dia- grammatic form in Figure 1.

To spell out the model in a little more detail, it is suggested that over the course of centuries there has been a general ten- dency in many countries for some of the more intelligent individ-

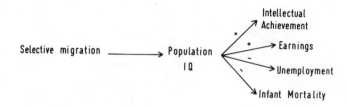

Figure 1

uals in the provinces to migrate to the capital city. Such individuals have been drawn by the attractions of wealth, status, intellectual stimulation and so forth which are available in capital cities. Many such individuals will have established homes and families in the capital cities and consequently their high intelligence will tend to pass down the generations through genetic and environmental mechanisms, leading over the course of time to significant differences in mean IQ between the population in the capital city and in the provinces. This is stage one of the path model shown in Figure 1.

In the next stage of the model it is suggested that the mean IQ differences between the regions are responsible for much of the variation in intellectual achievement, incomes, rates of unemployment and rates of infant mortality. It was first proposed by Galton that there would be a close association between the mean IQ of a population and its output of intellectually gifted persons and the expected association seems an obvious one. It is also proposed that a population with a high mean IQ would have higher average earnings and lower rates of unemployment and infant mortality. The reasons for these predictions are that intelligent individuals tend to have higher earnings, to be less prone to unemployment and to having an infant death in their families. Thus our expectations for the population differences are derived by regarding the populations simply as aggregates of individuals among whom these relationships are reasonably well established.

It is suggested that the model is applicable to regional subpopulations within nations, to districts within cities and possibly also across nations. There are thus quite a number of areas where the model could be tested. However, in this paper I shall be concerned only with data pertaining to the model from the regions of the British Isles, France and Spain.

Fitting Data to the Model: The British Isles

We turn now to the question of fitting data to the model and consider first the British Isles. Here we have thirteen regions whose mean population IQs range from 102.1 in the London area to 96.0 in the Republic of Ireland. The data are shown fitted to the

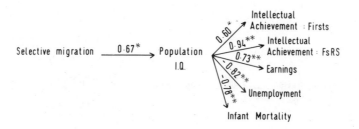

Figure 2

model in Figure 2, where it will be observed that all the predictions are fulfilled at statistically significant levels. The index used for selective migration is population increase over the period 1750--1950. This is considered a reasonable proxy for selective migration based on the assumptions that natural population increases are constant across regions and hence that differences in regional rates of population increase reflect migration in which there is a selective element. It has to be admitted that there are some assumptions in using this index and this is certainly the weakest of our variables.

The two measures of intellectual achievement are all first class honours graduates for 1973 expressed as a proportion of the total number of their age group in their region; and Fellows of the Royal Society, being all fellows born after 1911 expressed as a function of the populations in the regions recorded in the 1911 census. Data for income, unemployment and infant mortality are taken for the years 1959-61. A full description of the data is given in Lynn (1979).

2. France

The next case to be considered is France. The country is divided into 90 departments for which mean IQ data were reported by Montmollin (1958) derived from 257,000 male conscripts in the mid nineteen fifties. The index of intellectual achievement was membership of the Institut de France. The 253 members in 1975 were allocated to the regions where they were born and the numbers from each region expressed as proportions of the departmental population in 1974. Earnings, unemployment and infant mortality are taken for the years 1970-72. Selective migration was estimated in the same way as in the British Isles by taking the increase in population from 1801-54.

The French data are shown in Figure 3. All the predicted relationships are present at statistically significant levels with the exception of unemployment. It is suggested that the explanation

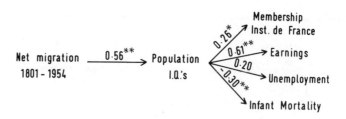

Figure 3

for this may lie in government subsidies to small farmers who would otherwise be unemployed, thus concealing the figures for natural unemployment in the French provinces.

3. Spain

Turning finally to Spain, our data is based on the 48 regions into which government agencies divide the country for the purposes of statistical compilations. IQ means for each region were calculated from the data of Nieto-Allegre et. al (1967) which gives results for approximately 130,000 Spanish conscripts into the armed services. An index of intellectual achievement was taken by using all Spaniards listed in World Who's Who and expressing these as functions of the populations in each of the regions. Data for mean regional incomes, rates of infant mortality and for illiteracy were also obtained from Spanish government statistics for 1970. Selective migration was estimated as in the case of the British Isles and France by taking population growth figures for the period 1900-1970.

The results for Spain are shown in Figure 4. It will be observed that they are less satisfactory than those for the British Isles and France in so far as there is no relationship between the measure of migration and mean population IQ, and the correlation between mean IQ and the index of intellectual achievement falls short of statistical significance. Possibly these less satisfactory results may arise because Spain does not have a single metropolitan city corresponding to London and Paris. While Madrid is of course the political and administrative capital, Barcelona is the most prosperous city in economic terms, characterised by both relatively high incomes and high mean population IQ. In spite of these possible distorting effects, relationships shown in Figure 4 between mean population IQ and income, infant mortality and illiteracy do appear consistent with the model.

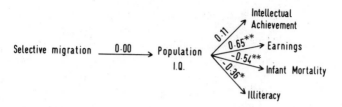

Figure 4.

References

Lynn R. (1979).   The social ecology of intelligence in the British
    Isles.   British Journal of Social and Clinical Psychology,
    18, 1–12.
Montmollin, M. (1958).   Le niveau intellectuel des recrues du
    contingent.   Population, 13, 259–268.
Nieto Allegre et al.   (1967). Differencias regionales en la
    medida de la inteligencia con el test M.P.   Revista Psicologia
    General y Aplicado, 22, 699–707.

VERBAL ABILITY, ATTENTION AND AUTOMATICITY

Steven Schwartz

University of Queensland

Queensland, Australia

Abstract

Analyses of the information-processing mechanisms underlying performance on verbal tasks have been relatively unsuccessful in identifying the components responsible for individual differences. A review of pertinent studies supports the notion that performance on practiced skills is more likely to be correlated with scores on psychometric tests of verbal ability than are performance measures obtained on new skills. With practice, skills become less attention-demanding and more "automatic." Such overlearned skills are not likely to be affected by transient situational variables but rather to reflect the limits of an organism's abilities.

Learning a new motor skill (driving a car, for example) requires conscious attention to details. Distractions such as the radio will influence performance as the novice driver struggles to keep attending to the car and the road. The expert driver, on the other hand, easily divides attention between the radio and the road without degrading driving performance. In a sense, driving becomes "automatic" with practice. An important question for students of learning and intelligence is at what point during skill acquisition will performance be best predicted by a standardized test? That is, are stable individual differences most likely to be found among practiced experts or untutored novices? The answer, at least for motor skills, is that psychometric tests are more highly correlated with performance after practice than with performance at the beginning of skill acquisition (Fleishman, 1965). The main thesis of this paper is that the same relationship holds true in the realm of verbal ability.

567

Recent efforts to study the component information-processing mechanisms underlying performance on verbal tasks have met with varying degrees of success in identifying sources of stable individual differences. The unstable pattern of findings may be the result of the varying degrees of practice and expertise subjects have with each component. If components require different amounts of practice before becoming "automatic" and if stable individual differences are only obtained when a component's attentional demands are minimal then highly practiced subjects should show different patterns of correlations (with psychometric tests) from those produced by novices. In addition, fast "maturing" components should be better predictors of psychometric scores than those requiring much practice. In this paper several sources of evidence supporting this relationship between performance on cognitive tasks and scores on tests of verbal ability will be examined. Before beginning, however, it should be noted that the distinction made here between novices and practiced subjects is similar to the one made by Schneider and Shiffrin (1977) with regard to "automatic" and "controlled" processes. There is no intent (nor is there any necessity) to adopt their model of information-processing in order to make the present argument. In this paper, practice is thought to lead to less attention-demanding (more automatic) performance at different rates for different information-processing components. No other assumptions are necessary. First, the relationship between verbal ability and "controlled" processing will be examined. Next, "automatic" processing will be reviewed. Finally, the implications of the various findings for the study of intelligence and learning will be discussed.

I. Attention Demanding (Controlled) Processes and Verbal Ability

A task with important implications for the present hypothesis is the memory search task described by Sternberg (1975). Subjects, in this task, have to decide rapidly whether or not a test item appeared in a previously memorized set of items. The function relating reaction time to the number of items in the memory set has two important properties--its slope and its intercept of the ordinate. The intercept is usually taken to indicate the overall speed of responding, whereas the slope represents the rate at which one scans items in short-term memory. The y-intercept is related to scores on test of verbal ability but the slope is not (Sternberg, 1975)--unless the task is given to mentally deficient subjects (Hunt, 1978, pp. 118-120). Scanning rate, then, which represents the speed with which one conducts a serial, controlled, search is generally unrelated to verbal ability scores, but total time to respond which includes such things as coding time and other highly practiced processes is related to verbal ability.

There have been several published exceptions to these general findings. Hunt, Frost, and Lunneborg (1973), for example, reported a negative correlation between scanning rate and verbal ability. This finding, however, has not been replicable (Hunt, 1978). Chiang and Atkinson (1976) found a positive correlation between scanning rate and SAT verbal scores for men but not women (the opposite relationship from the one found by Hunt, et al., 1973). This finding has not been explained or, to my knowledge, replicated. Keating and Bobbitt (1978) found scanning rate to be negatively related to verbal ability in children but the relationship disappeared by age 17. Hoving, Morin, and Konick (1970) failed to find a relationship even among children younger than 17. It seems safe and conservative to conclude that with few exceptions, scanning rate is unrelated to verbal ability in adults whereas total response time which includes "automatic" processes such as stimulus coding time and response key pressing time is related to verbal ability.

A variation of the Sternberg paradigm with implications for the present thesis was described by Hogaboam and Pellegrino (1978). They had their subjects judge whether a visually presented stimulus (picture or word) was a member of a previously designated semantic category. This task differs from the standard paradigm in that the memory set always equalled one and subjects scanned for semantic rather than physical features. Hogaboam and Pellegrino still found no relationship between reaction time and SAT verbal scores. This is as it should be. In the light of the present hypothesis, scanning speed should not be related to verbal ability. It makes no difference if one is scanning for physical or semantic features. (It should be noted that Hogaboam and Pellegrino take a much different view of their study from the one presented here.) One last confirmatory example will end this part of the discussion. One of the fastest scanning times in the literature, five times faster than average, belongs to a mnemonist studied by Hunt and Love (1972). Despite his speed, the mnemonist was of average verbal ability.

As part of his research into the information-processing components of analogical reasoning, Robert Sternberg (1977) engaged in what he called the "external validation" of potential models. Part of this procedure involved correlating measures of the various components with tests of reasoning, perceptual speed and vocabulary. These correlations are data relevant to the present discussion. Early in testing, components such as identifying attributes of the analogy, discovering (and generating) relational rules and so forth were hardly related to the psychometric measures. With practice, however, the correlations between the components and the psychometric measures improved markedly (Sternberg, 1977, pp. 210-211). This is as expected given the current thesis. The major exception was component C which was related to psychometric measures from the outset. C involves such things as preparation, preparing a response, and so forth. These

processes which are more-or-less constant across problems appear
to go on outside of awareness and may be "overlearned" and
automatic even before the experiment begins.

The components of complex information-processing models of
cognition are not highly related to psychometric tests of verbal
ability until they are practiced. Moreover, tasks which require
controlled, capacity-demanding search such as the Sternberg
memory scanning task are not good predictors of psychometric
test scores. Next, the situation with regard to highly overlearned
information-processing mechanisms is examined.

II. Practice, Automaticity, and Verbal Ability

Automatic encoding takes place when a stimulus directly
(without search) activates information resident in long term memory.
Such coding is an important part of reading (for practiced readers).
Hunt and his colleagues (Hunt, et al., 1973; Hunt, Lunneborg,
and Lewis, 1975) used the matching task developed by Posner
(Posner, Boies, Eichelman and Taylor, 1969) to study the relationship between verbal ability and automatic encoding. Subjects of
varying verbal ability were asked to determine whether two simultaneously presented letters were the "same" or "different" in
physical identity (PI) and name identity (NI) conditions. Everyone
took longer to make NI than PI matches but the difference (NI-PI)
was greater for low than for high verbals. In subsequent studies,
the relationship between verbal ability and automatic encoding has
been shown to apply to such diverse populations as university
students, children, the elderly, epileptics and non-college adults.
The relationship is obtained using a variety of psychometric
measures of verbal ability. Verbal ability is also related to speed
of access to long term memory codes when words are used rather
than letters (Goldberg, Schwartz and Stewart, 1977). This consistent pattern of results clearly supports the hypothesis that the
retrieval of highly overlearned materials from long term memory is
strongly related to measures of verbal ability. Memory for order,
an important aspect of understanding and using language has also
been studied in relation to verbal ability. Several experiments
were conducted by Hunt and his colleagues.

Nix (see Hunt, et al., 1973) performed an experiment based
on the release from proactive inhibition (PI) paradigm. In this
study, subjects of varying verbal ability were shown three words,
asked to count backwards for some period of time and then to
recall the words in their original presentation order. After
several repetitions of this procedure with words from the same
semantic category, recall accuracy began to decline. This is the
result of proactive interference. If the semantic category is
changed, say on the fourth trial, recall improves dramatically.
This effect is known as release from PI. Nix found a much

stronger release from PI effect for high than for low verbals
when recall was scored correct only when the subject duplicated
the original order of presentation. When order was ignored and
recall scored correct if the words were produced in any order,
high verbals did not differ from lows. These results suggest
that memory for the order of stimuli is related to verbal ability.
A similar conclusion was drawn from the results of another experiment reported in Hunt, et al. (1975).

    Subjects shadowed a sequence of four letters followed by a
variable number of digits. Following the final digit, subjects
were required to recall the letters in their correct order. In this
experiment, low verbals made more errors in order recall than
high verbals. Low verbals also recalled fewer letters. Although
the results appear to support the notion that order recall is
related to verbal ability, scoring is very tricky. Failure to recall
any of the letters, for example, results in no order errors. On
the other hand, the only way to make four transposition errors is
to recall all of the letters. Thus, memory for items and memory
for order are confounded. In order to explore the relationship
between memory for order and verbal ability, Schwartz and Wiedel
(1978) conducted a series of studies in which item and order
information were separated. These studies indicated that: (a)
Order and item information may be retained separatedly;   (b)
verbal ability is related to the recall (but not recognition) of
order;   (c) The relationship between verbal ability and memory
for order is most pronounced when the originally presented order
must be transformed at output (as in the "digits backwards"
task).

    Schwartz and Wiedel concluded that although verbal ability is
related to the recall of order, this relationship was not mediated
by any attention-demanding process not available to low verbals.
That is, high verbals were not better at rehearsal, chunking or
any other controlled process. Additional confirmatory evidence
for this conclusion was provided by Lyon (1977) who presented
digits very rapidly so as to eliminate grouping strategies and still
found verbal ability and order recall to be related and Cohen and
Sandberg (1977) who found the relationship between verbal ability
and memory for order to be largest at the end of a list. Attention
demanding organizational strategies should have their effects at
the beginning rather than at the end of a list. This pattern of
results makes a good deal of sense if memory for order is an
"automatic" process which all speakers of the language are thoroughly familiar with. Hunt (1978) suggested that the various findings
may actually represent indirect measures of attentional capacity
and that high verbals have a greater capacity than lows. Martin
(1978), in a direct test of this hypothesis, found memory for
order unrelated to measures of capacity. It would seem reasonable to interpret the various findings in the light of the present

hypothesis. Verbal ability is related to memory for order which seems almost certainly to be a largely "automatic" process.

One final source of data for the present hypothesis comes from recent studies of the hemispheric control of cognitive tasks. Studies in this area have been plagued by methodological and conceptual problems. Perhaps the most important issue is determining whether performance differences in the two hemispheres means that a particular hemisphere is specialized for a task (and that slowness in one, therefore, represents interhemispheric transfer time) or whether the task can be done by either hemisphere but one is merely slower than the other. Present views seem to have dealt with the problem by assuming the hemispheres to be interactive and load sharing; different parts of a problem are solved in different cerebral hemispheres. There is at least some evidence that the left cerebral hemisphere engages in serial processing (G. Cohen, 1973) and many have suggested that the right hemisphere is largely engaged in wholistic processing. It is tempting, but highly speculative, to associate the right hemisphere with highly practiced, automatic tasks (see Zaidel's chapter in this volume for some evidence in regard to reading). Recent findings suggest that the hypothesis may not be far off the mark. Poltrock (see Hunt, et al., 1975) found that memory for order in a dichotic stimulation experiment was better for high than for low verbals when the first stimulus was presented to the right hemisphere (left ear). High and lows were about the same when the first stimulus went to the left hemisphere. In an experiment conducted with Kim Kirsner, I found high verbals faster than lows in making NI and PI matches when stimuli are presented to the right cerebral hemisphere (left visual field). Highs are faster when stimuli are presented to the left hemisphere as well, but in the left hemisphere the difference is reduced by one-half.

Evidence from the Posner matching task, memory scanning studies, analogical reasoning, studies of memory for order and even studies of hemispheric specialization all seem to favor the hypothesis posed at the outset of this paper--highly practiced skills requiring minimal attentional control are better predictors of verbal ability test scores than unpracticed or highly attention-demanding skills.

III. Conclusions

Information-processing mechanisms underlying performance on verbal tasks may be of several types. Mechanisms which require a great deal of practice before they become "automatic" are not good predictors of psychometric test scores (until they have been practiced). Highly practiced components are better predictors of test performance and are more likely to yield stable patterns of individual differences. In some sense, it may appear counter-intuitive to say that measures of things such as decoding speed in the Posner task are more highly related to scores on test of verbal ability than such high level information processing components as those required to solve analogies. On the other hand, highly practiced skills are more likely to reflect the limits of one's information-processing ability than difficult new skills. This is because new skills requiring a great deal of concentration and performance may be easily affected by momentary distractions, motivation and other variables. As complex components become more practiced they are less affected by transient situational variables and better predictors of psychometric test performance.

References

Chiang, A. and Atkinson, R. C. Individual differences in inter-relationships among a select set of cognitive skills. Memory and Cognition, 1976, 4, 661-672.

Cohen, G. Hemispheric differences in serial vs. parallel processing. Journal of Experimental Psychology, 1973, 97, 349-356.

Cohen, R. L. and Sandburg, T. Relationship between intelligence and short-term memory. Cognitive Psychology, 1977, 9, 534-554.

Fleishman, E. A. The prediction of total task performance from prior practice on task components. Human Factors, 1965, 7, 81-127.

Goldberg, R. A., Schwartz, S. and Stewart, M. Individual differences in cognitive processes. Journal of Educational Psychology, 1977, 69, 9-14.

Hogaboam, T. W. and Pellegrino, J. W. Hunting for individual ences in cognitive processes: Verbal ability and semantic processing of pictures and words. Memory and Cognition, 1978, 6, 189-193.

Hoving, K. L., Morin, R. E. and Konick, D. S. Recognition reaction time and size of memory set: A developmental study. Psychonomic Science, 1970, 21, 247-248.

Hunt, E. Mechanics of verbal ability. Psychological Review, 1978, 85, 109-130.

Hunt, E., Frost, N., and Lunneborg, C. Individual differences in cognition: A new approach to intelligence. In G. Bower (Ed.), Advances in learning and motivation (Vol.7). New York: Academic Press, 1973.

Hunt, E. and Love, L. T. How good can memory be? In A. Melton and E. Martin (Eds.), Coding processes in human memory. Washington, D. C.: Winston-Wiley, 1972.

Hunt, E., Lunneborg, C., and Lewis, J. What does it mean to be high verbal? Cognitive Psychology, 1975, 7, 194-221.

Keating, D. P., and Bobbitt, B. L. Individual and developmental differences in cognitive processing components of mental ability. Child Development, 1978, 49, 155-167.

Lyon, D. R. Individual differences in immediate serial recall: A matter of mnemonics? Cognitive Psychology, 1977, 9, 403-411.

Martin, M. Memory span as a measure of individual differences in

capacity.   Memory and Cognition,   1978,   6,   194-198.
Posner, M. I., Boies, S. J., Eichelman, W. H., and Taylor, R. L.
    Retention of physical and name codes of single letters. Journal
    of Experimental Psychology: Monograph, 1969, 79 (1, part 2).
Schneider, W., and Shiffrin, R. M.   Controlled and automatic human
    information processing: I.   Detection, search and attention.
    Psychological Review, 1977, 84, 1-66.
Schwartz, S., and Wiedel, T. C.   Individual differences in cog-
    nition: Relationship between verbal ability and memory for
    order.   Intelligence, 1978, 2, 353-369.
Sternberg, R. J.   Intelligence, information processing and analogi-
    cal reasoning: The componential analysis of human abilities.
    Hillsdale, NJ: LEA, 1977.
Sternberg, S.   Memory scanning: New findings and current contro-
    versies.   Quarterly Journal of Experimental Psychology, 1975,
    27, 1-32.

ABILITY AND STRATEGY DIFFERENCES IN MAP LEARNING

Cathleen Stasz

The Rand Corporation

Santa Monica, California, U.S.A.

Abstract

This paper describes the influence of individual differences in abilities and subject-selected techniques for learning maps. Verbal protocols were obtained from 25 subjects who differed in psychometrically measured spatial restructuring and visual memory abilities. These protocols suggested a number of learning procedures and strategies that subjects used to focus attention, encode information and evaluate their learning progress while studying a map. High ability subjects differed from low ability subjects in the overall strategies they adopted to approach the learning problem, in their use of imagery for encoding spatial information, and in ther subsequent recall of spatial attributes of the map.

The study of intelligent behavior in any task domain requires an understanding of the sources of individual differences which influence task performance. Two typical and important sources of individual differences include basic abilities (e.g., Cronbach and Snow, 1977) and the strategies that people use to perform the task (e.g., Johnson, 1978; Hunt, 1978). This paper investigates how such differences influence knowledge acquisition from geographic maps. The research aims to understand expertise in map learning by analyzing differences between good and poor learners in terms of differences in both their basic information processing abilities and in their self-selected learning procedures and strategies.

Background

Map learning is a constructive process which produces a mental representation of the space depicted on the map. This internal knowledge representation stores many types of information, including names, shapes, locations, and distances. Since map learning is an active, intentional process, it may be viewed as a problem-solving task (Newell and Simon, 1972). The goal state corresponds to some memory representation of the map and the problem solving operators are the procedures and strategies the learner applies to produce the memory representation. These subject-selected procedures are specific techniques for selecting information from the map to study, and for determining how it will be encoded in memory. These procedures are of three types: attentional procedures restrict the map information which the learner attends to at any point in time; encoding procedures, such as rehearsal or imagery, elaborate the information in attentional focus and integrate it with information in memory; evaluajion procedures monitor the learner's progress by considering what information has been learned and what remains to be studied.

In addition to these procedures, people often adopt a global strategy for approaching the overall learning task. For example, an individual may decide to concentrate first on learning the spatial information on the map, then learn the verbal labels associated with the spatial locations. The subjects' strategy may determine, in part, the procedures they choose for accomplishing the learning task.

In previous studies of map learning, Perry Thorndyke and I (Thorndyke and Stasz, in press) collected verbal protocols from subjects attempting to learn fictitious, yet realistic, maps (see Figure 1). On each of six trials, subjects studied a map for two minutes and then attempted to reconstruct the map from memory. During study, subjects thought out loud, describing their attentional focus, their study procedures, and their evaluations of their learning progress.

Analysis of these protocols identified thirteen procedures that subjects employed during study for focusing attention, encoding information and evaluating the state of memory. We found large individual differences in subjects' use of these procedures and in their rate of learning of map information. A comparison of good learners (subjects correctly recalling at least 90 percent of the map information by the final trial) and poorer learners showed that subjects differed primarily in the use of a few study procedures. Of the procedures which differentiated good and poor learners, three required the encoding of spatial configurations of map information. These were imagery, pattern encoding, and relation encoding. Visual imagery involves subjects' construction of a mental image of the map. In pattern encoding, subjects would notice a particular shape or spatial feature of a map object, such as a street that curved to the east. Subjects employed the relation-encoding procedure when they studied explicit spatial relationships between two or more map objects, such as the intersection of two streets.

Our results and informal observations suggested that specific abilities might also influence the learning process. In particular, we conjectured that spatial ability, not procedure usage, might underlie the observed differences in performance. Since procedures comprise relatively low-level processes, subjects' choice of procedures might depend on their underlying abilities. For example, the best map learner reported that he had good visual memory and frequently used imagery to learn and remember information. By contrast, the worst learner reported that he had never experienced having mental images. He used primarily verbal learning procedures, such as associating map information with previous knwledge. This subject did not attempt to learn the more complex spatial configurations on the map.

Ability differences might also influence subjects' skill at using a particular procedure. For example, we observed that poorer learners were frequently inaccurate in their evaluations, during study, of what they had already successfully learned. The evaluation procedure requires subjects to retrieve knowledge from memory and compare it to information on the map. In this process, subjects might evoke a mental image of stored knowledge for comparison with the map. This image may be clearer or more accurate for subjects with better visualization ability.

Finally, abilities may influence the selection of global learning strategies. In the map learning task, all of the information to be learned is presented simultaneously rather than sequentially. Subjects must decide for themselves what information to learn first and how much time to spend studying each portion of the map. Individuals with spatial restructuring skill may employ strategies that subdivide the learning task. For example, subjects might adopt a divide-and-conquer strategy to help focus their attention on a subset of the information. They learn this information first, and then define and learn another subset. This strategy serves to structure the task into a sequence of smaller subproblems.

In sum, abilities appear to be a potentially important source of variation in map learning. The Thorndyke and Stasz (in press) results suggest how abilities and procedures might interact in the map learning process: both procedure choice and successful procedure use might depend on basic underlying ability differences. The present study was designed to directly investigate possible relationships between abilities, procedures, strategies, and map learning performance.

Method

Subjects and Ability Measures: Twenty-five subjects were
selected from an initial group of 94, based on their performance
on a battery of standard psychometric ability tests.   The tests
measured field-independence (Witkin and Goodenough, 1977),
which represents spatial restructuring ability, visual memory,
general intelligence, and verbal associative memory.   The selected
subjects differed in visual memory and spatial restructuring
skills, but had equivalent scores on tests of general intelligence
and verbal associative memory.

Procedure: Subjects were individually tested on a map-learn-
ing task.   For each of two maps, subjects alternately studied and
reproduced the map.   The Town Map is shown in Figure 1; the
Countries Map portrayed an imaginary continent with countries,
cities, roads, railroads, and large geographical features, such as
rivers and a mountain.   On each of six trials, subjects studied a
map for two minutes and then used as much drawing time as they
wished.   During study, subjects provided verbal protocols of
their study behavior, including the strategies and procedures
they were using to learn the map.   Following the final trial on
each map, subjects answered eight location and route-finding
questions from memory.

Results and Discussion

Although a variety of analyses investigated relationships
between abilities, procedures, strategies, and map learning, this
brief report focuses on analyses contrasting performance of extreme ability groups. Since tests of field-independence and visual
memory were highly correlated, ($r = .66$, $p < .01$), most subjects
fell into two extreme groups: relatively field-independent, high
visual memory (HIGHS; N = 10) and field-dependent, low visual
memory (LOWs; N = 10).

To determine the relationship between abilities and performance, recall scores between HIGH and LOW ability groups were
contrasted.   For each subject, map reproductions provided three
measures of recall performance: proportion of map objects correctly reproduced (both spatial location and verbal label correctly
specified), proportion of spatial information correctly reproduced,
and   proportion. of   verbal   information   correctly   reproduced.
Reproductions were scored at each trial.   For each subject, mean
recall was calculated across trials and maps.

Table 1 presents mean recall scores for the two groups.
Mann-Whitney U tests, with sample sizes of 10 and an alpha level
of .05, indicated that HIGHs recalled significantly more for complete elements and spatial attributes than LOWs.   The groups did
not differ significantly in recall of verbal attributes.   These

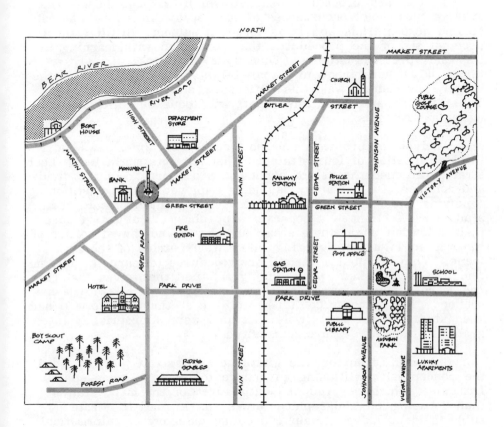

Figure 1.   The Town Map

findings replicate Thorndyke and Stasz (in press), who also found that good and poor learners differed in recall of complete elements and spatial attributes, but not verbal attributes. In general, subjects had little difficulty learning verbal information on a map. The present result extends those findings by demonstrating that subjects' visual-spatial abilities may underlie recall differences.

To compare procedure use between HIGHs and LOWs, the average number of occurrences of each study procedure was calculated across trials and maps for each subject. HIGH subjects used all six of the procedures that correlated with learning in this and previous studies (Thorndyke and Stasz, in press; Stasz, 1979) more frequently than LOWs. However, only for the imagery procedure was this difference statistically reliable. Thus, the remainder of this report will focus on differences in learning strategies.

Analysis of protocols and post-experiment interviews led to the identification of four strategies that subjects might use. Each strategy entailed the use of particular procedures. In the divide-and-conquer (DC) strategy, subjects employed spatial partitioning to divide the map into distinct sections. Subjects would then study each section as a separate subproblem. Subjects focused their attention on a single area, such as the northwest corner of the map in Figure 1, ignoring information outside of the area of focus. They adopted a variety of procedures to learn the information in the identified area. Having satisfied themselves that they had learned this information, they then moved on to study a new section. This process continued until all sections of the map had been studied. On final trials, sections were appropriately integrated to maintain feature continuity.

Subjects employing the global network strategy (GN) used the conceptual partitioning procedure to create a basic spatial framework which covered the entire area of the map. Rather than focusing on geographical areas, as in the DC strategy, subjects identified a certain conceptual category of information, such as streets, cities, or geographic features, to establish their initial framework. In the map shown in Figure 1, for example, a subject might first study vertical streets and large features, including the river, railroad track, and golf course. This initial framework acted as a point(s) of reference for learning new information. Subjects learned new elements by associating them to the the previously learned anchor points.

Progressive expansion (PE), the third major strategy, was characterized by subjects' systemmatic movement of attention across the map. Typically, subjects chose a starting point, such as the right side of the map, and studied as many adjacent elements

as possible in the alloted time. On successive trials they system-
matically focused on and learned new elements, moving across the
map in a slow progression and in a consistent direction.

A few subjects employed the narrative elaboration strategy
(NE). While the DC, GN and PE strategies relied on specific
attention-focusing procedures, the NE strategy did not. NE
strategists created verbal associations, such as a story or narrative,
to remember map elements and their spatial relationships. For the
map in Figure 1, for example, one subject invented and rehearsed
the following narrative: The butler went to church and saw
cedar trees in the park. Thus, he created an association among
Butler Street, Church, Cedar Street, and Park Drive.

To determine whether strategy use was related to subjects'
abilities, the study protocols were sorted into one of the four
strategy groups, or into the "no strategy" group. Table 1 shows
that 80% of the HIGH subjects' protocols exhibited one of the
three attention-focusing strategies. None of the HIGH subjects
used the NE strategy, and only four protocols were classified into
the no strategy group. By contrast, 50% of LOW subjects' protocols
showed no consistent strategy. Eight protocols contained attention-
focusing strategies, and two protocols were the NE strategy. To
test whether use of attention-focusing strategies versus no strategy
was significantly different for HIGHs and LOWs, Fisher's exact
test was computed separately for each map. The tests indicated
that the probability of observing differences as large or larger by
chance is .08.

Conclusions

These analyses suggest that both abilities and subject-selected
learning techniques are important sources of individual differences
in map learning. Visual-spatial abilities may underlie the use of
effective procedures for learning spatial information and the
adoption of attention-strategies. Both of these learning processes
contribute to successful map learning. Thus, three key characteristics identify good map learners: (1) they adopt an attention-
focusing strategy; (2) they use spatial learning procedures; and
(3) they have high visual-spatial ability.

References

Cronbach, L. J and Snow, R. E. Aptitudes and instructional
    methods: A handbook for research on interactions. New
    York: Irvington, 1977.
Hunt, E. Mechanics of verbal ability. Psychological Review, 1978,
    2, 109-130.
Johnson, E. S. Validation of concept-learning strategies. Journal
    of Experimental Psychology, 1978, 3, 237-266.

Newell, A., and Simon, H. A. Human problem solving. Englewood
    Cliffs, NJ: Prentice-Hall, 1972.
Thorndyke, P. W. and Stasz, C. Individual differences in pro-
    cedures for knowledge acquisition from maps. Cognitive
    Psychology, in press.
Stasz, C. Ability, procedure, and strategy differences in knowl-
    edge acquisition from maps. Unpublished doctoral dissertation,
    University of California, Los Angeles, 1979.
Witkin, H. A., and Goodenough, D. R., Field-dependence revisited.
    Research Bulletin 77-16, Princeton, NJ: Educational Testing
    Service, 1977.

INFORMATION-PROCESSING--"OLD WINE IN NEW BOTTLES" OR A CHALLENGE TO THE PSYCHOLOGY OF LEARNING AND INTELLIGENCE?

Herbert Geuss

Universität Osnabrück

West Germany

Abstract

Although pertinent research on learning and intelligence is guided more and more by information-processing models centering on an active subject, nature of this interaction is still dubious. A model is proposed to bridge this gap: 1. Learning usually demands at first far more information-processing capacity than the learner has available. 2. Only repeated and consistent intake and coding of the same feature-set stabilize stored information. 3. Repeated and consistent segmentation, selection, encoding, and storage are strategic activities. 4. Confusion arises if the learner has no, or many competing strategies available. Which, and how many strategies are used depends on situational characteristics, experience, and evaluation of task demands. Data obtained in several investigations dealing with reading and intelligence support this model.

Cognitive psychology and the majority of recently introduced information-processing models emphasize as their most central point the activity of the subject in interaction with environment. But current research on human development, learning, and intelligence largely ignores this point of view reducing subjects to passive and dependent objects (see e.g., training-studies on strategy development). Ahistoric experimentation only reveals cognitive structures by freezing in processes thus misleading to the assumption that all subjects perceive, process, store, and use information in the same way (see especially most of the computer-based models). Yet human learning and intelligent thinking and behavior do not emerge from merely passing information monotonously through unchanging structures; neither are they based on "general" strategies like

583

rehearsal being those of the experienced, not of the learner and thus only useful for discrimination but not to explain developmental differences. That means, a real paradigm-shift has not occurred, what we still have is the same old mechanistic world-view. Only labels have changed, and that is the old wine in the new bottles.

Dialectical approaches (e.g., Riegel, 1975) insist that subjects learn and develop in and through interaction with environment. But similar to above mentioned models dialectics do hardly more than describing this interaction, leaving its nature unexplained. What, then, is needed is not a merging of different models (Reese, 1976) but a reinterpretation of structure- and process-models.

Taken roughly together, structure-models reflect conditions and obstacles of learning, while process-models emphasize ways and strategies to overcome them. Within this very general framework the following assumptions are made: 1. New material in new situations is processed at first by surface characteristics or features which are numerous and ambiguous (see the levels-of-processing-model for this point). 2. Thus learning usually demands far more information-processing capacity than the learner has available, blocking transformation-space and preventing deeper encoding and chunking. 3. To overcome this gap input has to be restricted to only a few characteristics/features in order to save processing-capacity needed to form chunks. 4. Segmentation, selection, encoding, and storing of features have to be done in a consistent manner (repeated intake and coding of the same characteristics on the same coding level) in order to stabilize memory traces in long-term networks and form higher-order units being processed largely automatically. 5. Individual attempts of segmenting, selecting, encoding, and storing (either physical or other) characteristics consistently have to be interpreted as task-specific/knowledge-specific strategies. 6. In case the learner has either no, or (in early approaches) too many competing and thus planlessly alternating and weak strategies available, information will be processed in an inconsistent manner leading to weak memory traces, confusion, and decreasing confidence in one's own behavior.

Most tasks and situations allow adaptation and/or development of a variety of alternative strategies. Which and how many strategies the subject actually uses depends largely on his knowledge structure, metacognitive performance, evaluation of task demands, situational characteristics, and degree of self-confidence: Continuous information-overload, lack of confidence and independence, or anxiety, have deleterious effects on kind, number, and consistency of strategies to be adopted or developed, and consequently on the results of information-processing; lack of relevant knowledge or lack of knowledge about knowledge let the learner approach new material/situations in incorrect ways using inadequate and, often, too many competing strategies.

Amount and structure of knowledge, strategies, metacognitive competence, self-confidence, task demands, and situational characteristics are intimately interrelated and interdependent. Recording just some circumstances and prerequisites of learning and intelligent behavior in isolation can thus be of only little theoretical and practical value because there are many alternative ways the subject may select information, use his knowledge, and adopt strategies differing in appropriateness and effectiveness. In using strategies effectively they become skills and part of the knowledge base, changing actual ability-structure as well as subjectively experienced task demands hence requiring permanent adaptation of strategies and development of supplemental strategies. As, therefore, one characteristic of knowledge- and strategy-systems is permanent change in ongoing learning, and the other their task-specificity allowing study mainly in natural settings (call for ecological validity), the proposed model is consistent with dialectics focusing on the activity of the subject in interaction with environment.

Data obtained by the author in several investigations (Geuß and Schlevoigt, 1978) dealing with reading and learning to read support the model, also opening some new diagnostic and didactic perspectives. Systematic error patterns revealed four strategies concerning segmentation, selection, and encoding of written material by elementary school children (grades 2 and 3). Strategy use was determined from individually recorded reading errors and from performance on a task which required written reproduction of words presented tachistoscopically (0.5-1.0 sec. per word). Factor analysis of error patterns revealed five factors one of which represented attentional aspects. The remaining four factors were interpreted as different strategies to cope with information-overload, limited processing-capacity, and insufficient knowledge and metacognitive competence: 1. Visual translation strategy focusing on detailed feature-analysis (precise but slow); 2. Visual translation strategy with "sight-word"-emphasis (guessing from some features); 3. Visual translation strategy focusing on complete and precise information-processing but frequently ignoring spatial order of letters and letter groups; 4. Semantic translation strategy (a higher-order strategy following sight-word strategy; visual feature-configurations are rapidly translated into word-meaning displacing visual features from STM; fluent reading but bad spelling).

As has already been pointed out (Laberge and Samuels, 1974), the learner can selectively direct attention to any particular subprocess, ability, or strategy, but only by taking attention away from other possible strategies as long as a subprocess has not yet reached a certain degree of automaticity. In fact data obtained show clear evidence that at initial stages of learning to read less the kind than the number and consistency of strategies used are significantly related to reading achievement, in that an incompetent use of various competing and weak strategies leads to comparatively poor performance. This, and marked interindividual differences were found with respect to kind and number of alternative strategies used and class instructional method, grade, level of anxiety, and cognitive style. Some longitudinal studies designed to test the hypothesis that reading/writing performance may be improved by changing individual strategy-systems systematically proved to be very effective whereas no significant changes, despite repeated measurement effects, were found in the control-groups.

Focusing on intelligence, it is important to note that two of the above mentioned strategies seem to touch basic cognitive parameters being related to level of intellectual performance, namely "speed of information-processing" and "ability to retain order-information" (Hunt, 1976). Relating different reading strategies to scores on Primary Mental Abilities, moderate but substantial and interpretable correlations were found between some strategies and measures of intelligence (around .40), whereas correlations between test scores and more general reading achievement scores did not reach significance (grade 3, N = 62).

As correlations are rather rough indices, working patterns (distribution of omissions, correct and wrong solutions over all items) of some subtests of PMA (Reasoning, Embedded Figures; 40 items each test) were plotted for groups using different strategies in reading. Members of each group (N = 7) were "pure" strategy users in that they clung to only one strategy respectively. Groups were comparable with regard to age, sex, socio-economic background, and IQ. Results for Embedded Figures are displayed in Figure 1.

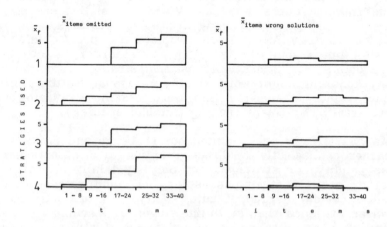

Figure 1. Mean number of omissions and wrong solutions (Embedded Figures) as a function of strategy used.

   As can be observed, there are some marked and predictable
differences between strategy groups. For example, in group 1
(detailed figure-analysis) omissions occur rather late compared to
group 2 (sight-word emphasis); group 3 (spatial order) tried to
work on all items but neglect of order information caused a steady
increase of wrong solutions.

   Comparing Embedded Figures- as against Reasoning-patterns,
the observed differences were predictable, too. Results for different
strategy groups and first and second halves of tests are shown in
Figure 2. For example, during first half of Embedded Figures group
3 (spatial order) shows a larger proportion of omissions than in first
half of Reasoning; during second halves it is just the other way. In
fact it appears from the nature of the Reasoning-items used, that the
observed differences are reliable: only about the last 20 items require attention to order-information for correct solution whereas
the first 20 items do not.

   Although correspondences seem to exist between structural
parameters associated with intelligence and optional information-processing strategies related to reading, some caution is necessary. In
the light of the present findings, further studies are needed to
investigate individual learning and thinking processes; subsequent
clustering of such processes may be more promising than the search
for "laws" of learning. Ann Brown's statement: "It is not how old
your head is but how much it has experienced in a particular cognitive domain" (1979, p. 253) may be modified: It is not only how
much one has experienced but how one has experienced it.

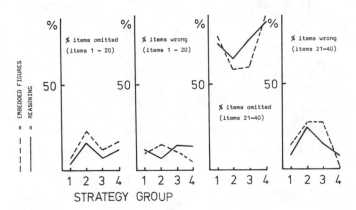

Figure 2

References

Brown, L. A. Theories of memory and the problems of development: Activity, growth, and knowledge. In L. S. Cermack and F.I. M Craik (Eds.), Levels of processing in human memory. Hillsdale, New Jersey: Lawrence Erlbaum, 1979.

Geuss, H. and Schlevoigt, G. Diagnostischer Lesetest für 2. und 3. Klassen DLT 2/3. Weinheim: Beltz, 1978.

Hunt, E. Varieties of cognitive power. In L. B. Resnick (Ed.), The nature of intelligence. Hillsdale, New Jersey: Lawrence Erlbaum, 1976.

Laberge, D. and Samuels, S. J. Toward a theory of automatic information processing in reading. Cognitive Psychology, 1974, 6, 293-323.

Reese, H. W. Models of human development. Human Development, 1976, 19, 291-303.

Riegel, K. F. Structure and transformation in modern intellectual history. In K. F. Riegel and G. C. Rosenwald (Eds.), Structur and transformation: Developmental and historical aspects. New York: Wiley, 1975.

GENERAL INTELLIGENCE AND MENTAL SPEED:

THEIR RELATIONSHIP AND DEVELOPMENT

Chris Brand

University of Edinburgh

Edinburgh, Scotland

Over the past century, differential psychologists have had to endure the embarrassment of having no compelling answer to the question "What is intelligence?" Early hypotheses that intelligence might be related to speed of reaction time proved unsuccessful; and more recent views (Eysenck, 1967) that intelligence might be related to the non-motoric, decisional, "information-processing" components of reaction time have at the time of writing found support merely at levels of correlation of "around 0.30" (see Jensen, 1973, pp. 32-45; Jensen, 1975, pp. 99-102).

But Nettelbeck and Lally (1976) report that Performance IQ on the WAIS correlates at around 0.9 with the speed at which two lines can be exposed to subjects who have to identify accurately their difference in length; so it seems reasonable to continue with the conjecture that measured intelligence may consist in or derive primarily from some kind of "mental speed" (cf. Spearman, 1923). The required modification to previous views of this kind may be that mental speed affects task performance at relatively early stages of perceptual input and registration rather than at stages that are "motoric" or even distinctively "central."

The experiments to be briefly described here were conducted by Anderson (1977), Hartnoll (1978) and Hosie (1979) and Grieve (1979) at the instigation of, and under varying degrees of supervision by the author. They investigate the relations between "inspection time" and various psychometric measures of intelligence in subjects at three different age levels.

Thirteen subjects of ages 16 to 26 and of good visual acuity were selected by Anderson to span an IQ range of 44 to 133 on

589

either Cattell's Culture Fair Test or on the Stanford-Binet (in the case of three hospitalized subjects of IQ≤77). Following Nettelbeck and Lally's (op. cit.) methods, subjects were asked on each trial to state the position ("left" or "right") of the shorter of two vertical lines presented tachistoscopically at a distance of 100 cm. within a visual angle of 1.6º degrees; presentations were succeeded by a backward mask. Once it was established that a subject made correct judgments at an exposure duration of ≤ 230 milliseconds, exposure time was varied over trials to discover the shortest duration at which the subject was correct in 95% of trials. This duration was called the subject's "inspection time" (IT); and such IT's ranged from 15 to 220 milliseconds.

The Pearson correlation between IQ and IT proved to be -0.88 (p ≤ 001, two-tailed) for the whole group. This IQ-IT correlation fell to -0.41 (nn.s.) when the six subjects of higher IQ's (99-133) were considered; but rose to -0.98 for those six subjects who spanned the range of lower IQ's (69-97). Since these last two correlations differ significantly (p≤05), it appears that, while there is clearly an overall association between IQ, and IT, this association is more substantial across the lower levels of intelligence.

Twelve of the above subjects (IQ's 69-133) were able to undertake similar comparisons involving selecting the position of the shortest of both three and four lines. For these conditions, IQ-IT correlations were respectively -0.78 (p≤01) and -0.66 (p≤05); and there was some (nonsignificant) indication that these correlations held up better over the lower IQ ranges within the sample. The four highest- and lowest-IQ subjects no longer showed the significant difference (by Mann-Whitney U-test) in IT that they had (p≤03) in the two-line condition: the low-IQ subjects had caught up to some extent - though there was no evidence from within-condition analyses that this could be attributed to practice. This may be some indication of high- and low-IQ subjects employing strategies for the task that were differentially adapted to increases in task complexity; but it would certainly appear to be evidence against the view that high-IQ subjects have any simple superiority at processing "bits" of information in this kind of task, since they had less of an IT advantage when selecting from four alternative positions than when selecting from two. It is as if the advantage of higher-IQ subjects lies in perceptual or attentional processes rather than in processes that are distinctively involved in decision-making or in whatever achievements of short-term memory and multiple comparison may have been involved in performance of the tasks involving three or four lines: their advantage might be said to lie in "initial processing speed" (IPS).

Since it might be held that the advantages in IT that are

associated with intelligence may be a cumulative product of the long period in which measured adult intelligence develops with age, it is of some consequence to ascertain by what age individual differences in general intelligence show any relation to IT. Hosie administered Raven's Coloured Progressive Matrices to twelve four-year-old children after pre-training them on comparable problems to ensure familiarity with the procedures. Scores ranged from 9 to 20 --i.e., approximately from IQ 95 to 123. The children were then asked to compare two line-lengths in conditions very similar to those of Anderson's (op.cit.) experiment. (Differences were that the lines were coloured red and blue so that the children did not need to refer to "left" and "right" and so that they associated them with teddy-bears of those colours who were made to "race" in between trials according to which colour had been associated with the short line on the previous trial). The IT's required by the children were established over several days and ranged from 200 to 600 milliseconds. It transpired that the IT-IQ correlation was 70.78 (p < 01); and, at this low level of mental age, the correlation did not differ between high- and low-IQ subjects.

Apparently, if the procedures can be considered comparable, IPS is associated with mental age: there was virtually no overlap in IPS between Hosie's and Anderson's subjects. But, if IPS is hypothesized to improve merely as a result of the development of general intelligence, it must be observed that this development has already occurred by age four to such a degree as to generate, amongst normal children of the same age, a strong relationship between IQ and IPS. It may seem more likely that IPS and its maturation provide one major psychological basis upon which general intelligence develops than that the large changes in measured intelligence over the course of development have improvements in such single tasks as comparing linelengths as a marked associated consequence.

The possibility that, once a certain IPS is attained, there are other influences that increase and sustain measured intelligence is suggested by the fact that the IQ-IT correlation was stronger over the lower ranges of intelligence (whether over lower IQ's amongst adults or at lower mental ages) in the above studies. Some confirmation of this possibility is provided by Hartnoll's study of 18 normal Dublin schoolboys of 11 to 12 years old whose IT's for words (five-lettered names of animals) were arrived at by successively increasing exposure durations until recognition occurred. Hartnoll's data show that the boys' ranked verbal intelligence (a composite of three measures of vocabulary, verbal fluency and verbal reasoning) correlated at 0.54 (t = 2.56, p < .02) with their ranked IT's. In its size, this correlation falls in between the high correlation of 0.78 for Hosie's normal four-year olds and the correlation of 0.41 for Anderson's brighter adults. Again, within the Dublin group , the IT-IQ correlation tended to be stronger (rho = 0.81, t = 3.2, p <.001) amongst the boys who had been selected for rather low verbal ability than it was (rho = 0.31, t = .86, n.s) amongst those of high ability. Thus the data from these several studies seem compatible with the view that the mental speed that is reflected in IT's might be particularly causal to intelligence up to some level of intelligence beyond which other factors also come into play.

In Hartnoll's study, there was no relation between "spatial ability" (on Thurstone's PMA) and IT's for either words or pictures; nor between verbal intelligence and picture IT's. The question of the relation of "spatial" ability to IT was pursued in Grieve's study. Ten subjects of ages 16 to 28 and of good visual acuity had IT's that ranged from 120 ms. down to 60 ms. according to a procedure resembling that employed by Anderson (op. cit.); their CultureFair IQ's ranged from 85 to 122. The results were as follows. (i) For these subjects, the correlation between IQ and IT was merely 0.61 (p. < 10); but the correlation for the five subjects of lower IQ (85105) was -.98 (p.< 01). This result adds further testimony to the conclusion drawn from Anderson's study that the IQ-IT relation is particularly striking across the lower ranges of IQ. (ii) Again, while scores for spatial ability on the Minnesota Paper Form Board showed little relation to IT (r = 0.11, n.s.), age-related percentile scores for Mill Hill Vocabulary correlated with IT at 0.88 (p< .001). Moreover, the relation between MHV percentile scores and IT was particularly strong across the lower ranges (percentiles 10 to 63) of Vocabulary: for n = 6, r = 0.97 (p. <002). These result are in line with the conclusion drawn from Hartnoll's study that IT is associated with "verbal" rather than with "spatial" abilites, and that this association is most marked across the lower ranges of verbal ability.

The results reported above cannot readily be attributed to "experimenter-expectancy" effects and they all constitute partial replications and extensions of each other and of Nettelbeck and Lally's (op.cit.) result. The relation between IQ and IPS appears to be both substantial and robust, particularly across the lower ranges of IQ.

References

Anderson, M. (1977) Mental speed and individual differences in
      intelligence. Final Honours Thesis: University of Edinburgh.
Eysenck, H. J. (1967). Intelligence assessment: A theoretical
      and experimental approach. British Journal of Educational
      Psychology 37, 8198.
Hartnoll, S. (1978). Verbal and spatial ability and their rela-
      tion to "inspection time" for word and picture stimuli. Final
      Honours Thesis: University of Dublin.
Hosie, B. (1979). Mental speed and intelligence: their relation-

ship and development in four-year-old children.   Final Honours
    Thesis: University of Edinburgh.
Jensen, A. R. (1973).   Educability and Group Differences.   London:
    Methuen.
Jensen, A. R. (1975).   Race and mental ability.   In F. J. Ebling,
    ed., Racial variation in man.   (Symposia of the Institute of
    Biology   No.   22).   London: Institute   of   Biology.
Grieve, D. (1979).   Inspection time and intelligence.   Final Honours
    Thesis: University of Edinburgh.
Nettelbeck, T. & Lally, M. (1976).   Inspection time and measured
    intelligence.   British Journal of Psychology 67, 1, 17-22.
Spearman, C. (1923).   The Nature of "Intelligence" and the Principles
    of Cognition.   London: Macmillan.

PRESENTATION MODE AND ORGANISATIONAL STRATEGIES

IN YOUNG CHILDREN'S FREE RECALL

Graham Davies and Anne Rushton

University of Aberdeen

Aberdeen, Scotland

Abstract

A series of experiments examined the free recall performance of five year-old children when the stimuli employed were actual objects, line drawings or photographs. In accordance with Sigel's "distancing hypothesis," clear effects due to semantic aids at presentation and recall were only found when objects were employed. Drawings and photographs showed continuing effects for aid at recall, but only weak and equivocal effects for aid at presentation. The implications of such mode effects for current theories of learning and intelligence are considered.

How well an adult remembers a list of words is largely governed by his ability to impose some sort of meaningful structure upon the disparate items. A good example of this comes from studies which have examined subjects' ability to recall word lists drawn from common taxonomic categories. These have reported that levels of recall are correlated with degree of "clustering": a measure of the degree to which recall order is organized in terms of the constituent categories. Support for the view that this relationship is causal is provided by the fact that techniques which emphasise the semantic structure of the material either at the time of learning (encoding) or at recall (decoding) increase levels of performance (see Baddeley, 1976, for a review).

However, this relationship between organisation and recall breaks down for children below ten years: increases of recall with age are not necessarily accompanied by concomitent rises in levels of clustering. Further, procedures designed to emphasise list structure do not appear to benefit consistently children under 7 years. This evidence has been used to support the claims of Piaget and others that the young children's memory processes differ qualitatively as well as quantitatively from their older peers and reflect the immaturities of his conceptual development (see Ornstein, 1978).

Davies and Brown (1978) presented five-year-old children with objects drawn from five common categories. The items were placed in five boxes. The child was required to label each object in turn, after which recall was requested. This procedure was followed for two cycles of presentation and recall. All items from a given category were either in the same box (blocked presentation) or distributed across the boxes (random presentation). Recall was attempted either with the experimenter asking the child to recall each category in turn (constrained recall) or leaving the child to recall as best he could (unconstrained recall). Four groups of children were tested corresponding to all combinations of the presentation and recall procedure.

Contrary to earlier findings with young children, aids both at presentation and recall facilitated performance, the effects being independent and additive. Further, increases in levels of recall were accompanied by corresponding increases in clustering. For the aided groups, this latter relationship held not only at the between-groups level, but also, more importantly, at the within-groups level. Only in the random-unconstrained condition were clustering levels low and unrelated to recall.

Similar effects on recall performance have recently been reported by Perlmutter and Myers (1979) using even younger children and a within-subject design. Taken together, these findings appear to indicate that even five-year-olds can and will show benefits to their levels of performance through the provision of semantic aids: it is premature to argue that previous negative findings necessarily reflect more primitive conceptual processes in the young.

Scrutiny of the literature suggested that one feature distinguishing these two studies from earlier experiments was the use of actual objects as stimuli. Sigel (1978) has provided evidence that young children perform at a significantly less mature level on concept induction tasks when the stimuli employed are pictures rather than objects. From Sigel's "distancing hypothesis," it was predicted that the substitution of pictures for objects in the Davies and Brown task would lead to a marked reduction in the effectiveness of semantic aids.

Rushton (1977) substituted uncoloured line drawings for the objects previously employed by Davies and Brown but maintained the same age of sample and experimental procedure. The change

in mode produced a marked alteration to the pattern of performance: the effect of constrained recall was still present but that of
blocked presentation was reduced from parity to technical insignificance (p < .10 > .05). Constraining recall also significantly
increased levels of clustering, but blocked presentation had no
effect whatever. This pattern of selective facilitation of recall
and organisation was also present when the relationship was examined at the within-group level: clustering and recall were significantly correlated for subjects in the constrained conditions but
not for those in the unconstrained.

A further study by Davies and Rushton (1979) paralleled the
earlier experiments in terms of procedure and subject population,
but used photographs of the objects as stimuli. From Sigel's
hypothesis, it was predicted that the experiment would produce a
pattern of facilitating effects midway between those of objects and
line drawings. Some support was found for this prediction: the
large facilitatory effects of constraint on recall and clustering
were replicated, but, on this occasion, there was a significant
residual effect for blocked presentation. However, this latter
effect was confined solely, in the case of recall, and mainly, in
the case of clustering, to the second trial. Within-group correlations were also consistent with this pattern: significant correlations were confined to the constrained conditions and the second
trial of the blocked-unconstrained group. These findings are
consistent with the view that photographs produce a delayed
effect upon conceptual organisation compared to the more immediate impact of objects and the weak effects associated with line
drawings. They are thus consistent with Sigel's hypothesis,
though it is noteworthy that the absolute levels of recall registered in the two picture studies were not significantly different.

These latter two experiments thus help to explain the discrepancy in findings between the studies of Davies and Brown
(1978) and Perlmutter and Myers (1979) on the one hand, and
much of the established literature on the other. The apparent
lack of effect of blocked presentation is seen to be limited to
conditions when photographs or drawings are employed as stimuli;
much higher levels of conceptual awareness are induced by objects
and this is faithfully reflected in levels of recall. Consistent with
other evidence cited by Sigel (1978), the child's primary difficulty
at this age lies not in deficiencies in conceptual knowledge, so
much as in applying that knowledge to pictorial representations.

Further evidence for the use of that conceptual knowledge is
provided by the continuing superiority of constrained over unconstrained recall. This latter result is surprising in the light of
the evidence accumulated by Tulving and his colleagues (Tulving
and Thomson, 1973) that a retrieval cue is only effective if it is
encoded at the time of original learning. It is possible that the

constrained recall instructions induce children to systematically search their extant semantic networks for relevant items in the manner of the "generation-recognition" procedure suggested by Bahrick (1970). Such a view would be consistent with the high incidence of category-relevant intrusions produced by subjects in the two studies using pictures. Significantly, the superiority of the constrained conditions in the two object studies was not accompanied by such intrusions, suggesting perhaps, that category prompts operated in a different manner in the latter studies.

More generally, these studies demonstrate the critical influence of task structure and demands upon the levels of cognitive sophistication exhibited by young children in cognitive tasks (see also Donaldson, 1978). Given optimal conditions of encoding and retrieval combined with a concrete mode of presentation, even five-year-olds will show a quality of performance characteristic of more cognitively mature individuals. However, unlike his older peer, such sophisticated behaviour is brittle and easily disrupted by minor changes in procedure, probably reflecting the lack of metamnemonic awareness of the purposes of the strategies elicited. Such findings are consistent with the arguments of Brown and Campione deployed elsewhere in this volume, that memory development involves the growth of access to available strategies as much as the development of strategies, as such.

References

Baddeley, A.D. The psychology of memory. London: Harper & Row, 1976.

Bahrick, H.P. A two-phase model for prompted recall. Psychological Review, 1970, 77, 215-222.

Davies, G., & Brown, L. Recall and organisation in five year old children. British Journal of Psychology, 1978, 69, 343-349.

Davies, G. & Rushton, A. The influence of presentation mode on organisational strategies in young children. Unpublished manuscript, 1979.

Donaldson, M. Children's minds. London: Fontana, 1978.

Ornstein, P.A. (Ed.). Memory development in children. Hillsdale, N.J.: Erlbaum, 1978.

Perlmutter, M., & Myers, N.A. Development of recall in 2 to 4 year old children. Developmental Psychology, 1979, 15, 73-83.

Rushton, A. The effect of blocked presentation and constrained recall on the performance of five year old children in a memory task using line drawings as stimuli: Unpublished M. Litt. dissertation, University of Aberdeen, 1977.

Sigel, I.E. The development of pictorial comprehension. In R.S. Randhawa and W.E. Coffman (Eds.). Visual learning, thinking and communication. New York: Academic Press, 1978.

Tulving, E., & Thomson, D.M. Encoding specificity and retrieval processes in episodic memory. Psychological Review, 1973, 80, 352-373.

QUALITATIVE AND QUANTITATIVE ASPECTS IN
THE DEVELOPMENT OF PROPORTIONAL REASONING

Gerald Noelting

Université Laval

Quebec, Canada

Abstract

An unsolved problem in stage-wise development is the dichotomy between sudden changes when passing from one stage to another, and more gradual changes between stages. This problem is studied here with an experiment on proportional reasoning.

An instrument, the Sharing Cakes Experiment, was devised for group questioning and administered to children between 9 and 16 years of age. It is made up of 24 items, each consisted of the comparison of two ratios presented graphically and included both multiple-choice and open-ended questions.

First-order analysis consisted of scalogram analysis, categorization of items on an ordinal scale, and chronological differentiation of subjects. Five stages were significantly differentiated, with structures integrating one another. Factor analysis performed on the results yielded seven factors which corresponded to the stages found.

A second-order analysis was then applied and consisted in searching for a pattern in the succession of structures: i.e., a structure of structures. Two periods of four phases each were found. These are described in terms of "increasing equilibration," or "adaptive restructuring" of the problem-solving scheme to new reality. This involves a "dialectical" interchange between subject and environment when the scheme is confronted with a new variable in a problem, with integration of novelty by the scheme.

The Experiment

Introduction

A problem situation involving a certain number of cakes divided among a certain number of people was devised with Luc Bégin. A 24 item Test was worked out with Gilbert Cardinal on the basis of an earlier experiment on ratios involving glasses of Orange Juice and Water, which had yielded significantly differentiated stages. This new situation was researched with R. Umbriaco. First results are given here.

Instrument: the Sharing Cakes Experiment

A group test comprising 24 items was devised. Each item compared two different ratios, presented as a certain number of cakes to be shared equally by a certain number of people.

In the experiment, the various items are presented graphically, with a three-choice answer ("each person receives more in group A, the same amount in A and B, more in group B"). Three lines are given for writing out an explanation justifying the choice.

Sample

The test was administered to subjects between 9 and 15 years of age. There were 30 subjects at each of the 8 age levels for a total of 240 subjects.

Treatment of results

a) A first-order analysis was first undertaken. This consisted of five steps:
i) Results were first corrected and submitted to scalogram analysis, with the program BMDO5S, Guttman Scale #1 (Dixon 1971). This program can treat up to 25 items. Items are thus ordered according to difficulty, and the scale obtained is directly analysed at the item level, to see if it forms a hierarchy. Items with their percentage of success are given in Table 1. Results of the scalogram analysis are given in the Note to the same Table. Thus a so-called "perfect hierarchy" is obtained at the item level. This was so because the universe of content was kept unique, with problems with exactly the same presentation given.

ii) Categorization is then applied to items which are close on the ordinal scale of difficulty. This leads to grouping items of the same category, and defining each category. Labelling of categories according to the Genevan chronology of stages was made on the basis of the type of problem involved. Results are given in Table 1.

Table 1

Items of sharing cakes experiment form C, ordered according to degree of success then categorized to form stages.

Stage	Item	Composition	% of success	Criteria for categorization
IA	1	4/1 vs. 1/4	100	Centration on numerator.
	3	1/2 vs. 2/1	100	
	2	3/1 vs. 1/3	100	
IB	5	2/3 vs. 2/1	97	Centration on denominator.
	6	3/1 vs. 3/2	97	
	4	1/2 vs. 1/3	96	
IC	8	3/4 vs. 2/1	91	Comparison between numerator and denominator.
	7	2/3 vs. 1/1	93	
	9	2/1 vs. 3/3	87	
IIA1	10	2/2 vs. 3/3	80	Equivalence class of unit.
	12	4/4 vs. 3/3	78	
	11	1/1 vs. 2/2	80	
IIA2	15	3/1 vs. 6/2	75	Equivalence class of digits, or unit fractions with lowest terms.
	13	1/2 vs. 2/4	73	
IIA3	14	2/4 vs. 3/6	69	Equivalence class of unit fractions without lowest terms
IIIA1	17	3/1 vs. 5/2	48	Fractions with two corresponding terms multiple of one another.
	16	1/2 vs. 2/3	51	
IIIA2	18	4/2 vs. 5/3	39	Same after reducing or extracting units.
	20	3/2 vs. 4/3	37	
	21	5/2 vs. 7/3	35	
	19	2/3 vs. 3/4	28	
IIIB	22	3/5 vs. 5/8	16	Fractions without multiple relation either within or between.
	23	7/12 vs. 4/7	12	
	24	8/5 vs. 5/3	11	

Note: Scalogram analysis gave a CR = .968, MMR = .803 and PPR = .838.

iii)   The subjects who had passed at least one item of a particular category, but none of the next, are then grouped.   The age-distribution of the various groups of subjects is then compared using the Kolmogorov-Smirnov test (Siegel, 1956).   If the difference is significant, this leads to a differentiation of stages, from both a qualitative (category) and chronological (age) point of view. Results are given in Table 2.   Five stages were significantly differentiated corresponding to IIA2: Lower Concrete Operations, IIA3:   Middle Concrete Operations, IIIA1: Lower Formal Operations, IIIA2:   Middle Formal Operations, IIIB: Higher Formal Operations. An age of accession was calculated for each.   Factor analysis performed on the results (Nie et al., 1975) yielded seven factors corresponding to the stages described (see Table 3).   When the succession of items (both within-stage and between-stage) was examined, changes were found of two types: Quantitative and qualitative.   A succession of quantitative changes or modifications in "extension" (corresponding to within-stage development of the scheme applied to solve problems) is followed by a qualitative change or modification in "comprehension" (corresponding to between-stage development, see Table 1).   This makes up an apparently linear process of development.   The quantitative aspect consists in accommodation to quantitative changes in the problem;   the qualitative aspect consists in differentiation of the scheme into anatagonistic subschemes to seize a new variable found in the problem.   The common aspect between quantititive and qualitative changes is mobility of the scheme when adapting to reality.

iv)   The particular items of each stage are then analysed in terms of structure.   The structure of the problem is defined as the relations between the four components involved: the two numerators and two denominators.   An analysis of successive structures showed that each preceding structure is integrated in the next.

v)   The strategy put into use to solve problems of each particular category is then analysed from the answers given by the subjects, and a mathematical expression of these strategies is attempted. At each level strategies are found of two types: within-state strategies, leading to the unit factor method: between-state strategies, leading to the Common Denominator algorithm.   Thus a qualitative description of each stage is given, with examples of success at items of the stage and failure at items of the next stage.   Adequacy of strategy to structure is investigated.

Space does not allow results of this analysis to be given here.

b)   A second-order of meta-theoretical analysis was then attempted.

Table 2

Comparison of age distribution of stages of sharing cakes experiment form C.

Age	N	Stage						
		I	IIA1	IIA2	IIA3	IIIA1	IIIA2	IIIB
9	30	21	4	2	3	0	0	0
10	30	11	2	6	8	2	1	0
11	30	7	4	3	12	3	1	0
12	30	5	1	2	12	5	5	0
13	30	2	1	0	4	9	11	3
14	30	1	0	0	3	6	9	11
15	30	1	0	0	0	3	16	10
16	30	1	0	0	2	1	12	14
Total 240		49	12	13	44	29	55	38
χ^2		–	–	–	5.495	8.565	10.398	5.618
p [a]		–	–	–	<.05	<.01	<.01	<.05
Age of accession [b]		–	–	–	10;8	12;5	13;3	16;0

Notes. – [a] Probability level of difference between age distribution of the stage, compared with preceding one, assessed by Kolmogorov-Smirnov Test.

[b] Age of accession to a stage is the age where 50% of Ss solve at least one item of the stage.

An analysis is made up of the succession of stages, both from the point of view of item-structure and problem-solving strategy. Comparison is made of the last strategy leading to an error, and the first strategy leading to success for each category of items. This allows one to infer what type of mechanism could explain the passage from one stage to the next. This leads us to adopt the concept of increasing equilibration (Piaget, 1975). Two periods of increasing equilibration were actually found, one consisting of the combination of numerator and denominator to construct the fraction concept (stages IA, IB, IC and IIA), and the

Table 3

Common factor analysis of sharing cookies experiment. (Principal factoring with iteration and varimax rotation of matrix.) Saturation of each item in each factor with hierarchical ordering of results.

| Item | | Percentage of succ. | | Factors, Eigenvalues and Saturations | | | | | | | Stage |
No.	Ident.	%	S-D.	Fact. 6 0.79	Fact. 9 0.26	Fact. 3 2.03	Fact. 5 1.05	Fact. 1 9.30	Fact. 2 3.52	Fact. 4 1.52	
1	4/1 vs. 1/4	99.6	6.5	0.607	0.599	0.190	0.043	0.030	-0.003	0.008	IA
3	1/2 vs. 2/1	99.2	9.1	0.914	0.241	0.122	0.084	0.071	0.029	0.010	
2	3/1 vs. 3/1	98.3	12.8	0.187	0.633	0.057	0.083	0.103	0.061	0.014	
5	2/3 vs. 2/1	95.0	21.8	0.042	0.036	0.954	0.226	0.146	0.077	0.026	IB
6	3/1 vs. 3/2	94.6	22.7	0.041	0.022	0.912	0.192	0.185	0.085	0.027	
4	1/2 vs. 1/3	92.9	25.7	0.177	0.134	0.693	0.246	0.231	0.119	0.031	
8	3/4 vs. 2/1	89.2	31.2	0.138	0.019	0.303	0.808	0.335	0.104	0.055	IC
7	2/3 vs. 3/1	88.8	31.7	-0.023	0.084	0.312	0.778	0.219	0.128	0.011	
9	2/1 vs. 3/3	85.8	34.9	0.075	0.144	0.259	0.531	0.465	0.162	0.058	
10	2/2 vs. 3/3	79.6	40.4	0.061	0.061	0.176	0.231	0.866	0.176	0.045	IIA1
12	4/4 vs. 3/3	78.8	41.0	0.076	0.034	0.115	0.154	0.891	0.229	0.068	
11	1/1 vs. 2/2	78.3	41.3	0.058	0.046	0.154	0.058	0.909	0.207	0.078	
13	1/2 vs. 2/4	74.2	43.9	-0.038	0.060	0.129	0.166	0.657	0.333	0.072	IIA2
15	3/1 vs. 6/2	74.2	43.9	-0.043	0.072	0.195	0.082	0.673	0.367	0.062	
14	2/4 vs. 3/6	73.3	44.3	-0.028	0.058	0.071	0.174	0.622	0.368	0.067	
17	3/1 vs. 5/2	48.3	50.1	0.017	0.048	0.092	0.108	0.296	0.797	0.105	IIIA1
16	1/2 vs. 2/3	46.7	50.0	0.022	0.051	0.112	0.004	0.334	0.632	0.190	
18	4/2 vs. 5/3	40.0	49.1	0.013	0.025	0.065	0.088	0.210	0.848	0.207	IIIA2
20	3/2 vs. 4/3	37.5	48.5	0.011	0.023	0.056	0.119	0.184	0.782	0.242	
21	5/2 vs. 7/3	35.4	47.9	0.015	0.019	0.061	0.075	0.158	0.770	0.305	
19	2/3 vs. 3/4	29.6	45.7	0.011	0.019	0.044	0.074	0.146	0.702	0.357	
22	3/5 vs. 5/8	17.5	38.1	0.008	0.010	0.030	0.042	0.085	0.449	0.728	IIIB
23	7/2 vs. 4/7	11.7	32.2	0.005	0.010	0.024	0.025	0.063	0.271	0.912	
24	8/5 vs. 5/3	10.8	31.2	0.006	0.011	0.023	0.020	0.058	0.272	0.868	

second of the combination of the equivalence class with the common denominator or unit factor to construct the Common Denominator or Percentage algorithms (stages IIA, IIB, IIIA and IIIB). (Items of stage IIB, e.g., 2/3 vs. 4/6, are missing in this version of the test.) However four phases were found in each period of adaptive restructuring of a scheme to new reality.

A general pattern of development was worked out, based on "dialectical processes" between a subject adjusting to new reality. Integrating this new reality, by means of a reorganization of existing schemes, leads to an adaptive process of development.

References

Dixon, W. G. Biomedical Computer Programs. Berkeley: University of California Press, 1971.
Nie, N. H., Hull, C. H., Steinbrenner, K., and Bent, D.-H. Statistical Package for the Social Sciences, Second Edition. New York: McGraw Hill, 1975.
Noelting, G. The development of Proportional Reasoning in the Child and Adolescent through Combination of Logic and Arithmetic. In Proceedings of the Second International Conference for the Psychology of Mathematics Education. Osnabrucker Schriften Zur Mathematik, 1, 1978, 242-265. An extended version of this experiment is due to appear in Educational Studies in Mathematics, Dordrecht, numbers 1 and 2, Vol. II.
Piaget, J. L'equilibration des structures cognitives, probleme central du developpement. Paris: PUF, 1975. Translated as: The Development of Thought. New York: Viking Press, 1977.
Siegel, S. Non Parametric Statistics for the Behavioral Sciences. New York: McGraw-Hill, 1956.